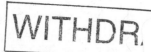

Praise for *Sustainable Market Farming*

Sustainable Market Farming is a great book! It is beneficial
grow a substantial part of their own food supply, in addition to those growing
for the markets. Besides knowledge of the soil and the crops, this book has
information for managing a crew (even if your crew is only you), trimming and
storing for the kitchen, and saving seeds. It is evident that Pam knows what it is
like to be in the fields and to truly feed people.

— Cindy Conner, permaculture educator, Homeplace Earth.

A great read full of solid, practical information. This book should be one of
the 'must-haves' on the bookshelf of every sustainable market farmer, whether
novice or long time grower. Great cultural information on most any crop a market
grower might be cultivating or contemplating. She includes outdoor production,
season extension, crop rotation, p   ases and much more in a friendly
easy to read style. Pam will be yo   urce for growing information for
years to come.

d Alison Wiediger, Au Naturel Farm

Whether you are new farmer just   an experienced producer looking
to move into year-round producti   acres or a home gardener who is
serious about food production thi   you. Pam has done a fantastic job
making the book well organized a   ead for the beginner but detailed
enough to be a useful reference f   ed growers who want to bring a
variety of healthy fresh delicious   rket year-round. What really gets
me excited about *Sustainable Market Farming* is finally having the details and
examples come from the mid-Atlantic and Southeast. *Sustainable Market Farming*
is a book I plan on referring to regularly for years to come.

— Ira Wallace, author, *Guide to Vegetable Gardening in the Southeast* and
coordinator, variety selection and outreach, Southern Exposure Seed Exchange

Pam Dawling's book is simply the best new cutting-edge resource for small organic
producers like us who grow for CSAs and farmers markets. I'll be spending the
winter off-season poring over everything from the detailed crop-planning to
organic insect control measures and high tunnel cropping, as the information
is both relevant and timely for our farm's evolution. A gold mine of practical
farming knowledge!

— Mark Cain, Dripping Springs Garden, Huntsville, Arkansas

If you are just getting started growing vegetables, or you are an experienced farmer or homesteader and still have unanswered questions, get this book! Thorough, accurate, and written in accessible language, *Sustainable Market Farming* covers all aspects of production from planning, planting, crop rotation, soil, and pest management to harvest and storage; and imparts in-depth knowledge of each crop and the care it needs to thrive. I can't wait to get my copy—it will fill a huge gap in my library.

— Mark Schonbeck, consultant in sustainable agriculture, Floyd, VA

After years of reading Pam's columns in *Growing for Market*, and visiting the Twin Oaks farm in the fall of 2011, I have been looking forward to her book. The integration of soil management, crop planning and production methods with hoophouse and cold cellar recommendations makes this book a valuable addition to the library of both beginning and seasoned vegetable farmers, and a unique text for vegetable production and urban farming classes. Thanks, Pam!

— John Biernbaum, Professor of Horticulture, Michigan State University

Pam Dawling is an experienced market gardener who employs her significant knowledge into feeding members of the Twin Oaks Community in Louisa, Virginia. For over a decade, Pam has led the Twin Oaks garden crew, providing year-round fresh produce to 100 people in a Community Supported Agriculture program. Pam is a wealth of practical know-how and has compiled her understanding of successful market gardening into this important manual. This is a must-read book for beginning gardeners and small farmers who are serious about market gardening, as well as for experienced market gardeners and students majoring in agricultural sciences.

— Reza Rafie, Ph.D., Associate Professor, Horticulture Extension Specialist, Small Fruits and Vegetables, School of Agriculture, Virginia State University

I've been using this book since receiving it, to plant my winter crops and plan for spring. Dawling explains the hows and whys, while also informing and encouraging the "what ifs." Balanced in depth, usefulness and inspiration; great for anyone wanting to grow (more) food for the table.

— Richard Moyer, Moyer Family Farm, Virginia

SUSTAINABLE
MARKET
FARMING

INTENSIVE VEGETABLE PRODUCTION
ON A FEW ACRES

PAM DAWLING

FOREWORD BY
LYNN BYCZYNSKI, *Growing For Market*

new society
PUBLISHERS

Cover design by Diane McIntosh.
Greenhouse: author supplied
All other images: © iStock
Chapter opener illustrations (unless otherwise noted):
Jessie Doyle (jessiedoylesstuff.blogspot.com/)

Printed in Canada. First printing January, 2013

Paperback ISBN: 978-0-86571-716-9
eISBN: 978-1-55092-512-8

Inquiries regarding requests to reprint all or part of *Sustainable Market Farming*
should be addressed to New Society Publishers at the address below.

To order directly from the publishers, please call toll-free
(North America) 1-800-567-6772, or order online at www.newsociety.com

Any other inquiries can be directed by mail to:

New Society Publishers
P.O. Box 189, Gabriola Island, BC V0R 1X0, Canada
(250) 247-9737

LIBRARY AND ARCHIVES CANADA CATALOGUING IN PUBLICATION

Dawling, Pam
Sustainable market farming : intensive vegetable production
on a few acres / Pam Dawling ; foreword by Lynn Byczynski.

Includes bibliographical references and index.
ISBN 978-0-86571-716-9

1. Vegetable gardening. 2. Organic farming. 3. Sustainable
agriculture. I. Title. II. Title: Market farming.

SB321.D39 2012   635   C2012-907150-1

New Society Publishers' mission is to publish books that contribute in fundamental ways to building an ecologically sustainable and just society, and to do so with the least possible impact on the environment, in a manner that models this vision. We are committed to doing this not just through education, but through action. The interior pages of our bound books are printed on Forest Stewardship Council®-registered acid-free paper that is **100% post-consumer recycled** (100% old growth forest-free), processed chlorine free, and printed with vegetable-based, low-VOC inks, with covers produced using FSC®-registered stock. New Society also works to reduce its carbon footprint, and purchases carbon offsets based on an annual audit to ensure a carbon neutral footprint. For further information, or to browse our full list of books and purchase securely, visit our website at: www.newsociety.com

Contents

Foreword by *Lynn Byczynski* vi

Introduction: Sustaining Community
and Agriculture ix

PART 1: TECHNIQUES

Planning

1. Year-Round Production 3
2. Create Your Own Field Manual . . . 7
3. Crop Review 14
4. Crop Rotations for Vegetables
   and Cover Crops 18
5. Seed Storage, Inventory
   and Orders 27
6. Scheduling Transplants 33
7. How Much to Grow 41

Planting

8. Crop Spacing 47
9. Transplanting Tips 53
10. Direct Sowing 61
11. Summer Germination of Seeds 65
12. Succession Planting for Continuous
    Harvesting 70

Sustainable Crop Protection

13. Season Extension 77
14. Cold-Hardy Winter Vegetables 83

15. The Hoophouse in Winter and
    Spring 90
16. The Hoophouse in Summer 99

Soil and Crop Quality

17. Maintaining Soil Fertility 108
18. Cover Crops 115
19. Cover Crops Chart 124
20. Sustainable Disease Management . . . 134
21. Sustainable Weed Management 141
22. Sustainable Pest Management 148

Harvest and Beyond

23. Manual Harvesting Techniques 155
24. Winter Vegetable Storage
    (Without Refrigeration) 161
25. Root Cellars 168

PART 2: CROPS

Legumes

26. Green Beans 175
27. Southern Peas, Asparagus Beans
    and Limas 182
28. Fava Beans 186
29. Edamame 191
30. Snap Peas and Snow Peas 196
31. Peanuts 203

Brassicas

32. Broccoli, Cabbage, Kale and
    Collards in Spring 208
33. Broccoli, Cabbage, Kale and
    Collards in Fall 217
34. Asian Greens 225

Other Greens

35. Spinach 233
36. Chard and Other Summer
    Cooking Greens 240
37. Lettuce All Year Round 246

Roots

38. Carrots, Beets and Parsnips 260
39. Celery and Celeriac 269
40. Turnips and Rutabagas 275

Cucurbits

41. Summer Squash and Zucchini 280
42. Winter Squash and Pumpkins 287
43. Cucumbers and Muskmelons 294
44. Watermelon 303

Alliums

45. Garlic 311
46. Bulb Onions 324
47. Potato Onions 334
48. Leeks 338

Nightshades

49. Tomatoes 342
50. Peppers 357
51. Eggplant 363
52. Potatoes 368

The Others

53. Okra 378
54. Sweet Corn 383
55. Sweet Potatoes 391

Seed Crop Production

56. Seed Growing 399
57. The Business of Seed Crops
    by Ira Wallace 408

Resources 415
Index 429
About The Author 435

Foreword

by Lynn Byczynski

Editor and Publisher, *Growing for Market* magazine

www.growingformarket.com

November 2012

This book could not have come at a better time. Vegetable gardening is more popular than it has been in decades. Across North America, people are trying to grow more of their own food. Young adults and retired people are starting market gardens, teenagers are working on urban farms, families are tilling up their lawns to plant gardens for themselves and neighbors. If you are among the vegetable growing population, whether you have a small backyard plot or many acres, *Sustainable Market Farming* was written for you.

The first thing you should consider when buying a book about growing food is the experience of the author. You want to be sure that the person whose advice you'll be taking is a successful grower with hard-won knowledge to share. Next you should ask if the author can convey that knowledge in a clear and readable manner, without a lot of irrelevant fluff.

I'll let the author tell you about her long experience as a vegetable grower, but I will fill you in on the second question.

I first encountered Pam Dawling's writing in 1999 when I mentioned in an editorial in *Growing for Market*, the magazine for market gardeners, that I would like to hear from people who were growing vegetables collectively. Pam sent in an article about her work as the garden manager at Twin Oaks Community in Virginia, and I published it. Years passed and Pam continued to grow vegetables. Then, in 2005, I heard from Pam again with a proposal for an article about choosing the right onion varieties based on latitude and daylength. It was an interesting, accurate, and lucid article about a topic that mystifies many growers, and my readers responded with appreciation. So I encouraged her to write about other topics of mutual interest. Thus began her long stretch of productivity as a farm writer. Since 2006, Pam has written an article for nearly every issue of *Growing for Market*—10 to 12 articles each year, more than 70 long articles in total.

Pam's articles are remarkable for several reasons. First, she is extremely well-organized both in her work as a farmer and in her writing. She proceeds logically and explains herself in an orderly fashion. That makes this book an excellent reference work; when it's time to start seeds of celery, for example, you can flip to the seed-starting section in the celery chapter and find what you need to know. Later, you can look up how to plant, weed, water, fertilize, harvest, clean, and save seeds. Pam's writing is also comprehensive. She doesn't assume a huge font of knowledge from her readers, but instead covers all the essential information you need to be successful. Even veteran vegetable growers can glean tips from her exhaustive treatment of each vegetable.

Another thing worth noting is that Pam is a meticulous record keeper. There are plenty of expert vegetable growers out there, but not all of them have recorded data to back up their advice. Pam does. She keeps track of everything she does in the garden. She can tell you where she bought her seeds, when she bought them, when she started them, what kind of pots she started them in, what temperature she ran the greenhouse, how many extras she planted, when she fertilized, and so on and so forth. Her records begin when she decides to plant a variety, and they end when the people she is feeding eat the food and voice their opinions about its flavor.

Another unique feature of Pam's work is the fact that she grows for 100 people. It's such a nice, round number — easy to multiple or divide, depending on the number of people you're growing for. For the vegetable farmer, Pam's data provides useful benchmarks that can help with the question of how much to plant. I also like the fact that she's growing in Virginia, which is a fine place to grow vegetables, but not ideal like California. She knows about heat, humidity, hail, bugs, drought, early freezes, and the many other challenges faced by the vast majority of growers in North America.

She provides information for growers in all climate zones.

Finally, the diversity of the gardens she manages is quite impressive. She grows almost everything you can think of — as long as it tastes good. If the residents of her community like it enough to have it on the menu, she works on her production system until she gets it right. And if they don't like it, well, maybe it's not worth your trouble. Let us not forget that vegetable gardening is, at its essence, about eating.

Several years into Pam's marathon of writing about vegetables, I suggested she was well on the way to writing a book. I knew that her work was timeless; that her knowledge about growing food would benefit many people in addition to *Growing for Market* subscribers. As an author myself, I believe that those who have the ability to write well in addition to having expertise on a topic should commit that knowledge to paper for posterity. But I also know that writing a book is a time-consuming, stressful affair, particularly for someone who has a tiring full-time job. I am so happy Pam took up the task. With the publication of this book, she has created a valuable resource that will help farmers and gardeners in their quest to grow more local, fresh, nutritious food. And that's a big accomplishment.

LYNN BYCZYNSKI is the founding editor and publisher of *Growing for Market*, a magazine for market gardeners. She is the co-owner of Wild Onion Farm, a small organic vegetable and flower farm in Lawrence, Kansas, and the author of *The Flower Farmer: An Organic Grower's Guide to Raising and Selling Cut Flowers* and *Market Farming Success*.

Introduction:
Sustaining Community and Agriculture

Since 1991, I have been living with a hundred people at Twin Oaks, an income-sharing, work-sharing intentional community (commune) and ecovillage established in 1967 in central Virginia. Before I moved here, I gardened for about seventeen years in the UK.

I decided to write this book after six years of writing monthly articles for *Growing for Market* magazine, at a time when my community was searching for more ways to earn money. I wanted to contribute by doing work I enjoyed. I also thought how valuable it would be for me and other gardeners at Twin Oaks to compile my shards of information for quick reference. I knew the exact book we (and probably many other growers) needed didn't yet exist and saw that as my opportunity to provide information for small-scale sustainable vegetable growers, on crop production, planning and organizing.

Many growers nowadays are producing a wide range of crops, over a long season. Here is information about a full range of vegetables, succession planting of popular crops and season extension techniques to provide food for the complete eating season. This book details varieties that are productive and disease-resistant, and techniques that are efficient. Relatively new methods such as the use of drip irrigation, biodegradable plastic mulch, plug transplants, farmscaping (the inclusion of flowers to attract beneficial insects), sustainable weed management, and computer software for record-keeping, calculations, planning, research and marketing are included. The organic seed movement and the popularity of baby salad mixes are exciting recent directions.

I am in USDA Plant Hardiness Zone 7 (out of 11), and the American Horticultural Society Heat Zone 7 (out of 12). Books from the Southeast are rare and Southern growers are traditionally under-served, as most US vegetable production books are written in New England or the Pacific Northwest. My book starts with a Southeastern "flavor" while remaining fully useful to growers in other regions. My experience growing in England gives me familiarity with cooler summers and milder winters.

I don't address finding and buying land, or USDA Organic certification. There are already books about those things. Our farming is sustainable: we work with awareness of limited resources, ecology and the long-term future of the planet. The methods we use are organic in spirit. Like many growers, we have decided that the all-round costs of Organic certification are not a good trade-off for our farm.

This book is intended for farmers growing vegetables sustainably on a few acres, using manual labor, hand tools and some machinery. These growers may be new or experienced and want to learn more. This includes CSAs (Community

Supported Agriculture farms), market growers, growers supplying restaurants or institutions, interns and apprentices on sustainable vegetable farms, students of courses in sustainable agriculture, urban farms, multi-plot city gardens and community gardens, schools with food gardens, intentional communities, hobby gardeners stepping up into commercial organic farming—or expanding their vegetable gardens to provide a larger proportion of produce for their own households—and people working on local food security and safety issues, all of whom are looking for information on small-scale food production.

At Twin Oaks I am the manager of three and a half acres (one and a half hectares) of vegetable gardens, part of our organic farm, which also includes dairy, beef, poultry, bees, herbs, tree fruit, mushrooms, seed growing, ornamentals and forestry. The work of the garden crew supports our community. And the community supports us. Our vegetable production is like a CSA with one very big box (the walk-in cooler at our dining hall). We're also like a grower supplying a restaurant or institution.

In very many ways gardening here is much like gardening on this scale elsewhere. We share the challenges most growers have with weather, pests, diseases, shortages and gluts. My managerial tasks are similar to those of other growers—getting organized, keeping up with the schedule, finding enough energy, keeping ahead of ten people, running back and forth to issue instructions and check how clearly they came across, recognizing problems and acting in a timely way.

Like a CSA, we need to grow variety, not just specialize in carrots and garlic because they do well here. We have a captive market, so we don't need to meet and greet customers and actively

sell each bunch of kale. (We don't even bunch our kale, we pick into five-gallon/19-liter buckets and deliver it to the kitchen just like that.)

You won't find much in this book about marketing. Production is my strength, not making sales. Our version of marketing is education, labeling and presentation. We need to be responsive to our diners and our cooks. We can grow lovely parsnips, but if cooks won't cook them, they go to waste. In that way, it's like supplying restaurants—we talk with our cooks, find out what each likes to cook with and supply information about unusual vegetables. We use the feedback we're given, and figure it out one vegetable at a time.

Like other farm managers, I consider how to distribute the available hours over the whole season for maximum productivity. I disappoint some people who would like to garden in April, because I know we'll need those hours in July for harvesting and hoeing. There's a balance to be found between having fewer people work longer hours so that they get faster and more experienced, and having a bigger pool of people learning.

In other ways my role as a manager at Twin Oaks is quite different from that of an owner-manager. I'm constrained from taking big risks and I'm cushioned from big calamities. My budgets, both money and time, are decided by the whole community of a hundred people, and balanced against other calls for resources. I hand in my annual budget requests and then I adjust my plans to fit the money and hours provided. Adjusting to fit money and labor available is probably familiar to all farmers.

As far as money goes, I alone can't decide to make a major investment, for instance, to double the greenhouse space. It means that the money I live on is independent (in the short term) from whether the garden has a bountiful or disastrous year. It means that I don't have a seasonal cash flow problem. It means that we don't have to focus primarily on growing vegetables that would bring in the most money if sold on the open market. It does however require careful thought in order to get best value for each dollar spent.

At Twin Oaks Community we each have a weekly work quota of around forty-two hours. Members can pick-and-mix jobs, and craft their own careers. Consequently we have different levels of involvement in the garden work. We have a group of six to seven Full Crew members, who commit to working a lot in the garden and helping out with other responsibilities to keep the whole thing running smoothly. Each season we form a bigger pool of people (the paracrew) willing to work four to six hours a week with us.

When it all goes smoothly, this is pretty nice. It makes sure we have enough people to do the work and lets people ease in and out of garden work to fit with their other commitments, stamina, health and ability to cope with the weather. We have extra help from visitors who are here for three weeks to check out the community. The visitors are a big unknown; they might be next year's crew in the bud, or they might be people with no skills in gardening. Those who operate CSAs and have sharers picking crops have similar experiences.

One of the biggest differences between my job here and that of most growers is that I very much "live above the store." I hear people's likes and dislikes over dinner; I see what gets thrown in the compost bucket. I might run into one of our "sharers" before I've even had my first cup of tea of the day. Sometimes a "customer" scurries

The entrance to the author's community. Credit: Bridget Aleshire.

from the carrot patch, hiding a blur of orange, in case they're helping themselves too soon. One of the best times was the night I stepped out with my flashlight to turn on the irrigation, and surprised some people sleeping out in the corn patch!

As you read the drier details of our planning systems, picture a motley bunch of people doing our best to keep it all together, grappling with the humidity, bugs and heat of central Virginia in the summer and winters that sometimes seem too short to get a good rest, and other times more severe than we feel we deserve!

Virginia Climate Summary

- Twin Oaks is in USDA Winter Hardiness Zone 7a: the average annual minimum temperature is 0–5°F (–18 to –15°C)
- The average rainfall for a year is 42" (107 cm). This is fairly evenly distributed throughout the year, at an average of 3.1–4.6" (7.7–10.8 cm) per month.
- The average daily maximum temperatures in December and January are 47°F (8°C) and 88°F (31°C) in July. The average night low temperatures in January are 26°F (–3°C) and 65°F (18°C) in July.
- The growing season, from last frost to first frost is around 167 days. The approximate date of the last spring frost is April 30th (later than May 14 only happens one year in ten); the approximate date of the first fall frost is October 14 (earlier than October 1st only happens one year in ten).
- Our climate is controlled by 3 weather systems:
  1) in the main by moisture from the Gulf of Mexico
  2) by the Bermuda High Pressure area in the summer
  3) by the recurrent waves of cold Canadian air in winter
  Rainfall peaks in January, February, March and early June and August. Due to the erratic movement of thunderstorms, some parts of our area may experience long periods of drought. September–November is the dry season but is also the hurricane season.
- Our latitude is 38°N, which is very relevant to onion growing and to daylight hours
- Little plant growth occurs between November 21 and January 21, when there are less than 10 hours of light each day.

PART I

TECHNIQUES

Planning
Planting
Sustainable Crop Protection
Soil and Crop Quality
Harvest and Beyond

Year-Round Production

Food production requires planning, and the stages of a good planning process are cyclical, with information from each stage suggesting changes for other stages. Before the cycle even starts, it is important to be clear about the goals of your farming. Here at Twin Oaks Community, the goal of our garden crew is to increase our self-sufficiency and reduce dependence on the cash economy. We aim to provide a diverse, year-round supply of tasty, fresh, organic vegetables and small fruit for our intentional community of a hundred people, and also to grow enough to process for out-of-season use.

The increasing interest in buying local food and eating organic is creating a need for a dependable supply of local, sustainably grown winter vegetables as well as summer ones. Before you wail with exhaustion at the thought of more work and no rest, let me emphasize that all farmers need time to rest, and this needs to be incorporated

into the farm's schedule. It might take the form of a complete shutdown of the farm's interaction with the public, or a slowdown that allows all the workers to take turns vacationing. Give careful consideration to what you can do to extend the season without overworking yourself, your crew, or your soil. If you decide to provide produce during the winter, you'll find that the pace is naturally slower: few weeds germinate and established crops need less attention. It's not a second hectic summer.

We've identified 16 factors that help us to keep good food on the table year round:

1. Planning: During the winter we spend a few hours each week working on some aspect of planning the coming year's work. This helps us make best use of our land, money, people, climate and crops. We also do some mid-year planning for the second half of the year, for the hoophouse and for the intensive raised bed area where we grow many different crops in quick rotation. We recognize that a plan is just a *plan* of what we intend to do: if the situation changes we can always change the plan. Having a map and a schedule helps us make the best use of the growing season and minimizes the need for last-minute, middle-of-the-field, brain-frying calculations in August. I create a field manual, with all the most important maps, schedules and crop specifications inside plastic sheet protectors. See the next chapter for all the details on this.

2. Caring for the soil: Compost, cover crops and organic mulches such as spoiled hay or old sawdust all improve the soil. Getting an annual or biennial soil test and amending with any needed lime, gypsum or other minerals will help increase yields. A good multi-year rotation schedule for the main crops will also help get the most from your soil, by varying what is drawn from it each year.

3. Gearing up: Having appropriate, functioning machinery and tools, as well as an ample irrigation system, ensures that productivity is not limited by your equipment. Implements need to fit the scale of the farming and the number of people available to do the work. We're growing on 3.5 acres (1.5 hectares). What we have is workable, although not ideal. We use a John Deere tractor for disking, compost spreading and bush-hogging, an 11-hp BCS "walking tractor" (roto-tiller), many scuffle hoes, an Earthway manual seeder, some drip irrigation, seven overhead rotary sprinklers (Rainbirds), seven wheelbarrows, six Garden Way carts and many stacks of plastic five-gallon buckets. We also have lots of helpers.

4. Research and information: One of the most important farm implements is the brain! Gathering (and retaining) information helps avoid silly mistakes in the field. Books, websites, seed catalogs, conferences and field trips to other farms can feed the farmer's mind and spirit and lead to better crops. A good filing system, in both paper and electronic formats, keeps the information accessible.

5. Choice of crops and varieties: Every year we try to introduce a new crop or two, on a small scale, to see if we can add it to our "portfolio." Sometimes we can successfully grow a crop that is said not to thrive in our climate. Rhubarb works, but brussels sprouts really don't. We like to find the varieties of each crop that do best for our conditions. We read catalog descriptions carefully and try varieties that offer the flavor, productivity and disease resistance we need. Later we check how the new varieties do compared with our old varieties. We use heirloom varieties if they do well, hybrids if they are what works best for us. We don't use treated seeds or GMOs, because of the wide damage we believe they do.

6. Maximizing plant health: Keeping plants growing well, by preventing and controlling pests, diseases and weeds, will lead to a longer productive crop life and a longer-running food supply.

7. Overwintering crops: Kale, collards, spinach, leeks and parsnips can all survive outdoors without rowcover in our climate (USDA Winter Hardiness Zone 7). We can harvest small amounts throughout the winter, and when spring arrives, the plants perk up and give us big harvests sooner than the new spring-sown crops. Arugula, mache (corn salad) and some other small greens are very winter-hardy too.

8. Season extension: The supply of a crop can often be extended at both ends of its normal growing season. Usually an extension of two or three weeks takes only a little extra vigilance and a modest investment in rowcover or shadecloth. Naturally, the further you try to extend the season of a crop beyond what is normal for your climate, the more energy it takes and the less financially worthwhile it becomes. We have recently discovered the wonders of biodegradable plastic mulch, such as Eco-One and Biotelo Mater-Bi. These mulches warm the spring soil and bring melons to maturity three or four weeks earlier.

9. Indoor growing: A hoophouse is a very good investment for winter crops, as the rate of growth of cold-weather crops is much faster inside, and the quality of the crops, especially leafy greens, is superb. Even though we had expected good results from a hoophouse, we were amazed at just how incredibly productive it was. Also, working in winter inside a hoophouse is much more pleasant than dealing with frozen rowcovers and hoops outdoors. Greenhouses and coldframes also offer opportunities for cold-weather cropping.

10. Transplants: Using transplants often makes multiple croppings possible in a bed in one season, because it reduces the length of time each crop needs to be in the bed. It also extends the season in the spring by allowing plants started inside in milder conditions to be set out as soon as the weather is mild enough, giving them a head start over direct-sown crops. And it means over-wintered cover crops can be left to grow longer (for example, until clovers, vetches or peas begin to flower), for improved soil nutrients.

11. Succession cropping: We plant outdoor crops here in central Virginia every month. Admittedly, in December and January the only things we plant are multiplier onions (potato onions). We grow nine plantings of carrots, six or seven plantings of sweet corn, five or six of cucumbers, squash, zucchini, edamame and bush beans. We do almost fifty plantings of lettuce! Cowpeas and limas get two plantings. This means as one planting is passing its peak, a younger one starts to be productive. Some crops grow here in spring and again in the fall, so we make the most of both seasons. Examples include broccoli, cabbage, spinach, kale, collards, turnips, beets, potatoes and many Asian greens. I recommend recording dates of sowing, first harvest and last harvest for each planting. You can use this information to determine the best sequence of planting dates for keeping up a continuous supply.

12. Interplanting and undersowing: Sowing or transplanting one crop (or cover crop) while another is still growing is a way of increasing the productivity of the land. Sometimes it enables a cover crop to get established in a timely way that would not be possible if we waited for the food crop to be finished first. We undersow our last sweet corn planting with oats and soybeans, which then become the winter cover. We interplant peas in the center of spinach beds in March, and plant lettuce either side of peanuts in April. We also undersow our fall brassicas with clovers in August, to form a green fallow crop for the following year.

13. Storage: We store potatoes in a root cellar; sweet potatoes, winter squash, pumpkins, garlic and onions in a basement; carrots, beets, turnips, rutabagas, celeriac and kohlrabi in a walk-in cooler; and peanuts in the pantry. Meeting the different storage requirements of various crops helps maximize their season of availability.

14. Food processing: We have a food processing crew who pickle, can, freeze and dry whatever produce we don't need to eat right away. They also make sauerkraut and jams. We make use of a solar food dryer and a small electric dehydrator. Processed (or "value-added") foods effectively lengthen the season, without requiring out-of-season growing.

15. Crop review: During the main growing season, we don't do a lot of paperwork. We record planting dates, and for our succession crops we note the harvest start and finish dates. We label each crop in the field with a row tag. When the crop is finished we pull up the labels and consign them to one of two plastic jars in the shed: "Successes" or "Dismal Failures." On a rainy day in fall, I transfer the information to the notes column on our Planting Schedule. At the beginning of the winter, we take time to discuss and write up what worked and what didn't, so that we learn from the experience and can do better next year. This is an example of those triangular cycles recommended in personal growth literature and management workshops, which rotate through three stages: "Plan–Execute–Review" or "Learn–Do–Reflect."

16. Lots of help: Last but by no means least, we arrange our work so that unskilled visitors and new community members can join in and be useful.

Protecting plants with a hoophouse or rowcover can extend the season. Credit: Kathryn Simmons.

Chapter 2

Create Your Own Field Manual

No one has the same farm you do! This chapter will give an overview of our winter annual planning process, and help you create a handy, customized reference file you can consult when anything seems unclear during the hot days of the busy season.

My dedication to winter planning came from the time I found myself standing in the full sun in the middle of the field with a tape measure, notebook and pencil, trying to figure out how many rows of sweet corn I could fit in, and how long to make them. My brain wasn't functioning at its best, and I was under pressure to get seeds in the ground. There had to be a saner way — ah, winter planning!

Once we have completed one step in our winter planning process, we print out a copy of the final spreadsheet, map or list and decorate it with a whimsical sticker. This lets us easily tell one sheet from another, and the final corrected versions from earlier drafts. The order in which we do these steps means that the information we need is gradually transferred along the chain, and we don't need to keep going back to consult the many different notes made during the previous year. At the top of the sheet we list which other charts, spreadsheets or maps are needed to compile the new one for that stage. At the bottom we list places to post or file copies, and which subsequent planning stages to pass that information on to.

Planning is circular, just like farming itself. For new farmers this can be daunting, but each year it becomes easier, as you are only tweaking the plan you used last year. My description of our planning process will include some pointers to different options and "starting" places.

Our planning sequence

1. Accounting — reviewing the year's numbers, planning next year's budget.
2. Crop Review Meeting. (See Chapter 3.)

3. An Annual Report for the community — what worked and what didn't, changes we plan for next year.

4. Plan the Main Garden Layout, following a ten-year rotation, noting changes suggested by the Crop Review. Fit in the smaller succession crops. (See chapters 4 and 7.)

5. Revise the Inventory Spreadsheet. Do the physical inventory of seeds left at the end of the year. (See Chapter 5.)

6. Prepare the Seed Order Spreadsheet. Decide what to order and place seed orders. (See Chapter 5.)

7. Spend any available end-of-year money on supplies or tools.

8. Revise the Hoophouse Planting Schedule and maps. (More in chapters 15 and 16.)

9. Revise the Greenhouse Seedling Schedule (before the first seeding date comes around!). (See Chapter 6.)

10. Revise the Outdoor Planting Schedule.

11. Plan the Labor Budget for the year — amount needed and when it will be needed.

12. Revise the Raised Bed Planning Chart and plan raised bed crops for Feb–June. (See Chapter 4.)

13. Revise the Garden Calendar (month-by-month list of tasks).

14. Revise the Harvest and Food Processing Calendars (what to expect when).

15. Revise the Lettuce List and Lettuce Log (a regular supply of lettuce every week is very important to us).

16. Update the Crop Planting Quantities Chart (gives us information on longer-term trends and choices).

17. Revise the Perennials Worksheet (a monthly checklist for each of the fruit crops and the asparagus).

18. Revise the Veg Finder (chart of succession crops and where to find each harvest).

19. Revise the Fall Brassica Spreadsheet (timing of sowing for fall brassicas is quite precise and complex).

20. Write up or update plans for specific crops we want to pay close attention to: onions, for instance.

21. Write a Seed Saving Letter to nearby growers, so we don't compromise isolation distances.

22. Revise and post a Paracrew Invitation (recruiting for casual help for the season).

23. Revise this list, file winter research notes, prune old files, discard junk.

A plan is just a plan!

Some people seem fatalistic about plans: "Oh, plans never work out!" or "I hate to be controlled by a plan." Probably most market growers see the value in planning, even if some days our plans unravel. A plan is just a plan! If a better idea comes along, or the situation changes, then you can change your plan. In agriculture we have to be ready to adapt, as many things are outside of our control. Having a good set of plans actually makes it easier to make changes and see how this will affect other parts of the farming year. A Notes column at the side of every sheet gives space to write in any alternative idea you already have. For example, our schedule for sowing winter squash reminds us that if sowing is delayed, we should not sow the slow-maturing Tahitian Butternut, but replace it with quicker maturing varieties. A Notes column is also useful for writing in anything different that you end up doing, and whether to make this a permanent change in future years or not.

Different styles of planning

Each farm will have its own style of planning. Some farmers prefer hardback notebooks or loose-leaf binders. We use spreadsheets. During the year we work off the printed sheets; we don't often go back to the computer. All our important sheets are in our Field Manual.

The main value for us in using spreadsheets is that the program will do calculations for us. We enter how many cabbages we want, the in-row spacing and the row length, and out pops the number of rows. We can quickly switch to a different number to make a whole number of rows. Then we can enter the harvest date and the days to maturity and out pops the transplant date and the sowing date, along with the number of starts and the number of flats to sow, allowing a percentage extra.

The second advantage of using spreadsheets is the ability to quickly sift out selected parts of the information and rearrange it to give us, say, a list with just the forty-six lettuce sowings in date order, or just the crops planted in the East Garden, or the seed orders sorted by supplier, so we can print these out separately for entering online.

Some farms are so well-organized they post their plans on their website. Jean-Paul Courtens of Roxbury Farm, a thousand-share CSA farm on three hundred acres in Kinderhook, New York, has posted an impressive array of information. Under the Farm Manuals tab on their website you'll find details of CSA share amounts, greenhouse schedule, and planting schedule for a hundred-member CSA. Courtens is also willing to send you their 1,100-member schedule. Posted are their manuals for each crop, their harvest and storage manual, soil fertility management plan, and crop rotation details.

If you would rather buy a set of spreadsheets than construct your own, a valuable resource is the Market Farm Forms package, which can be ordered in digital format or print plus DVD. See the Resources section for details.

Start somewhere!

If you have the time and energy, doing detailed field map layouts before you place your seed order is the best option. (We use hand-drawn photocopied maps. Some growers make maps using a spreadsheet. Databases can use aerial photography.) If you don't, make a rough plan for how much of what to grow for the following year and do the detailed maps later. (If you don't know how much of what to grow, see Chapter 7.) We prioritize getting our seed orders in early, so if necessary we postpone our maps. Making accurate maps after your seed order has the disadvantages that the space available for that crop might not be exactly as hoped, that ideas sometimes change, and the seed bought won't exactly match the need! But it will be close.

Some growers start instead with a Harvest Plan (how much of what they want to sell), then use yield projections to make a Planting Plan, followed by maps; then they revise the harvest and planting plans to make them fit the land available. After that they do a seed order. Then they look at desired harvest dates and extract field planting dates, and greenhouse sowing dates. The sequence you use depends partly on how drastic the changes are from last year, and partly on what makes best sense to you.

Here I'll describe the main planning steps of the sequence I use (that don't have their own chapter), and then the contents of my Field Manual.

Outdoor planting schedule

The outdoor (field) planting schedule is a list by date of what we intend to plant outside, how much, what spacing, and where. This is the plan we keep on a clipboard in the shed, and on which we enter details of crops we've planted at the end of each shift. It's assembled from information from the seed order as well as the maps made earlier. It has an open column to write in when we actually plant, a Notes column, and a "Success?" column to check off during the harvest season, all to inform the next Crop Review. We list each of the varieties, its row length and row spacing, and, for transplants, how many plants will be needed to achieve the stated in-row spacing (including 20% extra, so that we can select the sturdiest plants). The "Where" column lists the bed number or plot. If we make changes, we cross out the planned information and write in what actually happened. If we think our change is an improvement to keep for next year's plan, we circle the new information; otherwise we consider it a one-time anomaly.

Auditing accounts and planning the budget

By the end of the calendar year we need to sort out our money budget, request the budget we'd like for the next year, and then, once budgets are set, cut our cloth to make best use of the funds we get. Other farmers may have different timelines, but at some point it's necessary to look at the money, sort out any errors, assess the financial well-being of the farm, and decide if anything needs to be done differently. This is a point at which we see if we can afford to buy more tools or supplies. Our wish list is usually bigger than the budget, so we have to prioritize. At the end of one year, we had just the right amount for a Valley Oak wheel hoe, and what a boon it has been!

Outdoor Planting Schedule

Transplant or sow date	Actual planting date	Crop	Variety	Row length in feet	Plants per 100 ft	In-row space in inches	Inches between rows	Where	How many plants	Success?	Notes
Jan 25		Potato onions	Small Ones	540	240	5		RB	1555		Harvest ~ June 12
Sow: Feb 1		Fava beans	Windsor	30			16	RB			
Sow: Feb 14		Carrots #1	Danvers	360				RB			Germ ~ day 15
T/plant: Feb 21		Spinach	Tyee	1080	200	6		RB	2592		5 spd/bed or t/pl from hphs
Sow: Feb 28		Carrots #2	Danvers	360				RB			Germ ~ day 15
T/plant: Mar 1		Onions	Frontier 95d storer	720	300	4		RB	2592		2 beds, B, C
T/plant: Mar 1		Onions	Gunnison 102d storer	325	300	4		RB	1170		bed E
T/plant: Mar 7		Greens	Senposai	270	68	18	9	RB	220		1 bed. 3 rows/bed

Labor budget and recruiting help for the season

This step involves figuring out how much labor we are likely to need when, and then finding workers, whether interns, apprentices, full-time or casual labor. We have a spreadsheet for planning this too, slowly increasing the number of hours per week and the number of workers per day from January. We try to pace ourselves to ensure we get enough people during time-intensive harvesting and transplanting. We steel our hearts against the many people who want to help in April but will have gone indoors by July. Obviously, the issues are different depending on whether the labor is paid or voluntary. Even if it's free, you still need to organize everyone and find them tools.

Garden calendar (monthly task list)

This is a month-by-month list of tasks, taking a half to a whole page for each month. This calendar gives new workers an accessible overview of what we hope to accomplish each month, and reminds us of tasks we might otherwise forget. It includes not just the seeding and transplanting jobs, but also prompts to weed the strawberries, sort through the potatoes two weeks after harvest, look out for Mexican bean beetles, divide the rhubarb, and so on. It includes a section on climate information, latitude, and daylight length, and a list of books and websites I find useful. When we revise it, we include things we plan to do differently this year, to prevent automatic pilot taking over.

Harvest and food-processing calendars

These are lists of which crops to expect when. When I revise these calendars, I list a date halfway between what we had in print and what happened last year. This way we can gradually zero in on the likely date without wild pendulum swings to dates based on variable weather. I have the same list sorted by date and alphabetically by crop. I also make a version that shows when crops are abundant enough for our food-processing crews to start canning and freezing. An idea of when to expect lots of paste tomatoes can be very helpful in ensuring that people are available to process them and not on vacation.

Lettuce list and lettuce log

This is an example of a crop where sowing and transplanting dates need to be quite precise if you want a continuous supply. It has taken me several

Harvest Calendar, by Crop

Starting date	Crop	End date	Harvest frequency	Storage for out-of-season use	Notes
Oct 20	Arugula, fall & winter	Mar 15	As needed		We may not grow it.
Apr 5	Asparagus	Jun 6	Daily	Freezer	
Jul 10	Beans, asparagus	Oct 15	3 × week		Ends with frost
May 20	Beans, bush	Oct 15	3 × week	Freezer.	Die with first hard frost.
Jun 10	Beans, fava	Jun 10	One harvest		
Aug 15	Beans, limas	Oct 15	2–3 × week		Ends with frost
Sep 20	Beet greens, fall	Nov 15	3 × week	Freezer	
May 1	Beet greens, spring	Jun 30	3 × week	Freezer	

Dates are weather-dependent. See Garden Full Crew for current info. Note that quantities may be small and erratic at both ends of the season for that crop. Check against: Garden Task List, Hoophouse Log

Harvest Calendar, by Starting Date

Starting date	Crop	End date	Harvest frequency	Storage for out-of-season use	Notes
Jan 1	Lettuce	Endless	Daily as needed		
Jan 1	Scallions, spring	June 25	3 × week		
Jan 1	Turnips, winter hphs	Feb 28	As available		From hoophouse
Mar 10	Garlic scallions	April 20	3 × week	Freezer	
Apr 5	Asparagus	June 6	Daily	Freezer	
Apr 5	Sen po sai, spring	May 20	Daily as needed		

Dates are weather dependent. See Garden Full Crew for current info. Note that quantities may be small and erratic at both ends of the season for that crop. Check end of list for crops that carry over throughout the winter. Check against: Garden Task List, Hoophouse Log

years of fine-tuning to get an almost year-round supply without huge gluts. Any crop you are focusing on improving may warrant its own plan and recording sheet. See the lettuce chapter for more on this.

Onion plan

This is another crop we are paying close attention to. We have been trialing different varieties, so the planning is complicated and the record keeping important. We keep track of locations and monitor growth a few times during the season, then record the harvest of each variety separately to compare yields. Later we compare the keeping qualities.

Fall brassica spreadsheet

Planting dates for fall crops can be exacting, because cooling weather slows plant growth, and a day or two difference in sowing date can make a week or two difference in harvest date. With crops like broccoli, this can mean six weeks of harvest versus four, where the extra 50 percent of time means more than 50 percent extra yield because it takes place during warmer weather. We plant six to eight broccoli varieties with varying days to maturity, four to eight types of cabbage, up to eight types of Asian greens, two kohlrabis, one variety of kale and one of collards. All this means lots of seedlings to keep straight.

Crop planting quantities chart

This step is used for long-term planning/musing. It's simply a chart of how much of each crop we plant each year. We have about fifteen years of data collected. The chart helps us see if we are unconsciously drifting towards more or less of certain crops when we adjust our plans to fit available space. It's also useful to look at how we did things in "Olden Days" — different plant or row spacings, etc — and helps us be more intentional about what we do now.

Revise the planning schedule, file notes, prune old files, discard junk

This is the last task of our winter planning, a kind of house cleaning, so that once things get busy I can find what I need more easily instead of having important bits of information on tiny scraps of paper or lost in a sea of outdated trivia.

Compiling your own Field Manual

As our planning got more detailed, I needed to keep the information close by, in a relatively

Crop Planting Quantities

Crop	2006	2007	2008	2009	2010	2011
Beans, asparagus	0	30'	45'	38' × 2	38' × 2 = 76'	38' × 2 = 76'
Beans, bush and pole (double rows)	1,071' + (120' pole)	740' + 130' pole	895' +130' pole	180' + 170' + 180' + 180' + 140' + 180'	180' + 180' + 173' + 180' + 140' + 180' + 60' hphs = 1,093'	180' + 176' + 176' + 180' + 180' + 92' hphs = 1,044'
Beans, drying					340'	180'
Beans, edamame	2 × 180' + 90' hphs = 450'	670' + 90' hphs = 760'	5 × 90' + 90' hphs = 540'	90' + 65' + 65' + 60' + 60' + hphs	90' + 65' + 65' + 60' + 180' + 88' = 548'	45' + 88' + 90' + 90' + 90' + 9'2 hphs = 495'
Beans, fava	0	0	30'	45'	180' + 45' + 24' hphs = 249'	120'
Beans, lima	360'	140'	180'	180' × 2	120'	0
Beets	3 × 360' + 3 × 360'	3 × 360' + 1,285' + 80' bbb	3 × 360' + 3 × 360' + 100' bbb hphs	360' × 3 + 360' × 3 + bbb hphs	1,080' + 450' + 630' + 80' bbb hphs = 2,240'	1,080' + 1,080' = 2,160' + bbb 100'

(Check: Crop Review, Garden Plan, Raised Bed Plan, Veg Finder, Succession Crop Planning Sheet)

Note: hphs = hoophouse; bbb = Bulls Blood Beets

weatherproof and portable package. My solution is clear plastic "sheet protectors" — pockets you can slip a sheet of paper in, and file in a ring binder. The plastic keeps the pages fairly clean and dry, so I can take it out to the field with no worries. You can customize the manual to make crew versions, shed clipboard versions, customer versions, and website versions.

My current Field Manual includes spreadsheets, maps, charts and lists from our planning, along with other useful pages, such as:

- Tables of Soil Temperatures for vegetable seed germination, and Days to Emergence at different temperatures (for sowing and flame-weeding decisions)
- Ten-Year Rotation Pinwheel (see Chapter 4)
- Winter Cover Crops Maps
- Cover Crops Information and Chart
- Onion Planting Plan and Log
- Sweet Potato Slip Growing Plan and Worksheet
- Farmscaping Worksheet and suggestions
- Virginia Extension Vegetable Planting Guide for Spring
- Virginia Extension Fall Gardening Leaflet
- Sunrise and Sunset Timetable
- Map of the Blueberry Patches, and Monthly Care List
- Map of the Grape Rows and Monthly Care List and Log (I'm monitoring thirteen new varieties)
- Phenology Log
- Plastic Card Calendar (free from an insurance company).

Chapter 3

Crop Review

We have developed a tradition of having a Crop Review Meeting when each growing season slows down. (I had typed "at the end of each growing season," but it never really ends here in central Virginia; it just slows for a couple of months.) We encourage all the crew to attend by making it count as work hours, talking it up as interesting, and providing snacks halfway through. Our goal is to review how the season went while we can still remember it, and get the information in writing so we can use it in making our plans for the next season.

The format

Our basic format is to go alphabetically down a list of crops, using a spreadsheet we've prepared in advance. Anyone who remembers anything notable about that crop speaks up, and someone takes notes on a laptop. We usually time it so that while we talk we have seed garlic to separate into cloves, for planting the next day, or some other hand work. We allow five hours (all afternoon) with a break in the middle. We grow a lot of different crops and have a lot of different plantings; a farm that specializes in fewer crops would not need as much time. Naturally this is also a time for mutual congratulation and appreciation, reliving the highlights, and a few hilarious or rueful diversions when we recall the more disastrous events of the year. Among comments like "harder to pick," "not enough water," "Five blahs" (Red Sun tomato), and "don't remember," we have "Walla Walla: the big unkeeper," and "Zucchini Spineless Beauty: Fruitless! Pointless!" And then our silly mistakes: Broccoli Green Comet: "One row lost to sweet potatoes," and Acorn Squash: "Gap filling resowing hoed off, let's not do that again."

Preparation

Like most farming activities, the success of the crop review is partly related to the quality of the preparations. In the weeks before the Crop Review, I gather up all the miscellaneous scraps of information I can find about our plantings. Some of these are in my pocket notebooks, some on the backs of seed packets, some in more organized places like the Planting Schedule and the Seed Order. We also file our row labels (cut from Venetian blinds) after the crop is finished, in two plastic tubs, one for "Successes" and the other marked "Dismal Failures." If the crop wasn't very successful, we write a few words about why on the label before putting it in the Dismal Failures tub. It doesn't have to be a total disaster to qualify. I started out with tubs labeled "Successful" and "Unsuccessful," but the words looked too similar for a busy person to distinguish, hence the exaggerated wording we now use. We take each tub, sort the labels by month, then by date, and record the success or otherwise. We use a column on our Planting Schedule for this. It helps to consolidate the information in just a few places. I copy down any comments written on the label about the problem we encountered, bang the dirt off the labels and collect them together for cleaning and reuse (at the Crop Review Meeting, or on a rainy day).

I also post a paper on the community bulletin board asking the cooks and diners who have been eating our produce all year to write down any comments they have about what went well, and any requests for more/less/different produce. It's equivalent to asking CSA members in a newsletter or bag-note to send in their opinions. This input is taken to the meeting.

The next task is to prepare the spreadsheet for the Crop Review. This takes me two to three hours. I work from last year's Crop Review and Seed Order. I copy last year's Crop Review to a new worksheet, and empty the comments columns. We have a "General Comments" column for each vegetable and one for "Comments on Varieties." Referring to the Seed Order for the past year, I enter the row length and the variety names for each crop. I keep the previous year's information and reduce its prominence by reducing the point size and putting it in italics (it can be useful for comparisons). I also refer to any supplementary seed orders made during the year that were not recorded in the main spring Seed Order.

I tidy up the spreadsheet, print out a copy, then sit down in a comfortable chair and proofread. I might add details such as where and when that crop was planted, or divide the row length up between the spring and fall plantings. Then I print out enough good copies for us to look at while we talk.

The next important task is to line up the snackmakers! While we recognize the power of caffeine and sugar to keep us alert, we also cater for those avoiding these things. Our rule is that we have to get past Lettuce in the alphabet before we break out the snacks. It keeps us focused and prevents us from talking for too long about the crops early in the alphabet.

The Crop Review Meeting

This meeting replaces the outdoor work shift for the afternoon, providing a welcome change from the routine. Participation is not compulsory and many of the "paracrew" members who are less involved in the organizing and planning don't take part. We expect the main crew people to make a high priority of being there.

We hold the review in our dining hall. The facilitator keeps us on task and watches the time. Some people can concentrate better if they have hand work, and usually there is someone who can

Garden Crop Review

Vegetable	Row ft	General comments	This year's varieties	Last year's varieties	Comments on varieties
Beans: bush	180		Bush Blue Lake	Bush Blue Lake	
	675		Provider	Provider	
	133		Burpee's Stringless Greenpod		
	90		Burpee's Tender Pod	Strike	
	48		Jade	Jade	
Beans: asparagus	76		Purple Pod	Purple Pod	
Beans: favas	74		Windsor	Windsor	
Beans: edamame	480		Envy 75d	Envy 75d	
			Envy 75d		
			Envy 75d		
			Envy 75d		
			Envy 75d		
			Envy 75d		
Beans: limas	180		Jackson Bush	Jackson Bush	
Beets	630		Detroit Dark Red	Detroit Dark Red	
	900		Cylindra	Cylindra	
	100		Bulls Blood Beets	Bulls Blood Beets	
Broccoli: 3,550' total	758		Arcadia 94d	Arcadia s/f 94d	
	380		Blue Wind 72d	Blue Wind 72d	
	222		Diplomat 90d	Diplomat 90d s/f	
				Green King 85d	
	535		Gypsy 78d	Gypsy 78d s/f	
	265		Marathon 89d	Marathon f 89d	
				Packman s/f 73d	
	890		Premium Crop 82d	Premium Crop s/f 82d	
	222		Tendergreen	Tendergreen	
	488		Windsor 77d	Windsor 77d	

focus best if they are writing the notes, so they get that job. These days the use of a laptop to record the comments straight onto the spreadsheet saves us the time we used to spend transcribing. We bring all the crop records we have, plus the seed orders and the planting schedules (outdoor, seed starting and hoophouse).

In reviewing each crop, we consider yield, quality, flavor, ease of harvesting, pests and diseases, effects of whatever kinds of weather we had, as well as the popularity with our diners and cooks, and timing as far as space use and keeping up a continuous supply of that crop or a good flow of produce in general.

We write in any new crops we're interested in trying at the end of the spreadsheet. (We also consider new crops when we're doing the Seed Order in late December. We entertain ourselves by allowing a 30' × 4' (9 × 1.2 m) bed space for an experimental crop or variety for each person participating in putting the Seed Order together, but that's another story.)

After the Crop Review meeting, someone types up all the comments made (if we didn't use a laptop at the meeting). We use the information to plan improvements, and to fine-tune how we do things. We try really hard not to make the same mistake two years running! We file a copy and keep copies to work from when planning the crop layout for next year, composing the Seed Order, and making up the planting schedule. I also write up an Annual Report or "Informant" for the community.

The Crop Review starts us on a sequence of planning tasks that we do in the winter, with the information from each step going on to the next stage.

Popping garlic cloves in preparation for planting makes an ideal task to do at the Crop Review meeting. Credit: Southern Exposure Seed Exchange

Chapter 4

Crop Rotations for Vegetables and Cover Crops

Planning a crop rotation will help maximize the productivity and health of the land, reduce diseases and pests and make the planning work more predictable and easier on the brain. Before we made our Ten-Year Rotation Plan for our gardens at Twin Oaks, our basic scheme was to start all the early spring crops in one of our main garden areas, in order to simplify the tractor cultivations. All of the other gardens could then be disked up later in the spring. After that decision, the planning consisted of trying not to plant the same crop where it was planted in recent years, in an ad hoc way. Eliot Coleman's *New Organic Grower* inspired us to create a multi-year rotation for the main part of our garden. We now have a mixed system, with sixty permanent raised beds, each 4' × 90' (1.2 × 27.4 m), and ten plots of 9,000–10,600 ft² (836–985 m²), in three areas of "flat" garden.

Each farm needs to customize its own plan, depending on your goals. You might not want as much winter squash or potatoes as we do, for instance. You may want to have a nice long winter break, rather than practice season extension of greens. Perhaps there are certain crops you don't grow because they are too time-consuming to harvest in large quantities.

Permanent raised beds

We cultivate sixty raised beds manually or with a walk-behind tiller. For the raised beds we still don't use a permanent rotation plan: we like the extra flexibility of our ad hoc method. We fertilize the soil in these beds generously and never compact it. We use the space very intensively and get high yields. We plant a new crop as soon as we clear an old one. Some beds will get two or three crops in one season. If we have a four-week gap in the summer, we grow buckwheat; if six weeks, we add soy to the buckwheat.

A portion of the sixty permanent raised beds in early summer. Credit: Pam Dawling.

We use the beds for crops we grow in small quantities (celery, okra), very short-term crops (like lettuce), things we need to cosset (eggplant, because of the flea beetles, or early muskmelons), experimental crops we want to keep a close eye on and things that wouldn't fit anywhere else. The raised beds are more accessible for winter harvesting, and are a good place to grow crops such as kale, collards and leeks that we'll be keeping long after the rest of the gardens are in winter cover crops. We also use them for the very earliest crops such as peas, as the beds can be cultivated earlier than the flat gardens.

We plan the raised bed crops twice a year: once in the winter for the crops planted before mid-July and again in mid-June for the crops for the second half of the year. This two-part planning allows us the flexibility to respond to unexpected situations: crop failures, sudden needs for more of something sooner than we'd planned, something taking longer than expected to reach maturity, etc. A vital tool for this is our Colored Spots Plan, an outline map of the raised beds that shows the history of the crops planted in each one. Green spots indicate brassicas, orange spots umbelliferae (carrots, celery, celeriac in our case), purple spots show beets, chard and spinach, and so on. At the year-end, we put a vertical blue line after that year's spots: for overwintering crops we draw the vertical line through the spot. It's easy to fit many years' worth of spots on the same map, which helps us see at a glance which crops have been planted where in recent years. See the Colored Spots Plan in the first color section.

The crafting of these complicated little raised bed maps has become an enjoyable challenge for us. We sometimes call it Vegetable Sudoku—there seems to be only one right answer for each space! We gather the best minds on the crew and retire to armchairs for this job. We make a chart of how many beds of which crops we want, divided by family, along with the planting date and final harvest date. The amounts we hope to plant come from the previous planning stages, and the Seed Order sets the maximum we can plant (unless we order more seeds). We try to have two or three years' gap before the same crop family returns to a bed. One person reads from the Colored Spots Map which beds are suitable for the crop family this year, and others look at the chart to see what can work best time-wise. This way we manage to shoehorn in more crops during the season than we otherwise would. Sometimes this process leads to the sad realization that everything we want isn't possible, and we have to go that year with one bed of spring collards rather than two. Sometimes it leads to a more creative solution, such as trying five rows of carrots in a bed rather than four, or planting lettuce as we remove spinach, working along the bed. We always have to do a bit of back-tracking and some fudging of the rotation to a less-than-perfect match, but we do end up with a plan that uses the beds fully.

It might be a more efficient use of time to design a fixed rotation, but this flexible approach is a more efficient use of land.

The main gardens

The main part of the garden is in three patches, with rows 180, 200, or 265 feet (55, 61, or 81 meters) long. I recommend breaks or gaps every 100–200 feet (30–60 meters) for ease of access, especially with heavy loads. We do the initial cultivations in the flat gardens with a tractor and disks. We also have a manure spreader for compost, a seed drill for cover crops and a potato digger. For some crops we create temporary raised beds, but other crops are grown "on the flat." Here we use our ten-year rotation, planting out major crops and also most of our succession crops of beans, squash and cucumbers. As you'll soon see, although we refer to this as a ten-year rotation, it isn't ten years between corn plantings or potato plantings. It's a ten-year plan that rotates crops.

The first stage in setting up our rotation was to figure out how much area we needed for each of our major crops (the ones needing the largest amount of space). We already had records kept over many years, of amounts of various crops grown, and each year we have the Crop Review to remind us what changes to make.

Major crops

In terms of space occupied, our biggest crop is corn. In fact, sweet corn is such a space hog that we use three of our ten plots for it! Obviously, we love it!

Here's how the area for each of our major space-occupying crops adds up:

- Sweet corn: 6 or 7 plantings of about 3,500 ft² (322 m²) each
- Spring planted potatoes: about 7,000–9,000 ft² (644–828 m²)
- Summer planted potatoes: about 7,000–9,000 ft² (644–828 m²)
- Spring broccoli & cabbage: 4,000 ft² (368 m²)
- Fall broccoli & cabbage: 7,000 ft² (644 m²)
- Winter squash: about 8,200 ft² (736 m²)
- Watermelon: about 9,000 ft² (828 m²)
- Sweet potatoes: about 4,300 ft² (396 m²)
- Tomatoes: 4,000 ft² (368 m²)
- Peppers: 2,200 ft² (202 m²)
- Garlic: about 3,600–4,000 ft² (332–368 m²)
- Fall carrots: about 3,600–4,000 ft² (332–368 m²)

Some people can work with numbers in the abstract. Others prefer a more visual planning method. We modified Eliot Coleman's method, and wrote each major crop on a piece of graph paper cut to represent the relative area needed. You can use computer programs instead, even a simple spreadsheet program. To help get a crop sequence figured out, we looked at crop families. We have three major plantings of nightshades (Solanaceae): two of potatoes and one of tomatoes and peppers together. We have two of brassicas, six or seven sowings of corn, and two of cucurbits (winter squash and watermelons). We have one crop each of alliums (garlic), Umbelliferae (carrots) and Ipomoea (sweet potatoes). It became clear to us that the 7,000–9,000 ft² (644–828 m²) crops would each occupy one plot in our rotation. Then we grouped other crops together to use about the same area: for example, two or three of our corn plantings together in one plot; spring broccoli together with overwintered garlic; tomatoes together with peppers. We had an open mind about exactly how long a rotation we would have, but when our major crops fell into nine clusters and our area into ten plots, it suggested a ten-year rotation, with one plot left over for year-round cover crops (green fallow).

We then put the graph paper pieces in a circle, like a clock face with ten "hours," and set about imagining a good sequence. (Some directions for designing a crop rotation suggest setting crops out in a line, but crop rotations are a cycle, and a circular design makes more intuitive sense to us.) We started by spreading the three corn sections three or four years apart, and the three nightshade plantings likewise. We kept the winter squash three years after the watermelon.

We drew up our ten-year rotation on a piece of card with a small central disk attached by a brass paperclip so it can rotate each year to show which crops will be planted in which plots. We call this our Rotation Pinwheel (see next page). We are still using the same piece of card we made in 1996, even though we started our second ten-year sequence in 2006. It has seen quite a bit of "white-out" and small revisions!

Some crops are said to do better following certain other crops, although just how well this folklore has been tested I do not know. Some of that information is not relevant to the rotation for our major crops as it involves crops like onions, lettuce, or beets, which we grow in our permanent raised beds, or refers to beans or cucumbers, which we plant several times a year, and never a lot at once. We fit those crops in later, according to where space is available. Potatoes are said to do well after corn, so we put our spring potatoes after

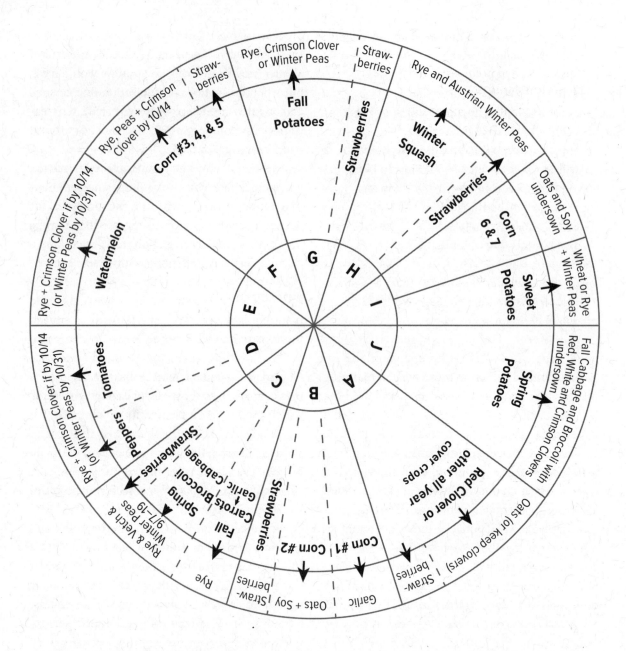

Garden 10 Year Crop Rotation Pinwheel. This shows the major crop groups with the winter cover crops that follow. Succession planting of beans, squash and cucumbers are fitted in according to space available and timing, keeping an eye on rotation too.

our late corn and our summer potatoes where the middle corn planting was in the previous year.

Including cover crops

Another one of our aims in devising our rotation was to improve our use of winter cover crops. For example, we wanted to add more legumes to our mixes. To get best value from crimson clover, for example, we need to wait until it flowers—mid-April at the very earliest—before turning it under. So after crimson clover it's best if the food crop we plant goes in later, such as later corn plantings or June-planted potatoes. Another factor is that crimson clover is best sown here before September 20, so it has to follow a crop that is finished by then. For early spring planting, a preceding cover crop of oats (maybe with soybeans) is ideal, as it winter-kills and is easy to incorporate. Oats need to be sown here in August or very early September, so they need to follow an early finishing crop. See the Cover Crops chapter for more detailed information.

For a few years prior to the change to the new rotation, we had been planting our tomatoes and peppers into a mowed cover crop of winter rye and hairy vetch. This reduces inversions of the soil, and the vetch (if plentiful) can supply all the nitrogen the tomatoes need. We like using this no-till method and wanted to incorporate it into our big scheme too. Rye and vetch is best sown here in early to mid-September, creating another restriction on which crops the tomatoes could follow. In fact these "restrictions" are more like the rules to a game, providing a structure to work within.

We formed several "strings" of a few crops that followed each other well. Spring broccoli can be followed by rye and vetch in good time

to grow a thick stand for the tomatoes the next year. Late corn can be undersown with oats and soy to provide a winter cover crop that is easily incorporated before a March planting of potatoes. Fall garlic planting can follow the early corn. Fall carrots can follow the garlic harvest. Mid-season corn is finished in time to establish rye and crimson clover, which will do well and produce lots of nitrogen and biomass before we need to plant the June potatoes. After some shuffling of paper pieces, we came up with a workable sequence for all the major crops, and also a transition plan for a couple of years to get us onto the new rotation.

Improvements

After a couple more years, we made a few improvements, and discovered Austrian winter peas as a leguminous cover crop. They can be sown as late as the end of October here, and are said to reduce the incidence of *Septoria* leaf spot in following tomato crops, so we now include them in our no-till rye and hairy vetch planting.

We also found that we could tighten up the rotation by having more than one vegetable crop within a year in a patch. We follow the spring planted potatoes with the fall broccoli and cabbage transplanted in August. This lets us keep a ten-year cycle round the ten plots while having one plot in cover crops all year round, to replenish the soil. We now undersow the fall brassicas with a mix of clovers about a month after transplanting. The following spring, we bush hog the old brassica stumps and let the clover flourish. If the plot is not too weedy, we let the clover grow all summer, mowing to prevent the crimson clover seeding. If the weeds are too bad, or the clover stand not thick enough, we turn the clover under once the warm weather has arrived, and sow sorghum-sudangrass hybrid with soy. This gets mowed to a foot (30 cm) when the sorghum-sudan is four feet (1.2 m) tall, to encourage deeper rooting for better soil drainage, and can stay until killed by the frost. If we still have the clover we may turn it under and sow oats in August. And so we get two food crops in one year and none the next from that plot. Plus it is ready early the following year for our first corn planting.

Another example of a tight rotation is that after the early corn, we sow oats, and divide the plot in two. We use half for the next year's spring broccoli and the other half gets mowed from time to time until late fall, when we disk and plant garlic there. We harvest the garlic in June, follow it with buckwheat and soy, then sow our fall carrots in late July or early August. That half-plot grows three food crops in two years.

A less elegant part of our rotation used to have one plot divided in two, for spring potatoes and cucurbits (cantaloupes/muskmelons and cucumbers). Because they are planted at very different times this involved some fussy tractor cultivations. We now grow more potatoes and fewer melons, which simplifies life.

Fitting in succession crops

The next task is to fit in the minor crops: the succession plantings of beans, summer squash and zucchini, cucumbers, cantaloupes and anything we didn't manage to find room for in the permanent raised beds. We start with a blank map of the plots and write in which major crops will go in which patch, using the ten-year Rotation Pinwheel. We look for any extra space in the plots. Off in the margin, we write when that patch will need disking or other tractor work. Usually we have four spring diskings:

1. As soon as possible in February, for the spring broccoli, cabbage and potatoes;
2. In March, for the early corn;
3. In mid-April, for middle corn, watermelon, winter squash and sweet potatoes;
4. In early May, for summer potatoes and late corn.

Factors we consider are: minimizing the number of disking occasions, to make tractor drivers' lives easier; waiting until legumes flower, if we can, before disking, to get maximum value from the winter cover crops; but also disking far enough ahead of planting to allow the cover crop to break down in the soil (dead oats are relatively quick, while rye takes more than three weeks).

Then on the map, we block in the area needed for each major crop (checking our Crop Review notes for changes in amount to plant) and calculate the remaining space in each plot. Our ten plots are not exactly the same size and our main crops don't all need equal space either, so some plots will have spaces of various shapes and sizes and others no gaps at all.

We have a special Succession Crops Planning Sheet (see the example here), where we list the spare spaces in the plots (in order of availability) and the crops we hope to plant (in date order). At the beginning and end of the season, and in mid-season when space in the main plots is tight, we also look for spaces in our raised beds. Then we pencil in arrows, fitting the succession crops into the spaces available. When we are satisfied we have the best fit, we draw up a "neat copy" of the map and transfer row length numbers to the Seed Order and any special notes to the Calendar and the Greenhouse Schedule. In this process we sometimes adjust the planting quantities from

Succession Crops Plan 2001

Plot	Where	Area Available	When	Date	Crop	#	Row feet
H	East	265' × 8' (88' × 6' × 2.7'	Late April	4/8	Cabbage	2	180' × 4'
				4/16	Beans 2 beds	1	180' × 4'
				4/20	Cukes 1 bed	1	90' × 4'
				4/20	Squash 1 bed	1	90' × 4'
				4/26	Edamame	1	90' × 4'
A	West	180' × 12.5' 2 rows 180' × 6.5'	Early May	4/30	Cowpeas	1	88' × 4'
				5/18	Edamame	2	88' × 4'
				5/12	Beans	2	176' × 3.5'
E	West	180' × 5.5'	Mid May	5/3	Cantaloupes	1	90' × 7'
				5/24	Cukes (sl + pkl)	2	180' × 6–8'
				5/24	Squash	2	90' × 5–7'
D	Cent	200' × 8'	Late May	6/7	Edamame	3	50' × 5.5'
				6/7	Beans	3	176' × 3.5'
				6/15	Drying beans		90' × 4'
D	Dogleg	60' × 5' × 5 beds	Late May	6/26	Edamame	4	50' × 5.5'
				6/23	Cukes	3	105' × 6–8'
K	East	11 beds 35' × 5'	Mid June	6/23	Squash	3	105' × 5–7'
				6/25	Cantaloupes	2	80' × 5.5'
				6/29	Beans	4	175' × 3.5'
I	East	Old Early Strawberry Beds 265' × 6'	Mid June	7/14	Edamame	5	65' × 5'
				7/15	Cukes (sl + pkl)	4	200' × 8'
				7/15	Squash	4	120' × 4'
				7/19	Beans	5	180' × 4'
				7/28	Extra carrots	9	200' × 5'
				8/3	Beans	6	180' × 4'
				8/5	Cukes 1 bed	5	90' × 4'
H	East	Foll Green Fallow	September	8/5	Squash 1 bed	5	90' × 4'
B	East	Follow Corn #3	September	9/1	Early strawberries		265' × 6'

Notes:

Beans: New patch to harvest every 20 days = 6 plantings. Pole beans need 2 weeks longer than bush beans to reach maturity.

Cukes and Squash: New patch to harvest every 25 days = 5 plantings

Cukes and Squash #2: Direct sow, unless spring is cold, in which case sow 5/17, transplant 6/7

Corn: #1 & #2 patch gets planted to oats in August, so no good for late succession crops. Corn #1 goes to garlic.

Spring Brassica: Patch gets planted in rye and vetch Sept 7; no good for late crops.

Strawberries: On the two year scheme: youngest patch each of Earlies and Lates gets used for providing new plants in July (ideally!). Older patch of each gets tilled in and becomes available in June. (If for Romas the following year, plant rye and vetch Sept 7–14.) New strawberry patches are planted in early-mid September, following part of corn#3, and part of the Green Fallow plot.

Tomatoes: To save Roma seed, need to isolate all other tomatoes by 75'–150' (75' with barriers, 150' without). Barriers: succession of sunflowers, cosmos, other white, yellow or blue flowers, planted every 3 weeks.

our original plan to make best use of the space available. This information will then go on to the Greenhouse Seedlings Schedule (for transplants), the Raised Bed Planning, the Outdoor Planting Schedule and the Crop Planting Quantities Chart.

See earlier chapters for our schedule of winter planning steps to keep us on track through the all-too-brief hibernation season, and see the Succession Planting chapter for details on timings and quantities of suitable crops for planting in successions.

Benefits

This tight crop planning might sound laborious, but for us it's very worthwhile. The division of the gardens into ten plots gives us mental and psychological advantages, in that we don't have to actively think about the whole of the area all of the time. In spring we "open up the rooms" one or two at a time to plant. By the beginning of July everywhere is in use, and in August we start to put the plots "to bed" with their winter cover crops. This annual expansion and contraction (breathing out and breathing in) of the space needing our attention helps us to stay sane and focused and keep perspective.

This system helps us get high productivity from our land, while taking good care of it. Some years ago, when I was researching plant spacing for watermelons, trying to plant as closely as pos-sible while keeping the melon size and yield up, I spoke with a farmer in the Midwest. He said if farmers in that region wanted more watermelons, they would just plow up more land, not try to plant them closer. That was a useful perspective for me to consider. Here on the East Coast, land for farming is less available and more expensive. I come from the UK, where land is even more expensive. At Twin Oaks the land available for vegetable gardening is finite. The whole community would want to consider any application from us to expand the area we use. All the other fields are fully used, for grazing or for hay. We're not enthused about clearing woodland. For growers using cultivated areas intensively, this sort of planning allows for that.

Probably the biggest snag for us in using this rotation is that it doesn't take into account that parts of the gardens with poor drainage are less suitable for some crops. Recently tomatoes went into one of our potentially wetter areas. My understanding of the El Nino Southern Oscillation system was that we could expect a wet spring, so the previous fall, before sowing our cover crop, we made raised beds. We mowed the no-till cover crop, crossed our fingers and planted. As it happened, no wet spring! For us, making temporary beds or planting on ridges in the wetter areas is easier than changing the crop rotation.

Chapter 5

Seed Storage, Inventory and Orders

In hard financial times, it's a waste to throw out all leftover seed at the end of the planting season and buy all new for next year. Yet it's a bigger waste to keep everything and risk poor germination of a vital crop. Here's a systematic method to minimize the chances of throwing out good seed, keeping bad seed, buying too little, or buying too much. The key components are:

- storing seed well, so you know that you can rely on it to grow the following year;
- planning the planting quantity of each crop for the following year;
- knowing how much seed is usually sown per 100' (30 m);
- calculating how much *your* farm uses per 100' (30 m) (optional fine-tuning);
- quantifying how much seed you have left;
- deciding which old seed to keep and which is unreliable;
- calculating how much seed to buy;
- deciding how much extra to buy in case a sowing fails;

- placing orders early, to increase the chances of getting exactly what you want.

Seed storage

Unless you store seeds well, there's no point in carrying them over for the following year. Good seed storage conditions are: dry, cool, dark and rodent-proof. Paper packets on a window ledge in the breezeway are fairly doomed! Seeds have an amazing capacity to draw in water from the air, soil, or damp fingers. Once partially hydrated, the seed respiration rate speeds up. The seeds revive from dormancy only to die when the conditions aren't good enough to really grow. When planting in our garden, we ask crew to keep one hand clean and dry for sowing the seed, and to avoid putting any seed that gets mixed with soil back in the packet. Soil always contains *some* moisture, so dirty = damp.

The paper packets which seeds are sold in are not designed for long-term storage. Foil packets are a lot better, but add to the cost of seeds, so many seed suppliers sell in paper, leaving growers to provide their own long-term storage containers. The storage life of seed is halved for every 1 percent rise in the moisture content of the seed above the very dry 5 percent. So it's wise to put freshly arrived seed packets in waterproof containers, away from humid air. Airtight conditions are generally the same as watertight: plastic, glass and metal boxes can all fit the bill. During the sowing season, we keep our seeds in plastic coffee "jars" with clip-on lids and built-in handles. Glass jars are probably more airtight, but we try to keep glass away from our garden because of the danger of cuts from broken glass. Also, the plastic jars are almost opaque, so they help keep the seeds in the dark. We group seed packets by type: one jar for nightshades, one for big leafy brassicas, one for rooty brassicas and so on. This means we take the summer squash jar to the field to sow the squash and if the Zephyr seed runs out, we can finish the row with Gentry, as we have the seed right there. Some other growers prefer to keep seed packets in alphabetical order in rectangular plastic shoebox-sized containers. Heavy-duty plastic ziplock bags are useful as a first layer of defense against the elements. We store bigger seeds like beans and corn in 2–5-gallon (8–20-liter) buckets with tight-fitting lids.

We store the seed jars and buckets at the back of the shed, where it is quite dark. At the end of April, May and June, I take the seeds we won't need again till fall (or next spring) to a cooler storage space in the basement. In April that's the spinach, peas, chard and beets; at the end of May it's the radishes, alliums, watermelon and winter squash; in June it's the nightshades, okra, peanuts and melons. The shed is dark, dry and fairly cool, but for summer, we'd rather keep the seed we won't need soon in a cooler place. For every 18°F above 32°F (10°C above 0°C), the life of seed is halved.

We put spinach seed (double-bagged for airtightness) in the freezer for two weeks or more before we need it for fall sowing. This helps break the heat-induced dormancy. I have heard some people keep lettuce seed in the fridge during the summer, and I often have good intentions, but so far I haven't managed to do that. Fridges are nice and cool, but also very humid, so seed containers for refrigerated seeds must be really airtight.

If you have seed you won't sow for many months (for instance, you deliberately bought enough for five years, or it's an heirloom from Auntie Doris that isn't available commercially), then you can store the seed in the freezer. Freezing doesn't kill seeds! Some growers believe plastic bags do let some moisture through, and they recommend freezing in glass jars, with the lids sealed with Parafilm self-sealing tape. Parafilm is a grafting tape available from Southern Exposure Seed Exchange and other companies who sell supplies for grafting or seed saving. Seeds that are dry enough to freeze are those at 5 percent moisture content or less. They snap or shatter when bent or hit—they don't mush. If you need to dry your seed further before freezing, put it in a jar with some desiccant such as silica gel until the seed snaps. Use a volume of silica gel equal to the volume of seed. In a pinch, far from town, you can use three times the volume of dried milk in place of silica.

Seed viability

Opinions vary a bit about how many years seeds of different vegetables are good for. The fuller story is that storage conditions make a big difference, as explained above. You can make your own decisions, weighing up the information supplied, your knowledge of how carefully you stored the seeds, the information on each packet about percentage germination when you bought it, and the economic importance to you of that particular crop. If you always transplant lettuce, as I do, you can risk one of your four varieties in that sowing coming up poorly, and just plant out more of the other three if it fails. Many seed catalogs include information about seed longevity, and so does Nancy Bubel in *The Seed Starter's Handbook*. Frank Tozer in *The Organic Gardener's Handbook* has a table including minimum, average and maximum longevity.

A simplified version is as follows:

- Year of purchase only: parsnips, parsley, salsify, scorzonera and the even rarer sea kale;
- 2 years: corn, peas and beans of all kinds, onions, chives, okra, dandelion and martynia;
- 3 years: carrots, leeks, asparagus, turnips and rutabagas;
- 4 years: spinach, peppers, chard, pumpkins, squash, watermelons, basil, artichokes and cardoons;
- 5 years: most brassicas, beets, tomatoes, eggplant, cucumbers, muskmelons, celery, celeriac, lettuce, endive and chicory.

Rather than deteriorating with age, some very fresh seed has a dormancy that needs to be overcome by chilling (lettuce). Other seed contains compounds that inhibit germination; these can be flushed out by soaking in water for about an hour (beets).

If you hate to throw seeds away, I have a couple of ideas for you, which I'll explain later. First of all, though, consider running some germination tests on critical packets of seed. Take a sturdy paper towel, folded in half, or a coffee filter, and dampen it by spraying with water. Count out 10–100 seeds and set them in a line in the fold or seam of the paper. The more seeds you can test, the more representative your result will be. Roll the paper loosely and put the roll in a plastic bag in a warm place, at a similar temperature to seed starting for that crop. One winter I made the mistake of trying to germinate watermelon seeds on the bathroom counter—it was too cold! Unroll and check the seeds twice a day. Initially your inspections will only be providing an air-change for the seeds, but soon you should see little sprouts emerging from the seeds. Spritz the seeds with water as needed. Count how many germinate and also notice if they are quick (vigorous) or slow (low-vigor seed, not for use in challenging conditions). When no more germinate, you have your final figure on germination. This will help you decide whether the seed is worth relying on next year, and whether you need to sow it thickly.

How much seed for a row?

Catalogs usually include guidance on how much seed to buy per hundred feet (thirty meters) of row, and you can use this information to help you decide how much to order. If you have kept good records, after you have inventoried your leftover seeds, you will be able to calculate (or have your computer calculate) exactly how much seed you use on your farm in real life, for each crop. Add the seed bought to the seed leftover from the previous year, subtract the seed left at the end of this season, and you have a figure for the amount of that seed you actually used this year. Add up the row length from all the plantings of that crop, then divide the total seed used by

Garden Seed Inventory

Vegetable	Row feet	Varieties	BOY 2010	Unit	Tot BOY 2010	Order Qty	Tot Seed 2010	EOY 2010 inventny	Plus/ Less	Adjusted 2010 EOY inv	Seed use 2010	2010 Use/ 100'
Beans: asparagus		Green Pod	17	g	17		17.0	17	−17	0	0.00	
	76	Purple Pod	64	g	64		64.0	33.1	−33.1	0	30.90	40.7
Beans: Bush	360	Bush Blue Lake	0.3	lb		5.0	5.3	3.72		3.72	1.58	0.44
		Mellow Yello	0.38	lb			0.4	0.38		0.38	0.00	
		Jade	0	lb			0.0			0	0.00	
	736	Provider	2.25	lb		5.0	7.3	2.28		2.28	4.97	0.68
	1096		2.93			2.93	10.0	12.9	6.38	0	6.38	6.55	0.60

⅛ oz=3g, 1 oz=28 g approx., 1 g=0.036 oz

the total row length. Your answer is the amount of seed for one foot or one meter, whichever set of units you are using. Record your figures to use when ordering new seed. This level of fine-tuning can help ensure you buy the right amount of seed, whether you are heavy-handed or parsimonious when sowing compared to the catalog average figures.

We have a spreadsheet for seed inventory and orders. It lists each crop in alphabetical order, with all the varieties of that crop in alphabetical order. There are columns for the row length, the amount of seed per hundred feet (thirty meters), and then a formula to calculate the amount of seed needed. Direct sown crops will have one kind of calculation, while for transplanted crops, we take account of the number of transplants per hundred feet (thirty meters) and the number of seeds per gram or ounce. Next is a column for the amount of seed on hand and another column with a formula to calculate the additional seed needed. The Inventory version of the spreadsheet refers to the year past, and then the seed-on-hand figures are carried forward to the Seed Order version of the spreadsheet.

Doing a seed inventory: weighing or guessing?

Depending on your preference, you can either weigh your seeds or eyeball the packets and estimate the proportion left. We used to just guess, but we weren't very good at it, so more recently we have started to weigh our seeds to help us get a better idea of how much we actually use. It's a nice indoor job once the weather has deteriorated. We have developed shortcuts so that we don't empty each packet out into the scale pan. We keep a list of the weights of the standard paper packets from each supplier, and weigh the packets and just subtract the weight of an empty packet of that size. We also keep a ziplock bag handy and use that for seeds in random bags.

At the same time we look at the age of the seed and decide whether to sow it next year or not. If we think it's too old, we consign it (still in the packet) to an Old Seeds Bucket. We make sure to subtract the weight of that seed from our calculations, so we don't confuse ourselves into thinking we sowed it. That would falsely increase the "Amount of Seed Used" figure and cause us to buy even more next year—the opposite of what we hope to do in making our orders closer to our actual needs.

Crop review and crop planning

As I wrote in Chapter 3, we conduct a Crop Review meeting each fall to record our impressions of that year's crops. This helps us make a rough plan for how much of what to grow for the following year (Less eggplant! More melons! Fewer sowings of beans!). With that information, we can go on to calculate how much new seed we need to buy.

Calculating the seed order

When we figure out how much seed to order we add in some extra for some things—crops that can be difficult to germinate, or we really don't want to cut too close. We add 20 percent extra for most crops, but only 5 percent for kale, 10 percent for onions and collards and 30 percent for melons. These numbers are based on our experiences—yours might be different.

We also know which seed we can buy in bulk and use over several years. This gives us an additional security against poor germination, or plagues of grasshoppers or caterpillars. For me, a big bag of broccoli seed for each of our main varieties gives some kind of warm glow of horticultural security!

This is the time to adjust the "seed rate" (seed/100' or /30 m) column using your own information from the past year, and feed in the next year's crop plan for varieties and succession plantings—everything you have decided so far about next year. Keep notes about any problems or questions you need to resolve later, and be sure to order enough seeds to cover these eventualities.

We have found it worthwhile to proofread our inventory and order form carefully before making our final decisions, as mistakes not discovered until planting day can be a big problem.

Formatting and placing your seed orders

On the Seed Order version of our spreadsheet, we include columns for the name of the supplier we buy each variety from (we just use the initial), the item number in the catalog, the packet size and the price. (Be careful though, if you carry this information over from year to year—prices change.)

Once we have composed our total seed order, we get the computer to sort the orders by the name of the supplier. Then we can separate the orders to each supplier onto different pages and calculate the total price for each supplier. This also gives us the opportunity to look at price breaks for large orders and move an item from one supplier to another, if that makes sense. At this point we usually make a cup of tea and reward ourselves with an "impulse buy" or two, if that doesn't push us up into a higher shipping cost bracket or blow the budget.

Each spreadsheet page can then be rearranged to put the columns in the order most compatible with the order form used by that supplier, if you are using mail order. Once you have done your calculations you can place your orders, nice and early, to increase the chances of getting exactly what you want. Online ordering is usually quickest, and sometimes offers extra discounts. But don't take that for granted. One year the best deal from one supplier involved inputting our orders onto their online form, printing it out and mailing it in!

Revisiting the Old Seeds Bucket

The Old Seeds Bucket may provide you with a backup plan if you somehow run out of seed during the growing season. You can reorder from a quick-to-respond seed company, and meanwhile make a sowing with your old seed. If the old seed germinates, you will be ahead of where you would be if you had to wait for the new seed to arrive, and you can keep the new seed for next year. If the old seed doesn't come up well enough, you can use the new packet and be no worse off than if you had composted the old seed back in the winter.

We sometimes use a mixture of old seeds for a short-term summer cover crop. We take into account the preceding and following crops and customize a mix that avoids those crop families (for a good rotation). Naturally, some small, slow-growing crops like parsley never make good cover crops, but peas, beans, corn and brassicas can all make quick growth. Be careful not to leave your free cover crops growing so long that they become problematic. One year we had a bed of cowpeas, corn and random old lettuce seed, sown beside our sweet corn. Our plans for that bed changed and before we knew it, we had cover-crop popcorn tasselling out next to our sweet corn. Luckily we noticed in time and cut them down before they cross-pollinated our sweet corn and ruined the flavor.

A third use for old seeds is micro-greens or salad mix. We have mixed leftover brassica, endive, beet and chard seeds to get a quick crop of salad mix from an unexpectedly empty spot in our hoophouse. Having these three possible uses for old seeds has helped us be more ruthless about buying new seeds at the beginning of each year.

Garden Seed Order 2011

							Put late orders, gifts, saved seed, under "Order"? Decided what to throw out?							Crop Review?				Plan?	
C	D	E	F (D*E* 0.012)	G	H	I (D*H/100) (D*H/100) f/g*(1-5)	J(I-AE)	K	L	M	U (P-R)	V (U/D* 100)	W	X	Y	Z	AA	AB	AC
						f/g*(1-5)	j(i-I)												
Vegetable	Row feet	Plants /100ft	Plants	Sd/ unit	Sd /100'	Plan	Need	Varieties	2011 BOY	unit	Seed use 2010	2010/ 100'	Order Qty	Pkts	C	#		Size	Price
Beans: Asparagus					64.0	0.0	0.00	Green Pod	0	gm					S	11,102		7 gm	
	76				50.0	38.0	38.00	Purple Pod	0	gm			50.0	1	S	11,103		50 gm	
Beans: Bush	776				0.8	6.2	0.09	Bush Blue Lake	6.12	lb	7.40	0.95			F	238	BB	D	
					0.8	0.0	0.00	Mellow Yello	0	lb					F	230	JD	C	
	351				0.8	2.8	(0.77)	Provider	3.58	lb	6.30	1.79	5.0	1	F	204	PR	D	$18.00
	1,127				0.8	9.0	(0.68)		9.7		13.32	1.18							

Scheduling Transplants

This chapter covers growing and managing transplants, as well as scheduling their production in line with your needs.

Why grow your own transplants?

Although it is becoming easier for growers to find vegetable starts for sale, finding organically grown ones is still not so easy. If you grow your own, you can get the mix of varieties you want, and keep them in tip-top condition until you plant them out. You can provide yourself with extra plants "just in case" at very little extra cost, and reduce your stress. You can bring the transplanting date forward if that seems best under your particular circumstances that year, which is not so easy to do if buying shipped-in plants.

If you want to harvest crops as early as possible, transplanting rather than direct-seeding is the way to do it. Also, if you want to let that crimson clover or Austrian winter pea cover crop flower, growing transplants allows extra growing time for the green manures. One year we transplanted all our winter squash, because we were having monsoons and couldn't get the plot prepared at the usual sowing date. So we retreated to the greenhouse, sowed them in cell-packs and didn't fret too much.

Basic equipment needed

To start your own transplants you will need a warm, well-lit space, a collection of flats and/or plug trays, some potting compost, an area for hardening off the plants before planting out, enough time in your days, and a plan. We have a greenhouse with a masonry north wall and a patio-door south wall. It has no heating apart from the sun (this is Zone 7). This space is (just) big enough and warm enough for all our seedlings once they have emerged. For growing-on the very early tomatoes and peppers, destined for our hoophouse, we put up a small plastic tunnel in one corner of the greenhouse.

Many seeds benefit from some heat during germination and are then moved into slightly less warm conditions to continue growing. This means it's possible to heat a relatively small space just to germinate the seeds in. Electric heat mats can be used, with the flats sitting on top. Many people construct some kind of insulated cabinet to start their plants, with fluorescent lights suspended above the flats. We use two broken refrigerators as our insulated cabinets, with extra shelves added. A single incandescent lightbulb in each supplies both the light and the heat (we change the wattage depending on what temperature we're aiming for). The extra light from the glass door can be an advantage, but most days it has a roller blind in front of it so the plants don't cook with the solar gain—it's in the greenhouse, facing south. The glass door has been a boon when we've had power outages following ice storms—then we can use all the solar gain we can get. I have also given the plants a rubber hot water bottle when the power has gone out!

We use two old style coldframes for "hardening off" our plants (helping them adjust to cooler, brighter, breezier conditions). They are rectangles of dry-stacked cinder blocks, with lids of wood-framed fiberglass. We set the flats directly on the ground in the frames. Having heavy flats of plants at ground level is less than ideal for anyone over thirty-five! Shade houses and single-layer

The greenhouse and traditional coldframes at Twin Oaks. Credit: Kathryn Simmons.

poly hoop structures with ventable sidewalls and benches for the flats are a nicer option. Some growers report that some pests are less likely to be trouble when flats are up on benches. Others say flats on the ground produce better quality plants. According to the nighttime temperatures, we cover the plants with rowcover for 32°F–38°F (0°–3°C), add the lids for 15°F–32°F (–9°C–0°C) and roll quilts on top if it might go below 15°F (–9°C). Once we get to hot weather, we put shadecloth over the coldframes, or rowcover for pest control (cucurbits and eggplant), and store the lids until late fall.

For a couple of years I started the first seedlings January 17, but came to realize that the extra freedom of a January 24 start was very important to me. And because plants grow so slowly when it's cold, the crops did not end up much earlier for another week of January growing. Seedlings need warmth, ventilation and watering, and you will probably need to check at least twice a day, even with some mechanization or an alarm system. If you are excited about starting a new growing season, don't let me stand in your way! But if you're still savoring sitting by the stove with knitting, novels, or even seed catalogs, consider waiting another week.

Having a complete plan is not vital, but it saves a lot of potential mistakes and stress later. Having enough time and attention for growing your own starts is essential, so think about that before you begin.

Greenhouse Seedlings Schedule

Some of the information we need before we venture far into the growing season is how many of which transplants we need and when we need them. When we have a rough idea of how much of each crop we want, we have the information we need for our transplant planning. Our Greenhouse Seedlings Schedule is a spreadsheet listing in date order all the transplants we need to produce and when we need to sow them to have them ready to go out in the ground at the right time. It is in date order and lists the crop, which varieties, and how many rows of what length we intend to plant. This information all comes from the work that has already been done at the Crop Review, Seed Order and Garden Layout stages. To create this we took all the crops we transplant, looked at the transplanting date we aimed for, and decided the sowing date. Finally, after setting up all the other columns, we sorted the spreadsheet by sowing date.

The spreadsheet contains the data on in-row spacing of each crop (plants per 100 feet or 30 m) and formulas that calculate how many transplants will be needed, allowing 20 percent extra of most things. For example, 1,080 row feet of spinach at 200 plants per hundred feet would be 2,160 plants. 2,592 is 20 percent more than 2,160, which allows us to choose the best ones when planting and have a margin of error to account for casualties.

The exceptions reflect the sturdiness or otherwise of the seedlings. For kale we only add 5 percent, because we grow lots and 20 percent of "lots" is more than we need, and because kale is so unlikely to die. For cipollini onions we add only 10 percent, because when we sow them in cell flats we sow clumps, with three to five seeds in each cell, and rarely get less than three germinating. For melons (cantaloupes and watermelons) we add 30 percent because they are so fragile.

After the "How many plants" column is one containing a formula that calculates exactly what to sow, whether in cells, open flats, or soil blocks.

Greenhouse Seedlings Schedule

Sow date	Crop	Variety	Row length in feet	Plants per 100ft	How many plants	Sow	Unit	Germ. date	Flats plan	Actual flats	Transplant date	Notes
Jan 24	Cabbage #1	Early Jersey Wakefield	90	67	72	0.3	flat		1.8		Mar 17	Half a raised bed
Jan 24	Cabbage #1	Farao 60d	90	67	72	0.3	flat		1.8		Mar 17	Half a raised bed
Jan 24	Lettuce #1	Tango/Red Salad Bowl	120	100	144	0.6	flat		3.6		Mar 17	Or sow in hoophouse
Jan 24	Onions: cipollini	Red Marble	120	200	264	1.5	speedling				Mar 21	Sow earlier if space
Jan 24	Spinach	Tyee	1080	200	2592	15.0	speedling				Feb 21	5 speedlings/bed or Jan 24 hoop-house or Feb 10 direct sow
Jan 24	Greens	Senposai	270	67	220	1	flat		5.5		Mar 7	1 bed, 3 rows
Jan 24	Tomatoes #1	Amy's Sugar Gem 75d indeterminate	12	50	7	9	cells				Mar 15	Hoophouse early. Deep nine-packs

We generally add 20 percent extra for each time a seedling will be moved.

Next we have a column giving the transplant date we are aiming for, taking into account how old the starts should be when we plant them out. We also have a Notes column and open columns to record actual sowing, germination and potting-on dates, as well as the actual numbers of flats, blocks or cells.

Open flats

For brassicas, lettuce and our paste tomatoes (a big planting), we use open flats—simple wooden boxes. The transplant flat size is 12" × 24" × 4" deep (30 × 60 × 10 cm). It holds 40 plants "spotted" or pricked out in a hexagonal pattern, using a dibble board. The "Flats plan" column tells us how many transplant flats we'll need: it's the "How many plants" number divided by 40. For sowing crops to be spaced out into these transplant flats,

we sow in shallower 3" (7.5 cm) flats. Usually we sow four rows lengthwise in each seedling flat. We reckon we can get about six transplant flats from each seedling flat, so the "Sow" column is the "Flats Plan" number divided by six. This allows for throwing out the wimpy seedlings. Naturally, we round up, when faced with instructions to sow fractions of a flat. This system allows us to start a higher number of plants in a smaller space.

Speedling flats

Speedling flats are Styrofoam cell or plug flats, which were originally made for the tobacco industry. We like them for onions and spinach especially, because the seedlings can be grown to full transplant size in them, they are light and easy to use, and the plugs can be pulled out (in the case of well-grown plants) or eased out with a butter knife, for the delicate approach. We use the 200-cell size, with a row of cells sawn off with a

Our greenhouse in late spring with seed flats resting on wood bars across blockwork. Credit: Kathryn Simmons.

bread knife, to make the flats fit in our germinating "fridge." So our customized Speedling flats are 190 cells.

Cell flats/plug trays

Because we transplant by hand, and because we hate to throw plastic away (or spend money when we don't need to), we use a range of different plastic plant containers. For crops where we are growing only a small number of plants of each variety, we use six- or nine-packs, or a tray that is divided into smaller units. We use deep nine-packs for tomatoes, which are later potted up into 3.5" (9-cm) pots; round 38-cell sheets (R38 in our spreadsheet); Winstrip vented square cells (W50), and some soil blocks. If we haven't made up our minds ahead of time, we calculate the amount to sow as individual cells. If we know from past experience

which containers work best for those plants, we record this information in the Notes column. For instance, for sweet peppers, we have a note saying "Cells 2"–2.25" diam. × 2.25"–3" deep (5 × 7 cm), nine-packs or R38, not W50," which alerts us that we have found the Winstrip 50-cell flats to be too small.

Soil blocks

These are less in fashion, perhaps, but we like them for early cucurbits (cucumbers, squash and melons) as these seedlings are harder to transplant and the soil blocks minimize the transplant shock. Soil blocks are made from a mixture of compost and coir (coconut fiber) or peat moss, and sometimes other plant nutrients. We use one part coir to one to one-and-a-quarter parts of good quality screened (sieved) compost. Plenty

Newly emerged okra seedlings in a Winstrip plug flat. Credit: Kathryn Simmons.

of water is added to make the mixture sloppy, and then it is compressed into special soil-block makers and ejected into a tray of some kind. Seeds are sown in the top and covered with soil. The blocks hold moisture well and are quite stable—they can be moved with a kitchen spatula. When the roots of the plants reach the edges of the blocks, they get air-pruned, so there is no root damage at all when the complete block is planted out.

Use of hoophouse for transplants in spring and fall

In the sample schedule on page 36, the Notes column for the lettuce says "or sow in hoophouse," which means that we might, instead of sowing in seedling flats, sow in the ground in the hoophouse. Increasingly, we are using this method for the hardier early spring seedlings. It takes very little space, and by January we have some of that in the hoophouse, where we have cleared some early winter crops. The plants grow sturdily, need little attention and transplant outdoors nicely when the time comes, perhaps because they have already been hardened off by the nighttime temperatures in the hoophouse (which is cooler than the greenhouse, where we grow most of our transplants).

In the Notes column for the spinach, it says: "5 speedlings/bed or Jan 24 hphs or Feb 10 direct sow." Three options here: on Feb 21, transplant 5 speedling flats of 190 cells per bed, or use bareroot transplants sown in the hoophouse Jan 24; or (if necessary) direct sow outdoors under rowcover Feb 10. We'd only do the direct sowing if we didn't have enough transplants from either source, as a later start means a later harvest.

Finally, in the fall, we sow onions in the hoophouse (around Nov 10–20), for outdoor transplanting as early in March as we can do it. These appear in the Outdoor Planting Schedule table below.

Outdoor bare root transplants

Later in spring, and during the summer, it might be easier to make an outdoor nursery bed and sow short rows of seedlings there. For us, this means we can ignore the greenhouse, which is off to one side of our main "circulation area." We use a short bed right at the front of the garden and keep a watering can there, so it's easy to remember to water and we can keep an eye on the plants. I know other growers who find it easier to continue producing lettuce transplants in cell packs,

Pepper transplants benefit from greenhouse warmth and should not be planted out small or in cold conditions. Credit: Kathryn Simmons.

because their setup means the greenhouse area is the best place for high-attention plants.

For several years we have sown leeks in an outdoor nursery bed on March 21, for transplanting at the end of May. Leek seedlings are very hardy so they don't need indoor "housing," which is just as well as our greenhouse and coldframes are full at that time. This works well if we pay attention to pre-emergent and post-emergent weeding, otherwise the slender seedlings get lost.

Age of transplants

As you can see from our sample greenhouse seedlings schedule, the first crops sown are not necessarily the first ones planted out. Our spinach gets sown Jan 24 and transplanted out 4 weeks later.

The early tomatoes get planted in the hoophouse at 6 weeks of age (the slower-growing peppers go in at 7.5 weeks with rowcover at the ready!). The lettuce goes outdoors after 6.5 weeks, the cabbage after 7.5 weeks, the cipollini mini-onions after 8 weeks. These are early season timings and as the days warm up and get longer, seedlings grow more quickly. The Transplanting Tips chapter has more on age and size of transplants.

Managing your transplants

If space is limited, you might need to schedule more tightly. Our germinating fridges can take a maximum of twenty-four flats at a time, so in some cases we wait until some sowings have germinated and can be moved on before we sow the

Outdoor Planting Schedule

Transplant or sow date	Actual planting date	Crop	Variety	Row length in feet	Plants per 100 ft	In-row space in inches	Inches between rows	Where	How many plants	Success?	Notes
Jan 25		Potato onions	Small Ones	540	240	5		RB	1,555		Harvest =June 12
Sow Feb 1		Fava beans	Windsor	30			16	RB			
Sow Feb 14		Carrots #1	Danvers	360				RB			Germ =day 15
T/plant Feb 21		Spinach	Tyee	1,080	200	6		RB	2,592		5spd/bed or t/pl from hphs
Sow Feb 28		Carrots #2	Danvers	360				RB			Germ =day 15
T/plant Mar 1		Onions	Frontier 95d storer	720	300	4		RB	2,592		2 beds, B, C
T/plant Mar 1		Onions	Gunnison 102d storer	325	300	4		RB	1,170		bed E
T/plant Mar 7		Greens	Senposai	270	68	18	9	RB	220		1 bed, 3 rows/bed

next thing. As noted earlier, being a few days later sowing something in early spring makes little difference, as later sowings can catch up by growing faster in the warmer weather. If the spring is cold and late, you may find your greenhouse packed to the gills with flats you don't want to take out to the coldframes. Packing more in takes creativity. We end up with flats entirely filling the south side of our greenhouse, leaving only the main path to the north. It takes extra planning to minimize the need to move large numbers of flats to get the ones you need. We try to put the faster-maturing crops near the doors and keep the open flats, which will need spotting-out, near the accessible north side. There are rolling benches, which can be very useful as a way to "move the path." Flats sit on mesh bench tops, which glide on rollers from side to side. I've also seen growers use the paths in their hoophouses to set flats down for a week or two. Navigating the paths involves stepping over flats every few feet. We have sometimes planted a coldframe with early peppers or eggplant, two or three rows in a staggered pattern, and then set flats diagonally between the plants.

But let's not complain about the bounty of so many plants! Spring is an exciting time of year, full of new growth and new potential. Working in the greenhouse with tiny plants on a sunny day when it's cold outside is a special treat.

Outdoor Planting Schedule

The information from this spreadsheet goes on to the Outdoor Planting Schedule, the Raised Bed Plan and the Calendar (Monthly Task List), unless we decided to do the Outdoor Planting Schedule before the Seedlings Schedule—it's all a big circle. See above for an edited beginning of year as an example.

Chapter 7

How Much to Grow

Planning is a cyclical process, and tweaking the plan for better results is an annual task. However, if you are starting from scratch, or if things are going really wrong, where should you start in planning your enterprise?

ATTRA's *Market Gardening: A Start-up Guide* has a set of questions to help clarify goals and develop a business plan, along with links to many other resources. An assessment of your available land, your preferred crops, your customers, your location and your financial situation will be the base on which you build your plan. Ellen Polishuk of Potomac Vegetable Farms links her success to decisions based on her seven core values: fun (a high quality of life); making a living; no or low debt; enjoying people; enjoying machines; continually investing in capital assets; and using organic practices.

Many growers will want to start with the money. Incoming money, that is. See *Crop Planning for Organic Vegetable Growers* by Brisebois

and Thériault, and ATTRA's *Holistic Management: A Whole-Farm Decision Making Framework* and *Sustainable Vegetable Production from Start-up to Market*. The *Organic Farmer's Business Handbook* by Richard Wiswall includes a CD you can copy and use to create your own budgets, timesheets, payroll calculator and more, compatible with Windows, Mac and Linux.

Set financial goals and then figure out how to achieve them. Plan your income (gross sales), then your profits (salary), then your expenses, which could be anywhere from 25 to 75 percent of your gross sales. Clearly, keeping expenses down will boost your income, so long as you don't make the farm nonviable.

To set your gross sales goal, consider how much produce you can grow and what the financial value of that is. If you are brand new, you will need to ask other farmers for help, study prices at the farmers' market, and see what other growers offer. The Roxbury Farm website in the

Resources section is helpful on this. A full-time farmer might work two thousand hours in a year and average $18/hour gross if things go according to plan. But a beginning farmer needs more slack while learning, and might expect to earn considerably less than the $36,000 of an experienced grower. Perhaps only $5,000–$10,000, according to Anne Weil, quoted in *Crop Planning for Organic Vegetable Growers*. Out of these gross sales come all the business expenses. So subtract your hoped-for earnings from the gross sales and look at how much will be available for covering the expenses. Consider if this is a reasonable amount, by listing all your expected expenses and then adding in something for contingency expenses (the unexpected but unavoidable).

SPIN-Farming (Small Plot INtensive vegetable growing) is geared to new growers using city plots and prepared to pay for the fairly expensive manuals. The website has a calculator to convert square feet into farm income using their methods.

Which crops to grow

Some crops require more skill or are less dependable. If your climate is marginal for okra, avoid relying on it for a large part of your summer income. *Gardening When it Counts* by Steve Solomon has a table of "Vegetables by level of care needed." His "Highly-Demanding Crops" list includes some greens that bolt easily in spring, crops which quickly go "over the top," and some which are plain old persnickety, requiring near perfect soil and conditions to perform well. He and I agree on the challenges of brussels sprouts, cauliflower, celeriac and bulb onions. Steve's list also includes asparagus, Chinese cabbage, early cabbage, cantaloupe, leeks, large fruited peppers, spring turnips and spinach. While those ones are

easy here, challenges in our climate include rutabagas, drying beans and shelling peas.

Some crops offer high yields or high market value for a small space. Do you have a lot of labor or a lot of land? In terms of yield per unit area, the best include carrots, summer squash, onions, potatoes, sweet potatoes and tomatoes. Peas, sweet corn, radishes and bush beans are among the worst. But in terms of tonnage per hour worked ("efficiency") the best crops are sweet corn, potatoes, cucumbers, cabbage, summer squash, peas and peppers. The worst include pole beans, radishes, onions, carrots, bush beans and lettuce.

Neither high retail price nor high yield is the same as most profitable. See Richard Wiswall's *Organic Farmer's Business Handbook* for twenty-five sample crop enterprise budgets that you can use to make a comparison of costs, sales and profit from each vegetable. This is a book about number crunching that's accessible and inspiring. (One of the author's main goals is to help create less stressed-out farmers.) Beware of preconceived notions on what is most profitable—get real numbers. His highest to lowest net profit per bed are: greenhouse tomatoes, parsley, basil, kale, field tomatoes, cilantro, dill, peppers, carrots, parsnips, celeriac, spinach, beets, lettuce, summer squash, bulb onions, cabbage, potatoes, cucumbers, broccoli, winter squash. Peas, beans and sweet corn all ran at a loss. Remember—your results may vary! One lesson from this list is the ability of long-season crops such as kale with an extended harvest to provide high yields for the time put into soil preparation, planting and cultivation. Another lesson is that while bunched herbs can bring a good profit, people will only use a certain amount, and a diversity of crops is needed to keep customers returning.

What the market wants and what it can take

Many CSAs, including Roxbury Farm, post the contents of their shares. Sometimes you will want to grow certain crops even if they are not the highest money-earners because they enhance what you have to offer. Perhaps they round out your market display, or your CSA boxes. Perhaps you'll grow a crop because it is extra early, or eye-catching. If you are growing for farmers' markets you can choose to only grow high-value crops, but if you are doing a CSA, your customers may expect to receive some of everything and you will need to grow some low-value crops. But don't be afraid to say no to growing a crop such as sweet corn or shelling peas that just doesn't work for your farm. A CSA has the advantage of money up-front and guaranteed customers, as well as avoiding the costs associated with going to market. On the other hand, a farmers' market booth can take a flexible range of produce, and so is easier if plans go awry.

Flow of decision-making on how much to grow

Producing crops when you want them and in the right quantities is a complex task, and the grower does not control all of the variables. However, to get the best chance of success, take decisions in a logical sequence. Once you've decided which crops you want to grow, here is a step-by-step process to determine how much to plant:

1. Figure out how much of each crop you'd like to harvest, how often and over what length of time;
2. Calculate how many plants you will need. This depends on the yield per plant and how long the crop will stand in the field;

CSA bags at Seven Springs Farm, Floyd, Virginia. Credit: Pam Dawling.

3. Add a percentage (perhaps 10%) to allow for culls;
4. Decide the dates for the sowings to meet the harvest date goals.

Likely yields

Charts of possible crop yields are available in the Roxbury Farm's *Field Planting and Seeding Schedule* and their *Greenhouse Schedule*. Some seed companies have tables of likely yields in their catalogs, although these are sometimes more for the home gardener than for market growers. The Center for Agroecology and Sustainable Food Systems at the University of California, Santa Cruz has a lot of useful information including a thirty-page Crop Plan for a hundred-member CSA, with planting requirements including total bed length for a range of thirty-six crops in its Unit 4.5 CSA Crop Planning. Their Appendix 9 includes the area requirements translated into fractions of an acre. A further source of this kind of information is *Sharing the Harvest* by Elizabeth Henderson and Robyn Van En. A two-page table includes yield per hundred row feet.

How much to grow

If the average person needs 160–200 pounds (72–90 kg) of vegetables per year and the average household (= 1 CSA share) is 2.5 people, then one share will be 400–600 lbs (180–270kg) per year, roughly ten pounds (4.5 kg) per week for a full year.

The table below lists forty-eight crops, along with likely yield; quantity required for a hundred CSA shares; and length of row needed to grow this amount. This fictional CSA (a blend of information gleaned from the various sources I've mentioned) runs for twenty-six weeks and has shares sized for two and a half "standardized" people. For comparison I have included how much of those crops we grow at Twin Oaks Community for fifty-two weeks for a hundred specific people. My point in including both is that every group is different, and no one else's table will reflect your group of customers exactly.

If all people were the same, the Twin Oaks list would total about the same amounts (twice as many weeks, less than half the number of people). You'll see some of our preferences come into play: we don't grow arugula in any quantity worth recording, and celery and mustard greens are not very popular. Even though we freeze and pickle green beans, corn, eggplant and okra, they're not as good as fresh crops, so we eat less than the fictional 250 people have fresh. On the other hand, beets and garlic store well, so we have more than Fiction Farm shareholders, as CSAs often don't supply for winter needs. Chinese cabbage, mizuna and pak choy bolt too readily to be worthwhile at Twin Oaks in the spring, so we only grow them in the fall, and most of that in a hoophouse, where yields outstrip those grown outdoors. Fiction Farm probably has cooler summers than we do! Kale, leeks and spinach overwinter outdoors here, so we grow lots more than a CSA supplying only in the warmer half of the year. I have to wonder how many of the hot peppers supplied by Fiction Farm get used? We make lots of salsa for winter use, but only plant 71 feet (22 m). Do your customers want attractive foods they might not actually eat? Maybe they do! Other differences are a matter of scale, and will be relevant to growers supplying institutions. For example, it's hard work to prepare scallions for a meal for a hundred, whereas a hundred separate cooks might enjoy adding them to the small meals they prepare. I wonder why the amounts of butternut squash and sweet potatoes are so much lower at Twin Oaks than Fiction Farm? Ours last through to May, quite long enough! I notice that we grow lots of paste tomatoes and fewer regular fresh eating ones. That might be because our quality standards can be lower because our tomatoes don't commute to market, and we're not so picky about looks!

Deciding sowing dates

It might be hard to orchestrate your annual start-up so that you have a generous bounty. It's OK to tell your CSA subscribers that the beginning-of-season boxes will be less full and the summer ones more bountiful. Johnny's Seeds website has a Harvest Date Calculator which you can copy and use to calculate sowing dates to meet a target date (e.g., first market of the year).

How Much to Grow for 100 CSA Shares

Crop	Weeks of Harvest	Number of Plantings	Crop Yield Pounds per 100ft	Annual Goal per Share (pounds)	Annual Goal for 100 Shares (pounds)	Row Length Needed (Feet)	Crop Yield Kilos per 30m	Annual Goal per Share (kg)	Annual Goal for 100 Shares (kg)	Row Length Needed (meters)	Twin Oaks Row Length for 100 People Year Round feet	meters
Arugula	20	several	17	3.7	370	2,176	7.7	1.7	167.8	663.4	0	0.0
Asparagus	8	1	35	3.5	350	1,000	15.9	1.6	158.8	304.8	1,400	426.7
Beans	10	up to 6	90	15.0	1,500	1,667	40.8	6.8	680.4	508.0	1,100	335.3
Beets	19	2	100	16.3	1,630	1,630	45.4	7.4	739.4	496.8	2,200	670.6
Broccoli	11	2	80	16.0	1,600	2,000	36.3	7.3	725.7	609.6	3,900	1,188.7
Cabbage	11	2	190	20.0	2,000	1,053	86.2	9.1	907.2	320.8	1,760	536.4
Cantaloupe	5	3	300	9.0	900	300	136.1	4.1	408.2	91.4	300	91.4
Carrots	18	9	100	66.0	6,600	6,600	45.4	29.9	2,993.7	2,011.7	7,800	2,377.4
Celeriac	2	1	80	3.0	300	375	36.3	1.4	136.1	114.3	360	109.7
Celery	1	1	150	2.0	200	133	68.0	0.9	90.7	40.6	44	13.4
Chard	14	3	90	3.3	330	367	40.8	1.5	149.7	111.8	300	91.4
Chinese Cabbage		1	75	5.0	500	667	34.0	2.3	226.8	203.2	43	13.1
Collards	13	2	100	2.7	270	270	45.4	1.2	122.5	82.3	1,080	329.2
Corn	6	up to 6	65	60.0	6,000	9,231	29.5	27.2	2,721.6	2,813.5	7,200	2,194.6
Cucumbers	11	up to 5	260	15.0	1,500	577	117.9	6.8	680.4	175.8	550	167.6
Edamame	6	2	20	1.0	100	500	9.1	0.5	45.4	152.4	540	164.6
Eggplant	6	1	140	10.0	1,000	714	63.5	4.5	453.6	217.7	180	54.9
Garlic	3		45	3.0	300	667	20.4	1.4	136.1	203.2	4,260	1,298.4
Garlic Scapes	3	1	1	0.3	30	3,000	0.5	0.1	13.6	914.4	3,180	969.3
Kale	16	2	100	7.0	700	700	45.4	3.2	317.5	213.4	3,500	1,066.8
Kohlrabi	5	2	65	2.0	200	308	29.5	0.9	90.7	93.8	540	164.6
Leeks	4	1	100	5.0	500	500	45.4	2.3	226.8	152.4	1,500	457.2
Lettuce	20	weekly	45	20.0	2,000	4,444	20.4	9.1	907.2	1,354.7	6,000	1,828.8
Mizuna	20	2	60	3.4	340	567	27.2	1.5	154.2	172.7	48	14.6
Mustard greens	21	2	85	2.0	200	235	38.6	0.9	90.7	71.7	0	0.0
Okra	6	1	75	5.0	500	667	34.0	2.3	226.8	203.2	90	27.4
Onions	1	1	80	8.0	800	1,000	36.3	3.6	362.9	304.8	1,800	548.6
Pak Choy	14	1	75	5.0	500	667	34.0	2.3	226.8	203.2	43	13.1
Parsnips	1	1	75	3.0	300	400	34.0	1.4	136.1	121.9	120	36.6
Peas, cow	3	1	40	4.0	400	1,000	18.1	1.8	181.4	304.8	300	91.4
Peas, snap	3	1	30	2.0	200	667	13.6	0.9	90.7	203.2	686	209.1
Peas, snow	3	1	30	2.0	200	667	13.6	0.9	90.7	203.2	180	54.9
Peppers. hot	10	1	75	1.5	150	200	34.0	0.7	68.0	61.0	71	21.6
Peppers., sweet	10	1	95	20.0	2,000	2,105	43.1	9.1	907.2	641.7	500	152.4
Potatoes	4	2	110	35.0	3,500	3,182	49.9	15.9	1,587.6	969.8	5,800	1,767.8

(cont'd.)

How Much to Grow for 100 CSA Shares (continued)

Crop	Weeks of Harvest	Number of Plantings	Crop Yield Pounds per 100ft	Annual Goal per Share (pounds)	Annual Goal for 100 Shares (pounds)	Row Length Needed (Feet)	Crop Yield Kilos per 30m	Annual Goal per Share (kg)	Annual Goal for 100 Shares (kg)	Row Length Needed (meters)	Twin Oaks Row Length for 100 People Year Round feet	Twin Oaks Row Length for 100 People Year Round meters
Radishes	12	9	33	1.7	170	515	15.0	0.8	77.1	157.0	712	217.0
Rutabagas	4	1	120	4.0	400	333	54.4	1.8	181.4	101.6	0	0.0
Scallions	14	about 6	40	6.0	600	1,500	18.1	2.7	272.2	457.2	206	62.8
Spinach		spr+fall	40	10.0	1,000	2,500	18.1	4.5	453.6	762.0	3,840	1,170.4
Acorn Squash	5	1	120	7.0	700	583	54.4	3.2	317.5	177.8	0	0.0
Butternut Squash	4	1	250	20.0	2,000	800	113.4	9.1	907.2	243.8	540	164.6
Summer Squash	13	up to 6	250	30.0	3,000	1,200	113.4	13.6	1,360.8	365.8	583	177.7
Sweet Potatoes	2	1	100	30.0	3,000	3,000	45.4	13.6	1,360.8	914.4	800	243.8
Tatsoi	18	2	20	3.0	300	1,500	9.1	1.4	136.1	457.2	240	73.2
Tomatoes	9	3	300	40.0	4,000	1,333	136.1	18.1	1,814.4	406.4	450	137.2
Tomatoes, paste	4	1	250	10.0	1,000	400	113.4	4.5	453.6	121.9	1,040	317.0
Turnips	6	3	70	4.0	400	571	31.8	1.8	181.4	174.2	1,263	385.0
Watermelon	4	1	700	35.0	3,500	500	317.5	15.9	1,587.6	152.4	1,080	329.2
Totals				580	58,040	64,971		263	26,326	19,803	68,129	20,765.7

Grow the amount you can handle and not more.
Credit: Bridget Aleshire.

Chapter 8

Crop Spacing

Most gardening books and many seed catalogs contain charts of ideal crop spacings. Often these repeat measurements that have traditionally worked well for many backyard gardeners. More recently, raised bed techniques, both for gardeners and commercial growers, have led to changes in crop spacing. Research has also revised spacing recommendations.

In this chapter we'll look at the best crop spacings to meet various goals, including consideration of the limitations of close planting. The best information I've found on research into spacing is a 1991 book from the UK: *The Complete Know and Grow Vegetables* by J. K. A. Bleasdale, P. J. Salter et al. John Bleasdale had noticed that information on in-row spacing and between-row spacing was inconsistent from one publication to another. Often the space recommended was found to be over-generous, which means growers have been getting less-than-top productivity from their land.

Research has determined the ideal area re-quired, according to whether you want to grow large individual plants, or the maximum total yield for the area, or the maximum early yield. Spacing alone can determine whether you grow jumbo-size onions or the size best for pickling or kebabs. Individual-sized cauliflowers and mini-lettuces are possible, and could be well suited to your market.

Knott's Handbook for Vegetable Growers (Donald N. Maynard and George J. Hochmuth) has a chapter on "Spacing of Vegetables and Plant Populations" that includes information on high-density spacing.

Winning the competition with weeds

Removing weeds is essential to getting good harvests. Row spacing has changed over time, depending on the methods used to deal with weeds. Farming with horses works best with long, narrow rows of each crop (maximum repetition, minimum complex instructions). This leads to a style of planting with relatively large bare (or weedy!) paths between rows. All this space is for the farmers, the horses and the machinery—the plants don't need it.

"Conventional" tractor agriculture uses even wider spaces than horses, to cultivate either side of long crop rows. Spraying herbicides works best when both sides of each plant are accessible and planting is not close-packed. Now we know that chemical herbicides are costing the earth.

Meanwhile in backyard gardens and small-scale mini-farms, biointensive planting proponents (using manual work only) have tightened up on planting spaces. When all the space will depend on hand work to keep it clear, wide rows or beds, rather than single crop rows surrounded by space, work best. In these raised beds, the in-row spacing and the between-row spacing are the same, for example, lettuces 12 inches (30 cm) apart in all directions, giving plants all the space they need to reach optimum size and no more.

For organic farmers, the use of cultivating tractors that can straddle a bed has made design changes possible in quite large fields. While the space needed for farming equipment is one factor in spacing crops, we can reduce that space to a minimum and then pay more attention to the space actually needed for the growth of the crop. The land can be used more efficiently when the growing space is as full of plants as possible.

Plant density and yield

There is a maximum yield per plant regardless of how much space is available. One cabbage alone in a field may grow large, but the yield is not high for the area. As more plants are put into a plot, the total yield goes up, until a point is reached where the plants are so crowded that they compete with each other and plant quality and yield go down. The key is to find the balance point at which the plant density provides the maximum total yield. This density will produce some plants that are too small to use, which is taken into account when calculating yield. The larger spacings recommended for home gardeners may take into account the fact that people (unless their livelihood depends on it) hate to throw crops away, even if the compost pile needs ingredients or the chickens are hungry. Gardening books recommend spacings that allow every plant enough room to grow to perfection, even though this does not lead to maximum yield.

Some crops such as carrots will provide the same maximum yield at very different densities. Although overcrowded carrots will take longer to size up, the weight of tiny carrots from very crowded beds can be the same as the weight of giant carrots from carefully cultivated and thinned beds (or from precision-sown pelleted seed). The response of your customers might be quite different, but it is reassuring to know that if your market is for giant carrots, you can still get as high a total yield as if you grew smaller ones. It also means that, if you have seed germination problems, the carrots can grow to fill the gaps and you'll still end up with the same tonnage, though the carrots will vary in size. Uneven emergence rate of carrot seed leads the first ones to germinate to hog the available space, so the final harvest contains a range of sizes.

Other crops, such as beets, behave very differently. While the tonnage of tops will be the same regardless of plant spacing, the weight of roots will not. This happens because the broad leaves of the beets shade out their neighbors. Beets will compete better with weeds than ferny-leafed carrots will, but they are their own worst enemy! So don't neglect singling and thinning of beets, unless you are working on a more valuable task and have made a conscious decision that the beets aren't worth losing sleep over (this is always a worthwhile question!).

Sweet corn is usually grown in rows about 36 inches (90 cm) apart. Research by Ray Samulis at Rutgers showed that with in-row spacing greater than 8 inches (20 cm), the yield-reducing effect of shading by the leaves of neighboring plants is much less. This confirms that 8 inches (20 cm) is usually as close as it's wise to go. Upper leaves get seven to nine times the light of lower leaves, and that is where photosynthesis is really happening. The lower, shaded leaves are not so vital.

Research has shown that reducing watermelon plant spacing from the usual 20–42 ft^2 (1.8–3.8 m^2) down to 10–12 ft^2 (0.9–1.1 m^2) per

Optimal Plant Spacing for Vegetable Crops for Various Goals

Crop	Row spacing	In-row spacing	Notes
Beets	7" (18 cm)	4" (10 cm)	For early harvest
	12" (30 cm)	1" (2.5 cm)	For max total yield (small). 2" (5 cm) in-row for bigger beets
Beans, fava	18" (45 cm)	4.5" (11 cm)	For tall varieties.
Beans, green	18" (45 cm)	2" (5 cm)	12" (30cm) × 3" (7.5 cm) gives the same area/plant
Broccoli (Calabrese)	12" (30cm)	6" (15 cm)	For equal amounts of heads and side shoots
Cabbage	14" (35 cm)	14" (35 cm)	For small heads
	18" (45 cm)	18" (45 cm)	For large heads
Carrots	6" (15 cm)	4" (10 cm)	For early crops, limiting competition
	6" (15 cm)	1.5" (4 cm)	For maincrop, medium size roots
Celery	11" (28 cm)	11" (28 cm)	For high yields and mutual blanching
Cucumber (pickling)	20" (51 cm)	3" (8 cm)	
Leeks	12" (30 cm)	6" (15 cm)	Maximum yield of hilled up leeks, average size
Lettuce	9" (23 cm)	8" (20 cm)	Early crops under cover
	12" (30 cm)	12" (30 cm)	Head lettuce
	5" (13 cm)	1" (2.5 cm)	Baby lettuce mix
Onions	12" (30 cm)	1.5" (4 cm)	For medium size bulbs
	12" (30 cm)	0.5" (1 cm)	For boiling, pickling, kebabs
Parsnips	12" (30 cm)	6" (15 cm)	For high yields of large roots
	7.5" (19 cm)	3" (8 cm)	For smaller roots
Peas, shelling	18" (46 cm)	4.5" (11.5 cm)	Can sow in double or triple bands, 4.5" (11.5 cm) apart
Potatoes	30" (76 cm)	9-16" (23–41 cm)	Depends on size of seed pieces; small pieces closer
Sweet Corn	30-36" (76–90 cm)	8" (20 cm)	Closer than 8" (20 cm) the plants shade each other.
Tomatoes, bush types	19" (48 cm)	19" (48 cm)	For early crops
Watermelon	66" (168 cm)	12–24" (30–60 cm)	For small varieties. 5–10 ft^2 (0.5–1 m^2) each
	66" (168 cm)	30–84" (76–215 cm)	For large varieties. 13–40 ft^2 (1.2–3.7 m^2) each

plant resulted in increases in yield of 37%–48% with only a 13% reduction in average fruit weight (1979, Brinen et al, see the watermelon chapter). With small varieties, halving the traditional spacing reduces the size by only 10%.

The chart on the previous page contains information from research by J. K. A. Bleasdale, Ray Samulis at Rutgers University, Scott NeSmith at Georgia Agriculture Experiment Station, Griffin, Warren Roberts at Oklahoma State University, and Oscar Lorenz and Donald Maynard in *Knott's Handbook for Vegetable Growers*.

Planting patterns

When I was studying crystallography at college, I didn't know the information about hexagonal close packing of atoms would be so useful later in life! When each lettuce is set out equidistant from all others, with one "row" staggered relative to the one next to it, the plants each have equal access to soil, sunshine and water, and can grow as big as conditions allow, without wasting any space. This, believe it or not, is the same as hexagonal close packing of atoms in crystals. Each lettuce is in the middle of a hexagon of neighboring let-

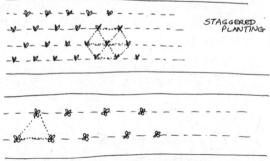

STAGGERED PLANTING SHOWING EQUAL SPACES BETWEEN ALL PLANTS

For best use of space, stagger the plants so that each one is equally spaced from its neighbors in all directions. Credit: Pam Dawling.

tuces, and there's no way to get them closer without overcrowding some of them. This staggered planting pattern makes better use of space than a square grid pattern, and the plants are easily hoed by working diagonal lines across the bed.

Crops that have large, wide leaves can make a closed canopy, which will then suppress weed growth. For such crops, close planting patterns help reduce the spots weeds can grow in. For crops with non-smothering leaves (onions are the best example), the crop will never smother weeds, but a consistent planting pattern will make it easier to use tools to cultivate.

An example of a consistent pattern (which is not an equidistant planting) from our farm has been beds sown with three to five rows using the Row Marker Rake from Johnny's. We set the pegs on the rake at the row spacing we want, drag the rake down the bed to make a set of marked lines, then run the seeder. (*Run with the seeder* might be a more accurate description—we use an Earthway, a very lightweight machine.) Now we can hoe between the rows as the seedlings are emerging, and do it at a fairly fast pace, because we know the rows are going to be exactly parallel and evenly spaced, so we only need to see a few seedlings to know where the rest will come up. Our old method involved just eyeballing the row spacing while using the seeder. That way it was too risky to hoe very early, because the rows wandered a bit. Growers focusing on large areas of crops use tractor seed drills and transplanters, and will get an even spacing as part of the job. But growers providing for CSAs or restaurants often need smaller amounts of a wide variety of crops, and tractors do not always prove to be the best tools.

Sometimes there are reasons to plant relatively close in the row, and have relatively wide spaces between rows, even though it means more open space for weeds. For most plants, it is the area per plant, not the ratio of in-row to between-row spacing, that's important in determining the yield. Research shows that broccoli yields best at a density of two plants per square foot ($21/m^2$). This could be with plants at 3" × 24" (8 × 61 cm), or 6" × 12" (15 × 30 cm)—both spacings provide the same area to each plant. This plant density gives the yield half in heads and half in side-shoots. Wider spacing gives more of the yield in side-shoots. Very close spacing gives small heads, all at once. There are situations where this could be the best choice. (Hmm, Local Broccoli Festival, anyone?)

For large plants such as okra or eggplant, it makes more sense to plant a single row in a bed and have the plants close together in that row, forming a "hedge." It is easier to cultivate plants that are 2 feet (60 cm) apart in 5-foot (150-cm) rows than plants at an equal spacing of 30 inches (76 cm) in all directions—an eggplant jungle!

Your own special circumstances may apply. For example, we grow no-till paste tomatoes. When the tomatoes have been growing for a while, the mow-killed cover crop starts to rot down, and we roll out round bales of spoiled hay between the rows to top up the mulch and control weeds for the rest of the season. So our row spacing is determined by the width of a hay bale. Additionally, tall plants get better air circulation and better light on the lower leaves if the rows are spaced further apart.

Some growers plant in clumps, as Eliot Coleman suggests for multiplant blocks of bulbing onions in *The New Organic Grower*. Onion seed is sown at four seeds per cell, then the clump is transplanted together in the row, 12 inches (30 cm) from neighboring clumps. This plant density is the same as individual onions at 4 inches (10 cm) apart. Transplanting and hoeing are much quicker with the clump planting method.

Water, nutrient, time and day length considerations

Dryland farmers use wider spacing so that their plants draw what they need from a larger area. The information about using a regular planting pattern still applies, even if your plants need to be four times further apart than where water is sufficient. The closer spacings discussed here will not work if your soil is poor in nutrients or if water is limited. If you plan on using close spacings, you need to ensure sufficient water.

For early-maturing crops, competition between plants must be minimal to allow each to reach its optimum size as quickly as possible. The higher prices paid for early crops can compensate for the lower yield. In the table on pg. 49, the space for early carrots and beets is more than the space for the maincrop plantings. This might not be intuitive—we might assume that because early crops are harvested small, they need less space. But competition will stunt them, which limits the size they can get to by their harvest date.

Flavor is another factor to consider when deciding spacing. Earlier watermelons are sweeter than later ones (this may be related to day length). And we all know there's not much demand for watermelons in October. Even though each watermelon plant might be capable of producing two or three melons over time if given enough space, it usually makes more sense to plant closely and be happy with one early melon per plant. More plants yield more early melons.

Senposai in a staggered planting pattern with three rows to the bed. Credit: Kathryn Simmons

For some crops, the market requires a certain size regardless of whether this gives maximum yield for the space or not. Onions are a good example. Onions give the maximum yield per unit area when they are 1¼" (32 mm) in diameter. Unfortunately that size isn't popular!

With potatoes, very wide spacing leads to large tubers, which get overcrowded in the clump and start to push out of the soil. Then they turn green. This is a specific case when too much space leads to bad results even beyond the uneconomic amounts of time spent dealing with weeds.

Intercropping

Growers who like to get the maximum from their farm are drawn to intercropping, where small or quick-growing crops are planted between slower-growing crops to use the space not yet needed by the slower crop. It is important that the intercrop not be allowed to outcompete the main crop, and that weeds not get any chance to compete with either crop. High soil fertility is needed to make this work. See examples in the crop chapters about intercropping (relay planting) of spinach and peas, okra and cabbage, and peanuts and lettuce.

Chapter 9

Transplanting Tips

Pros and cons of transplants versus direct sowing

Transplanting offers several advantages over direct seeding, but is not best for every situation or every crop. Controlled germination temperatures can provide a higher seed germination percentage and allow earlier sowing than would work outdoors in spring. It is easier to keep an eye on seedlings in flats, as well as protect them from pests and ensure optimal watering, because they are close together and in a protected environment. Better control of growing-on temperatures leads to faster-growing plants. The uniformity and larger size of plants results in earlier harvests and higher total yields. Growing the young plants elsewhere also allows you to use the outdoor space for another crop (either a food crop or a cover crop) earlier in the season.

Some crops are easy to transplant because they quickly grow new roots to replace those damaged in the transplanting and do not lose water at too fast a rate: brassicas, lettuce, tomatoes, beets. Some are moderately successful: eggplant, pepper, onion, okra, sweet corn. And some crops are tricky but still worthwhile (they do not quickly resume root growth): cucurbits are the main examples in this category. Other crops such as most root crops and most legumes are not usually worth transplanting.

Best of both worlds? Bare root transplants

Bare root transplants are plants dug up from a nursery bed outdoors or in a hoophouse and planted out directly. They offer the easiest option, if weather conditions are suitable. They need little extra care as they are less prone to drying out than seedlings in flats. However, they suffer more transplant shock. For us, onion seedlings overwintered in the hoophouse work very well, and spinach and kale sown in the hoophouse in January and planted outdoors in March are a lot less work than using flats. We have tried early lettuce transplanted from the hoophouse, but the plants were not as sturdy as those in flats. We may try collards, kohlrabi and broccoli this way.

Flats, plug trays and soil blocks

If you use a transplanter machine you will need to use plug flats of a size suitable for the machine. We do all our transplanting by hand, so we can vary the size and shape of our plugs. While we do use a lot of plastic plug flats, we also use many wooden flats, spotting (pricking out) the seedlings into more spacious flats from the smaller ones we germinate in (to make best use of our heated space). Moving up once into larger containers also allows us to select the best seedlings. We give most of our transplants a 4" (10 cm) depth of potting compost, and grow big, healthy plants. However, plastic containers are much lighter and easier to move, and sowing directly in cells reduces the transplant shock to a single occasion, resulting in earlier crops.

We use a motley collection of deep nine-packs, round 38-cell sheets, Winstrip vented square cells, Styrofoam Speedling flats, and 3-inch (7.5-cm) and 4-inch (10-cm) pots, as well as some soil blocks. See the University of Florida website listed under Resources for a comparison of Winstrip 72-cell (55-cc) trays, Speedling 128-cell (38-cc), and 51-cell round plug trays (filled to only 55-cc for comparison). The round cells grew the biggest transplants of the three and the earliest maturing fruit, although the other plants caught up and the final total yields for all three were similar. Which size cells we choose for which crops is a balance between what we know works and what we can find on hand.

We use the deep nine-packs to start our early tomatoes and peppers, and pot them up into 3.5-inch (8-cm) pots. It's easier for us to keep the smaller volume containers warm in January. These plants are destined for the hoophouse, so they do not have to get very big before they can be transplanted there. For our outdoor peppers, we sometimes germinate in the round 38-cell sheets, and they grow to transplanting size in these.

Winstrip and Speedling trays are designed to "air prune" the roots, provided the flats are suspended on rails or mesh or slatted benches to allow air under them. Our early cucumbers work well in Winstrip 50-cell flats, but those cells are too small for squash. For early squash and the second tomatoes, we use the 72-cell Speedlings. For the second squash, and for early melons, we use soil blocks. These have a slightly larger volume than the Winstrip 50s, and with careful handling, transplant shock to the roots is negligible.

Soil blocks are time-consuming to make, so we only use them for the cucurbits, which really do best with as little root disturbance as possible. Soil block making also involves purchasing a block-maker and peat moss or coir (coconut fiber), since some fibrous material is needed to hold the blocks together. I won't use peat moss because its rate of use is not sustainable, and if you've seen pictures of the gangs of giant machines tearing up large tracts of Canada, you'll know what I mean. Coconut fiber is imported, but it is an agricultural by-product, which makes more sense to me. My mix is one part of coir to one to one and a half parts of screened homemade compost and enough water to make a slumpy, not soupy, mix. It's important to dunk the block-maker into a bucket of water between fillings, to wash off the old remnants and enable the new blocks to slip out when ejected.

Winstrip 50s also work well for Swiss chard, okra and even watermelon or muskmelon (when we tire of making soil blocks) if transplanting is not delayed. I wish Winstrip made a larger-celled flat!

We use the 200-cell Speedlings for alliums, early sweet corn and spinach. For sweet corn and fall spinach we float the Speedling flats in a big tank of water. The sweet corn stays in the water until transplanting, but the fall spinach is floated during the day and drained overnight, as continuously waterlogged spinach roots won't be happy. The cooling effect of the water helps the spinach to grow while the weather is still hot. (Presprouting the seed in the refrigerator for a week also works well.)

Just about everything else in spring (brassicas, lettuce, celery, eggplant, paste tomatoes) gets started in 3-inch (7.5-cm) wood flats and pricked out into 4-inch (10-cm) flats when the first true leaf is 0.25–0.5 inch (0.6–1.3 cm) long. Once they grow close to the desired transplanting size, the flats are then hardened off.

Growing strong starts

Sterilization of seed flats has been a topic of lively discussion among growers. Some people believe strongly in sterilizing everything, others sterilize nothing. We're in the second school. There was research showing that wood harbored fewer diseases than plastics in a study of kitchen cutting boards, but I've heard that another study discredited the first one. So take your choice. Some plastics can react with bleach and produce compounds toxic to seedlings (and who knows what else?). Worm castings in potting composts can help reduce diseases. We screen compost in late summer and store it in our greenhouse over the

Transplanting spinach from Speedling flats, using butter knives. Credit: Denny Ray McElyea.

winter with lettuces growing in it. By spring it is full of worms and their castings. It is also there waiting, unfrozen and with a nice texture, for the use of growing seedlings in early spring. We only use our stored screened compost for seedlings, with no additions, and we get sturdy plants.

If you mix your own potting compost, ensure that you have a good pH and enough (but not too much) phosphorus. The Florida research website gives information on the best nutrient levels for various crops such as broccoli, lettuce, celery, cantaloupe, peppers and tomatoes. The Seedling Diseases section of the Florida website gives info on hot water seed treatments which can prevent the spread of seed-borne diseases.

Hardening off (or not)

Most plants started in the greenhouse need to be hardened off before planting out in the field. This is done by moving the plants to a transitional zone, such as a coldframe or a shade house, which has conditions midway between the outdoors and the protection of the greenhouse. The plant then experiences lower temperatures with greater fluctuations, more breezes and more direct sunlight. Less frequent watering can also be part of hardening off, but is not recommended for brassicas or lettuce. Too sudden a change of any type is to be avoided, as the stress may cause permanent damage. Plants respond to hardening off by reducing the size and number of their stomata (the holes in the leaf surface where air and water go in and out) and by developing a thicker leaf, perhaps with a waxier coating or a thicker skin and a sturdier stem. The slower growth causes the plants to accumulate carbohydrates, which are available for regrowth after transplanting.

If you use a coldframe, move the plants out to the frame when weather conditions are moderate, and use lids or rowcovers to moderate the microclimate for the first several days. You can also move the plants outside in the daytime and bring them back in for the night. Keeping the plants on a cart makes this easier.

A few crops have more particular requirements:

- Brassicas are sensitive to periods of chilling (2 weeks or more) below 45°F (7°C) at 4–5 weeks of age—these can lead to premature bolting. Younger plants are less sensitive.
- Celery and celeriac should be kept above 55°F (13°C), or they could bolt.
- Eggplant should be kept above 55°F (13°C). Avoid checking the growth of eggplant as it doesn't don't recover well.
- Melons should ideally be kept at 70°F (21°C) or above.

- Onions do better with moderate hardening at 40°F–45°F (5°C–7°C) nighttime temperatures for 7–10 days before transplanting. Yellow onion leaf tips at transplanting time are not a problem.
- Peppers seem to produce stockier plants if soil temperatures are 65°F–68°F (19°C–20°C) after germination. Avoid setting back peppers as they don't recover well. They should be transplanted before they are blooming.
- Tomatoes should ideally be kept above 45°F (7°C), but do OK with warm days (80°F; 26°C) and cool nights (36°F; 3°C).

Transplant quality

The transplanting stage is a good opportunity to practice some selection—plant the best you have. Avoid plants with spots, lesions, or pests. Also avoid over-lush plants that may lose too much water once in the field due to wind damage. Stems should be thick, not spindly—don't think the tallest are best. Roots should be unbroken, actively growing (not pot-bound) and a healthy white color. Ideally the root portion will be at least as big as the vegetative portion of the plant. Leaves should be plentiful and have a strong green color. Transplants with too few leaves can take ten extra days to reach harvest. Ensure that plants have adequate nutrition or they will be likely to be slower to mature and have lower yields.

Transplant age and size

There is quite a lot of flexibility about when a start can be transplanted, but there are accepted ideals to be aimed for. The University of Florida website (listed under Resources) has a wealth of research information. Transplants grown over winter or in

Veg Transplant Table

Vegetable	Notes	Ideal Age at Transplanting
Cucumbers/melons/squash	2 true leaves max (maybe less)	3–4 weeks
Watermelons	(older is OK)	3–4 weeks
Sweet Corn		3–4 weeks
Tomatoes	age is less important	4–8 weeks
Lettuce		4–7 weeks
Brassicas	5 true leaves is ideal	6–8 weeks spring/3–4 weeks summer
Peppers, eggplant	4 or 5 true leaves. Not yet flowering.	6–8 weeks
Onions (spring sown), leeks		10–12 weeks
Celery		10–12 weeks

very early spring in a greenhouse will take longer to reach a viable size than the times given in this table.

If transplanting will be delayed because of unsuitable weather, previous crops still occupying the space, or other parts of your life having priority, then examine the roots and, if necessary, move the plants up to larger containers. Tomatoes may have a few flowers open when you transplant, but there should not be fruit, which would compete with new roots for the plant's resources. Overly large plants can sometimes be pruned to reduce the transpiration losses. Removing one of the lower leaves from brassicas is possible, as is trimming the tops of the plants. I only do this if I feel sure that it's needed, as I hate to remove what might be productive.

If disaster strikes after transplanting (freeze, flood, groundhogs…) and you have more plants, it is definitely worth replanting as soon as you can. With good techniques and modern cultivars, transplant age is not such a concern. For important crops, I try to pot up the leftover transplants into larger containers, to hold in reserve for a couple of weeks after the initial transplanting.

Using larger cells for transplants can result in better yields. Certainly if you are going to be using older transplants, make sure the roots have enough room to grow. Research has shown that plants from larger cells grow faster in the field and are more resistant to pests and diseases. A larger cell allows the plant to make more and stronger leaves, shoots and roots. Field-grown transplants and those in generous cells can develop longer taproots compared to plants in smaller containers, where root growth becomes mostly laterals. Longer taproots help transplants survive drier conditions and erratic watering.

Transplant shock is generally less for plants with plentiful root systems, so harvest starts sooner for these plants. In watermelon, restriction of shoot growth can start as soon as four or five days after seedling emergence if the cell is small! Despite all these reasons to use larger cells, some crops show equal total yields regardless of cell size, in various research projects. The key is to minimize root restriction and transplant when conditions are right, potting up transplants if necessary to hold them longer. If your germination chamber is small, starting seeds in small cells or open flats and moving the starts up to larger containers may make the most sense. You'll still need to start cucurbits and other sensitive crops in larger pots or cells.

Planting depth

Deeper planting seems to reduce wind stress on young plants. Some plants, such as tomatoes, will grow extra roots along the buried stem. On the other hand, soil is cooler deeper down, and this may not be a good thing for warm-weather plants. One way to deal with this, e.g., for sweet potatoes and tomatoes, is to plant in a shallow horizontal or diagonal trench. Much of the stem can be buried in the soil, increasing the growth of extra roots and protecting the plant against wind damage, while keeping the roots in the warmer soil near the surface. With brassicas we usually bury the stem up to the first true leaves, deeper than the plant was in the flat. In general, planting to the depth of the first true leaves increases yields of many crops.

Suitable conditions for transplanting

The ideal conditions for transplanting are mild windless afternoons and evenings just before light steady rain. Transplanting late in the day gives the plant the chance to recover during the cooler night hours when the rate of evapo-transpiration is slower. Shadecloth or rowcover can be used to reduce the drying effects of wind and sun.

Damp soil is important before, during and after transplanting. If necesssary, in very dry weather, water the soil ahead of planting, either with overhead sprinklers or drip irrigation right on the planting row. Alison and Paul Wiediger taught me the wonderful method of setting out drip tape with emitters at the chosen crop spac-

Transplanting bare-root spinach plants. Credit: Jessie Doyle.

ing, watering for twenty minutes before planting, and then planting directly into the wet spots. No other measuring is needed. After transplanting, remember to keep the plants alive by watering frequently. For some reason, this is harder for me when the plants are out of sight under rowcover!

Use a soil thermometer to help you decide when to plant out tender plants. Once the soil temperature has reached 65°F (19°C), tomatoes and sweet potatoes can be planted. 70°F (21°C) is better for peppers and melons. As the climate (or at least the annual weather cycle) changes, basing decisions on reading a thermometer is smarter than using a calendar and planting on fixed dates.

Spacing and planting techniques

There are many possible ways to mark the appropriate spacing for transplants. Transplanting machines have automatic spacing, based on wheels with devices to punch holes in plastic mulch or soil at various spacings for different crops. Some growers have taken the wheel from their "Water Wheel" transplanter and rolled it manually over a bed for a small area of a crop, when it hasn't been worthwhile to set up the tractor for the job.

The Healthy Farmers, Healthy Profits Project in Wisconsin has instructions for making a rolling dibble, a drum roller with plastic cups attached at the required spacing, so that when the roller is run down a bed, it makes a series of divots, one for each plant at the desired spacing, both in-row and between rows. I set out to make one of these once and failed to read all the instructions through first. Consequently I made a handle for a three-foot (90-cm) roller, when I wanted a four-foot (120-cm) one. The dibble does sound like a good idea, but do read before you saw! Eliot Coleman in *The New Organic Grower* has a drawing of a mesh roller to which wood blocks can be attached for marking transplant spots.

For marking rows, we use the Johnny's Row Marker rake, while some growers use an Earth-way seeder without seeds, using the row marker to keep the rows parallel. Having parallel rows means you can hoe much faster, at a walking pace, with no casualties, using the widest possible scuffle hoe.

For accurate measuring of in-row spacing without sophisticated equipment, we have a collection of sticks marked at various lengths. Even workers who think they know measurements may not be as accurate as they think. I ask even the most experienced planters to check from time to time, as it's possible to lose yield by having fewer plants in the row than you intended. Using sticks can be inconvenient, so it helps if the desired spacing is the width of a hand, or a span from index finger to thumb. Sometimes we mark measurements on the tools we'll be using for the

job, using paint markers or nail polish. We have weeding tools marked for carrot thinning, and butter knives marked for spinach plug transplanting. Some trowels have marks imprinted, or are made in a length that happens to be a good plant spacing.

Some people have trouble keeping measurements good in two directions at once. They drift to the edge of the bed, or the two rows they are planting careen in towards each other. We try to get such people to measure in all directions. When using thumb-to-index-finger measure, it's possible to twirl the hand round like a geometry compass, measuring in all directions. Some growers make planting triangles of plywood, with cut corners, to show positions for plants in two rows at the staggered planting pattern.

For transplanting by hand we mostly use regular trowels. Some people like to hold them vertical, scoop facing towards the user, and pull forward to create the hole. Others prefer to push away from themselves. I discourage people from digging a sandbox-style hole, except for really large tomatoes. I like the Swiss plant hand (an ergonomic right-angled trowel), but no one else here does. For planting from Speedling flats we use butter knives. Insert the knife into the ground and wiggle it from side to side, opening a wedge-shaped hole the perfect match of the Speedling plug. Then slide the knife down the tapering side of a cell and twist upward, while grasping the plant near the base with the other hand. Usually this results in a plug sitting fairly neatly on the knife. Then slide it into the wedge-shaped hole and firm it up. Some other kinds of plugs I struggle to release from their cells. Sometimes squeezing the base of the cell works with thin plastic six- and nine-packs.

We grow 10-30 percent extra seedlings above what we plan to plant, to allow for some disasters and some selection of the sturdiest plants. Credit: Kathryn Simmons.

Aftercare: water, rowcover, shadecloth

My usual recommendation is to water your plants an hour before transplanting, and then also well after planting. You should also water them the next day, on days three, seven and ten after planting, and then weekly after that. When setting out a large number of plants, I recommend most people water every twenty to thirty minutes, regardless of the number of plants set out. If the person is skilled and moving fast, and the weather is not outrageously hot or windy, I might let an hour go by before pausing to water. The advantage of getting a lot of plants in the ground proficiently and quickly might outweigh the need to water more often, as the plants are not having their roots exposed to the air for as long when they are planted fast.

Shadecloth draped over recently transplanted crops can help them recover sooner from the shock in hot sunny weather. We use 50 percent shade, in six-feet (two-meter) width. I prefer to use some kind of wire hoops to hold the shadecloth above the plants. This improves the airflow as well as reducing the abrasion or pressure damage done to the plants. The airflow through shadecloth is better than with floating rowcovers. Rowcovers are useful in reducing wind damage, and in moderating cold weather. We use rowcover in summer when we transplant brassicas, to reduce the transpiration losses and also to keep harlequin bugs off our young plants.

Some people like to water newly transplanted crops with various "transplant teas." These may be compost teas or solutions of seaweed extract or fish emulsion. These solutions provide a boost of easily assimilated nutrients to feed the plant until the root hairs recover from transplanting damage.

Chapter 10

Direct Sowing

Many crops grow better from transplants, as explained in the last chapter, however, some don't. Two categories that are usually directly sown are those that do not readily recover when transplanted (tap-rooted crops such as carrots and parsnips, and delicate cucurbits), and crops that are needed in very large quantities. I once listened to an enthused friend who came back from a conference with details of making mini-soilblocks to grow carrot transplants, and special tongs (for handling the miniblocks) made from a couple of tongue-depressors! I was horrified at the very idea! I'm a big fan of transplants, but even I have limits. Some crops are just better off direct sown.

Getting the conditions right

Whether you sow by hand, with a push seeder, or with a tractor and drill, getting an even stand of plants is important. When you make up your planting schedules, you decide the date at which you expect conditions to be right to germinate each crop. The weather can vary, so be prepared to change your plans. It's a mistake to sow sweet corn when the soil is too cold—the seed will rot. I recommend getting a soil thermometer and a copy of Nancy Bubel's *New Seed Starter's Handbook*, which has excellent tables of the best soil temperatures for various crops, the number of days to germination at different temperatures, and the percentage of normal seedlings produced at different temperatures.

Most vegetable seeds need ample water to germinate. Peppers and cucurbits are two of the few exceptions. As these seeds easily rot if the soil is too cold, it's best to keep them on the dry side. Many flower seeds need light to germinate; most vegetables do not. Lettuce and celery are two that germinate better when exposed to light (sow only ⅛"/3 mm deep).

The germination rate of most seeds goes down as the seed gets older. See Chapter 5 for advice on seed viability, dormancy and germination needs.

Soil tilth and moisture also affect seed emergence. Sowing into clumpy soils can leave seeds "high and dry" in air pockets between the clumps, unable to draw on moisture and nutrients in the soil. Preparing a seedbed is a skill that develops with experience. To teach beginners, I find it helpful to explain that the soil particles need to be about the same size as the seed being sown. Tamping the soil over the covered seeds with a rake, hoe or foot helps the seeds stay in contact with the soil.

Some soils collapse and crust after heavy rains or overzealous watering. Under these conditions the washed-down silt can fill all the airspaces and seeds can drown for lack of air. Avoid irrigating after planting by watering before sowing. To reduce the compaction caused by heavy rain, lightly scatter some straw or hay over the row. The long-term cure to this problem is to increase the organic matter content of your soil.

Sowing by hand

Most growers mark where they want the row to be, using stakes and string, a row marker rake, or some kind of rolling dibble. Stake flags are useful as temporary markers when laying out wide-spaced rows. One person can measure and flag, while others hammer stakes and set up ropes. For close-spaced rows, we like the row-marker rake.

To make furrows (drills) we use pointed hoes, also known as Warren hoes. When we sow large seeds, such as corn, squash or beans, we often generously water the open furrows with a hose before sowing. This ensures the water and the seed are in close contact, and that the water is down at the level of the seeds. Usually this method provides enough water to get the seed to emergence.

The extra time at sowing is worthwhile, and saves work later. For small seeds, which we usually sow in many rows, it's easier to water later, with a hose and wand, drip tape, or overhead sprinkler.

For direct-sowing widely spaced crops, we "station-sow"—i.e., we put three seeds at each spot where we want a plant to grow. When the seedlings have three to four leaves, we thin to the strongest seedling. Station sowing can be used for parsnips, which often have a low germination rate, or for crops that will be widely spaced, such as melons.

Sowing at the right depth is important. If too deep, the seed will use up its reserves before the shoot reaches the surface. If too shallow, it can easily dry out.

Before sowing, we soak most legumes overnight and then sprinkle with inoculant. We also soak beets, but for just an hour or two, as they can drown if left too long. If sowing dry seeds, keep them clean and dry, so that any seed returned to the packet will stay in good shape. There are many special techniques for hot-weather sowing (see the next chapter).

Another factor you have to get right is the seeding rate. This can be a hard one for new people to get the hang of, as the human tendency seems to be to pour seed thickly in the furrows. Get new crew members trained on large seeds such as radishes, corn, beans and beets first, as these are easier to space accurately.

Push seeders

Push seeders come in three main types:
1. The cheap and cheerful Earthway, at around $90–$120;
2. Seed drills such as the Planet Junior for $600

new, $150 used; and precision seeders like the Jang JP-1 for $400 or Johnny's European Seeder for $300 (late 2011 prices);

3. The Pinpoint Seeder for $250.

I recommend trying before buying. Ask other growers, take the opportunity on farm field days to look in the back of the barn and ask the farmer about her or his seeder. Precision seeders give an even distribution of seeds, which saves time thinning and saves money on wasted seeds. But not all can handle corn, and some are devilishly difficult to adjust or to empty of leftover seed. Some have attitudes like princesses, demanding perfect soil conditions and jamming up in soil too damp or too fluffy.

Finding one seeder to tackle everything might not be possible. If you transplant a lot of your crops, or prefer to hand-sow large seeds, this reduces the range your seeder needs to deal with.

When we first bought a Planet Junior we dedicated our Earthway to carrot sowing (meaning we didn't need to change the seed plate or adjust the depth). Carrot sowing with the Earthway was so easy! Meanwhile the Planet Junior was so difficult! The soil was always too wet or too dry and fluffy. Changing plates was hard and the whole gadget was likely to fall apart. Sowing on raised beds while walking in the path was difficult. Emptying the hopper involved removing the plate and catching the seed in a dish. We found ourselves going back to using the Earthway for all our long row seeding, while the Planet Junior stayed in the shed. But some other growers love their Planet Juniors!

Johnny's European Seeder has a sturdy design, is lighter in weight than the Planet Junior, and has seed plates reputedly as easy to change as the Earthway. It doesn't work well for sweet corn or beets, which it drops in clumps. Because precision seeders are harder to empty than the Earthway, some people use a small cordless vacuum cleaner to empty the hopper.

Getting the best from the Earthway

Earthway seeders have a "Brassica Grinder" problem: they don't do well with small round seeds. Eliot Coleman has a fix for this in *The New Organic Grower*. A different one, involving a 3-inch (7.5-cm) PVC pipe cap, a long bolt and a wing nut, is to be found online (with photos) at wannafarm.com/earthway-seeder-fix/ posted by Chris Jagger, of Blue Fox Farm in southern Oregon. Saw the end cap down, removing about an inch (25-mm). Drill a hole in the center to take a quarter-inch (6-mm) bolt. Remove the bolt holding the seed plate assembly from the seeder and replace it with a longer, 2¼" (56-mm) one, from the outside to the inside and through the endcap. A wing nut secures the bolt. Adjust the tightness of the nut just enough to allow the plate to still spin freely. To change plates, remove the nut, endcap, and then the plate. (The function of the endcap is to hold the plate more tightly against the seeder and prevent seeds getting behind the plate.)

Apart from this problem, and the fact that the seeder drops out little clumps of seeds with much less precision than the more expensive seeders, the Earthway is wonderful. It works best if you lean it slightly to the right as you push it, so that the seeds are sure to be engaged by the plate. Watch the level of seed in the hopper and look for the twinkle of seeds exiting the seeder at soil level. Check that the seed-covering chain is not clogged with debris. Listen for the normal working sound it makes, so you don't walk along with a slipping belt, sowing nothing. It's relatively easy-going about soil conditions and easy to fill and empty (you can pick it up in one hand, turn it over and empty the seed hopper directly back into the packet).

The Earthway can also be used to sow cover crop seeds. See the Virginia Association for Biological Farming Infosheet listed in the Resources section.

It is possible to test out the sowing rate of different plates on the Earthway by rolling it over an area of concrete or a sheet of plastic, holding the back up so the furrower doesn't touch the ground, and tying up the chain. You can also fill or cover every other hole of one of the plates if you need a more spacious setting. Maintenance consists of washing the plates and hopper in water with dish soap. Instead of rinsing, let them dry with the soap still on. This helps prevent the seeds from sticking. Oil the wheel hubs. Careful owners de-tension the belt between uses, others just keep a spare belt on hand. Some growers buy several Earthways and bolt them together side by side, for faster sowing and perfect row spacing.

Dibbles and jab planters

Dibbles are rollers with knobs attached, used for making indentations in the soil at a regular frequency. They can be useful for large seeds such as cucumbers, squash, or melons, or if you find yourself trying to plant beans through plastic mulch. Their main use is for transplanting: see the previous chapter for more information.

Jab planters (sold by Johnny's) are useful for sowing large seeds, either in cultivated soil, plastic mulch, or no-till conditions.

Tractor seeders and drills

Tractor implements are good for big plantings, with precision sowing and evenly spaced rows. They make quick work of large areas, once the tractor is all set up. See Vern Grubinger's *Sustainable Vegetable Production from Start-Up to Market* for useful information on farm machinery.

Chapter 11

Summer Germination of Seeds

It can be hard to germinate some seeds when the weather is hot, because the temperature is just too high for that seed or the soil dries out too fast. Here are some techniques that can help. But the first tool in the kit is information: know the ideal germination conditions for your crop, the actual conditions, and the expected time to emergence under *your* conditions.

Temperature information

There are excellent tables in Nancy Bubel's *New Seed Starter's Handbook* and *Knott's Vegetable Grower's Handbook* for many of the main vegetable categories. There you can find not only the optimum temperature range for germination of each vegetable, but also the minimum and maximum temperature, the number of days to emerge at different temperatures and the percentage of normal seedlings produced at different temperatures. Get one of these books and a soil thermom-eter. This kind of information can save you wasted effort.

I recommend reading these tables carefully: you may find surprises! For instance, I knew that spinach does not germinate well at high temperatures. The tables say that the optimum temperature range is 40°F–75°F (4.5°C–24°C) and the maximum temperature is 85°F (29°C). One year, after a frustrating time trying to germinate fall spinach, I took a closer look and found that spinach will produce 82% normal seedlings at 59°F (15°C), but only 52% at 68°F (20°C), and a miserable 28% at 77°F (25°C). I hadn't realized how worthwhile it was to somehow get lower temperatures for spinach, rather than working at the top of the possible range.

Variety selection

Some varieties of some crops germinate better at higher temperatures than others. Consult the catalogs, especially ones from hotter parts of the country, and take a look at what growers in areas one or two zones warmer than you are growing.

Seed storage

Viability and seed vigor deteriorate if seeds are stored in warm places, especially if containers are not airtight and the air is humid. Lettuce seed suffers from heat-induced dormancy, but older seed less so. Even a few months of storage can make the difference, so you might consider buying seed of your favorite hot-weather varieties a year ahead. Chilling lettuce seed can help germination in hot weather. We make a practice of putting our spinach seed in double ziplock bags and putting it in the freezer for two weeks before we attempt late summer sowings. This gives the seed an extra "winter" so that it can be tricked into germinating better.

Soil temperature

Summer temperatures can make it hard to establish the crops that will grow well once the weather cools down. Crops which germinate best at soil temperatures below 80°F (27°C) include some beans, carrots, celery, lettuce, onions, parsley, parsnips, peas and spinach.

- Brassicas: It can be surprising to find that most brassicas will germinate faster at 86°F (30°C) than at 77°F (25°C). Turnips can be up the next day, and will still come through in 1.2 days at 95°F (35°C). Winter radishes and daikon have no trouble germinating at high temperatures. Don't expect much from the leafy brassicas (cabbage, broccoli, cauliflower) above 86°F (30°C) though.
- Beets prefer 50°F–85°F (10°C–29°C), and take 14.6 days to emerge at 50°F (10°C), 8.7 days at 59°F (15°C), 5.8 at 68°F (20°C), 4.5 at 77°F (25°C) and 3.5 at 86°F (30°C), their maximum temperature. Notice that if you can maintain a soil temperature between 77°F–86°F (25°C–30°C), you only have to do it for a few days. The percentage of normal seedlings drops off in hot weather above 86°F (30°C). Beets are notorious for spotty germination—their seed coats contain a germination inhibitor. Presoaking for two hours can help dissolve this compound. Room temperature water is better than cold water, and running water is the best, I've heard. Another option is to pre-sprout them just until small red shoots are seen. Getting good soil contact seems to be important, so tamp the row well after planting. It was a surprise to me to find beet seedlings transplant well, although they will be set back about two weeks.
- Carrots prefer 45°F–85°F (7°C–29°C), with 80°F (27°C) being optimum. They can germinate in as little as 6 days at this temperature. People who use pre-emergence flame-weeding for carrots and use a few beet seeds sown in the row as an indicator of when to flame, should note that beets are about half a day ahead of carrots at 50°F–68°F (10°C–20°C), but more than a day at 77°F–95°F (25°C–35°C). The challenge with carrots is to keep the soil surface damp until they come through. See the section on watering for more on this.
- Lettuce likes temperatures of 40°F–80°F (4°C–27°C), with 75°F (24°C) being the best. It germinates in only 2 days at 75°F (24°C). The maximum temperature at which lettuce will germinate is 85°F (29°C), and even then, only 12% normal seedlings will be produced,

in contrast to 99% at temperatures at or be-low 77°F (25°C). Research has shown that lettuce has better emergence in hot weather if sown between 2–4 pm, because the tempera-ture-sensitive phase of germination will then occur at night when soil temperatures are cooler. Even in the face of this research, I do it in the evening!

- Onions come up best at 50°F–95°F (10°C–35°C) with 75°F (24°C) being optimum. They will still germinate at 90°F (32°C), but more slowly, and the percentage of normal seed-lings will go down from over 90% to 73%. Most people wait for fall and won't be think-ing about sowing onions in hot weather, but perhaps you want scallions to fill your CSA boxes, or a quick crop of cipollini mini onions.
- Parsley can be hard to germinate any time, but 75°F (24°C) is the optimum temperature, though it will still take almost two weeks.
- Peas can make a good fall crop if started early enough to mature before frosts. 85°F (29°C) is the optimum temperature, and 95°F (35°C) the maximum. Peas are easy to pre-sprout, more on this later. Mature pea plants are more easily killed by frost than seedlings. In mild winter areas, peas can be fall sown for a spring crop. Sow 1" (2.5 cm) apart to allow for extra losses.
- Spinach is my most challenging crop in hot weather. Its optimum temperature is a mere 70°F (21°C) and its maximum 85°F (29°C). As noted earlier, it's worth getting down as close as possible to 60°F (16°C) to get a good stand of healthy seedlings. Swiss chard germinates best at 85°F (29°C), so consider if chard or leaf beet could substitute for spinach if the fall is being impossibly hot.

Pre-sprouting spinach seeds in fall is very worthwhile.

Reducing temperatures outdoors

If soil temperatures are too high for good germi-nation, look at options for cooling a small part of the outdoors. Options include shade from other plants, shadecloth, boards, burlap bags, or using cold water or ice. For crops you normally direct seed, in hot weather consider cooling a small nursery bed for your seedlings and transplant-ing later. Sowing lettuce from late afternoon until nightfall will give better emergence than morning sowings.

Sowing indoors

If outdoors is impossible, start your seeds indoors in flats, or simply pre-sprout the seeds. You may have room in your refrigerator or a cool room for a plastic flat of lettuce. If it's important to keep the space clean, put the flat in a clear plastic bag and open it a couple of times a day to let fresh air in. Indoor floors, especially in the basement, are usually cooler than shelves. Cool the flats by cov-ering them with wet newspapers until the seed-lings emerge. Plants in plug flats, cell packs or soil blocks transplant with less shock than plants from open flats.

Oxygen Requirements

Seeds need oxygen to germinate. Make sure your direct-seeded crops don't suffocate in over-com-pacted soils, or drown in an excess of water that pushes out all the air. Carrots germinate best in a high oxygen concentration—perhaps being near other growing plants in the garden helps them, as the plants put out oxygen during photosyn-thesis? Cucurbits are particularly sensitive to low oxygen/high CO_2. Would using a fan help them germinate?

Light or darkness?

Unlike flowers, most vegetables are indifferent to whether they have light or darkness when germinating. At high temperatures, celery and lettuce do better when they receive some light, therefore don't sow these crops too deeply if the weather is hot. Peppers may also need light.

Water

Getting enough water to the seed and maintaining that level can be tricky in hot weather. Read in the previous chapter about pre-watering furrows for large-seeded crops.

Soils that tend to "cap" or crust can make life difficult for small seeds that are slow to germinate. Here are some special techniques to minimize crusting:

- Avoid over-cultivating such soils before planting, as this makes crusting more likely;
- Water a day ahead of planting, to reduce the need to water copiously afterwards;
- Avoid hard tamping or rolling after planting;
- Some people sprinkle sand, grass clippings, straw or sawdust lightly along the rows;
- Fedco's seed catalog recommends sowing brassicas in trenches for better survival in dry conditions.

After sowing, watering should be shallow and frequent. Drip irrigation is a help for direct seeded crops, although it can be hard to get even watering all along the row unless the emitters are closely spaced. We use overhead sprinklers at night for close-planted small seeded crops. Chilled water,

night watering, and even ice on top of the rows can help reduce temperatures as well as supplying vital moisture.

Soaking

Some seeds benefit from soaking for a couple of hours or more before sowing, especially when temperatures are high and soils are dry. We generally soak beans and peas overnight before planting. This helps large seeds get all the water they need to absorb for the initial sprouting. After that the smaller amounts needed to help emergence are more easily found. Small seeds that have been soaked tend to clump together, so after draining off as much water as possible, mix them with a dry material like uncooked corn grits, oatmeal or bran, or use coffee grounds or sand. If you plan to put soaked or sprouted seeds in a seeder, dry off the surfaces of the seeds by spreading them out in a tray for a while. Experiment on a small scale ahead of a big planting, to make sure your seeder doesn't just turn the seeds to mush, or snap off any little sprouts. The length of time to soak a seed depends on its size: bigger seeds can benefit from a longer soak, say overnight. Don't soak legumes so long that the seed coat splits, as they then lose vital nutrients and may become vulnerable to attack by fungi. I suspect that when I've had failures with soaked beet seeds it is because I soaked them for too long and they suffocated from a shortage of oxygen.

Pre-sprouting

To pre-sprout seeds for out-of-season growing, first soak them. Then, drain off the water that has not been absorbed, and put the seeds in a suitably cool place. Rinse twice a day, draining off the water. Special plastic draining lids are sold for mason jars, for people who grow sprouts to eat. These are great to use, but you can also make your own with a piece of nylon window screen held on with the lid ring, or a rubber band. A pasta strainer or a sieve held upside down closely in the mouth of the jar will also work. For larger quantities of seed we use plastic one-gallon (4-liter) jars from catering sizes of mayonnaise and mustard.

Usually it's best to sprout the seed just until you see the seed has germinated. Seeds with long sprouts are hard to plant without snapping off the shoot. For most crops 0.2" (5 mm) is enough. For lettuce half that length is good, and one day may be time enough. If your pre-sprouting has got ahead of the weather or the soil conditions, slow down growth by putting the seed in the refrigerator. For fall-sown spinach we do the whole sprouting process in the fridge, and I confess that we don't rinse them very often at all! One week is a good length of time for fridge–sprouting spinach.

If you have leftover soaked or pre-sprouted seeds, you can store them in the fridge while the others come up, and then use them to fill any gaps in the row. One spring we had a crop failure of bush beans in our hoophouse, and we filled out the space with leftover soaked snap pea seed and grew a crop of pea shoots for salad!

Fluid sowing

A novel way of helping seeds get established in difficult dry conditions is to sow pre-sprouted seeds in a jelly, which protects them from damage. This method reduces the attention needed by slow-germinating seeds. Emergence will be earlier, spacing will be more even, and crops can get to maturity quicker, which can be an advantage if the winter arrives too soon after the summer in your region or you are extending the seasons to a maximum. Sprout the seeds, and once the shoots are 0.2" (5 mm) long, rinse and strain them gently.

Make up a starch or cellulose paste. In England I used wallpaper paste, although it is important to make sure this contains no fungicides. A cup of thick cornstarch paste is easily cooked up by making a smooth slurry with 1–2 tablespoons (15–30 ml) of cornstarch and an equal amount of cold water. Heat a cup of water, then stir in the slurry and boil for one minute to activate the cornstarch, stirring gently. If the paste is becoming too thick, stir in a little more water. Possible substitutes for cornstarch include the same quantity of xanthan gum, pectin, agar, gelatin, arrowroot, potato starch or tapioca. If using regular flour, use twice as much.

Allow the paste to cool, then gently stir the seeds in. They should not sink when the mix is allowed to stand. If they do, the mixture was not thick enough. Start again and decant the seed layer to mix into a thicker paste.

Pour the mixture into a plastic bag, and when you get to the garden, snip a small corner off the bag. Ideally the mixture should squeeze out into a damp furrow at about one tablespoon per foot of row (15 ml/30 cm), and should contain the number of seeds you want to plant in that distance. It may be worth experimenting on the kitchen counter first! After sowing in the garden, cover the seeds with soil in the usual way.

The Complete Know and Grow Vegetables, by J. K. A. Bleasdale, P. J. Slater et al., has a good description of this method, which is also used to speed emergence in cold spring weather.

Chapter 12

Succession Planting for Continuous Harvesting

Many vegetable crops can be planted several times during the season, to provide a continuous supply. Because the length of time from sowing to harvest varies according to temperature (and day length in some cases), simply planting squash once a month, for instance, will not provide an even supply. By keeping records and using information from other growers in your area, you can fine-tune your planting dates for better results.

Our first effort along these lines involved keeping a "Veg Finder," a chart where we note planting date, first harvest and last harvest ("veg" is the British abbreviation—you might prefer "veggie"). We use this for bush beans, summer squash, cucumbers, sweet corn and carrots. (We have a similar chart for lettuce, which is an exercise of its own.) The original aim of the Veg Finder was to alert crews about which plantings were currently ready to harvest and where to find

them. After keeping these charts for a few years we started to use the information to improve our planting dates.

Most growers are probably adept at planting as soon as possible in the spring. A lesson for me has been not to plant too early! Keeping old cucumber transplants on hold through cold early spring weather is just not worthwhile. I finally grasped this the year we planted out our first and second cucumber plantings side by side—the second ones did better than the first, and were ready just as soon!

It's also important to know the last date for planting each crop that has a reasonable chance of success. For this part we got help from a leaflet from the Virginia Tech Extension Service: "Fall Vegetable Gardening."

Here's the formula (for frost-tender crops) for figuring out the number of days to count back

Three sweet corn varieties sown on the same day provide over two weeks of harvests. On the right, fast-maturing Bodacious; in the middle, red-flowered Kandy Korn; on the left, Silver Queen still growing to its full height. Credit: Kathryn Simmons.

from the expected first frost date: add the number of days from seeding to harvest, the average length of the harvest period, 14 days to allow for the slowing rate of growth in the fall, and 14 days to allow for an early frost. For example, yellow squash takes maybe 50 days from sowing to harvest, and is good for 21 days, so the last date for sowing would be 50 + 21 + 14 + 14 = 99 days before the first frost. For us that means 99 days before Oct 14, so July 7. But with rowcover to throw over the last planting on frosty nights, the growing season is effectively 2 weeks longer, and we can ignore the 14 days for an early frost. So our last planting of squash is Aug 5.

Three pieces of information—the first planting date, the last planting date and the number of productive days of a planting—enable us to make a rough plan for how often to plant. For us the rough version is every two weeks for beans and corn, every three weeks for squash and cucumbers, once a month for carrots, and two or three plantings of muskmelons (cantaloupes) at least a month apart.

Record keeping

There are methods of succession planting that involve no paperwork. For example: sow another planting of sweet corn when the previous one is one to two inches (25–50 mm) tall; sow more lettuce when the previous sowing germinates; or sow more beans when the young plants start to straighten up from their hooked stage.

Another approach is to sow several varieties with differing days-to-maturity on the same day. We sow Bodacious, Kandy Korn and Silver Queen sweet corn on the same day, and get more than two weeks of plentiful harvests.

But to fine-tune for a really even supply of crops, nothing beats having real information about what actually happened, written at the time it happened! Our system for gathering crop records consists of a spiral pocket notebook I carry all the time, and a ring binder with alphabet dividers to store the information until the end of the year. I transfer the field notes from the small notebook to the binder sporadically during the season, and then have filing sessions on rainy s in late fall and early winter, storing the in-mation for each crop separately. We take the ...formation learned into account when we do our winter planning.

Graphs

Having graphs of sowing and harvest dates for each crop has been very useful for planning effective planting dates. Graphs can be done by hand or using a spreadsheet program such as Excel. Because the number of days to maturity varies as the season progresses, it is important to vary the interval between successive sowing dates to obtain equally spaced harvests for each crop. Typically, plants mature faster in warmer weather.

Here is a step-by-step guide to making and using these graphs:

1. Gather several years' worth of planting and harvesting records, into two columns (see the squash example). The date in the Harvest Start column is the first day of harvesting from a planting sown on the Sowing Date in the first column. Note that these aren't all in the same year!

2. Plot a graph for each crop, with the sowing date along the horizontal (x) axis and the harvest start date along the vertical (y) axis. In our squash example, the major grid lines are every 10 days apart on the x axis and every 20 days apart on the y axis. The small marks are 1 day apart on the x axis and 4 days apart on the y axis. For example April 18 on the x axis and June 1 on the y axis make the first point on our graph. Mark in all your points.

3. Draw a curving line joining all your points together. The line of the graph is often rather uneven, due to differences in weather from year to year, and to growing varieties with differing maturity dates. If you use a spreadsheet program as I did, the computer-generated line will be jagged and you will want to smooth it out a bit. To get a smooth line, which is more useful and more representative of typical reality, leave equal numbers of points above and below the graph line, but try to hit most of them. Sometimes there will be "outliers"—odd things happen. We once had an April 18 sowing that didn't produce till June 1. I guess the plants got cold and set back. Ignore atypical points. In the corn example later, the "blobs" marking the points are left in. Here they are hidden, so the line is clearer.

4. To use one of these graphs to plan future sowing dates, start from your first possible sowing date along the x axis. Draw a line up from this date to the graph line. Draw a horizontal line from the point on the graph line to the y axis. This is your first harvest date. (Yum!) Ours is around May 19.

5. Calculate your last worthwhile harvest date of the season. Draw a line across from this date on the y (harvest) axis to the graph line. Draw a vertical line from this point on the graph line to the x (sowing) axis to show when you need to sow this batch. Ours is sown Aug 5 (70 days before our average first frost) and we harvest from around Sept 24. From May 19 to Sept 24 is a 128-day harvest period.

6. Decide roughly how often you need a new patch coming into production, dividing the total harvest period into a whole number of intervals. If we want fresh squash every 32 days, we'll need four equal intervals between plantings, and five plantings ($32 \times 4 = 128$). The harvest start dates will be May 19, June 20, July 22, Aug 23 and Sept 24.

Squash Succession Crops

Sowing Date	Harvest Start
Apr 18	Jun 1
Apr 21	May 19
Apr 23	May 25
May 14	Jun 3
May 15	Jun 21
May 20	Jul 5
May 25	Jul 4
May 29	Jul 7
Jun 12	Jul 20
Jun 15	Jul 20
Jun 30	Aug 2
Jul 1	Aug 8
Jul 2	Aug 11
Jul 4	Aug 8
Jul 5	Aug 10
Jul 14	Aug 14
Jul 18	Aug 17
Jul 19	Aug 28
Aug 3	Sep 9
Aug 4	Sep 5
Aug 5	Sep 15
Aug 7	Oct 2
Aug 9	Sep 25
Aug 12	Oct 5

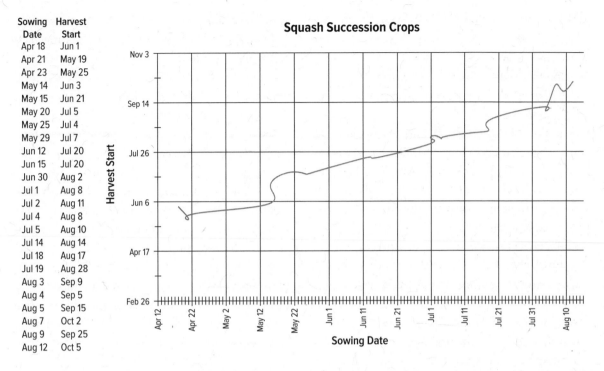

7. Draw horizontal lines from the y axis from the five calculated harvest dates to the line of the graph.
8. Then draw vertical lines from where these horizontals cut the graph line, down to the x (sowing) axis. This will give you the planting dates needed. The harvest dates above imply planting dates of April 21, May 17, June 21, July 16 and Aug 5. These sowing intervals are 26, 25, 25 and 20 days.
9. Note that our first squash is transplanted April 21 but sown March 25. Such early plantings grow very slowly.

If you don't have your own crop records yet, you could use our graphs as a jumping off point. You would need to change the first and last sowing dates to fit your own climate zone.

The spreadsheet program I use is Micro-soft Excel. It doesn't seem to recognize the word "graph," preferring "chart." I start the worksheet by creating two columns: for sowing date and first harvest. This type of graph with irregular intervals between entries is called a "scatter chart." I print out the computer-generated graph and just pencil in a more even line, using that to derive my sowing dates.

Save space and work: summer squash and cucumbers

Information from the graphs may indicate that you can plant less often than you thought, thereby saving space and work. We used to do six plantings of summer squash and cucumbers: March 25 (transplanted April 20), May 14 (transplanted June 7), June 13, July 3, July 19 and Aug 5. The intervals between these sowings were 50, 30, 20, 16 and 17 days. By using the graphs, we have been able to go down to five plantings: March 25 (transplanted April 21), May 17, June 21, July 16 and Aug 5, at intervals of 52, 25, 25 and 20 days. The sowing intervals decrease as the season warms up, as it takes fewer days for plants to mature. The first planting is very slow to mature—probably we could just start later still and lose nothing. By moving the second planting 10 days later than it used to be, we are able to direct sow rather than transplant, which saves us time. This revised schedule saves us from dumping zucchini on our neighbors' porches!

For us, squash plantings stay productive for around 40 days, while cucumbers sometimes only last 35 days. Cucumbers also take a little longer to mature than squash. These two features would suggest making more plantings of cucumbers than of squash, but after looking at the graphs, we decided to plant both on the same set of dates, for simplicity. We are often picking from two plantings on the same day—the overlap helps even out our supply, as the dwindling old patch augments the newly starting row. If we could be satisfied with a new patch coming on-stream every 36 days, we could sow only four times. In spring you can make a new sowing of cucumbers when the first true leaf appears on the previous sowing. In summer make the next sowing when 80% of the previous sowing has emerged.

Mexican bean beetles

Other factors can affect sowing dates for succession crops too. We used to be sorely plagued by Mexican bean beetles. This meant that pole beans were a complete waste of time (they didn't mature before the beetles ravaged them), and we needed a new patch of bush beans every 2 weeks to keep up supplies. We made seven plantings at 15-day intervals: April 16, May 20, June 9, June 24, July 9, July 22 and Aug 3. After two weeks of harvesting each planting, we would need to do "Root Checks," our euphemism for pulling up the beetle-ridden plants, picking off the last beans, and taking the plants off to our composting area.

Now we buy the parasitic *pedio* wasp. Once the parasites are established for the season, there's no more need for hand picking beetles, and the second and subsequent plantings will look very healthy. We now plan for a new patch to harvest every 20 days, sowing six times rather than seven: on April 16, May 14, June 7, June 29, July 19 and Aug 3. These sowing intervals are 28, 28, 22, 20 and 15 days. We also get more beans than previously. Much of the time we pick from two patches with an overlap period. See the Green Beans Chapter for more about beans.

Hoophouse crops

We now plant our earliest squash and cucumbers in our 30' × 96' (9 × 29 m) hoophouse. I think we could eliminate our chancy early outdoor plantings, and reduce our stress without losing much yield. We sow cucumbers Feb 14 (we like General Lee for its compact vines and prolific cucumbers) and squash March 1 (Zephyr has become a favorite), and transplant April 1 for harvests starting May 19 (squash) and May 21 (cucumbers).

Fast turnaround crops: lettuce

Succession planting of lettuce, because it is such a fast turnaround crop, is a subject in its own right. (See the Lettuce chapter.) The short version is that we sow twice in January, twice in February, every ten days in March, every nine days in April, every eight days in May, every six or seven days in

Sowing Date	Harvest Start
Apr 25	Jul 10
Apr 26	Jul 9
Apr 27	Jul 13
May 1	Jul 17
May 21	Jul 27
May 23	Jul 25
May 29	Aug 11
Jun 3	Aug 7
Jun 5	Aug 7
Jun 6	Aug 3
Jun 7	Aug 13
Jun 17	Aug 15
Jun 23	Aug 19
Jun 24	Aug 24
Jun 25	Aug 27
Jun 26	Aug 26
Jul 1	Aug 29
Jul 2	Aug 29
Jul 3	Sep 7
Jul 4	Sep 6
Jul 6	Sep 2
Jul 7	Sep 5
Jul 11	Sep 9
Jul 15	Sep 18
Jul 16	Sep 23
Jul 17	Sep 16
Jul 19	Sep 21

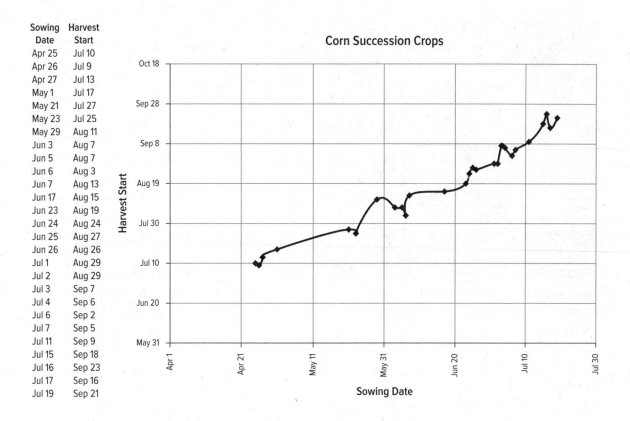

Corn Succession Crops

June and July, then every five days in early August, moving to every three days in late August and every other day until Sept 21. After that we ease back to every three days until the end of September. Those last plants will feed us right through the winter.

Avoid chancy sowings: sweet corn

We had been in the habit of making seven sweet corn plantings: April 26, May 17, June 2, June 16, June 30, July 14 and July 28. The intervals were 21, 15 and then 14 days. We eliminated the seventh planting because of shortage of space and increased the size of the sixth, sowing a range of three varieties, as noted above. The season does finish a bit earlier than previously, but we get no

complaints. The seventh sowing has always been risky because of deteriorating weather.

Using the graph of our corn sowing and harvest dates discussed earlier, I estimate that April 26, May 19, June 6, June 24, July 7 and July 16 would be good dates for six plantings to provide fresh eating every two weeks. The planting intervals are 23, 18, 18, 13 and 9 days.

Spring and fall crops: carrots

We sow carrots mid–late February, starting as early as possible. Then we sow every four weeks in March, April, May and, if needed, June and July. We scratch these hot-weather sowings if we still have spring carrots in the cooler, as the flavor of hot-weather carrots is not very good and

Have a new planting of cucumbers coming along before the old one passes its peak. Credit: Kathryn Simmons.

we can get Alternaria blight, which turns the leaves black and so reduces growth. We make a huge fall planting in late July or early August. If we miss those dates, we wait till late August to avoid the high numbers of grasshoppers here in mid-August. Late August sowings don't bring as heavy yields as the earlier ones, unless the winter weather lets us harvest later than usual. We don't do succession plantings for fall carrots, just one big one, because we are growing bulk carrots to store for use all winter and don't need multiple harvest dates. With fall crops, even a difference of two days in sowing dates can make a difference of two or three weeks in harvest date, because plants grow slower as days get shorter and cooler.

Season Extension

Growers wishing to produce crops over as long a season as possible use various methods to grow earlier crops in spring, extend the growth of cool-weather crops into summer, extend the survival of frost-tender crops beyond the first fall frosts, keep semi-hardy crops alive through the winter, and boost winter production of hardy crops.

ATTRA's publication *Season Extension Techniques for Market Farmers* addresses agricultural practices, use of plastics and the economics of season extension. It also includes a useful resource list. In this chapter I'll give an overview of the possibilities. Later chapters address some of the main techniques, and more detail on specific crops will be found in the crop chapters. The timing of succession plantings (see the previous chapter) is also an important method of extending the season of fast-maturing crops.

Growing earlier crops in spring

Careful planning can improve the potential early yield of some spring crops. Choose fast-maturing and hardy varieties to increase the chance of success. If you grow your own seed, you can select for cold tolerance in your plants. Sometimes the spacing of plants can influence the earliness of the harvest (see the Crop Spacing chapter).

Transplanting crops under rowcover is the main technique for earlier outdoor crops. Using transplants enables you to start the seedlings under warmer conditions and so start them earlier and grow them faster than if you direct seeded.

Look for nooks with a warmer microclimate than the rest of your fields, more protection from the prevailing winds, a slope to the south, or perhaps a barn wall to the north. Avoid frost pockets where cold air collects at the bottom of a slope. Providing top-quality rich soil for early crops will encourage fast growth. Raised beds and free-draining sandy soils warm faster than flat fields and clay soils. Raking the soil in a cold-frame to a five- to seven-degree slope towards the south gives conditions equivalent to moving to a warmer climate zone! Dark-colored soils absorb

A raised bed with a seeded crop protected by rowcover, held down by rolling it around sticks underneath the rowcover. Credit: Kathryn Simmons.

more sunlight and are less frost-prone. Black plastic mulch covering the beds will warm the soil, and can bring harvests of melons forward by as much as four weeks (see the Mulches chapter).

There are limitations to how early you can get harvests from some crops. Low light levels and cold soil temperatures can mean that plants started really early will not actually reach harvest any earlier than ones planted a month later. So experiment on a small scale at first and keep good records.

Rowcover

Rowcover is a wonderful invention: lightweight, easy to use, easy to store. The edges need to be held down by bags of rocks or sand, plastic jugs of water, or metal or wooden stakes lying along the edges. We use scrap wood bars about 5' (1.5 m) long and have found that a simple change to the direction of rolling of the rowcover edges made an enormous difference. (Thank you, Bri!) If you put the stick under the edge of the rowcover and roll the edge under several times, the stick is less likely to escape and leave the rowcover flapping in the wind. Any tendency of the stick to roll down off the edge of the bed causes it to lock itself further in the rowcover. When we use rowcover without hoops, we find we need to remind crew to leave some slack in the rowcover for the plants to grow.

To protect against frost, you need a heavy-weight rowcover—the thinner types are for protection from insects. We like Dupont Xavan 5131 (previously called Typar). It is a 1.25 oz/yd² (42 g/m²) fabric that can last for six years or more. It is spunbonded polypropylene with UV stabilizers, has a 75% light transmission, and provides about 6 degrees F (3.3 degrees C) of frost protection. We like its durability and reliability. We think polypropylene rowcover lasts longer and is tougher than polyester (Reemay). We have also liked Agribon 17 (or 19), spun-bonded polypropylene which weighs 0.55 oz/yd² (18.5 g/m²), transmits 85% of sunlight, and offers 4°F (2.2°C) of frost protection for winter use. It's usually available in 83" (2.1 m) width. Thinner rowcover can be used doubled up in severely cold weather, if you don't have enough thick rowcover available.

Rowcovers do have two disadvantages. One is that if you are growing on bare soil rather than plastic mulch, weeds will grow very well, secretly and out of sight. The second disadvantage is that rowcover reduces light levels. It is a good idea to ventilate covered crops in mild weather, so they don't lose their cold tolerance.

Hoops can be used to keep rowcover from sticking to frozen leaves and to reduce abrasion. We make hoops by cutting and bending 9- or 10-gauge wire. In winter we use double wire hoops—the outer hoops trap the rowcover so that it doesn't blow away. The microclimate under hooped rowcovers is very pleasant in chilly, windy weather. Some growers use fiberglass rods, or hoops made from scrap plastic piping. There are also spring steel hoops, for setting by machine or by hand. These are easy to store as they return to a relaxed bow shape when removed from the soil and don't get tangled. Their disadvantage is that they seem to come in just one length, 64" (1.63 m), which is fine for a single row of plants, but less good for our 48" (1.2 m) beds with multiple rows. It's possible that spring hoops could be cut from high-tensile fence wire, but I haven't tried that yet.

Plastic-covered tunnels

Once plants are established, if they can withstand cold nights, they may benefit more from clear plastic rather than rowcover over hoops. The plastic will let more daylight through, while still increasing the temperature and protecting from the wind. These covers may have slits to allow the plastic to curl open as it heats up, letting the hot air escape. If your plastic-covered tunnels are unvented, you will need to provide the ventilation yourself.

There are two versions of low tunnels that are taller, although not tall enough to walk under:

- Quickhoops cover more than one bed, and can be covered with rowcover topped with

Newly built hoophouse at Twin Oaks. A very worthwhile investment. Credit: Pam Dawling.

greenhouse plastic. Caterpillar hoops are similar (usually narrower), and have the plastic or rowcover held down by ropes;
- Hoophouses (walk-in plastic-covered hooped structures, also known as hightunnels, or confusingly, coldframes) are the most successful way to provide harvests in early spring and winter. Traditional low coldframes (beds with box sides and transparent lids) provide conditions between those of hoophouses and the outdoors. Frames with glass lids and insulating walls are warmer than those with wooden walls and plastic glazing. The latter don't offer any real advantages over rowcover low tunnels—in my experience they just cost more and offer less flexibility.

Frost avoidance

Overhead irrigation can be used to protect crops from late frosts in spring (or early frosts in fall). Commercial strawberry growers frequently do this once their plants are flowering. The water must be supplied continuously while the air temperature is below freezing, and continued until the sun is shining on the plants enough to keep them above freezing. This method works because water gives off heat to the plants as it freezes into ice, and the formation of an ice shell around the plant prevents the colder air reaching the plants. Commercial growers with large fields of strawberries may have an alarm system to wake someone if the temperature falls to a dangerous level. That person then starts up the irrigation. I prefer to start the sprinklers before going to bed, and then to sleep soundly.

If you wake early and find plants with that dreadful dark green glassy look of frozen death, you may be able to save them by watering immediately, either with sprinklers, or with a hose spray if the area you need to save is small. Just keep watering until the sun is shining on the plants and all may be OK.

Large-scale growers sometimes use wind machines to keep the air moving on cold nights and prevent a frost from descending. In the past, orchardists used small smoky fires to provide insulation over valuable crops. This is no longer recommended because of the air pollution and contribution to global warming.

Extending the growth of cool-weather crops into summer

Germinating seeds in hot weather and keeping the plants growing well are the twin aspects of taking on these challenges. The main methods of keeping cooler-weather crops growing once it warms up are to shade them and provide plenty of water. See the Summer Germination of Seeds and Lettuce chapters for more detail.

Shadecloth is available in many widths and different densities of shade, woven or knitted, in polypropylene or polyethylene. It's usually black,

sometimes green. We've bought from Gemplers, Wood Creek and Peaceful Valley. The general guidelines are:

- use 30% shadecloth for vegetables such as tomatoes and peppers in areas with very hot summers;
- for cool-weather crops such as lettuce, spinach and brassicas, use 47% in hot areas, 30% in cooler climates;
- 63% is for shade-loving plants;
- 80% is often used over patios and skylights to cool people as well as plants.

We generally use 45%–50% shadecloth to cool crops such as lettuce growing in summer. We also have a short roll of very dense (80% or 90%) shadecloth that we use to cover harvested produce before moving it to the refrigerator. And we have a giant piece of shadecloth that goes over our hoophouse mid-May to early September. It is knitted, black, 50% shade, UV-stabilized polyethylene. Knitted shadecloth is more flexible/drapeable than woven and will not unravel when cut. Polyethylene is the usual material for knitted shadecloth. Previously we bought woven polypropylene 50% shadecloth for the outdoor beds, and hemmed the ends on a domestic sewing machine using nylon thread. Now we buy knitted shadecloth for the vegetable beds too. We like to hold the shadecloth above the plants on the spring steel hoops, to improve airflow and let the plants grow unimpeded. Some people clip the shadecloth almost horizontal above the crop and leave the sides open for the best airflow.

We found some very good plastic clothespins to hold shadecloth onto hoops and to join short lengths of rowcover together at a hoop. They are Hurricane Grip Pegs from the UK, and come in hard-to-lose shades of yellow, purple and turquoise! I've also seen green ones sold in home gardening catalogs. These are made of a single piece of plastic (no metal), with an integral spring and a lock. They do break down in sunlight after a couple of years, but last outdoors a lot longer than the standard wooden kind and don't get tangled in the fabrics. Unfortunately I haven't been able to find a current supplier. For summer lettuce, we use one of these clothespins at each hoop, space the hoops about eight inches (20 cm) apart, and dispense with sticks to hold the shadecloth down. Because shadecloth lets air through better than rowcover, it's less likely to blow away. We keep it on our lettuce for two to three weeks after transplanting, then pull up the hoops with the shadecloth still attached, and four or six of us move it (like a giant accordion) on to the next planting.

For small areas, such as a lettuce nursery seedbed, we use scraps of window screen, old tent mesh windows, etc. Once when desperate, we covered a row of lettuce transplants with a collection of onion and cabbage nets for a few days. They were certainly better than nothing!

Extending the survival of frost-tender crops beyond the first fall frosts

The return of rowcover is a sign in our fields of the approaching frost. Some crops we simply harvest and store: Chinese cabbage, cowpeas, limas, eggplant, melons, okra and (if the frost is likely to be serious) winter squash. Cauliflowers can be protected by using clothespins to fasten the leaves together over the curds. Bright plastic pins help you locate the cauliflowers later! We cover late rows of summer squash, zucchini, cucumbers, bush beans, pak choy, lettuce and celery. Foliar sprays containing seaweed can be used a few days ahead of expected frosts, to toughen up cell walls and make frost damage less likely.

We use overhead sprinklers for our tomatoes and peppers if the forecast suggests a frost is likely. Tomatoes and peppers have big plants not easily covered by rowcover. Some growers take down the posts or cages supporting their tomatoes and lay the plants down on the ground so they can be covered. Here, and in many parts of the country, a frost or two will often be followed by a few more weeks of warm weather, so getting past the first few frosts is worth the effort. It's much easier to get extra harvests for a month or two from mature plants you already have, than it is to get harvests a week earlier in the spring.

With peppers, often only the top of the plant will get damaged by frost, so once we get close to our usual frost date, we change our harvest method. We remove all the peppers open to the sky, regardless of color. The peppers protected by foliage above them will often come through undamaged. The frosted top leaves die off, so before the next frosty night, we harvest another layer of peppers. This method allows us to get the highest number of ripe peppers possible, and also spreads out the harvest in a way that is more manageable.

Keeping crops alive into winter and boosting winter production

If you want to overwinter crops, start by believing it is possible, and look for the hardier cultivars available. If you can add some wind protection, do so. At our Zone 7 farm, we overwinter Vates kale without rowcover, but we've killed off Winterbor and Russian kales that way, while we were learning. Many greens and roots can survive some freezing, so it is worth experimenting to find how

Hoophouses do require new plastic every four-six years, but the space offers excellent value for season extension. Credit: Bridget Aleshire.

late you can keep crops outdoors. See the chapter on Cold-Hardy Winter Vegetables.

Consider the wisdom of season extension

The economics of season extension can be considered by using partial budgeting. Calculate the additional costs of using a particular kind of season extension for your chosen crop or group of crops, and the likely added income. Subtract any reduced costs (for example, using plastic mulches reduces time spent cultivating). Calculate the total change in projected income and decide if it is worthwhile.

Success with season extension is a matter of finding the balance point at which the time, money and energy you put in are still definitely worthwhile. The further we go from a plant's natural season, the more costly the efforts to keep it alive and productive. I sometimes imagine the pineapple growers in Victorian England—low-paid workers crafting thick straw ropes to wrap around tender plants for insulation, and stoking coal stoves day and night to grow a pineapple or two for the rich people in the Big House. Not a model to copy!

Money is not the only factor. A longer harvest season helps you retain and satisfy customers. Just how important this is will depend on your market: if students are your main customers, you may as well take a winter break (and a long summer break) too. But if you cater to snowbirds, then winter crops will be important. Season extension can help provide year-round employment for your crew, which helps you retain skilled workers. The flip side is that you get less free time. The issues of plastics disposal and the ecological impact of the use of fossil fuels to manufacture agricultural fabrics are both worth thinking about. However, people are going to eat year round, and if it isn't you providing local, sustainably grown produce, then someone else is trucking it in. The ecological costs of shipping are usually much higher than those of careful use of plastics.

Chapter 14

Cold-Hardy Winter Vegetables

While some growers are eager to get their fields and gardens into winter cover crops, others see the last quarter of the year, the "Second Spring," as an opportunity to sow and plant some cold-hardy vegetable crops. Some vegetables grow quickly and can be harvested before winter, others over-winter for early spring harvest, and some can even be harvested throughout the winter, depending on your climate. Because winter conditions can vary, success cannot be guaranteed. If a crop fails there is unlikely to be time to resow. I have plenty of respect for farmers who carefully guard their winter downtime to recover energy before next spring, or who use wintertime to build barns or greenhouses or to fix equipment. If you don't have big plans and are not desperate for a rest, fall planting is worth considering. As bonus points, weeds grow slower in colder weather, planting dates may become more flexible, and you have fewer crops to take care of.

Winter share CSAs will be much more in de-mand as the locavore movement grows, as will year-round demand for local foods in stores and markets. After all, people eat year round, not just for the duration of a twenty-six-week "summer" CSA. If you don't want to do the work yourself, perhaps you can rent out some land to a grower for the winter, or make some cooperative arrangement. There is a big potential for winter share vegetable CSAs to link up with bread shares, herb shares and other value-added products, strengthening the whole local economy. If you keep your customers through the winter, you'll have their loyalty next summer.

Before taking the plunge into winter gardening, know your climate, know your resources, know your market, know your crops, and when you don't know, experiment on a small scale.

Your climate

Useful data include first frost date, daylight hours (related to latitude north or south of the equator), average annual minimum temperature (USDA Cold Hardiness Zone) and wind-chill effects. Temperature is a big factor in the rate of crop growth, but it is not the only one. Excess soil moisture, wind strength, rodent and deer pressure also need to be considered. A minimum of three hours of direct sun per day is needed—in winter the sun arcs lower in the sky than it does in the summer, possibly not making it above the horizon for as long as you think, if you live among hills. On the plus side, deciduous trees will be leafless, so you may get more sunlight than in summer!

Your resources

If you have a hoophouse, you'll be able to produce an abundance of fresh crops; if you don't, fall is a good time to put one up. You can sow a nursery bed of transplants outdoors under rowcover to give yourself a head start, then transplant the seedlings into your hoophouse once you get the plastic on. (You'll also get the chance to experience an interesting "season reversal" in weeds if you cover your hoophouse in late fall: first the cool-weather weeds germinate, then the warm-weather weeds return as the space warms up—a harbinger of what you can hope for in vegetables too!) Whenever it is sunny, even if it's very cold outside, your winter hoophouse crops will be actively growing in the extra warmth.

Coldframes are useful in snowy areas, as they hold the snow off the plants and are easier to open when icy than rowcover, which can rip. Heavy-duty rowcover with hoops and something to hold down the edges is also valuable. Depending on your climate, some crops may not need rowcovers, or may only need covering for the coldest nights. Kale, for instance, may only need to be covered for a short time during the winter to keep

it alive for spring, when it can become very productive again.

Raised beds can help keep plants warmer and drier, and seaweed foliar sprays can increase their cold-toughness. Straw, tree leaves, pine straw or hay mulch around (not over) the plants can also help keep roots warmer.

What about your human resources? If your crew has slimmed down, the workload will need to reflect this.

Your market

If you have a CSA, find out how many sharers would be interested in a winter share, and how often they would want to pick up. Perhaps every other week? Or every other week in October, once a month in November, December, January, February, twice during March? Design something that works best for you and your CSA. It's unlikely to be the same as the summer schedule. If your farmers' market closes in the winter, perhaps deliveries to stores or restaurants would suit you? Also consider your market when choosing your crops—do your customers like the more bitter endives and chicories? Sour sorrel? Pungent mustards? Tiny leafy crops such as corn salad or claytonia? How much educational marketing are you willing to do to get people interested in the more unusual crops? Do you want to fuss with small-leafed crops? (We don't.)

Your crops

Winter vegetable crops fall into three categories: those that grow fast and mature before the coldest weather, those that can be harvested all winter, and those that sit in the ground and get off to an early start in the spring, allowing an earlier harvest. Look for especially cold-tolerant varieties or gene lines. Brett Grohsgal and Christine Bergmark of Even' Star Farm and Frank Morton of Wild Garden Seeds are working to produce extra-hardy varieties. Consider investing in cold-tolerant seed stock and then saving your own seed to select for cold tolerance as it is needed at your farm. Sometimes heirloom strains are more cold tolerant than modern hybrids—winter hardiness has fallen out of favor in the era of highway-traveling veggies.

Brett Grohsgal of Even' Star Organic Farm in Maryland has done extensive research and plant breeding to create extra-hardy greens. Brett and his family grow large fields of greens and root crops in the open, completely unprotected. They have started to sell seed of some cultivars for growers who want to do their own seed growing and develop locally suited lines. Their varieties will work for areas which get at least three hours direct sunlight each day and have minimum winter temperatures of 0°F (–18°C). Some of their lines are available through Southern Exposure Seed Exchange and Fedco Seeds.

Fast fall crops

Look for cool-weather crops that mature in sixty days or less. Mostly these are greens and fast-growing root vegetables. The fastest possibilities include leaf radish and mizuna, radishes (both the very fast small ones and the larger winter ones), kale and arugula. Also many kinds of Asian mustard greens: Chinese Napa cabbage, Komatsuna, Maruba Santoh, pak choy, Senposai, tatsoi, Tokyo Bekana and Yukina Savoy. Spinach, chard, salad greens (lettuce, endives, chicories) and winter purslane can all be ready in 30–35 days. Corn salad, land cress, sorrel, parsley and chervil can be ready in 35–45 days. Beets, collards, kohlrabi,

turnips and small fast cabbage (Farao or Early Jersey Wakefield) can take as little as 60 days.

If you use rowcover to get through your first frost of the season, there is often a spell of warmer weather before more frosts, effectively pushing back your first frost date. This may allow a late succession sowing of warm-weather crops such as bush beans, squash, or even cucumbers.

For crops to be harvested before killing conditions arrive, the formula to find the last workable sowing date is:

(1) Number of days from outdoor planting to harvest
(2) + Number of days from seed to transplant if growing your own starts
(3) + Average number of days of harvest period
(4) + 14 days "Fall Factor" to allow for the slowing rate of growth as the weather cools
(5) + 14 days "Safety Margin" from the date of killing temperatures for that crop
= Days to count back from the date of killing temperatures for that crop

It is possible to reduce the Fall Factor if you are providing protection. Pre-sprouting seeds will reduce the number of days from seed to transplant.

Here's an example: Early White Vienna Kohlrabi takes 58 days from sowing to harvest (line 1). You can direct sow, so line 2 = 0. You can harvest it all at once and store it in your cooler, so line 3 is 1 day. Assuming you don't want to use rowcover for this, line 4 = 14. Line 5 = 14. That all adds up to 87 days. Kohlrabi is hardy to maybe 15°F (–9.4°C). When is the temperature likely to drop to 15°F (–9.4°C)? Not before the beginning of November here, so counting back 31 days in October, plus 30 in September, plus 31 in August—that's 92 days already, more than enough. We could sow kohlrabi in early August and get a crop at the end of October.

While doing these calculations, consider sowing a winter cover crop such as winter rye between the crop rows a month before your harvest date. By then the cash crop will be big enough to withstand competition from the cover crop. This way, growing a late cash crop won't cause you to be too late to sow your winter cover crop.

Hardy winter-harvest crops

For crops to harvest all winter long, look for ones that will survive your lowest temperatures, taking any rowcovers or other crop protection into account. Consider arugula, some cabbage varieties, carrots in some areas, chard, collards, garlic scallions, horseradish, Jerusalem artichokes, kale, leeks, some onion scallions, parsley, parsnips, salsify, spinach and very hardy salad crops (corn salad being the hardiest). Winter purslane is extremely hardy, and protection will increase the growth rate. Fall is too late to plant the slower-growing winter hardy crops like parsnips or leeks.

In the middle of winter, when there is less than ten hours of daylight, little plant growth happens. At our farm, these dates are November 20 to January 20. Because the really cold weather arrives later than the shortest day, our period of little growth is more like mid-December to the end of January. If you want to harvest during that period, sow extra so you can harvest just part of the patch each time. Spinach is the hardiest green we grow over winter (we like Tyee). To grow big enough to harvest in the winter, we sow it at the very beginning of September. We use rowcovers on hoops to protect the leaves from the worst of the weather, and spinach will make new growth whenever the temperature rises above 37°F (3°C).

We grow seven beds, and use a marker flag to indicate which bed to pick next. If we arrive to harvest and find the spinach in the flagged bed isn't big enough, we wait.

Overwinter early spring harvest crops

If you can keep certain crops alive through the winter, they will start to grow again with the least hint of spring weather and be harvestable earlier than spring plantings. These include spinach, some Asian greens, kale, collards, some onion scallions, garlic scallions, chives, chard and arugula. Green-in-Snow is the hardiest Asian green. In spring the order of bolting of Asian greens is: tatsoi, pak choy, Komatsuna, mizuna, leaf radish, mustards.

Spinach, lettuce, chicories such as radicchio and Sugarloaf, fennel and cilantro seem to have the best cold tolerance when the plants go into winter half-grown. We usually have one or two beds of spinach, which we sow in late September, overwinter as adolescents and harvest in the spring. These plants bolt later than the ones we harvest leaves from all winter, and earlier than spring-sown beds, so we get a continuous supply. Notice that fall sowing dates are quite exacting: Sept 20 is the latest we can sow spinach for harvesting October–early April, and Sept 20–30 sowings will not get big enough to harvest until late February.

Some cabbage (Deadon, Brunswick and January King) can survive in Zone 7 for January harvest, but as they are slow-growing they need to be started in summer.

With alliums, such as bulb onions, multiplier onions and garlic, the harvest dates are regulated by day length, so the harvest cannot be earlier, but the bulbs will be bigger if you can overwinter the small plants.

Winter hardiness

Record how well your crops do in the colder season. My friend and neighboring grower Ken Bezilla of Southern Exposure Seed Exchange provided much of the information below, and has suggested the morbidly named 'Death Bed' idea: set aside a small bed and plant a few of each plant in it to audition for winter hardiness. Note when the various plants die of cold, in order to fine-tune your planting for the next year (and send me an email!). It's worth noting that in a hoophouse, plants seem able to tolerate lower temperatures than those listed here; they have the pleasant day-time conditions in which to recover. Salad greens in a hoophouse can survive nights with outdoor lows of 14°F (−10°C). All greens do a lot better with rowcover to protect them against cold drying winds.

Here are some starting numbers of killing temperatures, although your own experience with your soils, microclimates and rain levels may lead you to use different temperatures:

- 35°F (2°C): Basil.
- 32°F (0°C): Bush beans, cauliflower curds, corn, cowpeas, cucumbers, eggplant, limas, melons, okra, some pak choy, peanuts, peppers, potato vines, squash vines, sweet potato vines, tomatoes.
- 27°F (−3°C): Most cabbage, Sugarloaf chicory (takes only light frosts), radicchio.
- 25°F (−4°C): Broccoli heads, chervil, chicory roots for chicons and hearts, Chinese Napa cabbage (Blues), dill, endive (hardier than lettuce, Escarole more frost-hardy than Frisée), annual fennel, large leaves of let-

tuce (protected hearts and small plants will survive even colder temperatures), some mustards and Asian greens (Maruba Santoh, mizuna, most pak choy, Tokyo Bekana), onion scallions, radicchio.
- 22°F (−6°C): Arugula, tatsoi (both may survive colder than this).
- 20°F (−7°C): Some beets, cabbage heads (the insides may still be good even if the outer leaves are damaged), celeriac, celtuce (stem lettuce), perhaps fennel, some mustards/Asian greens (Tendergreen, Tyfon Holland greens), radishes, turnips with mulch to protect them (Noir d'Hiver is the most cold-tolerant variety).
- 15°F (−9.5°C): Some beets (Albina Verduna, Lutz Winterkeeper), beet leaves, broccoli leaves, young cabbage, celery (Ventura) with rowcover, cilantro, endive, fava beans (Aquadulce Claudia), Russian kales, kohlrabi, Komatsuna, some lettuce, especially small and medium-sized plants (Marvel of Four Seasons, Rouge d'Hiver, Winter Density), curly leaf parsley, flat leaf parsley, Asian winter radish with mulch for protection (including daikon), large leaves of broad leaf sorrel, turnip leaves, winter cress.
- 12°F (−11°C): Some cabbage (January King, Savoy types), carrots (Danvers, Oxheart), most collards, some fava beans (not the tastiest ones), garlic tops if fairly large, most fall or summer varieties of leeks (Lincoln, King Richard), large tops of potato onions, Senposai, some turnips (Purple Top).
- 10°F (−12°C): Beets with rowcover, purple sprouting broccoli for spring harvest, brussels sprouts, chard (green chard is hardier than multi-colored types), mature cabbage,

some collards (Morris Heading), Belle Isle upland cress, some endive (Perfect, President), young stalks of bronze fennel, perhaps Komatsuna, some leeks (American Flag), Asian winter radish, (including daikon), rutabagas, (if mulched), large leaves of savoyed spinach (more hardy than flat leafed varieties), tatsoi.

- 5°F (−15°C): Garlic tops if still small, some kale (Winterbor, Westland Winter), some leeks (Bulgarian Giant, Laura, Tadorna), some bulb onions (Walla Walla), potato onions and other multiplier onions, smaller leaves of savoyed spinach and broad leaf sorrel.
- 0°F (−18°C): Chives, some collards (Blue Max, Winner), corn salad, garlic, horseradish, Jerusalem artichokes, Vates kale (although some leaves may be too damaged to use), Even' Star Ice-Bred Smooth Leaf kale, a few leeks (Alaska, Durabel), some onion scallions (Evergreen Winter Hardy White, White Lisbon), parsnips, salad burnet, salsify, some spinach (Bloomsdale Savoy, Olympia, Tyee).

Scheduling your crops

Earlier I gave a formula for calculating sowing dates of crops to harvest before winter. Another, more general approach is this one from *The Heirloom Gardener* by Jeffrey Goss:

- 10 weeks before the average first frost date (Aug 1 for us): beets, cabbage (Brunswick and January King for January harvest), daikon, leeks, winter lettuce (Marvel of Four Seasons, Rouge d'Hiver, Winter Density), turnips, rutabagas, watercress
- 8 weeks before frost (Aug 15): winter radish, fall spinach

- 7 weeks before frost (Aug 22): kale
- 5 weeks before frost (Sept 7): spring spinach
- 3 weeks before frost (Sept 21): fall globe radishes

Eliot Coleman's *Four Season Harvest* has a table of fall planting dates for various first frost dates. Use the columns with a first frost date 20–30 days later than yours for coldframe sowings, and the columns for 40 or more days later for hoophouse sowings. Using this, in September with a first frost date of October 20, here in USDA Zone 7 where the temperature gets down to 5°F (−15°C) at some point, we can sow corn salad, claytonia, lettuce, endive, escarole, mizuna, onion scallions, radicchio, radish and spinach outdoors. In coldframes we can sow those crops into late October. In September we can still sow beets, early broccoli, Chinese cabbage, carrots, chard, chicory, kale, kohlrabi, leeks, mizuna, bulb onions, parsley, peas, sorrel and turnips in the coldframe. In the hoophouse we could also sow cabbage, celery, celeriac, rutabagas and tatsoi in September.

Fall planting techniques

It's important not to act as if it's April! Plant seeds deeper than you would in spring, as the soil is already warm and you want to save them from drying out. If the weather is still hot, for heat-sensitive seeds like spinach and beets, use all the techniques you know to cool the soil, the seeds and the water (see the Summer Germination of Seeds chapter). Good soil preparation is important, as good tilth lets the soil drain and breathe. Avoid compaction, which can lead to puddling. Consider making raised beds with permanent winter paths of grass or some other growing thing. Brett Grohsgal has found that direct-seeded crops are more cold-tolerant than transplants, probably because of the damage to the taproot with transplanting. Prepare to accept some losses, and to get some encouraging surprises.

Winter harvesting techniques

With fall sown crops the aim is often to keep the same plants alive all through the winter, as November, December and January are not good times to sow replacements. Don't harvest frozen crops—wait till they thaw. With leafy vegetables, the highest productivity is achieved with "Cut and Come Again" crops—i.e., the tops of the plants above the growing point are cut with scissors or shears every 10–35 days. Leaf-by-leaf is the method we use for kale, collards, chard and spinach (as well as outdoor lettuce under rowcover). Never remove more than 40 percent of the total leaf surface area of the plant—less than half of the leaves, with a safety margin. Whole Plant Harvesting works well for small plants like tatsoi and

Carrots are hardy down to 12°F (−11°C). Summer sowings can provide delicious early winter harvests. Credit: Kathryn Simmons

corn salad. A direct-seeded row can be thinned over time by harvesting out the biggest plants on each visit.

Chapter 15

The Hoophouse in Winter and Spring

The body of knowledge about using hoophouses for cool-weather crops has grown each year, and the options are many. Each grower will want to customize the mixture of crops to suit their climate and market. For example, our focus is on supplying a wide range of continuous fresh produce all winter and spring, to feed a hundred people. If your goal is to maximize harvests in December and then read novels by the woodstove till mid-February, you might rather plant a single large crop of spinach to overwinter and harvest after your time off.

We grow salad crops, cooking greens and some turnips, radishes and scallions in the hoophouse. We also use some of the space to grow bare root transplants for setting outdoors in February and March. In our climate, the outdoor beds can provide spinach, kale, collards and leeks all through the winter, although the rate of growth doesn't compare to what happens in the hoop-

house! We aim to be able to harvest greens in the hoophouse after the outdoor crops slow down, and turnips after the stored outdoor fall turnips have all been eaten, or as an occasional delectable alternative.

We have a 30' × 96' (9 × 29 m) Clearspan house from FarmTek, the gothic arch type, with bows 4' (120 cm) apart. We use an inflated double layer of plastic. Our hoophouse does not have roll-up or roll-down sides. We built it for maximum winter coziness, and so only the ends open. We don't use rowcover over our winter crops. We like that the insulation provided by the double plastic means we can move around freely and work inside the tunnel, unhindered by rowcovers. We don't need them for warmth, and in winter the amount of daylight is a limiting factor, and rowcover does reduce the light. If you are in a colder climate than we are, the extra benefit of rowcovers may pay off. Compared to using rowcover outdoors, inside the

Our hoophouse frame before covering with plastic.
Credit: Twin Oaks Community.

hoophouse is easy—no gusts of wind, no rain, no snow!

Inside, we have seven lengthwise beds. The two edge beds are 2' (60 cm) wide and the rest 4' (120 cm). The paths are 12" (30 cm). This is the same size we use outdoors, and works well for us. It really isn't much path space. (Do yoga! Or make wider paths.)

We only do hand work in the hoophouse—i.e., no tractor or rototiller. The lack of rain indoors means there are very few weeds, and by not inverting the soil at all, we get few new weeds. We enjoy the peaceful, enclosed, intensively planted miniature garden, protected from winter weather and heated by the sun.

Persephone days

If you want to harvest in the darkest days of winter—called Persephone Days by Eliot Coleman and the Solstice Slowdown by Paul and Alison Wiediger—you'll need to plan to have a good supply of mature crops to take you through. Be aware of the increase in days to maturity in winter. Here at latitude 38° North, the period when daylight length is less than ten hours and little plant growth happens, lasts two months, from November 21 to January 21. For most of the winter, our

Our first winter in the hoophouse. We still had weeds then. Credit: Pam Dawling.

plants are actively growing, not merely being stored for harvest (as happens in colder climate zones), so we can continue sowing new crops even in December.

Crop overview

Our winter crops are a mixture of direct sowings and transplants brought in from outside (in the fall), or grown inside and transplanted during the winter. We start in early September, sprouting some spinach seeds in a jar in the fridge, and sowing Bibb lettuces outdoors, to transplant later into the hoophouse. The Bibb lettuce will be harvested in December—it doesn't do well here any later than that. Meanwhile we clear and add compost to one of the beds inside to sow the one-week-old sprouted spinach seed, some radishes, scallions, Bulls Blood beet greens and tatsoi.

We make outdoor sowings on Sept 15 and Sept 24 for crops to later transplant into the hoophouse at 2–4 weeks old. The Sept 15 sowing includes about ten varieties of hardy leaf lettuce, pak choy, Chinese cabbage, Yukina Savoy, Tokyo Bekana, Maruba Santoh and chard. The Sept 24 sowing consists of Red and White Russian kales, another ten varieties of lettuce, Senposai, more Yukina Savoy, mizuna and arugula, and resows of anything from the previous week that didn't give a good stand of seedlings. We cover this outside nursery bed with hoops and rowcover to keep bugs off, and water it frequently. Using an outdoor nursery bed gives us cooler conditions for better seed germination, and allows our summer crops longer to finish up.

By the end of September we clear the summer crops from at least one more hoophouse bed, add compost and work it in. We transplant the Tokyo Bekana and Maruba Santoh at just 2 weeks old,

the Chinese cabbage, pak choy and Yukina Savoy at about 3 weeks and the Bibb lettuce at 4 weeks. (Plants grow so fast in September!)

By mid-October we clear and prepare another bed and transplant 280 lettuce at 10" (25 cm) apart, and the chard. We also sow various tender turnips and radishes. We transplant the second lettuce (300 plants), the kale, Senposai, mizuna, arugula and the second Yukina Savoy at about 4 weeks old, in the fourth week of October, after preparing the remaining beds.

We sow our first baby lettuce mix Oct 24, along with our second spinach and chard. At the end of October we sow some "filler lettuce" to transplant later to fill gaps. We try hard to keep all the space occupied, mostly using lettuce and spinach. See the Hoophouse Succession Planting Chart below. On Nov 10 we sow our second turnips, mizuna and arugula, more filler lettuce and spinach, and also our first bulb onions for outdoor transplanting as early as possible in the new year—we aim for March 1.

From Nov 10 on, we have a fully planted hoophouse, and as each crop harvest winds down, we immediately replace that crop with another. Our winter and spring crops come to an end in March or early April, when we dig holes down the centers of the beds to plant the early summer crops of tomatoes, peppers, squash and cucumbers. Usually we harvest the winter crops from the center rows first, plant the new crop down the center, then harvest the outer rows bit by bit as the new crop needs the space or the light. This overlap allows the new crops to take over gradually.

Varieties that work well for us

- Asian Greens: We like the big heads of Prize Choy or Joi Choi pak choy, as well as Tokyo

Bekana and Maruba Santoh, which have tender, bright yellow-green leaves. When small, the leaves are good in salad. When big, whole heads can be cut for quick-cooking greens. Tatsoi and its bigger cousin Yukina Savoy are good on flavor and strong dark color.

- Chard: Brite Lites, Rainbow. Although green varieties grow faster in low light, we don't need a lot, and we enjoy the varied colors.
- Chinese/Napa cabbage: We like Blues, but the Michihili types are taller and narrower and might make more sense in terms of space use.
- Kale: White and Red Russian kales do well in our hoophouse, but not outdoors, where they get weather-beaten. Inside, they do better than Vates, our outdoor kale.
- Lettuce: Oscarde, Merlot, Tango, Red Tinged Winter, Lollo Divino, Galisse, Salad Bowl, Red Salad Bowl, Rouge d'Hiver, Winter Wonderland, Outredgeous, Hyper Red Rumpled Wave, Revolution, Green Forest Romaine. Looseleaf varieties, mostly. We look for ones that do well in winter, not just in early spring. Romaines and Bibbs are good for harvest in December and late February onwards, but can get tipburn in midwinter.
- Lettuce Mix: Various mixtures, some homemade.
- Mizuna: Purple and green. Ruby Streaks looks very good. It adds texture to salad mixes.
- Radishes: Cherry Belle are the "industrial" radishes—all ready at once, quickly, looking perfect. We also like the multi-colored Easter Egg for variety and an extended picking season. It does well in the later plantings, staying mild in sometimes-hot conditions in March,

Rainbow chard is a cheerful winter hoophouse crop. Credit: Kathryn Simmons.

as does White Icicle, a long, crunchy, tender root.

- Scallions: Evergreen Hardy White. Sometimes we just use leftover seed from bulbing onion varieties that we tried and didn't like well enough to grow again.
- Senposai: Fast-growing, tender, delicious (the name translates as "Thousand Wonder Green"). Can grow very big, and will provide bunches of leaves many times over.
- Spinach: It's Tyee every time for us: reliable, and tolerates varied conditions.
- Turnips: Hakurei (short tops), Oasis, White Egg, Red Round (very tall).

Crops we tried and didn't like

Hot, bitter, or sour greens don't go down well with our crowd. We have tried and rejected chicories, endives and pungent mustards. We're also less wowed by crops that are tiny or have to be harvested very promptly: Chinese broccoli (Hon Tsai Tai), mache (corn salad) and claytonia. For us, it works better to use the space for more lettuce or spinach. In colder climates, lettuce grows more slowly, so smaller, hardier crops may be more worthwhile.

We tried Nero di Toscano/Dinosaur kale, but it didn't do well for us. It harbored aphids in the curled leaves and never developed the flavor we hoped for. We have also tried various large Asian greens and hybrids that ultimately seemed too rambunctious for our hoophouse: Komatsuna, Mizspoona, Tyfon Holland greens. In colder climates, these very sturdy greens may be just the ticket.

Crops other growers like

We don't grow carrots, beets, broccoli, collards, herbs, flowers, raspberries, blackberries, strawberries, dwarf cherries or potatoes in our hoophouse. However, these all have possibilities.

Daily tasks

We reckon on about two hours work each day in our tunnel in winter (only one hour in the summer). We aim to keep the temperature in the 65°F–80°F (18°C–27°C) range during the day, opening the big high windows, and some of the doors as needed. If the sun is shining we usually open the windows around 9 AM and close them around 2:30 PM (a few hours before dark) to store some of the warmth. On cloudy days we use the thermometer to help us decide if we can open the windows. Even in cold weather, plants need fresh air! While photosynthesizing, plants need carbon dioxide, so just walk around in there, breathing, and you are doing "carbon dioxide enhancement"! When it's below freezing and cloudy, you'll want to keep the place closed up.

High-density cropping can really use up the carbon dioxide in a closed hoophouse very quickly. When this happens, photosynthesis crashes and plant growth becomes limited. Soil high in organic matter contains high levels of organisms that produce carbon dioxide. Dense plant canopies can trap this near soil level, where it is most useful.

In spring, once the night temperatures are reliably above 45°F (7°C), we leave the windows open at night, to slow down the bolting tendencies of the cool-weather overwintered annuals. In March and April we have to balance this need with that of the newly planted tomatoes, peppers, squash and cucumbers for warmth. At this time, we leave the doors open only if the nighttime low will be above 55°F (13°C).

Our main task each day is harvesting. In the winter of 2009–2010, we had frozen soil or snow on the ground outside for a month (very unusual here). Despite this we were able to keep a hundred people in fresh salad and cooking greens (with turnips and scallions for variety) for the whole month.

Aside from harvesting, jobs include planting new crops, clearing old ones, spreading compost, hoeing, hand weeding and supplying water as needed. We have drip irrigation, with three lines of drip tape in a bed. We hand water new sowings and transplants using a hose. In the middle of winter, not much water is needed, and we try to only water when a relatively mild night is forecast. The idea is to minimize cell wall damage by not having turgid leaves in freezing conditions. For new hoophouse growers, it can come as a shock to find everything frozen in the morning, and

Hoophouse Succession Planting

Crop		Planting Dates	Harvest Dates	Notes
Chard	#1	transplanted Oct 15	Dec 11–April 9	
	#2	sown Oct 26	March 6–April 9	
Lettuce mix	#1	sown Oct 24	Dec 11–Feb 21	
	#2	sown Feb 1	March 20–April 20	3 cuts if we're lucky
Lettuce heads		until October	November to February	Harvest leaves from the mature plants
			Dec 6–March 31	Cut the heads
Mizuna	#1	transplanted Oct 24	Nov 1–Jan 25	
	#2	sown Nov 10	Jan 27–March 6	
Onions (bulbing)	#1	sown Nov 10		Transplanted outdoors as early as possible in March
	#2	sown Nov 22		
	#3 backup	sown Dec 6		
Radish	#1	sown Sept 6	Oct 1–Nov 15	
	#2	sown Oct 22	Nov 25–Jan 29	
	#3	sown Nov 27	Feb 12–March 13	
	#4	sown Dec 27	March 2–April 1	
	#5	sown Jan 27	April 2–April 15	
Scallions	#1	sown Sept 6	Dec 25–March 20	
	#2	sown Nov 13	March 19–May 15	Following radish #1
Spinach	#1	sown Sept 6	Oct 30–April 9	Sprouted seeds sown
	#2	sown Oct 24	Nov 20–May 7	
	#3	sown Nov 10	All these later sowings are harvested until May 7	We keep planting to fill gaps and pulling up finished plants
	#4	sown Dec 27		
	#5	sown Jan 17		
	#6	sown Jan 24	Until mid-May	To transplant outdoors in February
Tatsoi	#1	sown Sept 7	Oct 30–Dec 28	
	#2	sown Nov 15	Feb 15–Feb 28	
Turnips	#1	sown Oct 15	Dec 4–Feb 20	
	#2	sown Nov 10	Feb 25–March 10	Thinnings Jan 11
	#3	sown Dec 10	March 5–March 20	Only worthwhile if thinned promptly and eaten small. Greens are a very sweet and beautiful hoophouse crop
Yukina Savoy	#1	transplanted Oct 10	Dec 30–Jan 22	
	#2	sown Oct 24	until Jan 29	Only one week extra

Baby lettuce mix is a delightful winter hoophouse crop. Credit: Kathryn Simmons.

equally shocking, but much more pleasant, to find everything thaws out later and recovers just fine.

As well as these scheduled crops, we keep a supply of lettuce and spinach transplants growing and pop a few plants in whenever a gap opens up. Sowing small amounts of fast-growing, small-sized crops such as arugula, mizuna, cress and tatsoi is another option for making good use of every scrap of space.

Recent trial winter crops

Each year we try a few new ideas, or tweak something we tried the year before. Here's some we've tried recently:

- Spring cabbage greens, sown outdoors Oct 15, transplanted inside Nov 2, harvested March 13–27. In the UK these are sown in early August, grown outdoors and called spring greens. They are actually adolescent cabbages of particular varieties that will overwinter

there. They are transplanted about 4"–6" (10–15 cm) apart, harvested once by cutting alternate plants, then cleared as they start to head up. Some UK varieties include Offenham Flowers of Spring, Durham Early, Dorado and Avon Crest. I've tried Early Jersey Wakefield, because they are the right pointed shape, but they don't work as well as the real thing. This crop provides a very dark green, great-flavored, unhearted cabbage. It took a bit of work to convince people to try them.
- Celery grown outside for the summer, then transplanted big into the hoophouse Oct 24, 12 plants, at 10"–12" (25–30 cm). This keeps the plants alive, producing stems to add zest to winter stir-fries and salads (chopped). Harvested Dec 26–April 30.
- Wild Garden Seeds' Pink Lettucy Mustard salad mix, sown Nov 2. We're still experimenting to find the best sowing dates here.
- Bulb onions sown Nov 10, transplanted early March in a single row along the south edge of beds, for an early crop. We got good onions but they dwarfed the pepper plants behind them. Maybe planting them on the north side of the bed is better?
- Fava beans sown Nov 15, harvested mid-May. This worked well, so we may do a bigger planting next winter. Legumes add diversity to the rotation.
- Snap Peas Sugar Ann sown Feb 1, harvested April 21–May 15. Very successful. We've now added this to our repertoire.
- Peashoots—sometimes a "catch crop" crop if we have unexpected open space and leftover soaked seed from our spring outdoor planting, mid-March. Harvested April 10–May 5.
- Brassica salad mix, sown Feb 6, Feb 12 for

March and early April harvests. We mix our own from leftover random brassica seeds. For a single cut, almost all brassicas are suitable. Worthwhile if other crops fail, or outdoor conditions are dreadful and you need a quick crop to fill out what you have to sell.

Rotations

Rotating crops in the hoophouse is its own special challenge: we want to grow the same crops every year, in a very limited space. Our approach is to vary crops, so that even if we end up with lettuce in the same bed two winters running, we have had two other crops there in between, such as squash and cowpeas. We do keep track and do our best to rotate, but the task is like making a patchwork quilt, where the scrap of Aunt Rosie's wedding dress just has to be included somewhere.

Crop requirements

Because crops grow so fast in the hoophouse, the organic matter in the soil is consumed at a rapid rate. Each new crop requires a fertility boost. For fall planted crops, we spread one inch (2.5 cm) of compost across the beds and work it into the top of the soil with scuffle hoes or rakes. In general, for each crop, growers add compost at a rate of fifteen tons per acre. We use about four wheelbarrows (not full) per 4' × 96' (1.2 × 29 m) bed. A full wheelbarrow generally holds six cubic feet—about forty gallons or 170 liters.

For the early summer crops planted in March and April, we dig a hole for each plant and put a shovelful of compost in each hole. When we grow edamame or cowpeas in the middle of summer, we do not always add more compost at that point. See the next chapter for more on the early summer and high summer crops.

Sowing

We usually sow by hand, but other growers use Earthway type seeders or the more precise four- and six-row seeders sold by Johnny's Selected Seeds. Some growers buy several used Earthways and bolt them together for a multi-row seeder.

Transplanting

Because space is at a premium, it's good to measure plant spacings in some way, rather than drift to wider spacing and end up with fewer plants. We use wooden sticks. Rulers are cheap and convenient. Many growers turn on the irrigation and plant into the wet spots. Hoophouse crops grow bigger than they would outdoors, so sometimes the spacing needs to be up to 25 percent larger.

Cultivation

We use scuffle hoes, Cobrahead hoes and the hand cultivators that are commonly available for backyard gardeners. We call them "claws"—they have three tines and a short handle. Few weeds grow, but sometimes the soil texture looks in need of some help, because of the lack of rain—any big lumps of soil just get drier rather than being broken down by the elements, and salts can deposit out around the drip tape.

Pests and diseases

Voles are our most frequent winter mammal visitor. We use mousetraps, the plastic Intruder type. Our worst winter insect pests are aphids and vegetable weevil larvae. For the aphids, we bring in hibernating ladybugs (they wake up with warmer temperatures!) or spray with soap. For the vegetable weevil larvae, which love turnips and Chinese cabbage (but will move on to spinach), we use Spinosad. We also try to attract beneficial insects by planting flowers—you can put them in pots and move them around to where they are needed. We have used alyssum, echinacea and buckwheat.

By far our worst disease problem is the fungus we call solstice slime; its scientific name is Sclerotinia drop. It attacks lettuce when the days are short and temperatures chilly, causing the whole plant to collapse into a soggy gray pancake. So far, our approach has been to finish our shift by collecting up the slime and taking it outside, then washing our hands before returning. Prompt action does seem to limit the spread. Summer solarization will kill the spores. See the next chapter for more details.

Tipburn (brown leaf margins, including internal leaves) is caused by quick drying of the soil when the weather makes a sudden switch from overcast to bright and sunny. Be ready to irrigate when the weather suddenly brightens.

Harvest

Our winter hoophouse harvest starts in November, with spinach, lettuce leaves, mizuna, arugula, beet greens, tatsoi and brassica mix for salad, as well as radishes and scallions. We harvest lettuce by the leaf, leaving the center to keep growing, and switch to harvesting the heads in late January, when growth begins to pick up.

From December we also have baby lettuce mix, Tokyo Bekana, Maruba Santoh, chard, kale and turnips. Kale grows whenever it is above 40°F (5°C).

The bigger greens, including Senposai, pak choy and Chinese cabbage, feed us from January till mid-March, if we plant enough. Yukina Savoy starts in the new year.

We harvest spinach by the leaf, using scissors, until the plants start to look a bit past their peak;

then we "crew-cut" or buzz-cut them. This brings me to another nice thing about hoophouse growing. For winter crops, even kale, there is no need to remove finished plants to the compost pile if they are not diseased. We simply leave them on the surface to dry up and disintegrate, improving the texture and nutrients in the surface layer of the soil. The higher temperatures and the lack of rain ensure that crop residues will dry up rather than rot.

Nitrate accumulation

In winter, when light levels are low, there is an increased likelihood of high levels of nitrates in leafy vegetables. This is generally thought to be a health hazard (although recently this has been questioned)—nitrates can be converted in the body into nitrites, which reduce the blood's capacity to carry oxygen and may be further converted into carcinogenic nitrosamines. To keep nitrate levels as low as possible:

- Grow varieties best suited for winter;
- Avoid fertilizing with blood meal or feather meal;
- Water enough but not excessively;
- Provide fresh air so that carbon dioxide levels are high enough;
- Harvest after four or more hours of bright sunlight in winter;
- Avoid harvesting on very overcast days;
- Harvest crops a little under-mature, rather than over-mature;
- Use crops soon after harvest;
- Store harvested greens at temperatures close to freezing;
- Spinach contains about twice as much nitrate as lettuce, so mix your salads; don't just eat spinach.

Chapter 16

The Hoophouse in Summer

When we first put up our hoophouse we, like many growers, were focused on winter greens and early tomatoes and peppers. "What to use the hoophouse for in summer" has been a high-level problem we've happily been experimenting with and researching since. We're in the humid Southeast. Those of you in colder zones will have no doubts about what to grow in your hoophouse in summer: all those hot-weather crops that struggle outdoors in your climate! But if you already grow melons, limas and okra outside, you may be left wondering how to make good use of your prime real estate in hot weather.

Below are some ideas along with thoughts about timing and rotation of crops.

Removing the plastic

This option allows existing warm-weather crops to continue growing, or new ones to be planted. Personally I can't imagine doing this every year, but you might! The two layers of plastic could be pulled up and over, part way down the north side, then rolled up and covered in black plastic and tied up along the hipwall until fall, in the manner of the Haygrove seasonal-use tunnels.

A related cooling option is to increase airflow by using big vents and roll-up or drop-down sides. We don't have separate sidewalls on ours, because as I said, our primary goal was to keep the space warm for winter crops. Drop-down sides have the advantage over roll-up sides in colder weather, as they protect small plants at soil level.

Another related option is to move the hoophouse by building it on sleds or rollers. However, doing that doesn't answer the question of what to do with the space; it just moves the space over. It does allow you to get an early start sowing heat-loving crops inside the hoophouse; you then move the hoophouse and grow those crops outdoors, while you start yet more crops on the new hoophouse site. Eliot Coleman writes about various options for mobile hoophouses. Rolling

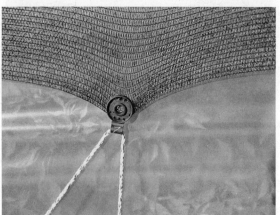

Thunder is one manufacturer of mobile tunnels. The challenge of these mobile houses is anchoring them well so they don't take off in a strong wind.

Shading the plastic

Cooling the hoophouse means you can continue to grow crops you would otherwise grow outdoors, but with the benefit of enriched soil and protection from wild weather. We cover our hoophouse from mid-May until early to mid-September with a single large piece of shadecloth, with clip-on grommets every two feet (60 cm) along each long edge. We set large hooks every two feet (60 cm) along the baseboards on the north and south sides, and thread ropes through the grommets. It takes four of us about an hour to put the shadecloth on. We throw over a tennis ball in a sock tied to a rope, or a plastic flip-flop sandal, and then pull the shadecloth over. We then feed the ropes along through the grommets until the shadecloth hangs evenly, and hook it on. This is much easier to do than putting on new plastic!

The shadecloth makes a real difference. The

Shadecloth on our hoophouse, showing details of the clip-on grommets, ropes and hooks in the baseboards. Credit: Twin Oaks Community.

hoophouse becomes much cooler than outside, where the sun is relentless! Removing the shadecloth in mid-September takes thirty to forty minutes with three or four people. Using shadecloth enables crops to thrive later into the summer than they otherwise would, and lets the crew harvest in there without dying. We keep sweet peppers growing in our hoophouse from April until November. Harvesting them can be done at either end of the day. With shadecloth it is possible to grow crops such as lettuce mix or arugula, that are harvested young and don't get the chance to bolt. The hoophouse conditions help these crops stay attractive. That said, you may find that pests prevent you growing some crops (such as brassicas) indoors or out in summer.

Our shadecloth is 34' × 96' (10 × 29 m), about the same size as the footprint of the hoophouse. It's knitted polyethylene 50% shade from Gemplers. Shadecloth is available woven or knitted, polyethylene or polypropylene, in white, black or other colors and a range of shade factors. Polypropylene is longer-lasting than polyethylene, but only available woven. Avoid PVC as it will degrade the poly sheeting of the hoophouse. We chose knitted fabric as it is stronger, lighter in weight, more flexible and doesn't unravel when cut. We went with 50% shade. Some growers recommend 40% for vegetables, 50% for flowers and 60% for cool-weather vegetables in hot weather.

A related option is to use KoolLite 380 plastic. I haven't tried this, but reports say it reflects back the infrared, ultraviolet and green parts of the spectrum, leaving the hoophouse about 10°F (5°C) cooler in hot weather, while keeping it just as warm as regular infrared-blocking plastic on cool overcast days. It does cost almost twice as much as IR/anti-condensate plastic. Another relatively new covering is Solaroof material (from Robert Marvel), a four-layer woven material that diffuses the light and reduces overheating.

Warm-weather food crops

Suitable candidates are crops that you'd like more of, but earlier; crops that grow in warmer climates; ones that grow better in drier climates; and, ideally, ones that are not in the same families as your main crops in other seasons. Hot peppers, ginger, cowpeas, soup beans and edamame are successful summer food crops in our hoophouse. Crops that have not been successful for us are melons, jicama (yam beans), eggplant and peanuts.

The melons failed because the earlier yellow squash and cucumber plants had brought in lots of striped cucumber beetles, which gobbled up the melon transplants before they grew their fourth leaf (demonstrating once again the importance of crop rotation!). We had sown the melons (muskmelons/cantaloupe and watermelons) in pots June 1 and transplanted July 1, if you want to try.

Jicama, a crunchy tuber, is not a quick-maturing crop, but it is vining and we grew it at the "back" (north side), where it would not shade anything. We bought the seeds from Pinetree, and they are also available from Baker Creek, who warn: "Takes a very long season, these must be started very early in all areas except the Deep South. Caution: the seeds and pods are poisonous." We came to the conclusion that we did not have hot enough conditions for long enough. If you are in zone 8 or 9, it might work for you.

The eggplant was killed by droves of flea beetles. When we plant outdoors we keep the eggplant under rowcover until they flower, but because the hoophouse is warmer, we didn't use rowcover. So eggplant could work, if you guard against flea beetles.

The peanuts (sown July 1) didn't produce well; perhaps they were too hot and dry, perhaps the soil was not fertile or loose enough, perhaps the voles found them. I'm still not sure what went wrong—I think they have potential to be a good hoophouse crop. As peanuts can grow OK outdoors where we are in central Virginia, and take about 120 days, we decided they didn't deserve the space inside. In general, though, we have found legumes (green beans, cowpeas, edamame) to be a good group of summer hoophouse food crops.

Now onto the successes: We have found the hoophouse a good place to grow unusual hot peppers. Compared to outdoors, we have a small group of people with more training who harvest in our hoophouse, so unusual varieties do better there, as there's less chance of them being harvested at the wrong stage.

Our humid climate, and possible heavy September rainfalls from hurricanes, can make for ugly spotty bean pods outdoors. Growing beans under cover in the hoophouse means we get beautiful pods. This matters particularly for edamame, where the diners see the pods. Edamame have the advantage of being picked all at once—take it from me that you do not want a crop that requires daily harvesting in high summer in your hoophouse, unless you live in the far north. We like Envy edamame, a short bush type that matures quickly. We sow several plantings between June 2 and July 27 and harvest after 90 days for fresh eating, or 110–120 days for seed.

Cowpeas can be sown as early as April 1 or as late as July 23, later than they would be outdoors, to provide a succession following on from outdoor crops. We really like the flavor of Missis-sippi Silver. We harvest the July 23 sowing in late September through early October for fresh eating. Leaving them till late October/early November provides a seed crop (about 95 days from seeding). We sow 2 rows in a 4-foot (1.2-m) wide bed. We support the cowpeas by hammering in a 5-foot (1.5-m) wooden stake each side of the planting every six to ten feet (2–3 m) and using twine in a modified single-sided Florida weave to make a "fence" around the planting, with a row of twine every foot (30 cm) up the stakes. We've found that we don't like putting up big trellises, so we mostly choose bush-type cowpeas. Bush varieties also allow the sunlight to better reach the north side of the house.

Thanks to a SARE grant, ginger was trialed as a hoophouse crop in Massachusetts in 2007 and 2008. Melissa Bahret and others cut a mother root into budded pieces in late March and planted the pieces in nursery crates. After about six weeks at 80°F (27°C), they transplanted into passive solar double-skinned hoophouses. Ginger needs lots of water and foliar feeding. They hilled up three times and harvested in September. In 2007 they got yields of 5:1. Fresh ginger retails at about $20/pound ($45/kg). The mother root remains usable for grating, and young roots can be grown over-winter in a frost-free place to become next year's mother roots.

We tried growing ginger on a small scale in 2010, transplanting in early April from roots potted in a warmer greenhouse. Our own little experiment had a 4:1 yield. In 2011, we planted about 20 lbs (9 kg) and got a 7:1 yield. We started the plants in a warm greenhouse in mid-March, planted them in the hoophouse in late May, and harvested in the middle of October. Fresh baby ginger is very popular and eye-catching. See the

Our 2011 baby ginger trial. Credit: Kathryn Simmons.

websites of Alison and Paul Wiediger (aunaturel farm.homestead.com) and East Branch Ginger (eastbranchginger.com).

Timing of summer crops

Don't get so excited about summer crops that you forget to keep your fall planting dates and crop rotations in mind, especially if the winter greens and salads are the main purpose of the hoophouse. Essentially we have three crop seasons in our hoophouse: winter crops planted in the fall, early warm-weather crops planted in March and April, and high summer crops mostly planted in July.

We plant the bulk of our winter crops from mid-September to mid-October. Some are direct-sown, but many are transplanted into the hoophouse from sowings made in an outdoor nursery bed. This makes good use of space and allows the summer crops to grow until then.

In spring, four to six weeks before the last frost date, we transplant tomato, pepper, squash and cucumber plants in the middles of the beds. With this plan in mind, we harvest the winter crops from the centers of the beds first, then plant the early summer crops. Soon we harvest the greens on the south side, which are blocking light to the new crop. Later we harvest the old crops on the north side, before they start impacting the new crop. The other spring crops we grow are early dwarf snap peas (sown Feb 1) and bush beans (sown March 15). The peas finish May 31, the beans mid-June, the cucumbers in mid-July, the squash and tomatoes at the end of July. We keep our peppers through early fall until cold weather arrives (November).

High summer crops have from mid-July at the earliest to early November at the latest to be in the ground, to fit in with our all-important winter and spring crops. Obviously we can only grow crops that mature quickly, unless we either give up some space from our spring/early summer crops or our fall/early winter crops, or get very creative! Crops we squeeze into this summer period are cowpeas, drying beans and edamame, and soy and buckwheat as cover crops. An example of being creative might be growing a tall trellised crop in the northernmost bed and planting a ground-hugging winter crop below it on the south.

Rotations

Our hoophouse is divided lengthwise into five main beds with a half-width bed along each edge. Initially, we worked out an ad-hoc plan for each

Planting Schedule for Early Summer and High Summer Crops.
Seven longitudinal beds, A–G. Beds A and G are the two-foot (60-cm) wide edge beds.

Bed	Date	Task	Date	Task	Date	Task
A: 2' (60 cm) wide	Feb 1	Sow snap peas along north edge of bed	May 31 / July 12	Pull snap peas, sow buckwheat. Sow cowpeas	Oct 15– Oct 25	Pull cowpeas, spread compost, rake in. Transplant Senposai. Sow radishes, scallions.
B: 4' (1.2m) wide	April 1	Transplant cucumbers with compost in holes	July 12	Pull cucumbers, sow drying beans	Oct 23	Pull beans, spread compost, rake in. Transplant kale, chard, mizuna, arugula, Yukina Savoy. Sow tatsoi, chard
C: 4' (1.2m) wide	March 15	Transplant tomatoes with compost in holes	July 30	Pull tomatoes, sow buckwheat & soy cover crop	Sept 8	Pull cover crop (use to mulch peppers); spread compost, rake in. Sow spinach, radish, scallions, Bulls Blood beets, Tatsoi
D: 4' (1.2m) wide	March 15	Transplant tomatoes with compost in holes	July 30	Pull tomatoes, sow buckwheat & soy cover crop	Oct 1– Oct 10	Pull cover crop (use to mulch peppers); spread compost, rake in. Transplant Tokyo Bekana, Maruba Santoh, Chinese cabbage, pak choy, Yukina Savoy. Sow turnips
E: 4' (1.2m) wide	April 1	Transplant peppers with compost in holes			Oct 15– Oct 24	Pull peppers, spread compost, rake in. Transplant lettuce. Translant mature celery. Sow mizuna, arugula, lettuce mix
F: 4' (1.2m) wide	April 1	Transplant summer squash with compost in holes	July 22	Pull squash, sow edamame	Oct 1	Pull remaining edamame for seed. Spread compost, rake in. Transplant lettuce, chard
G: 2' (60 cm) wide west	April 1	Transplant ½ bed hot peppers with compost in holes			Oct 23	Pull peppers, spread compost, rake in. Sow spinach
G: 2' (60 cm) wide east	March 15	Sow ½ bed bush beans	June 15	Pull beans, sow edamame	Nov 5	Pull edamame, spread compost, rake in. Sow onions

year, juggling rotation, timing, height and shading. At first we grew two beds each of tomatoes and peppers, then added in half a bed of hot peppers and a bed of late tomatoes. Clearly this is a lot of nightshades (*Solanaceae*)—five and a half beds out of seven! We've cut back to two beds of early tomatoes, half a bed of bell peppers and half a bed of hot peppers. It's still quite a lot of nightshades—almost half of our growing space. For the rotation, we look at the sequence of crops rather than the number of years since a nightshade was grown. Because everything happens faster in a hoophouse, we are growing multiple crops in each bed each year. We conjecture that the part of the year spent growing other crops contributes to a speeded-up version of the time needed away from that crop in an outdoor rotation. Perhaps the generously enriched soil also helps prevent diseases.

Early April cucumber transplants down the center of a bed where we have removed spinach. The wire hoops are for rowcover on frosty nights. Credit: Kathryn Simmons.

Cut flowers

I know almost nothing about flowers, but research on growing cut flowers in hoophouses is being done at Virginia State University and elsewhere. Hoophouse flowers arc protected from battering winds, rain, deer and, to some extent, Japanese beetles. In the Southeast, from July to September, cut flowers can be scarce. In the hoophouse, to take one example, you can have *Campanula* (bell-flowers) blooming at the end of July. Hoophouses can boost production and improve quality. See the Association for Specialty Cut Flower growers and *The Flower Farmer* by Lynn Byczynski.

Seed crops

We were growing a late food crop of Mississippi Silver cowpeas in our hoophouse, and noticing how fast they were maturing and how nicely the unpicked ones were drying, we decided to make it a seed crop instead. We now also grow edamame seed in the hoophouse. Legume seed crops provide a nice crop rotation break away from brassicas, lettuce, cucurbits and nightshades, and may add some nitrogen to the soil from the nitrogen-fixing bacteria on the roots. They mature much more quickly than outdoors, with the seed most of the way ripe about three weeks after the edible stage.

We sow these crops in early to mid-July and pull up the almost-dry beans in late October or early November, taking them to an upstairs barn to finish drying out of the sun. (This way, we also avoid seed shedding in the hoophouse.) We sow a range of greens in an outdoor nursery bed, with rowcover, on Sept 15 and 24, to plant in the hoophouse following the beans.

A hoophouse will not provide space for a large-scale crop, but it is ideal for specialist commercial varieties as well as for home use seeds. And it can make good use of the space in summer when you don't need to grow food crops in there. Experiment on a small scale at first, and learn about seed growing and the isolation distances required to guarantee purity for commercial standards for your particular crop. Also be sure to have a large enough population of plants to ensure a diverse genetic pool.

An early lettuce, cucumber or squash crop could be left for seed, once outdoor crops are supplying plenty of food and provided the hoophouse plants are still healthy. Another possibility might be buckwheat seed, or other fast-growing cover crop seed, for use on your own farm. A bed of short sunflowers for seed at the north side of the hoophouse might be worthwhile, but only if you get plenty of pollinators coming in. Some growers have reported difficulty getting pollinators to come inside, although we haven't had this trouble.

See the Seed Growing chapter for more about our experience growing seed crops in the hoophouse and the chapter by Ira Wallace for more about the Business of Growing Seeds.

Cover crops

Hoophouse growing burns up the organic matter in the soil at a fast rate. If you don't need to grow another cash crop in the summer, you could grow cover crops to replenish the organic matter, ready for your next fall plantings. Buckwheat, soy and cowpeas all work well. Because of harlequin bugs, we avoid growing brassicas as cover crops, but if you don't have this pest they may be a good choice, as they have a biofumigation effect on the soil, tackling some soil-borne diseases.

If you grow lots of winter greens, brassica cover crops might not work with your rotation, even without harlequin bugs. Avoid any cover crops that harbor pests you have trouble with. Short-term, fast-growing cover crops are needed, and clovers are too slow. Winter cereal grains won't grow in the summer. Beware of huge hot-weather grain crops such as sorghum-sudan or tall millets. We grew sorghum-sudan one year and it grew all too well! We cut it down repeatedly and ended up pulling it up by hand. It did make great mulch for the pepper plants, and it all disappeared into the soil without any tilling, but it's not an experience I want to repeat. I still have the scar of a deep cut from the edge of a leaf!

Soil solarization

If you have pest or disease problems in your winter crops, and can live without your hoophouse for a hot month, you could solarize the soil. Ideally, prepare the soil before solarization by cultivation and watering to germinate weed seeds. For best results, grow a brassica crop just before your planned solarization and leave the residues, which will be especially effective as a pesticide. When heated in the solarization process, they release volatile biofumigation compounds, which are trapped in the soil under the plastic. Leave the drip tape in place and irrigate the soil regularly throughout the solarization process.

Cover the soil tightly with old hoophouse plastic, burying the edges in the soil. Close up the whole house, make sure the cat is not sunbathing in there, and let the soil heat up to kill off the bugs and fungal spores. Control is generally obtained down to 4" (10 cm). After solarizing, minimize any disturbance of the soil, which could bring up new weed seeds and pests.

Desalinization

Salts can build up and damage crops in hoophouses, especially if your soil doesn't drain well or if you use animal manures or synthetic fertilizers. Salinization can cause symptoms like drought stress, interfere with seed germination, and inhibit plant growth. It can also encourage some pests. You can test salinity with an electrical conductivity meter. If your salinity is too high, you can scrape off and remove the top layer of soil or cultivate and flush out the salts by flooding, using sprinklers. Alternatively, remove the plastic for a while and let rainfall solve the problem. Avoid a repetition by improving soil drainage and switching to vegetable-based composts and more cover crops. See the Resources section for more information.

Drying

The hoophouse can also be used to dry seed crops grown outside, gourds and luffas, or flowers or ornamental grasses for winter arrangements (experiment first to make sure they don't bleach too much in the sun, or use thick brown paper bags). Don't cure onions or garlic in your hoophouse unless you can be sure the temperature won't go above the temperature at which they cook, around 90°F (32°C). At Southern Exposure Seed Exchange, they covered a hoophouse with 80% shadecloth and use the space to dry peas and beans, and to cure and sort garlic and potato onions.

Some growers use hoophouses in the fall to cure winter squash or sweet potatoes. (Curing improves the sweetness and flavor.) We don't use our hoophouse for curing because we are very actively planting in the hoophouse in October when our squash and sweet potatoes come in. We never actively cure squash. Lisa Dermer pointed out to me that in *The Resilient Gardener*, Carol Deppe says that curing conditions for squash are the same as storage—you cure squash simply by waiting to eat it! Sweet potatoes need temperatures of 80°F–95°F (27°C–35°C) to cure, and need to stay above 55°F (13°C) for storage. Our nights get too cold for keeping sweet potatoes in our hoophouse.

We have bundled up plastic trellis net with attached bean vines, and hung the whole lot from the purlins until the vines were brittle. Then we took the bundle down and danced on it to break up the vines and shake out the pieces, so we could store the trellis for the next season.

Chapter 17

Maintaining Soil Fertility

One of the keys to sustainable farming is to feed the soil so that the soil can feed the crops. Most of us probably do this by some combination of adding compost, organic fertilizers, cover crops, organic mulches, and some soil amendments to change the pH or fix some shortfall of a certain element. I can't do justice to this important topic in a few pages. The best resource I know on the subject is *Nature and Properties of Soils*, fourteenth edition, by Nyle Brady and Ray Weil. The many kinds of soil are explained, and also the differences in approach needed with various kinds of soil and climate. It's a complex world we farm in, and various authors, farmers and researchers have set out to simplify approaches to maintaining soil fertility.

Three very specific approaches to building fertile soil are: the Kinsey Albrecht system of soil nutrient management; the Ingham Soil Foodweb approach via microbiology; and the Jeavons Bio-intensive Mini-Farming system. It is also possible to take something from each of these methods and create your own system. I am grateful to Mark Schonbeck for his valuable comparison of these approaches.

Fixing the soil chemistry

The Kinsey/Albrecht system seeks to balance soil nutrients to improve the chemistry of the soil and make conditions right for plants. This is the theory behind soil nutrient testing and the use of soil amendments to improve the base saturation ratio with the goal of maximizing the cation exchange capacity. The cation exchange capacity is a measure of the soil's ability to hold cations (positively charged atoms of elements which act as plant nutrients). Higher levels indicate a greater nutrient level available. The base saturation ratio shows the relative amounts of calcium, magnesium and potassium, and this theory recommends a ratio of Ca:Mg:K of 20:4:1. This is sometimes expressed as 65%–80% Ca, 5%–15% Mg, 2%–5% K. See the section on Soil Tests for more details.

It is widely recognized that excesses or insufficient supplies of certain elements can cause problems for plants. It is not really proven, however, that one ideally balanced soil will suit all crops. We know that one pH level does not work best for everything. Critics of the Kinsey/Albrecht system are skeptical of the cation levels chosen as ideals. Albrecht insists that some native soils under natural vegetation are a long way from a good balance and must be extensively modified for farming to work well. This belief does not sit well with other soil scientists or with the experience of some farmers. From a perspective of sustainable farming, a system that often requires extensive mining and hauling of large volumes of minerals from one part of the world to another has to be questioned.

Fixing the soil biology

Elaine Ingham's Soil Food Web approach suggests that improving the biological life of the soil will lead to improved soil chemistry. The idea is that if we encourage and maintain a diverse, balanced and well-fed underground world of aerobic soil microorganisms, this soil life will convert organic residues to humus and release nutrients to the crops. The Soil Food Web approach is inspiring in microscopically mirroring the Gaia Hypothesis. Plainly, everything is connected, and changes in one component affect all others in some way. Most growers can see the value in improving the soil biology, even if they don't embrace the more esoteric aspects of this approach.

Soil Food Web management practices recognize three stages in the reclamation of soil by plant species. Asparagus, brassicas and chenopods are in the first "early grass stage" group of plants. They do best on bacterial-dominated soils and don't form fungal mycorrhizae. The second "late suc-cessional grass stage" group includes most vegetables and row crops. These benefit from a 1:1 ratio of bacteria:fungi. The third group, "trees," includes berries and nuts. These do best in fungal dominated soils with minimal soil disturbance. Within the Soil Food Web system, management practices are specific to the crop being grown. It is possible to add purchased mycorrhizal fungi. Some soil scientists disagree with this microbial succession model.

Make "Foodweb" compost from 30%–45% green materials, 30%–45% dry brown materials and 25% high-nitrogen materials (such as manure or alfalfa meal), aerobically under close monitoring. The final product should be the color of 70% cocoa chocolate. Then make this compost into tea following their very precise method. Modest applications of the tea or the compost are used to inoculate the soil, rather than as food for plants or soil. There have been spectacular results from some of the work done under this system. It does, however, rely on a lot of experience and intense management. The warning "Your results may vary" will surely apply to new practitioners! Prescribing the best remedial action for any soil is a tall order, given the enormous complexity of soil ecosystems.

Biointensive mini-farming

John Jeavons' system of intensive farming employs double-digging and close planting on raised beds. It focuses on affecting the physical environment of the plants, rather than the chemical or biological. The aim is to supply a complete and balanced vegan diet for humans with minimal outside inputs. A skilled practitioner can grow a complete diet for one person on 4,000 square feet (371 m²). This is a fairly radical kind of sustainability, to address worldwide land and food needs. 60 percent of the land is used to grow cover crops that are cut to make compost. 1–4 cu ft/100 sq ft (3–12.3 litres/m²) of compost is added before each crop. Under this system, compost is made from a mix with a C:N ratio of as high as 60:1 (30:1 is more usual elsewhere). This approach uses a slow, cool process, to avoid heating the pile above 120°F (49°C) in order to reduce carbon dioxide losses.

Maintaining high soil organic matter is done in ways easy to understand: simply apply compost and mulches, and mineral supplements when needed. Crop yields can be very high. This system is quick to comprehend and try out and its theories are not complicated. It makes intensive use of a relatively small area of land. The double-digging to incorporate the compost and aerate the soil does, however, require large inputs of human energy.

Some critics suggest that these biointensive methods, which originate in California, may not be suitable in other regions. Such close planting in more humid climates may cause disease problems. Likewise, close planting may be risky in hotter climates with limited irrigation. Organic matter decomposes more rapidly in humid climates, quickly leading to a shortage available for crops. Colder winters may cause more frost heave (which loosens and damages plant roots) in soils in loose deep beds—perhaps traditional rows would work better at preventing this.

Reducing erosion and other causes of soil loss

Across the globe, agricultural soil is being lost at eighteen to eighty times the rate it can replenish. Conventional farming loses six to twenty-four times as much soil as the weight of food produced. Organic farming on average loses a "mere" three to eight times the weight of food produced. See David Montgomery's *Soil: The Erosion of Civilization* for more on soil loss. To avert global disaster we need to reduce the rate of soil loss to zero, and build up healthy soils with our farming practices.

Making compost

Many farms make their own compost, as this improves the soil while using up materials that could otherwise be a waste disposal problem. The Center for Environmental Farming Systems at North Carolina State University has good information on compost–making. Farms which are USDA Organic certified will need to follow Organic rules.

Compost is central to our soil fertility program. One of our community businesses is making and selling tofu. Okara is a high-nitrogen waste product from tofu making, the part of the soybeans that doesn't go into the soymilk. We mix in high-carbon sources such as sawdust (waste from our hammock-making business) or woodchips that we trade for with a neighbor. We also add kitchen scraps from our dining hall, and sometimes weeds or crop refuse from our garden. In the summer we don't collect up the weeds, just let them die in place, as that is easier. We don't have specialized compost-turning machines or screens. We use the tractor bucket to lift and turn the piles.

Compost–making is both art and science and there are practitioners of several methods and recipes. Most people strive to make hot (aerobic) compost, by combining 1 to 3 parts high-carbon materials with 1 part high-nitrogen materials in a 25:1 to 40:1 C:N ratio, with enough water to make the piles damp and enough air to keep the bacteria alive. The initial mesophilic stage lasts for the first

Medium- or large-scale compost making needs its own space, with easy access. Credit: Twin Oaks Community.

two to three days after the pile is made. Bacteria which are active at 90°F–110°F (32°C–43°C) begin to break down the sugars, fats, starches and proteins. The pile then heats up and moves into the thermophilic stage, which lasts several weeks. Temperatures in the middle of the pile can reach 120°F–150°F (48°C–66°C). Thermophilic bacteria increase, and keep working as long as decomposable materials remain available and the oxygen supply is adequate. Pathogens, weed seeds and fly larvae are destroyed, and the particle size of the compost becomes smaller.

A decrease in temperature shows that more oxygen or more water is needed. The pile benefits from turning during this stage to provide more oxygen and remix the material, so that all of it can be composted. Turning also prevents the pile from overheating—above 150°F (66°C), the thermophilic bacteria can be killed. During turning, more water can be added if needed to keep the pile damp but not dripping. Too much water

will cause air to be excluded, and the pile will slow down, go anaerobic, and emit foul-smelling by-products. High-nitrogen mixes are likely to lose (waste) nitrogen by volatilization as ammonia. Some loss is inevitable, but compost-makers strive to minimize loss by getting a good balanced mix. If there is not enough nitrogen in the mix, the pile will not heat up and the process will move slowly.

After the compost materials have all been consumed by the bacteria and the nitrogen mineralized (converted to nitrates which will be available as plant nutrients), the pile cools to around 100°F (37.7°C). It can no longer be reheated by more turning, and once it reaches ambient temperature, it is left to cure for about thirty days. This allows beneficial microorganisms to recolonize the compost.

The carbon in mature compost is resistant to further breakdown, and the nitrogen, initially contained in the bodies of microbial soil life forms, slowly becomes available to the plants. Finished compost ideally has a C:N (carbon:nitrogen) ratio of 10:1. See the CEFS publication *Composting on Organic Farms*.

If the C:N ratio is greater than about 25:1, almost no nitrogen is available from the compost and it is unable to mineralize. Between 16 and 20:1, about 10 percent of the N is available. Even at a C:N ratio of 10:1, only half of the nitrogen is available in the near term. Compost is a long-term plan!

When I made compost on a smaller scale I experimented with turning the compost three times every five days. This caused an impressively fast process. In twenty days we had usable compost. It wasn't finely broken down, but it was usable, although it was better after time to mature. (If immature compost is applied to crops, seed germination and plant growth may be inhibited.)

Because it breaks down slowly, generally about 10 percent of the nitrogen will remain after harvest for the next crop (assuming an adequate amount of good finished compost was used). Some growers aim to build the soil to a high overall fertility level, and then maintain that level with smaller applications of compost each season. Others aim to apply a consistent amount each year. Whatever the aim, it is generally agreed that the occasional shortfall in compost application will not be too dire if the soil fertility is high from previous applications.

Compost enhances the soil organic matter and humus, and improves soil structure. Its effects last longer in the soil than cover crops and crop residues, especially in humid conditions where the breakdown of plant material is very rapid. In addition, compost can add a range of beneficial bacteria and fungi to the soil, which can inocu-late plants against diseases by inducing systemic acquired resistance in them. The plants produce antibodies and other protective compounds before any infection occurs.

In his *Winter Harvest Handbook*, Eliot Coleman recommends spreading compost at 5 gallons/25 ft^2 or 15 tons/acre (8.6 l/m^2) of raised beds, for each successive crop.

Compost recipe software is available from Cornell University to help with compost recipe calculations when you are using three or more ingredients. User-friendly spreadsheets are at cfe.cornell.edu/compost/science.html

Compost teas and other "watery extracts"

There are several kinds of compost extracts that sometimes get called compost tea, but they are distinctly different products. See the ATTRA publication *Notes on Compost Teas* for more information.

Compost watery extract is a traditional product. A sack of compost is suspended in a barrel of water for one to two weeks, then the solution is used as a liquid fertilizer, diluted to the color of weak black tea. The same practice is used in England with cut comfrey leaves or stinging nettles.

Compost leachate is the dark-colored liquid that seeps out of the bottom of compost piles. It is usually rich in nutrients but can contain pathogens if it comes from immature compost. It is not suitable for use as a foliar spray until it has had time to mature, and for the pathogens to be killed.

Compost tea, as the name is used today, is a fermented product. Compost is brewed aerobically for 24–36 hours, using a pump to introduce fresh oxygen into the solution. The compost provides microbes, and microbial food sources and catalyst amendments are added to feed the microbes and encourage their multiplication in the tea. Some examples of microbial food sources are molasses, kelp powder and fishmeal. Microbial catalysts include humates, rock dusts and yucca extract. The resulting compost tea is high in biologically active microorganisms including beneficial bacteria, fungi, protozoa and nematodes, as well as plant nutrients. When compost teas are sprayed on foliage, the beneficial organisms consume leaf exudates and outcompete pathogenic organisms for the food. Other microbes directly kill pathogenic organisms.

Organic fertilizers

Sustainable growers who cannot make enough compost use purchased organic fertilizers, such as seed meals, or organic waste by-products from food or fiber manufacturing, sold as commercial nutrient sources. Examples of the first category include cottonseed meal and soybean meal. The second category includes dried poultry manure, feather meal, bonemeal, fish meal and kelp (seaweed). In addition there are products such as earthworm castings, biodynamic sprays, endomycorrhizal inoculants, humates and biochar, which are intended to restore soil balance. The *SoilFacts* publication from NCSU has a valuable table of the nutrient content of organic materials as well as minerals and inorganic fertilizers.

Growing cover crops

Cover crops can improve soil organic matter and crumb structure, thus increasing the moisture-holding ability, porosity and drainage. Leguminous cover crops can provide nitrogen and, if densely planted and turned under at the early flowering stage, may be all the nitrogen the next crop needs. An above-ground biomass of 4000 lb/acre (4480 kg/hectare) of cool-season legumes can provide 60 lb/ac (67 kg/ha) of nitrogen. Cover crops are another big part of our soil fertility program at Twin Oaks. See the chapters on Cover Crops for more about growing them.

Using organic mulches

We use a lot of hay mulch in our gardens. In the past we used even more, but our supply has been reduced and we were having high phosphorus (P) and potassium (K) readings on our soil tests. Using a lot of hay mulch will eventually increase potassium levels, even into the very high level, so watch your soil test scores if you tend towards high K. If your soil has very high or excessive K levels, cut back on the addition of animal manures, hay, or soil amendments high in K (such as wood ash), and use legume cover crops to increase the nitrogen and organic matter without increasing the K readings.

Soil testing and amendments

Soil tests measure the pH and the content of various nutrients. We take soil samples every October (it's good to do it at the same time of year each time). Having some solid figures makes some decisions about improving soil fertility easier. It can also make some harder, if you get too lost in the details. Remember, the overall goal is to feed the soil, not just the next crop! If your soil needs lime, use a hi-cal (calcitic) lime unless magnesium (Mg) is also low, when you need to use dolomitic lime. If you need to add calcium without altering the pH, spread gypsum. If you need to make the soil more acid (such as for blueberries) add elemental sulfur.

Large scale compost making requires specialized equipment, such as this compost screener. Credit: Pam Dawling.

Next, be sure you are adding at least enough of whichever nutrients are low to bring the level up to a sufficiency for the yield desired of the following crop. With nitrogen, be aware that if you only provide the amount of nitrogen the next crop will remove, there is a danger that there will be none spare for the microbe pool, and that crop roots will not find all the applied nutrient. The organic approach is to build up a plentiful soil nutrient bank account, feeding the microbes and being ready for anything. Organic fertilizers release nutrients slowly, particularly if the weather is cold or very wet or dry. In warm, moist conditions, about 75 percent of the nutrients from organic materials are generally available during the first month, with the remainder released by microbial activity in the following two months.

The phosphorus content of soils varies widely, and while too little or too much can cause reduced yields, phosphate ions in the soil are "unlikely to move a greater distance than the diameter of a root hair" in one growing season (Nic Lampkin, *Organic Farming*). In other words, phosphorus does not leach readily, especially from flat fields. Some phosphorus is removed with each crop. Long-term cropping without inputs of animal manure will lead to a decrease in soil phosphorus: about 60 percent of the total P in the manure is available in the year after application and the rest remains in the soil. Compost generally contains more P than N, so if you have high-P soils, find other sources of N besides compost (such as leguminous cover crops) or you will be creating higher P levels. Be sure you have sufficient calcium as this can help lock up excess P as calcium phosphate. Iron and aluminum can also tie up excess phosphorus. On the other hand, if phosphorus is low, rock phosphate is the most common addition.

Greensand is one organic source of potassium. Rock potash is available for non-USDA Organic farmers.

If soil tests show boron is low, add borax at 10–20 lb/acre (11–22kg/ha). This provides 1–2 lbs elemental Boron/acre (1.1–2.2 kg/ha). Be cautious—don't add too much as it could be phytotoxic. Boron deficiency can show up in tomato plants as small, distorted new leaves (looking like virus symptoms). If you see this, give a foliar spray of borax solution at 0.1–0.2.lbs elemental Boron/acre (0.1–0.2 kg/ha).

Chapter 18

Cover Crops

There are many advantages to using cover crops in vegetable production, and a wealth of information is available on them. Cover crops can add biomass and nutrients to the soil, smother weeds, reduce erosion, absorb and "store" rain, salvage excess nutrients from a previous crop, and also — in some cases — attract beneficial insects. No single solution suits all situations and all times of year, so it's good to experiment with different ideas, take notes and be flexible about your plans, to take account of the weather, the crops, the weeds and your schedule. See the Resources section, and the charts that make up the next chapter for more details.

The first step

Begin by identifying a possible niche for a cover crop, and think about which of these benefits are your priorities at that site:

- Prevent weeds growing and seeding

- Add organic matter to the soil and increase the biological activity
- Improve the soil and sub-soil structure, tilth, drainage and water-holding capacity
- Prevent erosion by keeping something growing (roots anchor the soil)
- Add nitrogen to feed the next crop (leguminous cover crops)
- Absorb any nitrogen and other nutrients left over from feeding the previous crop (non-leguminous cover crops)
- Encourage beneficial insects (flowering cover crops)

The second step

Look at the "cover crop window" you have:

- *Winter cover crops* are usually sown after harvesting a major summer vegetable crop. Sometimes they can be undersown in the vegetable crop, while it is growing. They will

115

then continue to grow after the vegetable is finished. Winter-killed cover crops are another option for planting after your main crop. The dead mulch covers the soil and is easy to till in as soon as spring arrives. Winter cover crops include grasses, legumes and brassicas that grow in cool and cold seasons.

- *Spring gaps* are locations where you plan to grow a crop later in the season. If you have six weeks or more in spring you can till in the winter growth and sow oats. Plan ahead and order oats along with winter cover crop seed. Quick-growing cool season grasses and brassicas may be suitable.
- *Summer gaps* may occur between the end of one vegetable crop and the planting of the next, or if a cash crop fails. Cover crops will reduce weeds. You could sow a quick-growing cover like mustard, buckwheat or soy, to till in before the next cash crop. Or you could plant a warm-weather cover crop, then till strips to plant fall vegetables, and perhaps till the rest of the cover crop later. (Winter can offer time to reconfigure planting schedules to make more future windows for short-term summer cover crops.)
- *Full year cover crops* (green fallow) can be used to rebuild fertility, using perennials or biennials such as clovers. When we set up our ten-year rotation in a more organized fashion than we had been using, we discovered to our surprise that we had a plot "spare," which we now use to grow cover crops to replenish the soil and reduce annual weeds.
- *Undersow a cover crop* between rows of a food crop. Perhaps undersow a spring vegetable crop with buckwheat, white clover, or sorghum-sudan, which will take over after

the spring food crop is finished. Summer vining crops like winter squash, pumpkins or watermelon, if not on plastic mulch, also make good use of this technique, provided the cover crop between the rows is progressively mow-killed or tilled in, or is low-growing and will not compete with the crop.

The third step

Identify suitable cover crops for each situation. Ray Weil divides cover crops into six groups:

- Cool-season grasses
- Cool-season legumes
- Cool-season brassicas
- Warm-season grasses
- Warm-season legumes
- Warm-season broadleaved crops

See the next chapter for details on cover crops in each of these groups. Then choose good candidates for your situation.

How we use winter cover crops at Twin Oaks, in Central Virginia, Zone 7

We sow oats in August (or early September—later is too late to make enough growth), in the areas where we plan to plant the early spring crops next year. The oats will be killed by hard frosts, creating a mulch that will be very easy to till under, come spring. We do this before peas, cabbage, broccoli, March-planted potatoes, spinach and sometimes the first sweet corn. This necessitates a crop rotation that clears those patches before the end of the previous August. For us, that's early sweet corn, spring broccoli and cabbage and spring-planted potatoes.

Most of our crops are *not* finished before then, so in order to have more areas in winter-killed cover crops, we undersow some crops. We have good success undersowing our last sowing of sweet corn four weeks after seeding, with oats and soy. The next year this area is disked in early spring for potatoes. It's at least theoretically possible to transplant into the winter-killed oat mulch in the spring without tilling, although weeds may be a problem and the soil will be colder than bare soil—this may work for cabbage and broccoli.

In addition we have the option of tilling in all or part of our green fallow plot, which has been in a clover mix, and sowing oats there in August. If the clover is growing well and the weeds are not too bad, we leave the clover to overwinter, and disk it in February. But if the weeds are gaining the upper hand in August, sowing oats (perhaps mixed with soy) is a better bet. If the weeds are bad in July, we disk in the clovers and sow sorghum-sudan hybrid mixed with soy. While this deals effectively with the weeds, it is a poor crop rotation, as the next year's crop there is early sweet corn, which is related to sorghum-sudan.

During August and September, we sow other warm-weather cover crops that will be winter-killed, into areas that will be planted early next year. Examples include sorghum-sudan hybrids, buckwheat, soy, cowpeas, Miami peas and millet.

The spring broccoli and cabbage finish in early July. We follow them with a round of buckwheat summer cover crop. Then, on September 7–14, we sow winter rye, Austrian winter peas and hairy vetch for the following year's paste tomatoes and peppers. The next year we do not till in this cover crop but mow it very close to the ground and transplant into the dying mulch. The vetch supplies all the nitrogen the tomatoes need, and the dead mulch keeps weeds away for weeks. The peas are thought to help limit *Septoria* leaf spot,

Sweet corn undersown with soybeans and oats. The soybeans will die with the first frost, while the oats will grow later into the winter before winter-killing. In early spring this patch will be easy to work up for planting. Credit: Kathryn Simmons.

a troublesome disease of tomatoes in this region. (We do eventually roll hay between the rows, in July, to top up the mulch). We have also used this no-till cover crop technique for watermelons, planted out in soil blocks, but watermelons like warm soil, so plastic mulch is a safer bet.

We plant crimson clover with oats or winter rye where we plan to have the late crops, which will not be planted till late May or June

A winter cover crop mix of rye, hairy vetch and crimson clover. Credit: Kathryn Simmons.

next year (chiefly the later sweet corn plantings, June-planted potatoes, winter squash and watermelons). This gives the leguminous cover crops time to flower before we need to disk them in. Once again, the legume supplies all the nitrogen for the following crop. Oats will winter-kill and be easier to incorporate, but rye will make more biomass, most of it in early spring. In practice we usually go for the rye, as we are too late for oats. In recent years we have moved towards bringing our watermelon, sweet potatoes and winter squash to a timely end, in order to get good winter cover crops established. We used to hold onto those crops until the last minute to maximize crop yields, but realized we would do better with a perspective on productivity that went beyond one season.

The introduction of biodegradable black plastic mulch revolutionized our watermelon crop! The melons really do ripen three to four weeks earlier than with organic mulch. So now we harvest until we have enough melons to see us through to early October, then disk in the plot. We store our watermelons outdoors in the shade of trees or the eaves of a building, where they store quite well for several weeks. When the cover crop sowing happens before mid-October, we use rye and crimson clover. If we're later we use rye and winter peas.

We plant winter rye and Austrian winter peas (which can be sown later than crimson clover) during late October or even early November after the late finishing crops (winter squash, sweet potatoes, tomatoes and peppers, middle sweet corn). We will not disk these areas until mid-April the next year. Examples in our rotation include the watermelons to middle corn mentioned above; the winter squash to sweet potatoes and late sweet corn; the tomatoes and peppers to watermelons; and the June potatoes to winter squash. We go through the winter squash field for the last time just before Halloween and pull out the large semi-mature squash and give these away for lantern carving. Then we disk that patch and sow rye and winter peas. We aim to harvest the sweet potatoes in the week we expect the first frost (mid-October), then disk that plot and sow the cover crops.

We use winter rye alone after early November, as it is too cold for anything else to make enough growth. Rye needs three to four weeks after tilling in, in spring, to break down and to disarm the allelopathic compounds that prevent small seeds from germinating. Rye is our Last Chance cover crop, and we reckon that we can sow until mid-November, although it's only worth sowing in November if it will have time to make growth in spring. For example, we harvest our fall carrots (sown in early August) in November, and if we are too late for rye, we simply spread the tops over the area to protect the soil and let weeds grow. Rye can be sown up to a month after the first frost and also in the early spring. If we are sowing in early spring, though, we prefer to sow oats, as they break down quicker.

How to simplify your options

Bart Hall once published a very useful table in *Growing for Market* which works back from your farm's first frost date, to see what options you have. If you are more than 120 days before frost, you would most likely plant another food crop. If you are 80 days before your frost date, you could sow buckwheat, oats, Japanese millet, sorghum-sudan, or possibly another vegetable crop. If only 60 days remain before frost, sow oats, Austrian winter peas, crimson clover, or red clover; or soybeans, buckwheat, Japanese millet, sorghum-sudan, winter barley, Miami peas to winter-kill. When about 40 days from frost, use oats, Austrian winter peas, crimson clover, hairy vetch, red clover, fava beans, or winter barley; or soy beans or Miami peas to winter-kill. When only 20 days remain before frost, sow winter rye with or without hairy vetch, winter wheat, winter barley, Austrian winter peas, red clover or crimson clover. It is too late to usefully sow crops that are not frost-hardy. Up to 10 days past the frost date you can sow winter rye, winter wheat, or perhaps red clover or crimson clover. After that, options

become fewer, but it is still possible to sow winter rye up to a month past your average frost date.

How we use summer cover crops at Twin Oaks

Unlike our winter cover crops, we usually only plant our summer cover crops in small areas each time, so we broadcast and till in with our walk-behind BCS tiller. It is also possible to use an Earthway-type seeder for small areas of cover crops; good information about this, including which plates to use for which crops, can be found in the Virginia Association for Biological Farming Infosheet *Seeders: Using Manually-operated Seeders for Precision Cover Crop Plantings* by Mark Schonbeck and Ron Morse.

We sow oats in February or March, when we have winter weeds and will not be planting a food crop for eight weeks. Once we get to March 31 here, it is too late in the year for oats (they will quickly head up after making very little growth) and too soon to rely on frost-tender cover crops. We sometimes sow winter rye, which "languishes" here once it gets hot. I did this one year when our spring potatoes got flooded in April—we actually transplanted potatoes from the flood margins to the drier end of the patch and sowed rye in the lower end, once the floods had subsided. This kept the ground covered, and was easy to deal with in July at potato harvest time.

Once we get to late April (close to our average last frost), we are ready to sow frost-tender cover crops like buckwheat or soy, if mixed with a grain such as winter rye or wheat for insurance and some shielding from harsh weather. Warm-weather cover crops we use include sorghum-sudan hybrid, buckwheat, soy, cowpeas, Miami peas and the millets. After a bad experience

We undersow our fall broccoli with a mix of red and white clovers. The following spring we mow the old brassica plants, and leave the clovers growing all year to replenish the soil. Credit: Kathryn Simmons.

battling oversize millet with small-scale equipment, I avoided growing them, but some are better behaved than the giants I had. German foxtail millet grows to 3'–4' (1–1.3 m), Japanese millet to 3'–5' (1–1.6 m) and pearl millet (my nemesis) gets taller, at 5'–10' (1.6–3.2 m).

We like to include legumes in our cover crop mixes whenever we can, to add nitrogen to the soil. Soy has been a cheap legume for us, because Twin Oaks also has a business making and selling organic tofu. We can get the beans at wholesale price and be certain they are organic. Almost all non-organic soy grown in the US is genetically modified, so if you don't want to add to the problems caused by GMOs, buy organic, or if you are not certified organic, Identity Preserved. See the Organic & Non-GMO Report's *Non-GMO Sourcebook* for a searchable database of non-GMO suppliers.

Full-year cover crops

You can plant long-term cover crops in a section of your crop rotation, to replenish the soil. They can also be planted as perennials in areas that are too challenging to use for production: edges, slopes, tight corners. This is *farmscaping*. These cover crops can attract beneficial insects, and the pollen and nectar can offer an alternative food source to beneficial insects when their insect prey is scarce. This can also be achieved by leaving an unmowed section when most of the crop is mowed, so that the beneficials don't all die.

Undersowing (inter-seeding)

The timing of undersowing is critical. Sow the cover crop late enough to minimize competition with the food crop, but early enough so that it can survive and be big enough to endure foot traffic. The leaf canopy of the food crop should not yet be closed. Often the best time is at the last cultivation (known in the UK as "lay-by"). With vining food crops, it's important to sow the cover crop before the vines run. Choose vigorous *food* crops, but *cover* crops that are only moderately vigorous. Buckwheat, millets and cowpeas all have their fans. Ensure the seedbed is clean and the soil crumbs small enough. Use a high seeding rate, whether broadcasting or drilling. And irrigate sufficiently—this is the second critical component of successful undersowing.

Kale can be undersown with rye and hairy vetch, rye alone, or oats in mid-late August in New York, but in Virginia, winter rye sown in August will go to seed the same year. And perhaps more importantly, we expect kale to overwinter, and cover crops would compete too much.

Peas, eggplant, and peppers can be undersown with oats. Living mulches cool the soil, which is not usually wanted for early plantings of warm-weather crops, but works well for cool-weather crops like peas. It is important to keep mowing the living mulch so that it doesn't outcompete the food crop.

We tried undersowing winter squash and pumpkins with red clover, which works well in New York State, sown when the vines are just starting to run. The clover germinates OK in the low light levels under the squash, and survives foot traffic. Financially, this is probably only worthwhile if the clover grows for a full year. We also tried undersowing corn with crimson clover, which also works in NY. The corn can be mowed after harvest, and the clover left to grow over the winter. In the South, corn and squash grow too fast compared to clover for this scheme to work. Our vines very quickly cover the whole field, once they start running. It is difficult in Virginia to get the clover to germinate in the heat and dryness of June and July. Soy is much easier to deal with, and cheaper.

Buckwheat can be undersown with winter squash or sweet potatoes, and mowed as soon as the vines start to run. One year we tried buckwheat between our squash rows to keep the weeds down, but failed to mow or till in the buckwheat, and had to wade in and pull it by hand—the crew hasn't yet forgiven me! One trial of undersowing buckwheat in corn reported that if sown the same day as the corn, the buckwheat outcompeted the crop; if sown at the eight-leaf stage of the corn, the buckwheat did not get enough light. Find the happy medium!

We tried a forage brassica (canola/rape), before a new strawberry planting, but it encouraged too many brassica pests, harlequin bugs being the worst. We decided brassica cover crops are not for us. In areas where they work, try daikon, forage radish, mustards or canola. Do this when the soil is 45°F–85°F (7°C–30°C). That's up until early October for us. Aim to get six to eight leaves before the killing frost. The brassicas produce allelopathic compounds that inhibit weeds and biotoxins (glucosinolates) that kill pests.

Organic no-till cover crops

Some cover crops can be killed by mowing, rather than having to till them in; food crops can then be planted in the dying residue. Roll-killing is another option, but usually requires special equipment. Mow-killing and roll-killing can be challenging to do successfully, so it's best to experiment with a small area first.

Advantages of no-till include:

- Reduction of soil disturbance and the associated burn-up of organic matter;
- Prevention of new weed seeds being brought to the surface;
- Smaller number of tractor passes across the field, saving fuel and preventing soil compaction;
- Improvement of soil life diversity in the top layers of soil;
- Cooler soils in summer;
- And reduced need to bring in other mulches to control weeds.

Disadvantages include:

- Cooler soils in spring, meaning slower crop growth and later harvests;
- Slower rate of nitrogen release from the cover crop;
- Possibly more fungal diseases and slugs;
- Possibly problems with regrowth of the cover crop;
- And possibly weeds growing in the cover crop.

Summer vegetables, especially tomatoes, can do very well after a legume-grass mix no-till cover crop, needing no additional source of fertility. Fast-maturing spring vegetables will not do well with no-till cover crops as they need nitrogen more quickly than can be got from no-till. It is sensible, especially if you haven't got much experience with no-till, to have a backup plan in case cover crop stands end up too thin, or the weeds are too plentiful for the no-till system to work.

Usually this means tilling in the poor cover crop and adding compost or some other source of fertility.

A related approach is Reduced Tillage: take an existing plot of cover crops and till out strips in summer and plant the next food crop. Sometimes the rest of the cover crop is tilled under later, as the food crop needs the space.

How to clarify your options for summer gaps

If you have only 28 days until the patch is needed for a food crop, you can grow mustards or buckwheat; or weeds, if you're careful not to let them seed! If you have at least 45 days, you can grow soy or Japanese millet. If you have 50–60 days, Browntop millet is possible. In the right climate, sunn hemp can mature in 60 days. With 60–70 days, German/foxtail millet, pearl millet and some cowpeas will mature.

In high-moisture years, grow the most weed-suppressing crops, e.g., alfalfa. If fall moisture is low, sow peas, or wait until spring and sow a fast-maturing legume. Spring-planted peas can produce more nitrogen than fall-planted Austrian winter peas.

Cover crop mixes

A 2011 ATTRA webinar advocates mixes of cover crops. Using mixes can give the plot the advantages of each of the components, and also insure that regardless of the weather or rainfall, some cover will grow. Additionally, most mixes include some crops that attract beneficial insects and some legumes to add nitrogen. Mixes can generally be sown at a depth of one inch (2.5 cm), regardless of seed size.

Major ingredients for a summer mix could

include soy, cowpeas and buckwheat. Lesser ingredients could include pearl millet, proso millet, radish, turnips, sunflowers and sunn hemp.

A spring mix could have oats and peas as the main ingredients, with hairy vetch, radish, turnips and red clover as minor ingredients.

Nitrogen availability

How much nitrogen becomes available to the next crop depends on the C:N ratio in the biomass of the cover crop. C:N ratio is as important here as in compost-making. The soil microbes that digest the cover crop have a C:N ratio of 10:1. When a cover crop is incorporated into the soil, the microbes use the carbon and some of the nitrogen in the cover crop to build more microbes, tying it up until they die, meaning it is not immediately available to the next crop. If the cover crop has a C:N ratio of 50:1 (like sorghum-sudan), the microbes will need to find extra N to make use of all the C. They will use N from the soil, tying it up until they die. Hence any crop following *immediately* after a high C:N ratio cover crop will need a different source of nitrogen.

Legumes have a lower C:N ratio (from 30:1 down to 12:1) and when they are incorporated, soil N is unlikely to be tied up, so is available for the next crop. This process is explained in the North Carolina Extension Service Horticulture Information Leaflet 37 cited in the Resources section. This explains the advice to incorporate high-C cover crops a few weeks ahead of planting the

Winter rye and hairy vetch. Credit: Kathryn Simmons.

following crop. It also allows time for any cover crop allelopathy to subside.

If cover crop residues are left on the surface rather than incorporated, the rate of decomposition is slowed. Some nitrogen is lost to the air (denitrification), but the increased organic matter can boost the diversity of microorganisms at the surface. Some of the carbon from cover crops is below the top eight inches (20 cm), where almost all soil data are collected. Remember the value of the roots!

Chapter 19

Cover Crops Chart

This chart is designed to help growers find crops suitable for their climate, time of year and time-window. It does not include all possible cover crops, nor Latin names, information on planting depth, inoculants, soil pH requirements, effect on soil compaction, biomass and N production. It is arranged by six crop types: Cool season grasses, Cool season legumes, Cool season brassicas, Warm season grasses, Warm season legumes, and Warm season broadleaf plants.

Notes:

- When legumes and grasses are mixed, sow on the date for the grass.

- When 2 grasses are mixed, reduce the seeding rate of each by a third.
- Do not reduce the seeding rate of legumes in mixtures.
- "Frost Seeding": Early in the morning after a hard frost, broadcast clover seed on prepared soil. The thawing will wet the seeds and pull them down into the soil.
- Dates are for zone 7, adjust as needed.
- D = drilled, B = broadcast, # = pounds, ac = acre, ha = hectare, EW = Earthway type manual seeder, ft^2 = square feet, m^2 = square meters

Type	Cover crop	Sowing date if last frost is 4/30 first frost 10/14. Adjust dates as needed.	Sowing rate (US units)	Sowing rate (metric)	Uses and Cautions	Notes
Cool season grasses	**Winter Rye/ Cereal Rye** 5-7' (1.5-2.1 m) #22 EW plate. *Wren's Abruzzi* is better for SE. *Wheeler* matures a week later than *Abruzzi*	14 days before first frost to 28 days after last frost. April is OK for zone 7. Spring sowings in warm climates, with a soil temp >50°F (10°C) will usefully "languish" and be easy to incorporate. Don't sow in August in zone 7—may set seed.	D:60–120 #/ac B:90–150 #/ac B:4–5 oz/100 ft²	D:67–134 kg/ha B:100-170 kg/ha B:13 gm/m²	To be followed by anything from mid-spring onwards. Could become a weed in wheat fields. Can be undersown in tasselling or silking corn or fall brassicas in late August, and left as a winter cover crop.	Very cold hardy: −30°F (−34°C). Can be sown later than anything else. Weed suppressant (especially lambsquarters, redroot pig-weed, ragweed). Turn under at least 3 weeks before sowing small seeds. If tilling rather than disking, till in at 8" (20 cm) tall or mow if tilling will be delayed. Mow-kills at flowering, not earlier. Cheap.
Cool season grasses	**Winter Wheat** 3' (0.9 m) #22 EW plate	Around first kill-ing frost date: mid-September to early Novem-ber in zone 7.	D:60–120 #/ac B:70–150 #/ac B:3–6 oz/100 ft²	D:67–134 kg/ha B:90–170 kg/ha B:9–20 gm/m²	Good nurse crop for winter legumes.	Winter-killed in zone 4. Matures a week or so later than barley. Less vigorous than rye (easier to incorpo-rate with small equipment). Cheap.
Cool season grasses	**Oats** 2–3' (0.6–0.9 m) #22 EW plate "Feed" oats are acceptable for cover crops but beware inclu-sion of GMO rapeseed	2/15–3/31 or August/early September	D:80–110 #/ac B:110–140 #/ac B:4–6 oz/100 ft²	D:90–123 kg/ha B:123–155 kg/ha B:12–20 gm/m²	Spring sowings to be fol-lowed by late spring/early summer planted crops. Winter-killed late summer sowings to be followed by early spring planted crops: peas, potatoes, broccoli, cabbage, spinach, onions. Good nurse crop for winter legume cover crops.	Takes only 6 weeks in spring to make a worthwhile stand. Winter-kills in zone 7, 10–20°F (−12 to −7°C). Cheap. Can mix with peas, 3 oats:7 peas by weight, sown in August.
Cool season grasses	**Winter Barley** 2–3' (0.6–0.9 m) Shorter than rye. #22 EW plate	14–44 days before first frost in fall: in zone 7, 9/1–9/30.	D:50–100 #/ac B:80–125 #/ac B:3–5 oz/100 ft²	D:56–112 kg/ha B:100–140 kg/ha B:9–16 gm/m²		Quick growing. Winter-killed in zone 6, 0–10°F (−18 to −12°C). Hardy zone 8. Tillers (grows multiple stems from one plant), so provides high biomass. Tolerates dry soils, but not acidic soils. Cheap.
Cool season grasses	**Black Oats** 5' (1.5 m)	September-October in Zones 8b–10a. Not suitable for colder zones.	50–70 #/ac	56–80 kg/ha		Winter-killed at 12–15°F (−11 to −9°C). Tillers profusely. Allelopathic—sup-presses broadleaf weeds very well. Resistant to root-knot nematodes.

Type	Cover crop	Sowing date if last frost is 4/30 first frost 10/14. Adjust dates as needed.	Sowing rate (US units)	Sowing rate (metric)	Uses and Cautions	Notes
Cool season grasses	**Annual Rygrass/ Italian Ryegrass** 2–3' (0.6–0.9 m) Don't buy Perennial Ryegrass by mistake.	In spring or late summer/early fall, at least 40 days before the average first frost.	D:10–20 B:25–35 #/ac B:1.5–3 oz/100 ft^2	D:11–22 kg/ha B:30–45 kg/ha 5–9 gm/m^2	Good nurse crop for clovers. In colder areas, can be a good choice for pathways Has the potential to become a weed in warmer climates. (We think the risk at our farm is not worth the possible benefits.)	Fast-growing. Winter-kills at −20°F (−28°C). If fall-sown, tillers profusely, producing a dense mat of dead mulch in the spring (when winter-killed). Easy to establish, tolerates flooding. Self-seeds readily. Cheap.
Cool season legumes	**Hairy Vetch** 2' (0.6 m) 6' (1.8 m) if mixed with tall cover crops #22 EW plate	From spring, up till 35 days before the first freeze. Needs min soil temp of 60°F (15.5°C). In zone 7 9/7–10/10 is ideal.	D:15–20 #/ac Mix:50 # rye + 25 # hairy vetch/ac B:20–60 #/ac B: Hv alone, 2 oz/100 ft^2. Mixes: 1–2 oz hv + 2–3 oz rye/100 ft^2. Or 1.5 oz hv, 2.5 rye, 0.5 oz crimson clover/100 ft^2. Or 1.5 ox vetch, 1.5 oz AWP, 2.5 oz rye/100 ft^2.	D:17–22 kg/ha Mix:56 kg rye + 28 kg hairy vetch/ha B:22–67 kg/ha B: Hv alone, 6gm/m^2. Mixes: 3–6 gm hv + 6–9 gm rye/m^2 Or 5 gm hv, 8 gm rye, 1.5 gm crimson clover/m^2. Or 5 gm hv, 5 gm AWP, 8 gm rye/m^2.	To be followed by crops planted later than 6/1 (late corn, summer potatoes, late successions of beans, cucumbers, zucchini, possibly winter squash). Before early spring food crops, use Purple Vetch (which winter kills). Suppresses yellow nutsedge, lambsquarters. Vines can be very tangly. Can be invasive if it sets seed. Seed is poisonous to poultry. May harbor pest nematodes.	Very cold hardy legume, to −15°F (−26°C). (*Lana Woolypod* to 10°F (−12°C), *Common* to 0°F (−18°C), *Purple* to 20°F (−7°C). Drought tolerant once established. Not tolerant of shade or flooding. Can be tilled in, or mowed for no-till transplanting. (Use nylon line trimmer for small patches). *Lana Woolypod Vetch* is earlier and more heat tolerant. *Common Vetch* is cheaper, *Purple Vetch* is more vigorous. Best killed at early flowering (late April/ early May for zone 7) for maximum N.
Cool season legumes	**Fava Beans/ Bell Beans** 3–5' (0.9–1.5 m) *Banner* survives 10°F (−12°C)	As early in spring as possible. Fall in mild areas.	80–140 #/ac 3–5 oz/100 ft^2	90–160 kg/ha 10–16 gm/m^2	To be followed by summer or fall vegetables (after 6/1). Suppresses tomato wilt.	Fast growing. Frost tolerant, hardy to 15°F (−9°C). Tall. If organic seed not needed, Greek groceries may be cheapest source.
Cool season legumes	**Austrian Winter Peas** #14 EW plate	In zone 7, 8/10–10/24 ideally (11/8 may be possible). Up to 35 days before first hard freeze. May not do well if sown in spring (requires a cold dormant spell).	D:60–120 #/ac B:5–6 oz/100 ft^2 in mix	D:67–134 kg/ha B:16–20 gm/m^2 in mix	Best used when the next crops will be planted in June/July. Suppresses Septoria leaf spot in tomato. May increase pest nematodes.	Winter-killed in zone 6, 0°F (−18°C). Hardy in zone 7. Rapid spring growth. Blooms in late April at Twin Oaks, before hairy vetch. Not tolerant of shade or of long cold spring weather below 18°F (−8°C). Can mix with oats, rye or barley.

Type	Cover crop	Sowing date if last frost is 4/30 first frost 10/14. Adjust dates as needed.	Sowing rate (US units)	Sowing rate (metric)	Uses and Cautions	Notes
Cool season legumes	**Lupins** Indeterminate OP white lupins or determinate hybrids	Aug–Nov in the north. September or April in the southeast.	70–120 #/ac	90–134 kg/ha	May suppress nematodes and other diseases and pests. Attracts beneficial insects. Sponges up phosphorus.	Very cold-tolerant. Zones 4–9. Deep roots, breaks compaction. Good weed control. Not flood tolerant. Need relatively high P and K.
Cool season legumes	**Canadian Field Peas** 5' (1.5 m) #14 EW plate *Trapper* is most common variety	Early spring, or late summer or fall. Need cool soil to germinate, and 100 days to reach maturity from spring sowing.	D:50–80 #/ac B:90–150 #/ac B:3–5 oz/100 ft²	D:56–90 kg/ha B:100–170 kg/ha B:9–16 gm/m²	To be followed by summer crops planted June–August	Fast growing. 100 days. Hardy to 10–20°F (–12 to –7°C). Smothers weeds better than spring sown clovers. Mow-kills. Incorporate at flowering, 5–8 weeks after spring planting. Can mix with oats.
Cool season legumes	**White Clover** 7" (18 cm) *Ladino*: 12–15" (30–38 cm), fine stems. *Dutch/NZ*: low-growing 12" (30 cm) dense, drought tolerant. *Alsike*: white-pink flowers, fine stems, upright.	2/15–3/15— can be frost-seeded. Or Sept–mid Oct. Clovers should be sown at least 6 weeks before a hard fall frost, to ensure winter survival. Can sow in summer if soil can be kept damp.	D:5–9 #/ac B:7–14 #/ac B:1.5 oz/100 ft²	D:6–10 kg/ha B:8–16 kg/ha B:5 gm/m²	Provides an all-year cover crop for the next year. Re-seed by frost seeding every two years. Tolerates foot traffic, so good for orchards and vineyards. Needs regular mowing to kill weeds. Doesn't mix well with grasses.	Generally hardier than red clovers, smaller. *Dutch White* aka *New Zealand* is hardy to –20°F (–28°C). Has stolons, so good at re-growing after mowing. Perennials used as winter annuals in the South. Water frequently until established. Competes poorly with weeds until well established.
Cool season legumes	**Crimson Clover** 18" (46 cm) or twice that #5 EW plate	Usually sown in the fall, mid-July to mid-Sept. In zone 7, 9/1–10/10 is ideal. Also 2/15–3/15 in zone 7 Spring sowings don't make much growth before flowering.	D:15–18 #/ac B:22–25 #/ac B:1–2 oz cc + 2 oz oats or rye/100 ft² B alone: 2–3 oz/100 ft²	D:17–20 kg/ha B:25–28 kg/ha B:3–6 gm cc + 6 gm oats or rye/m² B alone: 6–9 gm/m²	To be followed by crops planted later than 6/1 (late corn, summer potatoes, late successions of beans, cucumbers, zucchini, possibly winter squash). Can undersow in fall crops for winter cover. If seed is not raked in, triple the seeding rate and ensure adequate irrigation. Attracts beneficials, including assassin bugs which eat Colorado Potato Beetle. Suppresses Italian ryegrass.	Hardy to 10°F (-12°C). Establishes earlier than hairy vetch in fall, so suppresses weeds better. Fall sown crops make fast growth in spring, to 18" (45 cm) or more. Deep rooting annual. Shade tolerant. Provides lots of nitrogen. Mow at early bloom when N-fixing has peaked. Reduced tillage: sow in September, and in spring strip-till rows and plant eggplant.

Type	Cover crop	Sowing date if last frost is 4/30 first frost 10/14. Adjust dates as needed.	Sowing rate (US units)	Sowing rate (metric)	Uses and Cautions	Notes
Cool season legumes	**Red Clover** *Medium* or *Mammoth*. 2–3' (0.6–0.9 m)	In the south fall or spring, ideally 30 days before first fall frost, and 30 days before last spring frost. In zone 7, frost seeded 2/15–3/15. In the north., from 45 days before last frost, up till 30 days before first frost. Spring sowings do not make much growth until the nurse crop is harvested. In warmer weather, sow with nurse crop (rye, buckwheat or oats). Ensure sufficient water, especially for summer sowings.	D:8–10 #/ac B:15 #/ac B:1–2 oz/100 ft^2	D:9–11 kg/ha B:17 kg/ha B:5–9 gm/m^2	Provides an all-year cover crop for the next year. Best incorporated after 2nd season of full growth (needs time to produce enough growth). Shade-tolerant, so good for under-sowing, for example, fall brassicas at 30 days after transplanting. Attracts beneficial insects.	Easy to establish. Fairly high cold tolerance (not as hardy as white clover). Short-lived perennial. Tolerates shade and poor drainage (not flooding). Good weed suppression. Add crimson clover for earlier bulk. *Mammoth* is easier to establish in dry soils than *Medium*, faster growing, more biomass, but less good at re-growing after mowing. Cheap.
Cool season legumes	**Sub-terranean Clover/Sub Clover** Low-growing	Best suited to areas with dry summers, mild wet winters. In hardiness zone 7 and warmer, sow in the fall. In zones colder than 7, sow in spring. Can mix with crimson clover.	D:10–20 #/ac B:20–30 #/ac B:3 oz/100 ft^2	D:11–22 kg/ha B:22–34 kg/ha B:9 gm/m^2	Long term orchard floors, living or dying mulch, erosion control. Survives close mowing.	Re-seeding cool season annual. Strong seedlings, good weed suppression. Thick, low growth. Dormant in winter, regrows in spring. Seeds, then dies in summer. New plants grow in the fall.

Type	Cover crop	Sowing date if last frost is 4/30 first frost 10/14. Adjust dates as needed.	Sowing rate (US units)	Sowing rate (metric)	Uses and Cautions	Notes
Cool season legumes	**Berseem Clover/ Egyptian Clover** *Bigbee, Multicut* 2' (0.6 m)	Spring (frost seeding not very successful). Late summer if water is adequate (leaves a friable seedbed needing minimal spring tillage).	D:15–20 #/ac B:15–20 #/ac B:2oz/100 ft²	D:17–22 kg/ha B:17–22 kg/ha B:6 gm/m²	To be followed by no-till early spring vegetables.	Hardy to 15°F (−9°C). Very productive, vigorous tall annual white clover. Does best in hot moist conditions. Tolerates wet soil. Mow to 3" (7.6 cm), not less, when 7–20" (18–50 cm) tall. Excellent weed suppression. Easier to incorporate than hairy vetch. Sow Berseem clover or Lana Woolypod Vetch with oats as a winter cover before early spring vegetable crops. It will provide nitrogen and then be winter-killed.
Cool season legumes	**Sweet Clover** 2–6' (0.6–1.8 m) Yellow is earlier but less productive than white	Sow April, May or late summer. Can be frost-seeded in late winter	D:6–10 #/ac B:15–20 #/ac B:1–4* oz/100 ft²	D:6.5–11 kg/ha B:17–22 kg/ha B:3–12* gm/m²	Sow late summer for heavy growth next spring. Spring sowings make growth by early summer. Mow high for maintenance, e.g., if inter-seeded between crop rows.	White annual *Hubam* variety winter-kills at 19°F (−7°C). Other sweet clovers are biennial. Deep rooted. Requires 17" (43 cm) of water per season. Yellow sweet clover seed is cheap, white annual is expensive. *Yellow sweet clover is not easy on a small scale. It can be difficult to incorporate. (White is much easier.)
Cool season brassicas	**Brassicas** Daikon, forage radish, turnips, canola/ rapeseed or specific cover crop mustards 2–3' (0.6–0.9 m) #5 EW plate for radish seeds	Sow when the soil is 45–85°F (7–30°C). Until early October in zone 7. Oilseed radish and daikon are best sown in late summer, for winter-killing	D:5–14 #/ac B:10–20 #/ac B:0.5–1 oz/100 ft² Use low rates for turnip, medium rates for rape, mustard, high rates for radish (large seeds)	D:5.5–16 kg/ha B:11–22 kg/ha B:0.5–3 gm/m² Use low rates for turnip, medium rates for rape, mustard, high rates for radish (large seeds)	Winter-killed radishes can leave a very mellow seedbed for early spring crops, such as onions, lettuce, spinach, peas, (although soil can get crusty if left too long before spring planting). Attracts beneficials. Suppresses pest nematodes. Good in preparation for new strawberry beds. Not recommended where lots of brassica food crops are grown or where brassica diseases are a problem.	Fast growing, cover 90% of surface in 22 days. Brown and black mustards are hardy to 0°F (−18°C), oilseed radish and daikon to 20°F (−7°C). Tolerate heat and drought in late summer and fall. Aim to get 6–8 leaves before the killing frost. Mustards produce allelopathic compounds that inhibit weeds, and biotoxins (glocosinolates) that kill some pests and diseases. They also inhibit legumes, so do not mix, in large amounts.

Type	Cover crop	Sowing date if last frost is 4/30 first frost 10/14. Adjust dates as needed.	Sowing rate (US units)	Sowing rate (metric)	Uses and Cautions	Notes
Warm season grasses	**Sorghum-Sudangrass** hybrid/*Sudex* 6' (1.8 m) Sow ½-1 ½" (1–4 cm) deep. #5 EW plate, or #22 plate if weed pressure is high	From 2 weeks after corn-planting date (needs warm soil). Less deep than corn. In zone 7, mid-May to late August.	D:25–40 #/ac, less if rows 36–42" apart B:40–50 #/ac B:2 oz/100 ft²	D:28–56 kg/ha, less if rows 90–110 cm apart B:45–56 kg/ha B:6 gm/m²	Excellent as part of a soil improvement year in a rotation. Good in preparation for new strawberry beds. Winter-killed sorghum-sudan can be followed by early spring crops. Good ahead of fall crops—needs 8–10 weeks of growth. Suppresses root-knot nematodes and annual ryegrass. Can be hard to incorporate once tall.	Not frost hardy. Fast growing, deep rooting, especially if mowed to 6" stubble when it reaches 4' (1.2 m) in height and allowed to re-grow. Large biomass. Allelopathic, allow one month for decomposition. Cannot be mow-killed. Does well in mixes with buckwheat, soy, and/or viney legumes. Can mix 10 # with 50 # cowpea/ac (11 kg with 56 kg cowpea/ha).
Warm season grasses.	**Japanese Millet** 3–5' (1–1.6 m) Easier to incorporate than sorghum-sudan-grass—good for small equipment	From corn planting date. May and early June are best. Day-length sensitive. Can be sown until early July, but growth is considerably less if planted after summer solstice.	D:15–25 #/ac B:35 #/acIn mix with soy:54 # soy:12 # millet/ac B:2 oz/100 ft²	D:17–28 kg/ha B:40 kg/ha In mix with soy: 60 kg soy: 14 kg millet/ha B:6 gm/m²	To be followed by summer and fall vegetable crops. Good choice for cleaning weedy areas after early vegetable crops: mow repeatedly for the rest of the year.	Fast growing annual, 45 days. Not frost hardy. For best re-growth, mow at 60 days when 3' (1m) tall, and before heading. Cut to 3–8" (8–20 cm) and repeat every 40 days after that. If cut when older, it will rapidly set seed. Tolerates drought and wet soils, including *cold* wet soils. Cannot mow-kill or roll-kill reliably: it re-grows, and if cut after heading up, will flower again in 2–4 weeks, on short stems.
Warm season grasses	**Browntop Millet** 2–5' (0.6–1.5 m)	May–August	D:20–30#/ac B:30–40 #/ac B:3 oz/100 ft²	D:22–34 kg/ha B:34–45 kg/ha B:9 gm/m²		50–60 days. Tolerant of acidic soils, low fertility and flooding. Regrows after mowing
Warm season grasses	**German/ Foxtail Millet** 3–4' (1–1.3 m) #24, #5, or #10 EW plate, depending on weed pressure	2 weeks after last frost, onwards, even until late summer. May–August here. May and early June are best. Growth is considerably less if sown after summer solstice—day-length sensitive.	D:20 #/ac B:30 #/ac B:1–1.5 oz/100 ft²	D:22 kg/ha B:34 kg/ha B:3–4.5 gm/m²	To be followed by late summer and fall crops. Fairly well-behaved—unlikely to become a weed.	Fast growing annual (60–70 days). 75–90 days to seed formation. Not frost hardy. Needs warm soil. Small seeds: need good seedbed and few weed seeds. Fairly drought tolerant once established. Shallow roots Easier than most other millets to incorporate. Mow-kills or roll-kills reliably after heading.

Type	Cover crop	Sowing date if last frost is 4/30 first frost 10/14. Adjust dates as needed.	Sowing rate (US units)	Sowing rate (metric)	Uses and Cautions	Notes
Warm season grasses	**Pearl/ Cattail Millet** 5–10' (1.6–3.2 m) #24 or #10 EW plate, depending on weed pressure	From corn-planting date until 60 days before fall frost. Soil temperatures of 75–90°F (24–35°C) are ideal. Late April to mid-August for zone 7. (Later than other millets). Not day-length sensitive, so produces good growth in spring or summer.	D:15#/ac in 18" rows (use stale seedbed) B:25–40 #/ac B:1–1.5 oz/100 ft²	D:17 kg/ha in 45 cm rows (use stale seedbed) B:28–44 kg/ha B:3–4.5 gm/m²	To be followed by summer and fall crops. Millet can be sown 60–85 days before expected frost in order to be winter-killed and avoid seed formation. This can be followed by early spring crops.	Fast growing, 60–75 days.. Not frost hardy. Annual bunchgrass. Does OK in poor soils. Does not tolerate water-logging. Excellent biomass, even in late summer. Suppresses weeds better than sorghum sudangrass. Mow before heading for fast regrowth. After heading it is fairly easy to mow-kill, although not as easy as German and Japanese millets.
Warm season grasses	**Proso Millet** More than 5' (1.6 m) #5 EW plate		D:20 #/ac B:30 #/ac	D:22 kg/ha B:34 kg/ha	To be followed by summer and fall crops.	Cut after 60 days before it gets tough. Generally sold for cattle forage or game bird feedlots. A much finer textured grass than Pearl or Japanese millets. Makes a good mulch.
Warm season legumes	**Miami Peas** #14 EW plate	May–August	70–120 #/ac B:4–5 oz/100 ft²	80–134 kg/ha B:12–16 gm/m²	Winter-killed, to be followed by spring crops. Or sow in spring to be followed by summer crops.	Not frost hardy. Can mix with oats for a winter-killed mix.
Warm season legumes	**Soy Bean** 2–4' (0.6–1.2 m) #22 EW plate Beware GMO non-organic seed.	From last frost date (late April for zone 7), until late August	D:60–90 #/ac B:100 #/ac B:4–6 oz/100 ft²	D:67–100 kg/ha B:112 kg/ha B:12–20 gm/m²	OK in shade, and withstands foot traffic, so good for undersowing corn at 30 days. Can be followed by summer and fall crops. Winter-killed soy can be followed by early spring crops.	Not frost hardy. Establishes quickly, competes well with weeds. Moderately drought tolerant. Roots mostly in top 8" (20 cm). Use soybean inoculant. Grow 11–14 weeks for maximum biomass, but 6–8 weeks is still worthwhile. Can mow-kill. Residue has some allelopathy. Cheap, widely available. Can mix with oats for a winter-killed cover. In warm weather, can mix with Japanese Millet 54 # soy:12 # millet/ac. 60 kg soy:13 kg millet/ha.

Type	Cover crop	Sowing date if last frost is 4/30 first frost 10/14. Adjust dates as needed.	Sowing rate (US units)	Sowing rate (metric)	Uses and Cautions	Notes
Warm season legumes	**Cowpea/ Crowder/ Southern pea** 2' (0.6 m) Taller if mixed with a tall crop (SS or millet). #22 EW plate	From 2 weeks after corn sowing date until late August. Soil temperature above 65°F (18.3°C). Allow 9 weeks before first killing frost. Sow ½–1" (1.2–2.5 cm) deep	D:30–40 #/ac in 6–8" rows B:70–120 #/ac B:5 oz/100 ft²	D:34–45 kg/ha in 15–20cm rows B:80–134 kg/ha B:16 gm/m²	To be followed by summer and fall crops, or may be sown after early vegetables. Can undersow sweet corn at 30 days. May reduce root-knot and soybean cyst nematodes. Susceptible to reniform nematodes.	Fast growing, 60–90 days. Not frost hardy. Deep tap-root, drought tolerant. Does better in poor soils than soy. Tolerates some shade. May raise the pH of the soil. Mow-kills. Mung and azuki beans are also worth trying. Can mix with foxtail millet, oats or sorghum sudangrass. 10 # SS with 50 # cowpea in mix/ac 11 kg SS with 56 kg cowpea in mix per hectare.
Warm season legumes	**Velvetbean** 30' (7.6 m)	For southern areas only	D:25–35 #/ac in 40" rows	D:28–40 kg/ha in 1m rows	To be followed by late summer or fall crops.	190 days to maturity, but can till in after 67 days. Huge vigorous vine. High biomass. Suppresses yellow nutsedge and chickweed. Can be hard to incorporate. Decomposes quickly. Needs inoculant. Mow-kills. Large seed, so not cheap.
Warm season legumes	**Lab Lab** #2 or #4 EW plate	For southern areas only	20–50 #/ac	22–56 kg/ha		
Warm season legumes	**Sesbania** 6–8' (1.8–2.4 m)	For southern areas only	20 #/ac B:2 oz/100 ft²	22 kg/ha B:6 gm/m²		90 days. Not frost tolerant. More tolerant of flooding and drought then sunnhemp. Vigorous, fibrous.
Warm season legumes	**Sunn Hemp/ Crotalaria** 9' (2.7 m)	For southern areas only From corn-planting date. Can be sown 3 months before fall killing frost for a large bio-mass of dead mulch.	D:5–7 #/ac in 40" rows B:30–50 #/ac	D:5.5–8 kg/ha in 1 m rows B:34–56 kg/ha	Can sow in late summer after early corn. Or before early corn, if winter-killed. Can reduce nematodes.	60 days. Frost-killed at 28°F (−2°C). Vigorous—stems become fibrous with time, and if widely spaced. Needs specific inoculant. Allelo-pathic—inhibits weed germination. Relatively expensive and not widely available—could be mixed with white rice in seeders to reduce sowing rate.

Type	Cover crop	Sowing date if last frost is 4/30 first frost 10/14. Adjust dates as needed.	Sowing rate (US units)	Sowing rate (metric)	Uses and Cautions	Notes
Warm season broadleaf plants	**Buckwheat** 2–4' (0.6–1.2 m) Sow ½-2" (1–5 cm) deep. #22 EW plate	From a few days before last frost date until 60 days before first frost. Late April-late August.	D:40–70 #/ac B:80–120 #/ac B:2–4 oz/100 ft^2	D:45–80 kg/ha B:90–112 kg/ha B:6–12 gm/m^2	To be followed by crops planted from early summer to fall (quick to incorporate). Can be used whenever there are gaps of 4 weeks or more between crops. Useful cover in case of crop failure or early finish. Attracts beneficial insects. May harbor root lesion nematodes.	Takes only 4–5 weeks to grow to full size. Not frost hardy. Can be left to self-seed if time allows. Scavenges calcium and phosphorus. Not tolerant of drought or shade. Can mow-kill or roll-kill if soil is too wet to disk or till.

Crimson clover flower. Credit: Kathryn Simmons.

Chapter 20

Sustainable Disease Management

Three factors affect whether or not a disease takes hold. Diseases need a susceptible host, the presence of a pathogen, and suitable environmental conditions. The grower can affect all three of these conditions, to reduce the likelihood of diseases. Disease-resistant varieties can be planted, crops can be rotated, and planting and harvesting dates can be planned to reduce vulnerability to diseases. Disease organism populations can be monitored and prevented from reaching economically damaging levels. Soil fertility and crop nutrients can be optimized to provide the best environment for the crop and for beneficial organisms.

How diseases spread

Plant pathogens can be soil-borne, foliar-borne, or seed-borne, or a combination of seed-borne with one of the others.

Soil-borne pathogens can live in the soil for decades, so long crop rotations are needed to eliminate them. I had some friends who managed to eliminate clubroot, a fungal-like disease of brassicas, by stringently keeping all brassicas out of their garden for ten years. They ate a lot of chard, and found other land to grow brassicas on. *Fusarium oxysporum* and *Verticillium dahliae* are two soil-borne fungi. *Fusarium* survives a long time in soil without a host, and can also be seed-borne.

Foliar pathogens die in soil in the absence of host plant debris, so practice good sanitation (prompt incorporation of plants after the end of harvest, effective hot composting), good crop rotation, removal of volunteers, and avoidance of cull piles. Late blight (*Phytophthora infestans*) is a good example of this type of disease: it does not carry over in the soil, on dead plants, the seeds or the stakes. Cucurbit angular leaf spot (ALS) bacteria (*Pseudomonas syringae*) overwinter in diseased plant material and on the seed coat. Disease spreads when splashes of rain or irrigation water carry bacteria to healthy plants from infected material in the soil. People, insects, plant contact and the wind can also transfer ALS.

Lettuce mosaic virus is an example of a disease in which the seed is the main source of the pathogen and if seed infection is controlled, the disease is prevented. Other seed-borne pathogens may start life as a foliar-borne or a soil-borne pathogen. Infected seeds will produce infected plants even in clean soil. Pathogens can infect the seed via several routes:

- The parent plant can become infected by drawing soil pathogens through its roots up into the seed;
- Pathogenic spores can float in on the air (*Alternaria solani*, early blight of tomatoes; *Anthracnose* fungus that affects nightshades, watermelon and cucumber);
- Insects that feed on the plant can transfer the disease (striped cucumber beetles vector bacterial wilt, which is caused by *Erwinia tracheiphila*);
- Insects that pollinate the plant can bring infected pollen from diseased plants. Honeybees fly more than a mile (1.6 km).

Biointensive integrated pest management (IPM)

The basis of IPM in biointensive systems involves a pro-plant, rather than anti-problem, approach to creating healthy plants and soil. Vigorous plants are better able to withstand damage. The aim is to prevent disease organisms finding the crops, keep damage to a minimum, pay attention to what is happening, and control diseases by the least ecologically damaging method. This approach is site-specific and must be adapted to each crop and situation. The sequence of steps is basically: prevent problems; monitor for problems; then, if needed, treat problems.

1. Cultural controls

Focus on proactive (preventative) strategies, starting at the planning stages, to minimize opportunities for diseases to become a problem. Restore and maintain vibrant health in your crops. Increase biodiversity to provide greater stability. Specifically:

- apply good compost and maintain healthy, biologically active soils;
- use rotations to minimize disease and improve the environment for natural enemies of diseases;
- practice good sanitation of tools, plants and shoes;
- optimize nutrients, moisture and planting dates for crop vigor;
- enhance natural disease control strategies, such as by providing good airflow;
- use mulches to reduce splashback from soil to plants;
- use drip irrigation to reduce moisture on foliage;
- plant locally adapted, resistant varieties;
- use farmscaping to encourage beneficial insects;
- time plantings to avoid peak periods of certain diseases;
- practice good soil management (eg timing of tillage) to preserve maximum diversity of microorganisms.

2. Monitor crops for problems

It's a good practice to make a regular tour of your crops once a week to monitor growth and health, and also to keep a "weather eye" open whenever you are working in an important crop. When you notice a problem, you can take a picture, mentally or with a camera, and then consult a variety of resources to identify the problem. The Cooperative Extension Service, part of the USDA, is an educational program provided by the land-grant universities. It operates at a county or regional level within each state, and includes a Plant Diseases Diagnostic lab that is free for farmers. It can be a big help when you are really unsure, or when a serious disease would require drastic action and you want confirmation of your suspicions. There are some print publications with excellent photos, although growers are increasingly using the Internet to identify problems. A mnemonic I was taught goes: "Fungi are fuzzy, viruses are mottled, bacteria are slimy." Bacterial diseases are not as easily controlled by modifying the environment as fungal diseases are. Viral diseases can interact with each other and become hard to distinguish. Viruses are often spread by aphids, so reducing aphid populations is a key early step.

For some major diseases there are web-based sites offering regional forecasts. They combine scouting and reporting of outbreaks with meteorological forecasting of temperature and air movement to come up with a trajectory map and risk assessment. An example is the cucurbit downy mildew forecast from NCSU: ces.ncsu.edu /depts/pp/cucurbit

The mere presence of a disease does not automatically require spraying. The economic threshold (ET) or action level is the point at which losses from the disease warrant the time and money invested in applying control measures. The results of any preventive and control measures taken should also be monitored. Good record–keeping helps in adapting and fine-tuning biointensive IPM.

3. When control measures are needed

When this is the only way to deal with an outbreak, use the least toxic materials and the least ecologically disruptive methods. Design a series of complementary steps that work together, so that one tactic does not interfere with another.

Physical controls: Removing diseased plant parts, protecting vulnerable plants with rowcovers or sprayed kaolin barriers, mulching to isolate plant foliage from the soil, tool and shoe sanitation, soap washes for foliage, hot water or bleach seed treatments, and soil solarization to kill disease spores are all methods that reduce problems without adding any new substance into the mix.

Biological controls: Beneficial animals and insects are more common in insect pest reduction than in disease control, but the use of milk as a fungicide qualifies as a biological control. Plants in danger of developing powdery mildew can be sprayed weekly with a mix of one volume of milk with four volumes of water. When exposed to sunlight, this is effective against development of fungal diseases.

Microbial controls: Homemade microbial remedies employ liquids (simple watery extracts and fermented teas) made from compost. For a simple compost extract, mix one part mature compost with six parts water. Let it soak one week, then strain and dilute to the color of weak black tea. Fermented compost tea can deal with many maladies. If your strawberries are prone to *Botrytis*, apply fermented compost tea every two weeks, starting when the berries are still green. See ATTRA or the Soil Foodweb site for how to make fermented compost teas. Specific fungi, bacteria and viruses can be brought in to help control specific diseases.

Botanical controls: Using plant-based products to reduce disease. Neem oil, as well as being a pesticide, forms a barrier on foliage that prevents some fungal diseases from establishing. It degrades in UV light in four to eight days and must be reapplied if the disease organisms are still around. Like all broad-spectrum insecticides, neem can kill beneficials as well as pests, so caution is needed if it is used. Garlic can be used against fungal diseases: blend two whole bulbs of garlic in one quart (one liter) of water with a few drops of liquid soap. Strain and refrigerate. For prevention, dilute 1:10 with water before spraying; for control, use full strength. Kelp sprays are also used to generally boost the resistance of plants to pest, disease and weather-related problems.

Inorganic controls: Also known as biorational disease controls, these include bicarbonates, oils and soaps, and copper and sulfur products. Several of these need to be used with caution if the plants and the planet are to survive the treatment. Details below.

Reduction of diseases by maintaining healthy soils

Management of soil-borne diseases focuses on making the soil environment less welcoming to pathogens. This approach is the basis of Elaine Ingham's inspiring work on the Soil Foodweb. Soil-borne diseases are the result of a lack of biodiversity of soil microorganisms. Restoring beneficial organisms can make the soil disease-suppressive. Compost is a food source for beneficial microorganisms and encourages a more diverse soil population. Mature compost can induce resistance in plants by acting as a source of

weak pathogens. The plants develop a response to the pathogens and then are "vaccinated" against stronger pathogens. In microbially active soils there is fierce competition for mineral nutrients (from the bodies of other microorganisms), and pathogens may lose out. Compost can be made more disease-suppressant in three ways: ensuring the compost is properly matured before applying it; incorporating the compost in the soil several months before planting; or inoculating the compost with specific biocontrol agents, such as *Trichoderma* species or *Flavobacterium balustinium*. Beneficial organisms introduced via the compost have a better chance of thriving and being effective in the soil than they do if they are introduced directly to the soil, where there may not be suitable food or a congenial environment.

The suggestion to incorporate compost months ahead of planting might lead to the idea of providing compost for the preceding cover crop, then incorporating the cover crop. However, undigested organic matter in the soil may decompose anaerobically, leaving compounds that reduce the soil quality.

Bacteria may experience a dramatic bloom in the presence of large amounts of organic matter (OM), then die, leaving the OM less accessible (in the form of their bodies). Meanwhile many of the other soil microorganisms have died off during the fierce competition, creating an unbalanced soil population. When incorporating cover crops or other raw organic matter, do your best to ensure the decomposition is aerobic. For instance, avoid incorporation in cold wet weather.

Both compost and compost teas sprayed on plants have been found to activate disease resistance genes in plants. Once activated, the disease-prevention program is already running if a disease pathogen arrives, and the plant can withstand the attack. Without compost, the plants turn on the disease resistance gene when a pathogen arrives, but it takes time to develop a full defense and by then it may be too late. Damp composts (40%–50% moisture) are more disease-suppressant than drier composts, as they are more likely to be colonized by a broader spectrum of organisms. Composts with less than 34% moisture are more likely to be colonized by fungi unbalanced by bacteria and more likely to permit growth of *Pythium* fungi.

Attention to soil nutrient balance can also help prevent some diseases:

- Soils with sufficient calcium are less likely to harbor *Fusarium* or *Pythium* (some damping-off fungi);
- There is a decreased chance of *Fusarium* and *Verticillium* in soils with enough potassium;
- *Fusarium* is suppressed by high carbon to nitrogen ratio composts;
- Potato scab is less severe in acidic soils with pH below 5.2;
- Brassica clubroot is inhibited in neutral to slightly alkaline soils with pH 6.7–7.2;
- Silicon is an important nutrient with beneficial effects on disease suppression and stress tolerance;
- Wollastonite is a mineral source of calcium silicate that can delay the onset of powdery mildew in cucurbits.

Plant roots may be protected from disease organisms by symbiotic associations with mycorrhizal fungi, which physically cover roots with a fungal net, produce antibiotics and other compounds toxic to pathogens, and increase the supply of nutrients to the plant. Mycorrhizal fungi may al-ready be present in biologically active soils and are also available commercially. They may be used as a seed dressing, or added to growing plants, to improve yields and grow healthier plants. Symbiotic bacteria can also release nutrients to plant roots.

Crop rotation can reduce the chance of some diseases but will not prevent diseases that blow in on the wind, are brought in by flying insects, or arrive with infected seeds or dirty tools. It is helpful to group vegetable rotations by family rather than by the part we eat—grow all the brassicas in one patch, with the spinach and chard (non-brassica leafy greens) elsewhere. This makes it easier to rotate a patch out of crops susceptible to a particular disease. Crop rotation is most effective for soil-borne diseases that survive five or fewer years without host plants. Examples are root rots of beans and peas, blackleg and black rot of brassicas, parsnip root canker and sweet potato scurf.

Damping-off in seedlings can be reduced by inoculating the growing mix with 1%–4% fish emulsion, 28 days before sowing seeds. The fish emulsion produces a biological environment that suppresses the disease.

Hot water and bleach seed treatments

Hot water or bleach seed treatments are most easily done immediately before sowing the seed, as this removes any need to dry the seed.

Hot water treatment destroys seed-borne fungi and bacteria. First put the seed into a small cloth bag or tea strainer, and warm it. Then soak seed at 122°F–125°F (50°C–52°C) for twenty to twenty-five minutes. Exact details vary from one seed to another. Use a thermometer and nearby supplies of hot and cold water to regulate the temperature. At the end of the time plunge the hot seeds into cold water, drain and dry on paper towels. Fresh seed withstands heat treatment better than one- or two-year-old seed.

Bleach treatment kills pathogens on the seed surface only. For one pound (500 g) of seed, make one gallon (four liters) of a solution of commercial bleach in water, with 24 oz/gall (0.19 l/l). Put the seed in a cloth bag, submerse it in the solution, and agitate continuously for forty minutes. Rinse under running water for five minutes, then dry. This process can stimulate germination, so use the seed within three weeks of treatment, or risk reduced germination later.

Good detailed information on seed treatments can be found at the Cornell Extension website: vegetablemdonline.ppath.cornell.edu/NewsAr ticles/All_BactSeed.htm

Soil solarization

This is a very effective way to control soil-borne diseases and pests. It involves taking a patch of land out of production in the height of summer, and using solar heating through clear polyethylene to cook the top layers of soil. For additional benefit, first grow one of the mustard varieties known as "biofumigants," such as Ida Gold and Pacific Gold, then till it in. (If you are solarizing without mustards, simply till the soil.) Next, water thoroughly or, better still, run drip tape over the surface to keep the soil continuously damp. Then cover the soil tightly with clear plastic, burying the edges so that hot air does not escape. Keep the plastic in place for a minimum of six weeks, preferably eight or more. For more details on soil solarization, see the chapter on the Hoophouse in Summer.

Our first trial of solarization, to fight nematodes in our hoophouse. Ideally, the plastic should be stretched tight across the bed. It is important to seal the plastic to the soil by burying the edges. Credit: Kathryn Simmons.

Biorational disease controls

Several leaf spot diseases and powdery mildew (PM), caused by several species of fungus, can be controlled using inexpensive materials you likely already have, or that are cheap and easy to buy. To prevent the fungi building up resistance to any one remedy, spray only once a week and rotate remedies. A simple remedy uses baking soda: one teaspoon (5 ml) in one quart (one liter) of water, with a few drops of liquid soap as a spreader-sticker. Household soaps include ingredients that can burn plants; insecticidal soaps are safer. Having said this, I've used Murphy's Oil Soap at 3 Tbsp/gall (12 ml/l) sprayed in the early morning or evening as a pesticide with no plant-burning problems.

Baking soda is especially effective when used in rotation with horticultural oil, which, when sprayed on foliage, can make the surface inhospitable to fungi. Commercial horticultural oils may be based on soy, canola, neem oil or mint oil, with an emulsifier. You can make your own, using vegetable oil at 2½–3 Tablespoons per gallon (9–12 ml/l) water.

Hydrogen peroxide is a broad-spectrum fungicide/bactericide, with curative as well as preventative properties. Oxidate is one brand. While approved by OMRI, and having a zero pre-harvest interval, it is corrosive and needs care in use.

Sulfur has been used as a fungicide for more than two thousand years. It must be applied before the disease develops, as it works by preventing spores from germinating. Do not use sulfur if you have used an oil spray within the last month, as the combination can kill the plants. Also, do not use sulfur when temperatures are expected to exceed 80°F (27°C), to reduce the risk of plant damage. Cucurbits should never be treated with sulfur as their leaves are sensitive, and the cure may be worse than the disease!

Copper fungicides can kill fungi and bacteria. Care must be taken to prevent copper from damaging the host plant. Bordeaux mixture combines copper sulfate with lime (calcium hydroxide), which reduces the chance of the copper damaging the plants. Copper products accumulate in the soil, and are toxic to humans, fish and other animals, so many sustainable farmers avoid using them.

Some microbial controls:

- *Ampelomyces quisqualis* is a parastic fungus effective in preventing (not curing) powdery mildew. AQ10 is one brand;
- *Bacillis pumilis* tackles powdery mildew, downy mildew, *Cercospora*, *Phytophthora*, *Alternaria* and fire blight. Sonata is one brand;
- *Bacillus subtilis* targets *Pythium*, *Rhizoctonia*, *Phytophthora*, powdery mildew, *Sclerotium rolfsii* (southern blight) and *Fusarium*. Serenade and Companion are two brands;
- *Coniothyrium minitans* fungus tackles the two lettuce drop fungi, *Sclerotinia sclerotinia* and *Sclerotinia minor*. Contans and Intercept WG are two brands;
- *Gliocladium virens* fungus tackles *Rhizoctonia solani*, and *Pythium* species (both damping off diseases), *Sclerotium rolfsii* (southern blight) and *Fusarium* wilt. Soilgard is one brand;
- *Pseudomonas* bacterium targets *Botrytis*, *Penicillium*, *Erwinia*. Bio-Save and BlightBan are two brands;
- *Reynoutria sachalinensis*, a plant extract rather than a biofungicide, tackles powdery mildew, downy mildew, *Botrytis*, *Alternaria* (early blight), *Phytophthora* (late blight) and bacterial leaf spot. Regalia is one brand;
- *Streptomyces griseoviridis* and *S. lydicus* bacteria targets *Fusarium* spp, *Alternaria brassicola*, *Phomopsis*, *Botrytis* (gray mold), *Pythium* and *Phytophthora*. Actinovate (*lydicus*) and Mycostop (*griseoviridis*) are two brands;
- *Trichoderma harzianum* beneficial fungus tackles *Sclerotium rolfsii* (southern blight), *Pythium*, *Rhizoctonia*, *Fusarium*, *Sclerotinia*, *Thielaviopsis*, *Botrytis* and powdery mildew. Plantshield and Rootshield are two brands;
- *Trichoderma viride* fungus also controls *Sclerotium rolfsii* (southern blight).

Chapter 21

Sustainable Weed Management

Do you need a justification for having some weeds visible among your crops? Do you crave a system to help you get a grip on your to-do list, so you're not overwhelmed? Sustainable (or ecological) weed management does all this! In the earlier days of organic farming, maximum use was made of frequent cultivation to kill weeds. Now we know that too-frequent cultivation can cause soil erosion, and that each tilling or deep hoeing stirs air into the soil and speeds combustion of organic matter. Sustainable weed management is about effectiveness—including removing weeds at their most vulnerable stage, or at the last minute before the seedpods explode—and ignoring weeds while they are doing little damage. Work smarter, not harder!

A holistic approach to organic weed management

As always, strive to restore and maintain balance in the ecosystem. Develop strategies for preventing weeds and controlling the ones that pop up anyway. An obvious way to prevent weeds is to avoid adding new kinds of plants to any part of your fields. Remove the hitchhikers from your socks out on the driveway, not when you notice them as you squat to transplant onions! We use our driveway as a convenient place to "roadkill" particularly bad weeds by letting them die in the sun. Beware of Trojan plant swaps (nice plants in soil concealing nasty weed seeds)!

Weeds are not a monolithic enemy but a diverse cast of characters. Applying biological principles is not an attitude of war but more like jujitsu, using the weaknesses of the weeds to contribute to their downfall. This chapter aims to develop our understanding of weeds and the different types: annuals and perennials; stationary perennials and invasive perennials; cool-weather and warm-weather types; quick-maturing and slow-maturing types; and what Chuck Mohler (*Manage Weeds on Your Farm: A Guide to Ecological*

Strategies) refers to as "Big Bang" types versus "Dribblers" (see below).

One factor to consider is how vulnerable the crop is to damage from that weed at that time. Weeds that germinate at the same time as a vegetable crop usually do not really affect the crop's growth until they become large enough to compete for moisture and nutrients. If allowed to grow unchecked, however, these early weeds have the greatest potential for reducing crop yields. We need to cultivate or otherwise control weeds before this two-to-three-week grace period is over.

The critical period for weed control is the interval from the end of the initial grace period until the end of the minimum weed-free period, which is approximately the first third to one half of the crop's life. In other words, the most important time to weed a crop is from two weeks after sowing until the crop is halfway to being finished. For vigorous crops like tomato, squash and transplanted brassicas this is four to six weeks; less vigorous crops like onion or carrot need weed-free conditions for eight weeks or more. During that time it is essential to control weeds to prevent loss of yield.

Weeds that emerge later have less effect, and ones that emerge quite late in the crop cycle no longer affect the yield of that crop, although their seeds give a reason for removing them to improve future crops.

Know your weeds

Learn to identify the major weeds on your farm, and any minor ones that suggest trouble later. Observe and research. Start a Weed Log with a page for each weed. Add information about your quarry's likes and dislikes, habits and possible weak spots. Find out how long the seeds can remain viable under various conditions, and whether there are any dormancy requirements. Note when it emerges, how soon it forms viable seed (if an annual), when the roots are easiest and hardest to remove from the soil (if a perennial), what time of year it predominates, and which plots and which crops have the worst trouble with this weed. Monitor regularly throughout the year, every year. Look back over your records and see if anything you did or didn't do seems to have made the problem worse or better.

Next think about any vulnerable points in the weed's growth habit or life cycle, or responses to crops or weather that could provide opportunities for prevention or control. List some promising management options. Try them, record your results, then decide what to continue or what to try next.

Most weeds respond well to nutrients, especially nitrogen. If you give corn too much nitrogen, even as compost, its productivity will max out and the weeds will use the remaining nutrients. Sun-loving weeds like purslane are more likely to thrive among crops (like carrots and onions) that never cast much shade at any point of their growth. They won't be a problem for crops that rapidly form canopies that shade the ground.

A few weeds, such as giant ragweed, emerge only during a three-week interval, while others, such as pigweed and velvetleaf, can germinate during a two-month period, if temperatures are warm enough. Galinsoga seeds are short-lived and germinate only near the soil surface, but velvetleaf seeds can lie dormant for years deep in the subsoil and germinate whenever they are brought close to the surface. Clearly, different strategies work best with different weeds.

Red root pigweed is a "Big Bang" weed—the plant grows for a long time, and then all its seeds ripen at once as it starts dying. Most seeds come from a few large plants—pigweed monsters that mature late in summer can shed four hundred thousand seeds! Pulling the largest 10 percent of the weeds can reduce seed production by 90 percent or better. We used to ignore pigweed growing in our sweet corn, once it escaped two cultivations, believing anything that big must already have done damage. Now we pull while harvesting. Some pigweeds are as tall as the corn, but most don't have mature seed heads. Since starting this a few years ago, we have noticed a considerable drop-off in the number of pigweeds we have to deal with. This is different from the "seed dribblers" like galinsoga, which mature seed while still quite small plants, shed some, make some more, and can carry on for a long seed-shedding season.

Another useful piece of information is that a constant percentage of the seeds left from one year's shedding dies each year. For lambsquarters in cultivated soil it's 31 percent per year (only 8 percent in uncultivated soil). The number of seeds declines rapidly at first, but a few seeds persist for a long time. The percentage varies widely among species.

While seeds survive better deeper in the soil, they don't germinate better down there. Larger seeds can germinate at deeper levels than small seeds. If you are trying to bury seeds deep, use inversion tillage; don't rely on rotavating, as seeds somehow manage to stay near the surface with rotary tilling. Chuck Mohler has verified this with colored beads.

Most of the weeds in cultivated soils are annuals, but some of the worst ones are perennials, ei-

Red (or purple) dead nettle is one of our most common winter weeds. At least this one is feeding a honeybee. Credit: Kathryn Simmons.

ther stationary (taprooted) perennials like docks and dandelions or wandering/invasive perennials with tubers, rhizomes, or bulbs (Bermuda grass, quackgrass). Stationary perennials in their first year act like biennials—leaves, roots, but no flowers or seeds. In annually tilled areas, they get killed in year one and don't often establish. Wandering perennials are a more difficult problem, and understanding apical dominance is important in tackling them—see "Reducing the strength of perennial weed roots and rhizomes," below.

Preventing weeds from germinating

Any technique that helps fill the growing space and the growing season continuously with thriving crops will leave less opportunity for weeds to germinate. Options include:

- Keeping soil fertility high, helping the plant canopy to close quickly, out-shading and outcompeting weeds;

- Growing vigorous crops, providing the right soil conditions, rowcover or shadecloth if needed;
- Choosing locally adapted varieties with good disease and pest tolerance;
- Using crop rotations that switch between spring and summer crops, so that cultivation takes place at different times of year;
- Using drip irrigation (rather than overhead watering), leaving aisles relatively dry and inhospitable;
- Using mulches, especially for slow-growing or vertical crops such as garlic that have a poorer chance against broad-leaved weeds;
- Planting promptly after cultivation to give the crop the most advantage over sprouting weeds (I had not realized how much difference it can make to plant immediately after cultivation rather than two or three days later. With large areas, if we end up needing to wait before planting (with sweet corn, for example), we'll scuffle hoe along the planting rows just before planting, knowing that we can till weeds between the rows later);
- Using stick seeders or easy-plant jab planters to sow large crop seeds into a seedbed that has already had the weeds removed (by flaming or stale seedbed technique), avoiding bringing new weed seeds to the surface;
- Transplanting, rather than direct sowing, giving the crop a head start;
- Choosing close crop spacing, leaving less space for weeds;
- Practicing rapid multiple cropping (where one crop is planted immediately after the previous one);
- Interplanting/relay cropping (where one crop or cover crop is planted in the spaces between the standing crop before it is finished);
- Planting cover crops wherever there are no food crops (in some cases, cover crops exude compounds which inhibit germination of small-seeded weeds);
- Using no-till cover crops grown to maturity (heading/flowering) then rolled or mowed to create an in-situ mulch (plant vegetable crops into this mulch, reducing new weed seed brought close enough to the surface to sprout);
- Practicing reduced tillage/strip tillage, thus reducing the new weed seeds brought to the surface.

Reducing weed seeding

It's not always possible to prevent weeds germinating, so the next step with annual weeds is to remove them before they seed. It is possible to reduce weed seed banks to 5 percent of their original levels when weeds are not allowed to produce seeds for five consecutive years. Extreme energy might be needed to apply this across your whole farm, but in a hoophouse the improvement is greater for the amount of work. Techniques for controlling weeds before they drop seed include:

- Timely cultivation, when the soil is dry enough, and the weather warm and/or breezy. This can kill weeds by the million. Additionally, the loosened soil is not conducive to more seed germinating;
- Mowing, by cutting the flower stems, prevents weeds seeding. Use big mowers for big areas and nylon line mowers, scythes, or sickles in small or closely planted areas;

- Flaming (pre-emergence of the crop, or once the crop is big enough to take the heat);
- Grazing by cattle, chicken tractors or goslings, if you are not prevented by organic rules;
- Precision cultivation, whether that's tractor-mounted or a good hoe. Close in-row spacing, and wider row spacing to provide the same planting density, means that most of the weeds will be between the rows, easily wiped out by cultivation, rather than between plants in the row, where they are harder to remove. Plan your row spacing to fit your cultivating tools;
- Using post-emergence organic weed killers: corn gluten, vinegar. A fairly new area of research.

Reducing seed viability

Most weed emergence happens within two years of the seeds being shed. Not all seeds that are produced will ever get to germinate (I was very pleased to learn that seeds have many ways of not succeeding!). You can help reduce their chances:

- Farmscaping: plant habitat areas to encourage seed-eating birds, insects, earthworms, mice;
- "Grow" a healthy soil, with abundant soil microorganisms to keep the biological activity high. This can give vegetable crops which benefit from mycorrhizal fungi (legumes, alliums and nightshades) an advantage over some weeds (such as lambsquarters, pigweeds, smartweeds, nutsedges);
- Mow the crop immediately after harvest (to prevent more weed seed formation), then wait before tilling to allow time for seed predators to eat weed seeds already produced. Seeds lying on or near the soil sur-

face are more likely to deteriorate or become food for seed predators than buried seeds, so delaying tillage generally reduces the number of seeds added to the long-term seed bank (short term, they may germinate, only to be killed when you do till!);
- If they do not get eaten, dry out or rot, seeds on top of the soil are more likely to germinate than most buried seeds, but longer-lived seeds like pigweed, lambsquarters and velvetleaf, if buried, may remain viable and dormant in cold dark storage for years, and any tillage during that time can bring them back to the surface, where they rapidly germinate and grow. Avoid deep tillage if you struggle with these long-lived-seed types of weeds;
- On the other hand, small, short-lived seeds of weeds with no dormancy period, such as galinsoga, will almost all die within a year or two if they are buried a few inches. Till and mulch to bury short-lived weed seeds;
- Stale seedbed techniques draw down the "wealth" of the seed bank in the soil. Prepare beds ahead of time, water them, perhaps even rowcover them, to germinate a flush of weeds. Then remove the weeds by shallow cultivation or flaming;
- Soil solarization can be used in hot weather for particularly difficult weeds. This kills anything in the top layer of soil that is unable to move out of the way. See the Disease Management chapter for details. For weed control, solarization can also be combined with allelopathic cover crops, such as brassicas, rye, or sorghum–sudan, grown to full size, mowed green and tilled in, before the soil is covered with plastic.

When sowing carrots, sow a few "Indicator Beets" at the same time. As soon as the beets germinate, it's time to flame the carrot bed. Credit: Kathryn Simmons

Reducing the strength of perennial weed roots and rhizomes

Perennial weeds can often regrow from pieces of roots or rhizomes left in the soil after tillage. Rhizomes are modified stems, usually growing underground, that are capable of growing new shoots and roots from each node. When a rhizome grows a shoot, chemicals from that shoot prevent other nearby nodes from sending up shoots. This is called apical dominance (the node at the apex dominates the other nodes into not sprouting). These chemical messages get weaker over distance, so that on long rhizomes, after a certain length, the dominance effect is too weak and another node can grow a shoot. When rhizomes are cut into little pieces during tillage, the apical dominance is lost and each piece can grow a shoot. This is not necessarily as disastrous as it sounds, because these are small, weak plants with only a small source of nutrients. The danger is in walking away at this point and leaving them to grow. If we cultivate again before the new shoots have grown enough to be sending energy back to the roots, or manually pull out the pieces to dry out on the surface, the depleted pieces of root or rhizome may die.

Even simply removing top growth whenever the weeds reach the three- to four-leaf stage can be quite effective in further weakening invasive perennial weeds. This may need to be done several times at three- or four-week intervals to knock out a bad infestation. For quack grass the three-leaf stage is the time to act. It's more effective to wait each time until the new top growth has drawn down the plant's reserves (in the roots) before hoeing or pulling, than to go almost daily after every sprig. Thickly planting buckwheat or other smothering cover crops immediately after tilling helps put extra pressure on the weed, and can reduce tilling passes.

Putting it together: two examples

Galinsoga can be a troublesome weed for vegetable growers, as it can produce seeds in as few as thirty to forty days after emergence and its seed has no dormancy period—it sprouts after the next cultivation. Fortunately, the seeds are short-lived and have to be in the top quarter inch (6 mm) of the soil to germinate. Strategies include:

- Inversion tillage such as moldboard plowing (seeds will die off deep in the soil within a year or so);
- Mulching—the seeds will not germinate or be able to grow through the mulch, and will

be dead by next year. Be sure to rotate the mulched crops around the farm;

- No-till cover crops, with summer crops transplanted into the dying mulch;
- Stale seedbed techniques, including flaming.

Nutsedge is a "wandering perennial." Most of the seeds are sterile, so they are a distraction, not the main threat. Those seeds that are viable can persist for three to four years deep in the soil, but if they are near the surface they germinate or die within three years. The main sources of new plants are the small tubers (edible "chufa nuts") that form on the roots and the rhizomes that grow in late summer. Tubers need a month of cold dormancy before they start spring growth. Strategies include:

- Tilling in late spring and early summer after the tubers have sprouted, but before new tubers or daughter plants from rhizomes have had a chance to form. Repeated cultivations are needed to kill all the young plants, and this can work with a late spring/early summer crop that needs frequent cultivation, such as sweet corn;
- Growing food crops that finish in early summer and follow by deep tillage to disrupt tuber formation, which mostly occurs in late summer;
- Maximizing soil fertility and making dense plantings of crops that can overshadow and outcompete the nutsedge may work. Nutsedge is not very responsive to soil fertility, so it is possible to boost crop growth without boosting nutsedge growth;
- Sweet potatoes can suppress the growth of

Galinsoga grows quickly in warm weather and sheds seed in a short amount of time. Credit: Wren Vile.

nutsedge, by releasing growth-inhibiting substances (allelopathy);

- Allelopathic cover crops such as rye (overwinter) and sorghum-sudangrass (through the frost-free growing season) could be used if the area can be (needs to be?) taken out of crop production;
- Ducks or pigs are the action of last resort for organic growers. Land will need to be taken out of vegetable production (and put into meat production?) while the livestock roots out the nutsedge.

Knowing and understanding the particular weeds that are giving you the worst problems enables you to design an approach that includes removing weeds at the most important point in their life cycle, before they do their worst damage. While focusing on that you can relax and ignore weeds that are not doing much harm.

Chapter 22

Sustainable Pest Management

Biointensive integrated pest management

Integrated pest management (IPM) is a systematic approach to pest management that tailors strategies for each situation. The goal is to reduce pest damage of crops to an economically viable level, while minimizing the risks associated with use of chemical pesticides. This approach considers the effects of pesticides on not just the crops and workers but also the wider environment. Organic and sustainable farmers who use biointensive IPM go further in avoiding use of all synthetic pesticides. They (we!) focus on restoring and enhancing natural balance and resilience in agriculture to create healthy plants and soil that are better able to withstand attacks. Biointensive IPM includes controlling pests by physical and mechanical means (preventing the pest reaching the crop, or removing the pest from the crop), and incorporates environmental quality and food safety as both ecological and economic factors in farming decisions.

Conservation stewardship program and PAMS

The Natural Resources Conservation Service (NRCS) has a Conservation Practice Standard on IPM (Code 595), which as I write is under revision to "make it more relevant and applicable to organic and sustainable farmers, and to reflect the conservation benefits of organic and sustainable farming," quoting the instructions to reviewers of the new draft.

The NRCS uses the mnemonic PAMS (which I find especially easy to remember!) for the four steps of IPM: prevention, avoidance, monitoring and suppression. According to the likely severity of each particular pest, prevention alone, or prevention and avoidance, may be enough action. If the trouble is likely to be worse, monitoring is used to assess if and when the action level has been reached and suppression is needed.

Certified Organic producers are required by the National Organic Program (NOP) regulation to use multiple prevention and avoidance strategies before using any suppression strategy (and then to use only approved inputs).

Prevention

Start in your yearly plan to focus on preventative strategies and minimize opportunities for pests to get out of control. Preventative actions are also known as "cultural controls" and include such strategies as:

- Caring for the soil, in order to grow strong plants better able to fend off attack. Regularly add organic matter of different types to enhance the biological, physical and chemical properties of the soil. Use a diversity of cover crops, keeping the soil covered with growing plants or crop residues whenever possible;
- Minimizing soil compaction and hardpan layers. Adjusting pH to suit the crops being grown;
- Planting resistant, pest-tolerant, regionally adapted varieties when available;
- Growing strong plants by careful spacing, planting at the right time, providing optimum water and nutrients, managing weeds to reduce competition for nutrients, and improving airflow;
- Maintaining good sanitation of equipment and clothing;
- Promptly removing and destroying pest-infested plants;
- Ensuring that transplants are pest-free before planting out.

Encourage ladybugs and other insectivorous insects. Colorado potato beetle eggs and small larvae are eaten by ladybugs. Credit: Kathryn Simmons.

Avoidance

After growing the healthiest plants possible, the next stage is taking actions to reduce the chances of a specific pest taking over. These actions are also known as physical controls. All these methods reduce problems without adding any new compounds into the soil:

- Use good crop rotations so that it's hard for pests in the soil to find new host crops;
- Plant successions of the same crop at a distance from each other, and harvest the newer crop before the older one to minimize transfer of pests to the new planting. We do this with beans, cucumbers and squash/zucchini. Before we started buying the *Pediobius* wasp to deal with Mexican bean beetle, we used to plant more successions of beans, and flame the old plants when the pest count got too high;
- Remove pest habitat, by mowing, pruning, clearing debris, grazing with livestock, hand weeding, or flaming;

- Use organic mulches (e.g., spoiled hay or straw on potato plantings will greatly reduce the number of Colorado potato beetles that find the plants). Use reflective plastic mulches or those of a certain color to deter other pests (e.g., reflective mulches will deter aphids, thrips and whiteflies);
- Use deterrents such as garlic spray or hot pepper spray against rabbits; soap bars against deer;
- Grow transplants in a protected greenhouse and plant out once the starts are large enough to withstand most pests;
- Use rowcovers to physically exclude pests. White polypropylene rowcover is available in lightweight forms for insect protection in summer. We have found it very fragile and easily torn, so we often use thicker rowcover, even in summer. If you have a serious insect problem and don't mind single-use plastics, lightweight rowcover might suit you fine. Hungry rabbits, though, chew holes in rowcover, and are smart enough and big enough to chew out the squares between the reinforced grid of Agromesh type reinforced rowcover! For warm weather use, we usually drape the rowcover directly on the plants, without hoops. We don't need to keep the cover from touching the foliage as we do in winter. For our fall broccoli and cabbage, we support rowcover on ropes above the crop. The 83" (211-cm) wide rowcover will cover two rows at 34" (86 cm) spacing. Johnny's now sells squarish hoops with special loops at the top to hold twine to support rowcover.

We use rowcover against cucumber beetles when we transplant or direct sow cucurbits. We keep it on until the crop is flowering, then pack it away so that the flowers can be pollinated. By then the bigger plants are better able to resist the pests. Rowcover is one of the defenses against the brown marmorated stink bug, a devastating pest recently and accidentally introduced to the US.

We use rowcover in summer on brassica transplants to protect against harlequin bugs and to reduce transplant shock. We remove it after about three weeks. We also use rowcover on young eggplant transplants to protect against flea beetles, removing it when the plants flower and need pollination. We keep the seedlings covered in the coldframe while hardening off, then do our best to cover the transplants promptly after setting them out. Sometimes we hose the plants with a jet of water to dislodge any flea beetles already there, while someone else follows immediately behind to spread the rowcover. (By the way, brassica flea beetles and nightshade flea beetles are not the same species, and don't care for each other's food, so don't worry about setting out eggplant next to broccoli.) Continuing our list of physical controls:

- Use polyethylene sheeting. In the UK, carrot rust fly/carrot root fly is a big pest, and prior to the invention of rowcover, organic growers made three-feet- (one-meter-) high fencing around carrot beds, usually of framed panels of polyethylene sheeting. This works because the carrot rust fly flies in close to the ground. It also provides wind protection and a mini-greenhouse effect, but is more work than floating rowcover;
- Use netting. A few years ago I brought back from the UK a small roll of Enviromesh insect netting to try out. Something similar is available online or by mail order

from Purple Mountain Organics in Maryland: ProtekNet insect/pest netting. A roll 82" × 328' (2 × 100 m) sells at $339.99 in 2012. Made from UV resistant 80 g/m² knitted polyethylene, it transmits 83% of the light and has a lifespan of 7–9 years. The mesh size is 0.0394" × 0.0335" (1 × 0.85mm). They also sell a lighter weight ProtekNet Insect/ Pest Control Netting 25 g/m², with a lifespan of 3 seasons. The price in November 2011 is $432 for 6.9' × 820' (2 × 250 m), also available in smaller pieces. They recommend burying the edges in the soil. Farmtek has an insect screen fabric for doors, vents and hoophouses;

- Use collars around each transplant to repel cutworms and cabbage root fly;
- Use mixed plantings (intercropping) to maximize diversity of insects and other beneficial organisms;
- Use hedgerows, field borders and sections of planting beds for farmscaping to attract beneficial insects—plant 5 percent of the crop area to insectaries. We plant sweet alyssum in our spring broccoli patch to attract predators of aphids and caterpillars;
- Provide habitat for bats, insectivorous birds, spiders, birds of prey and rodent-eating ground predators (snakes, bobcats). We encourage black snakes to live in our root cellar so they eat any mice that come in. We have installed bat-boxes on barn walls near our garden to reduce the number of mosquitoes;
- Plant perimeter trap crops to lure pests from the food crop. Destroying the trap crop in a timely way, before the pests move on to the food crop, is an essential part of this strategy. Trap crops of zucchini may be the best im-

This baby black snake on our hoophouse pepper plants was probably eating hornworms. Credit: Pam Dawling.

mediate hope for catching the brown marmorated stink bug. Cleome attracts harlequin bugs, which can then be shaken off the tall flower heads into a bucket of soapy water;
- Physically remove pests, by handpicking, spraying with a strong water spray, flaming, vacuuming, or by using a leaf-blower to blow bugs into a collecting scoop;
- Solarize soil in the summer to kill soil-dwelling pests, as well as diseases. Cover cultivated damp soil with clear plastic for four to eight weeks in high summer, during which time the area is out of production.

Monitoring

Just because a pest appears does not mean you have to kill it. The "action level" (or economic threshold) will be the point at which the losses from the pest warrant the time, money and ecological disruption needed to apply control measures. The following monitoring practices are crucial in determining your action level:

- Make a habit of walking the fields regularly (weekly) to scout for trouble as well as for natural enemies of pests. Learn to recognize bugs of all kinds, and understand their lifecycles and enemies. Record what you find where and when. Take a hand lens. Use a red LED headlamp for nighttime scouting. Monitor the results of any measures previously taken;
- Scout vulnerable crops frequently (daily) at critical times;
- Photograph or capture insects you are unsure about and identify them from websites or books such as Whitney Cranshaw's *Garden Insects of North America*;
- Use phenology to predict the likely arrival of key pests. Phenology is the study of seasonal and yearly variations in climate by recording regular plant and animal life cycle events each year. A visible change can be used as an indicator that another species is also likely to be changing. For example, we expect brassica flea beetles when the redbud blooms, and bean beetle eggs to hatch when the foxgloves bloom;
- Consult weather forecasting as it helps determine when pest outbreaks are more likely. Mexican bean beetle eggs seem to prefer hatching in cloudy weather. Use any available online pest-specific forecasting sites. These sites combine scouting and reporting of outbreaks with weather forecasts of temperature and wind direction to make a forecast map;
- Record and use growing degree days (GDDs) to help predict pest outbreaks. The develop-

ment rate of many insects, from hatching to maturity, normally depends on temperature. By keeping temperature records and calculating accumulated heat, it is possible to predict when pest outbreaks may occur during a growing season, even though temperatures vary from year to year. GDDs are the accumulation of each day's growing degrees above the baseline level. I haven't started using this method yet, but Japanese beetles appear at a GDD level of 850, on a base of 50°F (10°C), and corn earworm somewhere between 150–490 GDD on a base of 54°F (12°C). Some weather forecast websites have a GDD calculator, such as weather.com/outdoors /agriculture/growing-degree-days;
- Set traps and lures (sticky traps and pheromone traps). Usually these will not significantly reduce the pest population, but are useful for indicating whether or not a certain pest is on your farm, so that other measures can begin as early as possible. Some, such as the pheromone lure for cucumber beetles, can make enough difference on some farms;
- Practice good record keeping. Record the pest name and population, date, which fields, which input (if any), relative success of action, date repeated, equipment used, cleanup method, comparison with methods used previously and overall evaluation. Keep learning, and applying what you learn. "Never make the same mistake two years running" is a mantra on our farm.

Suppression

When the established action level for a particular pest has been reached, and prevention and avoidance strategies have been exhausted, bio-

Carrot weevil damage is usually in the top third of the root and consists of relatively open tunnels. Carrot rust fly damage consists of smaller, closed tunnels, usually in the lower third. Credit: Jessie Doyle.

The culprit attacking our carrots. Our state entomologist said this looks more like the weevil larva than rust fly, as it has an obvious head capsule. Credit: Jessie Doyle.

logical, microbial, botanical, mineral (and for non-Organic growers, chemical) control measures can be used to reduce or eliminate that pest or its impact while minimizing environmental risks. Design a series of measures that work together, to ensure that one action does not mess up another. Choose the least toxic materials and the least ecologically disruptive methods. Precision equipment, calibrated sprayers and focused nozzles all help minimize the collateral impact on the environment. Be aware of sensitive areas, such as waterways, slopes, woodlands and habitat of vulnerable species. Be aware of better times to spray, e.g., for materials that kill honeybees, spray in the very early morning or at dusk when bees are not flying. For compounds that break down in sunlight, spray at dusk for maximum effectiveness and minimum amounts.

There are five types of control measures to choose from:

1. **Biological control** involves either introducing beneficial predators or parasites of the pest species, or working to boost populations of existing resident predators and parasites. Most years we buy the *Pediobius* wasp parasite of the Mexican bean beetle. Some years populations are not high enough for us to need the parasites.

2. **Microbial controls** refer to the use of fungi, bacteria, and viruses to kill pests. Bt (*Bacillus thuringienisis*), a species of bacteria, is the best-known microbial control. One strain kills caterpillars, another mosquitoes, another Colorado potato beetles. Spinosad is a fermentation product (no longer live material) of a soil microbe containing an enzyme that kills thrips, leaf miners, most caterpillars and some other insects. It is effective at one-eighth the rates recommended, if you don't need to annihilate every last bug. It breaks down in sunlight, so spray at dusk. We use it against Colorado potato beetles on our March-planted crop. The June-planted potatoes don't need any CPB control beyond the hay mulch we

Spiders are an asset in pest control. Zipper spiders are especially large and attractive. Credit: Wren Vile.

roll out to keep the soil cool and moist. We have also used Spinosad against the vegetable weevil larva when they took over our winter hoophouse turnips and greens. Among insecticides available in OMRI-approved forms, Spinosad, insecticidal soaps and pyrethrins are the three currently most promising for dealing with brown marmorated stink bugs.

3. **Botanical control** uses plant-based products for pest control. An example is neem oil, which degrades in UV light in four to eight days and must be reapplied if the organisms are still around. Rotenone and pyrethrin are two botanical insecticides effective against aphids, stink bugs, beetles, stinkbugs, squash bugs, cucumber beetles, vegetable weevils and other pests. They are generally strong broad-spectrum insecticides with short-term results.

4. **Inorganic (mineral) controls**, also known as biorational disease controls, make use of oils and soaps. Several of these need to be used with caution if the plants and the planet are to survive the treatment. Petroleum-based horticultural oils control soft-bodied insects such as aphids, spider mites and whiteflies (rather than hard-bodied insects such as beetles). Most are not approved for Organic use, sometimes because of secondary ingredients. As an alternative, vegetable oils and fish oils are available in some Organic formulations. Insecticidal soaps are made from potassium salts of fatty acids. If used too heavily they can damage foliage. They are effective on soft-bodied insects, and are one of our choices for dealing with early spring outbreaks of aphids in the hoophouse. After reducing the pest numbers, we collect hibernating ladybugs and bring them in for a feast. Sulfur is sometimes used against spider mites. Kaolin clay, besides being a protective layer on crops, can suppress feeding in striped cucumber beetles, flea beetles, thrips and grasshoppers.

5. **Synthetic/chemical control** is not allowed on USDA certified Organic, transitioning-to-Organic or Certified Naturally Grown farms, and cannot be used on biointensive or sustainable farms. That said, some other farmers may choose to use chemical controls when facing a plague and after trying and failing with sustainable methods.

Non-insect pests

We tend to think mostly of insects when we use the word "pests," but of course there are also microscopic pests and vertebrate pests. Vertebrate pests include mammals such as deer, rabbits, groundhogs, armadillos and voles, as well as birds such as pheasants, pigeons and crows. Physical barriers, strong deterrents such as vigilant dogs or cats, and hunting are the three main options for dealing with vertebrate pests.

Chapter 23

Manual Harvesting Techniques

Every year early in the season our crew watches an old video, *Efficient Harvesting Techniques*, made by CSA Works (Food Bank Farm) in Hadley, MA. It's out of print, sadly (Michael Docter says "We've retired from the film industry to do farming!"). Check sustainablemarketfarming.com for progress on making this work available again. The film has cult movie status with us, and old hands join in on key punch lines and heckle the workers on the screen. It's packed with good information. It took me many screenings before I no longer took away a new tip each time I watched. Much of what we have learned is from them.

We have also compiled our own tips for harvesting crews, which I describe below. Another good resource is the Roxbury Farm website, which includes a detailed Harvest Manual.

Planning

The seed for smooth efficient harvesting is sown before the season starts, when you plan your field layout. Locate the giant winter squash, pumpkins and watermelons near a road, or at least a path wide enough for a garden cart, if not a truck. Plant long rows with access gaps every hundred feet (thirty meters) or so, to reduce load carrying distance.

Our mixed system of beds and row crops means that small plantings of delicate crops are always located where we will pass by them often. (Permaculturists call this close-at-hand area Zone 1). It is poor planning to have your lettuce at the bottom of a long field, "over the horizon" from the top of the row.

Plant tall but closely spaced crops (corn, broccoli) in paired rows so that one person can pick two rows for the same travel distance. This works best for crops where each plant offers up only one piece for harvest at a time. With peppers, tomatoes, beans and the like, we find it more efficient to pick just one row at a time, or else the picker "loses their place" and has to scan over the plant again when switching from one row to another.

Don't harvest more than you need — aim to have the right amount coming in each week.

Organization

Before you set out, make a picking list and gather the containers and knives you will need. Watch the weather forecast — harvest more or sooner if a big rain is coming, and preemptively pick crops that should not be handled when the leaves are wet.

Plan who will pick each crop. It is often more efficient to have regular crew members specialize in certain crops. However, cross-training is also important, so pairing up a newbie with an experienced person makes the crew more resilient. Explain the signs of maturity: have people gently squeeze eggplant of different sizes, for instance. I recently learned from the High Mowing catalog that cabbage is fully mature when the biggest leaf on the head (not the loose outer leaves) curls back on itself. I had struggled for years to explain exactly how I was determining maturity — paleness of the center, firmness, size? I couldn't clearly say how I knew! The curling back leaf is a simple sign, easy to explain.

We generally divide our harvest into three categories: daily, three times a week and twice a week. How often to harvest each crop depends on its shelf life as well as your markets. Daily harvests include salads, okra, cucumbers, zucchini and summer squash, and also topping up supplies of potatoes and sweet potatoes from storage, according to the season.

For the every-other-day crops we have developed an ingenious phonetic system. On Monday, Wednesday and Friday we harvest crops beginning with a k/c/g sound; on Tuesday, Thursday and Saturday we harvest b and p crops. This works almost perfectly, with just a few crops we force into place: eggPlant not eGGplant! sPinach, senPosai! This system ensures we harvest some cooking greens each day: kale, collards, cabbage some days; broccoli, pak choy, spinach on the other days. Beans take over from peas as the spring heats up. Corn gets picked on the days we don't pick labor-intensive beans.

The twice-a-week crops include some with a reasonable shelf life, like peppers, and those that are at the end of their season when production has dropped off. At some times of year we have once-a-week harvests, such as winter squash in September and October.

Blue flags mark which bed of spinach or kale to pick next. We always pick a whole bed, to save confusion. We move the flag after harvesting, in a clockwise rotation (bird's eye view, not earthworm's!) round the area of raised beds. We harvest by the leaf, and aim to give each bed a week between pickings to regrow.

Potatoes, sweet potatoes, garlic and onions mature all at once and are bulk harvested as the major task for the day after the regular harvesting is done.

Sequence

Temperature considerations are a major factor in deciding the order of harvest. Leafy greens benefit from cooling as soon as possible after picking. Corn and lettuce need to get to the cooler quickly. Pick before the day heats up, and when the cart or truck has just enough room left, so that a trip to the cooler follows loading of the most perishable crops.

Other crops, including cucurbits (cucumber, squash, melons), tomatoes and beans, must wait until any dew has dried from the leaves before working with them, to minimize the spread of disease. On dewy mornings we'll often start with hoeing rather than harvesting.

Avoid harvesting in midday during extremely hot periods. Roots and fruits can be harvested in the afternoon and stored overnight for the next day's sales.

In winter, nitrate accumulation in leafy vegetables is a health issue, and a reason to delay harvest of greens until the sun has been up for at least four hours. See the Winter Hoophouse chapter for more on this.

Management

Every day, one of our most experienced people acts as crew honcho. The honcho makes the pick list, matches people with crops to pick, and then tries to ensure a smooth, happy and efficient harvest. She or he shows the workers good techniques, sets the pace, and creates a pleasant atmosphere.

Honchos keep an eye on who has picked where, when the cart is full, and if more buckets are needed. They make sure no one leaves tools or buckets of produce behind (yes, it has happened). They also make the decision on whether a plot is worth returning to next time, or if it should get tilled in, and pass this message on to the other honchos.

Teaching is an important part of the job. The honcho trains pickers as needed and, importantly, watches to see if any worker needs remedial training. Often I find it best to fine-tune my instructions, once people learn the basics, rather than giving a deluge of detail at the beginning. Some useful phrases if things go wrong are: "Oh dear, that's not what I meant" or "I see I didn't explain that clearly enough."

When we have large harvest far from the cooler we use the truck rather than multiple trips hauling a garden cart. Credit: Twin Oaks Community.

Get people to start as far from the road as possible, so when their buckets are full they'll be nearer the road. When picking long rows, we get people to mark the place they started or left off, for example to take a full bucket to the cart. The Sign of the Crossed Beans (two beans laid across one another on the ground) is one of our favorites. If we can't finish one of the big harvests on a particular day, the next time we start at the opposite end of the patch, hoping we can pick it all, but using a fail-safe system in case it is again too much.

Keep yields up by sowing often enough — how many times will you pick the same bean plants? It depends on the cost of labor and how much land you have. Pick newer plantings before older ones of the same crop to reduce the spread of disease. If you have enough after picking the younger sowing, stop picking the old planting and till it in.

Weed-free rows reduce time spent searching for the crop. I encourage fast workers to pull out a few weeds every time they harvest as their hands are right there. On the other hand, newer, slower people are told not to weed — we don't want the kale wilting in the buckets!

On Saturdays we harvest some crops smaller than usual to avoid Monday Monsters: zucchini and summer squash, of course, and also okra (we take Sundays off). If some labor-intensive crops (peas, beans, tomatoes, herbs) are getting away from you, you could consider allowing U-pick for them, or calling in the Society of St. Andrew, a volunteer group who glean to feed the hungry.

Weekly scouting will help you to plan when the harvest of each planting will start and end. Records from previous years can provide expected start and end dates.

Tools of the trade

We use Garden Way-type carts, sometimes a pickup truck, and lots of five-gallon (19-l) buckets. Customize your buckets — we have some with holes drilled in the bottom. We keep these and the non-holey ones separate, so we can easily find buckets to hold water or to drain washed produce in, as needed. Train your crew not to pour from a full wash bucket into a holey one (greens with grit sauce), but to lift the produce out of the gritty water and put it in the holey bucket. We also have "berry buckets" made from cut-down plastic one-gallon (four-liter) jugs, with rope loop handles. Long roped ones go round the neck to free up both hands for picking blueberries or cherry tomatoes. Short handled ones go over the wrist or are moved along the ground, for strawberries, snap peas and snow peas.

Some crops do better in shallow or ventilated crates rather than in deep airless buckets. We use sheets of plastic Bubble Wrap as cushions in the crates we use for slicing tomatoes. Open-topped backpacks are another idea for a harvest container. They are more ergonomic than carrying a heavy bucket on one side. Broccoli heads can be tossed over the shoulder — I imagine this takes a bit of practice.

Small tools are easily lost, so find some kind of portable container to take to the field. The Universal Container, the five-gallon (19-l) bucket, is often the answer. We use dishpans for trowels and three-pronged weeding claws, and have made a pouch for our pruners with the exact number of pockets for the number of pairs, making it less likely we will leave tools behind. Although special harvest knives can be bought, and we have some, we get most of our knives at yard sales and thrift stores. Great value for the money! Serrated bread knives are excellent tools for cutting cabbage and kohlrabi. We use pruners to harvest okra, eggplant and winter squash, and scissors for spinach leaves.

We have made size-cards to hang from the okra-harvesting pruners, after spending too much time debating size, during which I actually forgot that we harvest Cow Horn okra at 5" (12.5 cm) rather than 3" (7.5 cm)! To protect arms against spiny crops like okra, CSA Works suggests cutting the ends off of cheap crew socks. After years of grumbling about so-called "gardening gloves" full of seams and far too stiff for delicate vegetable farming, I started buying Gempler's latex-coated work gloves, sold by the dozen pairs.

Some kind of barn, packing shed or staging area makes a useful headquarters for communication, especially if supplied with a chalkboard, bulletin board, and perhaps a stand-up desk. For ideas on the layout of packing sheds see Healthy Farmers, Healthy Profits, and the Growing For Market articles in the Resources section.

As the day draws to a close, you might need a

pond, river or outdoor solar shower, followed by some shade trees and a hammock.

Harvesting whole heads

For leafy greens, we have three main ways of harvesting: heads, leaf-by-leaf, and "buzz-cutting."

We cut whole heads of mature crops such as cabbage, lettuce, heading oriental greens (napa Chinese cabbage, pak choy, Maruba Santoh, Yukina Savoy, tatsoi), and older spinach. We bend the head to one side, cut through the stem, trim off a few outer leaves if needed, and put the heads in a bucket or crate. With crops that can get bitter in warm weather (lettuce, broccoli), harvesters break off a piece and taste test for bitterness, rejecting culls in the field. In late spring and early summer, we cut celery bunches about an inch (2.5 cm) above the ground, which allows secondary bunches to develop later. Crops that wilt quickly are set like a bunch of flowers in the bucket, initially held on its side, making it easy to add an inch (2.5 cm) of water when the bucket is full. We do this with celery, chard, Russian kale, turnip greens, mustard greens and also leeks.

Harvesting leaf-by-leaf

For many leafy crops, we extend the harvest period by harvesting leaf-by-leaf. For adolescent lettuce, arugula and young spinach plants, we use scissors to cut individual leaves or several leaves in one snip. At the same time, we snip off old yellowed leaves. I like to cut the leaf stems near the main stem, allowing the new leaves to spread for maximum sunlight and preventing a "cage effect" from near-vertical stalks. With bigger leaves, such as chard, collards, kale, senposai and adolescent oriental greens, we snap leaves off. To avoid over-picking, we tell harvesters to aim to leave at least 60 percent of the leaf area, or, if it is easier to understand, count the leaves from the center outwards and leave "eight for later" — the eight youngest leaves in the center.

Buzz-cutting

A quicker variation on cutting a few leaves at a time is what we call "buzz cutting" — giving the plant a crew cut. When the plants are growing fast or we're short on time, we do this with mizuna, spinach, lettuce mix and arugula. Simply gather the leaves of the plant into a bunch and, using scissors or a sharp knife, cut above the growing point, so the plant can continue growing. During the winter, we often cut our hoophouse mizuna and tatsoi on one side of the plant only, leaving the remaining leaves to make the most of the limited light available.

Roots and alliums

How you harvest roots and alliums depends on your equipment and the scale of your farm. For example, with carrots, you can mow or tear off the tops, undercut with machinery, then lift. Or you can use the tops to help get the carrots out of ground, as we do, loosening them with a digging fork, then trim. When we harvest carrots for immediate use, we snap the tops off right at the junction of the foliage and the root. When we harvest for storage, we trim with scissors to leave a small length of greens. We don't do bunched carrots. If you do, you'll know to band them and wash, keeping the greens in good condition.

We plant our early scallions in bunches, which makes for quicker transplanting and easier harvesting: we loosen the soil, pull up the bunches, band them, trim the roots off, trim the tops with scissors, wash, then stand the scallions in small buckets with water. For tiny roots, like radishes and baby turnips, we harvest directly into a small bucket of water, after trimming tops and tails. This way the roots wash themselves during harvesting and very little cleaning needs to happen afterwards.

It is important to avoid bruising of alliums and fruits — it may be invisible at the time, but cause trouble later. Onions dropped a foot (30 cm) or more suffer interior bruising. Don't cram too much into one container: this causes bruising too, and reduces air circulation.

When picking peas, beans, tomatoes and peppers, we encourage harvesters to use two hands, and use their eyes to look ahead. Because we deliver our peas and beans directly to a kitchen, we trim the pods as we pick. Once we have gathered a handful of pods, we break off the stem end caps before dropping the pods in a bucket. Meantime, we look back to see if any have been missed. We pick and discard over-mature pods.

We pick our cucumbers by hand, encircling the cucumber with the palm and pushing against the stem with the thumb. This helps reduce scratches (to the cucumber, not to the harvester, unfortunately!). For zucchini and summer squash we use knives. We harvest cantaloupes at "full slip," when the melon leaves the vine after a little nudge. Some people separate ripe watermelon from the vine with their bare hands, but I prefer using pruners or scissors. It is a good idea to cushion vulnerable crops, especially if the ride to the packing shed is bumpy.

Corn is another crop we trim in the field, because we deliver within the half-hour direct to the cooler in our kitchen. We shuck the corn in the patch, throwing the husks right back on the ground. This provides immediate feedback for the harvester on whether the ear has reached full maturity. It also saves cooler space and minimizes soil nutrient removal. Our field trimming produces "kitchen ready" crops, saves time later, and reduces the number of times the crop is handled. Obviously, this technique won't work for most growers, but it is an example of tailoring harvesting techniques to fit the uniqueness of the market.

Field washing and food safety concerns

It is important to avoid bacterial contamination. Wounds and abrasions can lead the crop to pick up new bacteria from the environment. Crops can be punctured by the sharp edges of containers as well as the more obvious knives and fingernails.

At the washing station, crops may be washed by spraying down on a mesh table, or by dunking in troughs or buckets of clean water. Not all crops require washing: for some, such as basil and zucchini, it is a poor idea, as quality suffers.

Draining is important. Some crops can drain on a mesh table or in a holey bucket, suspended mesh bag or laundry basket. On a field trip with local growers, I learned Marlin Burkholder's method of filling a laundry basket with salad crops, hanging it from a tree, and twirling it round to spin out the water. No electricity! Barrel root washers and salad spinners have the draining stage built in.

After washing — or perhaps before — comes cooling (washing itself can also act to cool the crop). Make full use of all possibilities, such as damp burlap, high-percentage shadecloth, or the shade of trees, buildings, or a truck. In the shed, setting buckets or crates of produce on a concrete floor will keep them cooler than on tables, especially if the floor is splashed with water periodically.

Chapter 24

Winter Vegetable Storage (Without Refrigeration)

As more farmers aim to sustainably produce local food year-round, the storage of vegetables for sale over the winter becomes important. Understanding the needs of different crops can help reduce your electricity bill and carbon footprint, and maximize the amount of produce you can store for later sale. Only critical crops need refrigeration. Many others may be stored without electricity, perhaps in buildings that serve other uses at the height of the growing season.

A publication from Washington State University Extension, *Storing Vegetables and Fruits at Home*, is a good introduction to alternatives to refrigerated storage, using pits, clamps and root cellars. There is also good information in USDA *Agriculture Handbook 66*. Nancy and Mike Bubel's book *Root Cellaring* has a wealth of information.

They identify five different sets of storage requirements:

- Cold and Very Moist: 32°F–40°F (0°C–5°C), 90%–95% humidity—refrigerator conditions. For most roots, greens, leeks;
- Cold and Moist: 32°F–40°F (0°C–5°C), 80%–90% humidity—cold winter root cellar conditions. For many fruits, greens;
- Cool and Moist: 40°F–50°F (5°C–10°C), 85%–90% humidity—root cellar. For potatoes;
- Cool and Dry: 32°F–50°F (0°C–10°C), 60%–70% humidity—cooler basements and barns. For garlic and onions;
- Warm and Dry: 50°F–60°F (10°C–15°C), 60%–70% humidity—basements. For sweet potatoes and winter squash.

161

Reasonable expectations

Only store the good stuff. Damaged and poor quality vegetables will not store well. All crops for long-term storage should always be handled gently, to avoid bruising. For long-term storage, make sure crops are fully mature but not over-mature when you harvest. Potatoes need firm skins that don't rub off when you apply thumb pressure. Other crops, such as beets and sweet potatoes, don't have a "ripe" stage, but are ready when they reach the size you like. Very small or skinny roots don't store well. Expect that a small percentage of your crops will go bad in storage—it's not a sign of failure, just a reminder that life has limitations. Some warm-weather crops get chilling injury when refrigerated or stored in cold outbuildings, so don't expect a one-shed-fits-all solution to crop storage. For instance, cucumbers and eggplant, like peppers, are better above 45°F (7°C).

Some varieties store better than others, so advance planning will help achieve good results. Scrutinize the small print in the seed catalogs before your next seed order.

Cure, then store

Some vegetables need to cure before storage and the curing conditions are different from those needed for storage. Curing allows skins to harden and some of the starches to convert to sugars. Sweet potatoes need curing at high humidity (80%–95%) and high temperatures: 4–7 days at 85°F–90°F (29°C–32°C), 10 days at 80°F–85°F (27°C–29°C), or 20 days at 70°F (22°C), until the skins remain undamaged when you rub two of them together. Then they can be safely stored in warm, dry conditions for up to seven months. Sweet potatoes stored below 50°F (10°C) will be damaged beyond usefulness.

Regular (white, Irish) potatoes need curing in moist air (90% humidity) for one to two weeks at 60°F–75°F (15°C–24°C). Wounds in the skin will not heal below 50°F (10°C). We sort our potatoes after curing and find this usually reduces the chance of rot so that we don't need to sort again. Remember to keep white potatoes in the dark while curing as well as during storage.

Some people cure winter squash by setting the freshly harvested crop in a windrow with the stems facing the sun or the breeze. I've also read recommendations to cure at room temperature for ten days before storing. When ready, skin should be undamaged by the attempt to puncture it with a thumbnail. We never cure ours, and have seen no adverse effects. Perhaps in cooler climates it's important to cure at a warmer temperature, with good airflow, especially if storage will be in deep crates. Our storage is on open shelves, with reasonably good airflow. Certainly the squash sweetens with length of storage, and as with potatoes, the rate of deterioration drops right down after a few weeks. Different squash species take varying amounts of curing or storage time to sweeten fully.

Cabbage for storage benefits from having the cut stems exposed to the sun for an hour or so to dry, before bagging. We usually do this by setting the cut cabbage upside down on the outer leaves, and coming back later to bag up.

Onions and garlic need curing or drying with the tops and roots on, for about two weeks, before trimming and storing.

Peanuts will cure in two to three weeks indoors at ambient temperatures, or as little as six days with perfect conditions, including good airflow. It's important not to heat them higher than 85°F (29°C), as this causes skin slippage and kernel splitting. To find out when peanuts have finished curing, taste some. If they are still watery and crunchy like water chestnuts, they are not ready. For long-term storage, they need to be very well protected from mice and kept very dry.

Biocontrols to reduce storage losses

There has been some USDA ARS (Agricultural Research Service) research into biological disease control for stored fruit and vegetables. It takes three directions:

- Using biologicals such as Aspire yeast, Bio-Save bacteria (*Pseudomonas syringae*) or chitins to form a semi-permeable film over the surface of the roots and fruits;
- UV light to induce rot resistance. Primarily used for fruit;
- Natural fungicides derived from jasmine and peaches, which induce disease resistance in the crop itself.

Currently these methods are only used by large operations, but in the future, they may be useful to small growers.

Preparation for storage

Plan your storage sites, buy a thermometer for each site, and gather suitable containers. Wood crates are good for nostalgia and agritourism, but plastic is kinder on aging backs and less likely to harbor diseases. Containers for produce should not be set on bare concrete floors but rested on shelves, pallets or blocks of some kind. This helps improve ventilation and reduce condensation.

For root cellar storage, roots are traditionally packed in layers in damp sand, sawdust or wood ash in open boxes. The goal is to buffer temperature changes and prevent the roots drying out.

Extra water can be sprinkled on the media if the crops are getting too dry. For traditional storage without refrigeration, most roots store best unwashed (less wrinkling), though this can make them harder to clean later. A modern alternative is perforated plastic bags. We perforate our own and, if necessary, add more holes during storage if too much moisture is building up. If you are using perforated plastic bags for cold storage rather than boxes of sand, you can wash the roots first, as the bag reduces the water losses that lead to wrinkling. Plastic bags are less flexible about temperature conditions than open boxes, so if you might not be able to keep temperatures low enough, choose boxes rather than bags.

When you have choice in the matter, try to harvest roots from relatively dry soil, so they are less likely to grow mold. Trim leafy tops from root vegetables, leaving very short stems on beets and carrots. For non-refrigerated storage, unless using outdoor pits or clamps, several smaller containers of each crop are often a safer bet than one giant one, in case rot sets in.

The packing of your containers should allow for airflow, but you don't want the produce to shrivel up, so be observant. Celeriac needs more ventilation than beets, for instance. Sometimes night ventilation offers cooler, drier air than you can get in the daytime. Keeping root cellar temperatures within a narrow range takes human intervention, or sophisticated thermostats and vents. If needed, electric fans can be used to force air through a building.

Ethylene

Ethylene is generally associated with ripening, sprouting and rotting. Some crops produce ethylene gas while in storage—apples, cantaloupes and ripening tomatoes all produce higher than average amounts. Environmental stresses such as chilling, wounding and pathogen attack can all induce ethylene formation in damaged crops.

Some crops, including most cut greens, are not very sensitive to ethylene and so can be stored in the same space as ethylene-producing crops. Other vegetables, however, are very sensitive to the gas and will deteriorate in a high-ethylene environment. Potatoes will sprout, ripe fruits will go over the top, carrots lose their sweetness and become bitter. When storing ripe fruit, ventilate with fresh air frequently, maybe even daily, to reduce the rate of over-ripening and rotting. Ethylene also hastens the opening of flower buds and the senescence of open flowers.

Other produce can affect your crops—don't be tempted to set that bargain box of very ripe bananas you bought on the way home near anything you don't want to sprout or ripen further. Conversely, ripe bananas can be a cheap ethylene source if you *do* want to speed ripening.

Inground storage

Depending on the severity of your winter temperatures, some cold-hardy root crops (such as turnips, rutabagas, carrots, parsnips, Jerusalem artichokes and horseradish) and also leeks can be left in place in the ground, with about a foot (30 cm) of insulation (such as straw, dry leaves, chopped corn stalks, or wood shavings) added after the soil temperature drops to "refrigerator temperatures." Hooped rowcovers or polyethylene low tunnels can keep the worst of the weather off. There could be some losses to rodents, so experiment on a small scale the first winter to see what works for you. We have too many voles to do this with carrots or turnips on our farm, but

horseradish survives without protection, as do some winter-hardy leek varieties. Besides being used as a method for storage of hardy crops deep into winter, this can be a useful method of season extension into early winter for less hardy crops such as beets, celery and cabbage, which would not survive all-winter storage this way. Access to crops stored in the ground is limited in colder regions—plan to remove them all before the soil becomes frozen, or else wait for a thaw.

Clamps, pits and trenches

Cabbage, kohlrabi, turnips, rutabagas, carrots, parsnips, horseradish, Jerusalem artichokes, salsify and winter radishes (and any root vegetables that can survive cold temperatures) can be stored with no electricity use at all, by making temporary insulated outdoor storage mounds (clamps). The general idea is to mark out a circular or oval pad of soil, lay down some straw or other insulation, pile the roots up in a rounded cone or ridge shape, and cover them with straw and then with soil, making a drainage ditch round the pile. As a chimney for ventilation, leave a tuft of straw poking out the center or one side. Slap the soil in place to protect the straw and shed rainwater. For the backyarder, various roots can be mixed, or sections of the clamp can be for different crops. Those growing on a large scale would probably want a separate clamp for each crop. It is possible to open one end of a clamp or pit, remove some vegetables, then reseal it, although it takes some care for it to be successful. There is a balance to be found between the thermal buffering of one large clamp and the reduced risk of rot that numerous smaller clamps provide.

To store in pits or trenches dig a hole in the ground first, lining it with straw, lay in the vegeta-

We built a rodent-proof cage in our basement for winter squash. Credit: Pam Dawling.

bles, then cover with more straw and soil. To deter rodents, it is possible to bury large bins such as (clean) metal trash cans, layer the vegetables inside with straw, and cover the lid with a mound of more insulation and soil. Trenches can have sidewalls made with boards to extend the height. Another alternative is to bury insulated boxes in the ground inside a dirt-floored shed or breezeway. A new life for discarded chest freezers! Insulated boxes stored in unheated areas need six to eight inches (15–20 cm) of insulation on the bottom, sides and top.

Basement storage rooms and root cellars

Walk-in root cellars are larger versions of the "buried box" idea. Traditional root cellars are made by excavating a large hole near the house, lining it with block or stone-work walls, casting a well-supported and well-insulated concrete

roof, then covering the top with a big mound of soil. The doorway may have bulkhead doors or an entryway with additional doors. The more modern version is to construct an insulated cellar in the basement of a building such as a CSA distribution barn or your house. See the Bubels' book, the Washington State publication, and the next chapter for drawings and instructions on making these. Provide wide doorways with ground-level access if possible (roll that garden cart right in!). Good lighting and drainage are important, so you can see if everything is storing well, or hose the shelves and floor down if it isn't. Mouse-proofing is worth considering upfront.

Refrigeration and the CoolBot

For those needing some electrically boosted cooling, an alternative to full-on refrigeration is the CoolBot, as advertised in *Growing for Market*. It is a device that enables you to run a standard window air-conditioner at a lower temperature. Any well-insulated small room or building can be used.

Green tomatoes

In a category of their own are green tomatoes stored for ripening. They'll never taste as good as vine-ripened fruit, but a few weeks after your local frost date they'll be better than jet-lagged vegetables shipped in from afar. Fruit from late-planted vines still in good condition will ripen best. Only unblemished fruit that have passed a certain magic point will ripen. Unripenable fruits are matte, dark green (although unripenable Romas are paler, not darker) and often fuzzy. If they are no longer fuzzy, they will ripen if nothing goes wrong. Set them in trays such as those in boxes of apples. It helps prevent rot if the tomatoes don't

Table of Storage Conditions. The Summary column indicates the general conditions needed for each crop, and allocates each crop to one of four groups: A=Cold and Moist; B=Cool and Fairly Moist; C= Cool and Dry; D=Warm, Dry to Fairly Moist. (Compared to the Bubels' five groups, I don't distinguish between Very Moist and Moist in Cold storage conditions.) By providing storage spaces with these four types of conditions, these twenty-five crops can be stored.

Crop	°F	°C	% Humidity	Need for Ventilation	Summary	Storage Life in months	Notes
Apples	30–40	–1–4	90–95	Low	B: Cool and Fairly Moist	2–7	
Beets	33–40	1–4	95–100	Low	A: Cold and Moist	4–6	Perforated plastic bag in cellar. Short–term inground. Clamp. Temperatures above 45°F (7°C) cause sprouting.
Cabbage	32–34	0–1	90–95	Low	B: Cool and Fairly Moist	5–6	Net bag in cellar. Or dig up and hang upside down. Short-term inground. Clamp. Light reduces yellowing and weight loss.
Carrots	32–41	0–5	90–95	Medium	A: Cold and Moist	7–9	Perforated plastic bag in cellar. Inground. Clamp. Temperatures above 45°F (7°C) cause sprouting.
Celeriac	32–40	0–4	97–98	Medium	A: Cold and Moist	4–8	Well-perforated plastic bag in cellar. Short-term inground. Temperatures above 45°F (7°C) cause sprouting.
Celery	32	0	95–100	Medium	A: Cold and Moist	1–3	Can dig up and replant in buckets in cellar.
Chinese cabbage	32–41	0–5	99–100	Medium	A: Cold and Moist	1–3	Perforated plastic bag, or dig up and replant in buckets in cellar
Daikon radish	32–34	0–1	95–100	Low	A: Cold and Moist	4	Perforated plastic bag in cellar. Clamp. Temperatures above 45°F (7°C) cause sprouting.
Garlic	32–38 or 65–86	0–3 or 18–30	60–70	Low above 50°F (10°C)	C: Cool and Dry. Or D: Warm and Dry	6–7 (1–3 months at warm temps)	Net bag. Keep warm or keep cold.. Never 40°F–56°F (4°C –13°C), or they will sprout. Never warm after cold either.
Ginger root	54–57	12–14	85–90	Low	D: Warm and Fairly Moist	4–6	If stored with other warm-storage crops, will need extra humidifying.
Horseradish	30–32	–1–0	90–100	Low	A: Cold and Moist	10–12	Perforated plastic bag or plastic bucket, no lid, in cellar. Inground. Clamp. Temperatures above 45°F (7°C) cause sprouting.
Jerusalem artichoke	32–34	0–2	90–95	Low	A: Cold and (Fairly) Moist	4–10	Paper or plastic bag in cellar. Inground. Clamp.
Kohlrabi	32	0	95–100	Medium	A: Cold and Moist	2–3	Perforated plastic bag in cellar. Clamp.
Leeks	32	0	95–100	Medium	A: Cold and Moist	3	Perforated plastic bag, or plastic bucket with small amount of water, or dig up and replant in buckets in cellar. Inground.
Onions (bulbs)	32–40 or 60–90	0–4 or 16–32	60–70	Low	C:Cool and Dry. Or D: Warm and Dry	1–8	Net bag. Keep warm or keep cold.. Never 45°F–55°F (7°C –13°C), or they will sprout. Never warm after cold either.
Parsnips	32–34	0–1	95–100	Medium	A: Cold and Moist	6	Perforated plastic bag in cellar. Inground. Clamp. Temperatures above 45°F (7°C) cause sprouting.

Crop	°F	°C	% Humidity	Need for Ventilation	Summary	Storage Life in months	Notes
Pears	29–31	-1.5 to -0.5	90–95	Low	B: Cool and Fairly Moist	2–8	Open trays. Temperature requirements vary with variety, and are critical.
Potatoes	40–45	4–7	90–95	Low	B: Cool and Fairly Moist	5–10	Plastic or wood crates, paper bags in cellar. Protect from light. Below 40°F (4°C) the flavor deteriorates.
Radish, winter	32	0	95–100	Low	A: Cold and Moist	4	Perforated plastic bag in callar. Clamp. Temperatures above 45°F (7°C) cause sprouting.
Rutabagas	32	0	95–100	Low	A: Cold and Moist	4–6	Perforated plastic bag in cellar. Inground. Clamp. Wax unnecessary. Temperatures above 45°F (7°C) cause sprouting.
Salsify	32	0	95–98	Low	A: Cold and Moist	2–4	Can dig up and replant in buckets in cellar. Clamp. Temperatures above 45°F (7°C) cause sprouting.
Squash, winter	50–60	10–15	50–75	Med	D: Warm and Dry	2–12	Storage life varies widely with variety.
Sweet potatoes	55–65	13–18	70–80	Low after curing	D: Warm but not too Dry	4–10	Never below 50°F (10°C). Ideal temperature 55°F–59°F (13°C–15°C). Temps above 65°F (18°C) hasten sprouting.
Tomatoes, ripening green	55–70	13–21	75–85	Low	D: Warm but not too Dry	1–3	Egg trays, apple trays
Turnips	32	0	90–95	Low	A: Cold and Moist	4–5	Perforated plastic bag in cellar. Inground. Clamp. Temperatures above 45°F (7°C) cause sprouting.

touch each other. Stack the trays in a warm place. Light is not necessary. Sort through once a week, removing the ripe ones and the composting ones. It probably helps ripening to mix in a half-ripe fruit in each tray (because of the ethylene). At 65°F–70°F (18°C–21°C) they will ripen in two to three weeks; at 55°F (13°C) they will take about a month. Temperatures cooler than this will not produce reliably good results.

"Unstoring" vegetables

Out of sight should not mean out of mind! Keep a record of what you have stored where. Watch the thermometer readings and adjust temperature and humidity as needed. We have a natural tendency to "check out" once the daily vigilance of the busy season is over for another year. Look over your stored crops regularly, removing damaged specimens. Monitor the rate of use and notice if you'd benefit from more or less of each crop next winter.

After long storage, some vegetables look less than delicious, and benefit from a bit of attention before the public gets them. Cabbages can have the outer leaves removed, and can then be greened up by exposing to light for a week at 50°F (10°C). This isn't just cosmetic—the vitamin C content increases ten-fold. Carrots can lose their sweetness over time, unless frequently exposed to fresh air, by ventilating well.

Chapter 25

Root Cellars

Storage is not a one-solution-fits-all project. Produce in storage might need to be frozen, refrigerated, or kept cool, warm, moist or dry. As a sustainable alternative to refrigerated storage for crops needing cool, damp conditions, consider the traditional root cellar. Here I will address the requirements for the group classified in Nancy and Mike Bubel's book *Root Cellaring* as "cold" and "moist." See the previous chapter for details on their classification.

Since their book was published, more evidence suggests that potatoes are better stored at 40°F–50°F (5°C–10°C) than 32°F–40°F (0°C–5°C). Below 40°F (5°C) the starches start to convert to sugars, giving the potatoes an unpleasant flavor and causing them to blacken if fried. Potatoes which have become sweet can be brought back to normal flavor by holding them at about 70°F (21°C) for a week or two before eating. This chapter will look at root cellars, primarily for potatoes but also for apples, cabbage, or root vegetables.

Just be careful what you mix, because ethylene from the apples, for example, will cause potatoes to sprout! With a good inground root cellar, potatoes can be stored for five to eight months.

Crop maturity

The life cycle of potato plants comes to a natural end when the leaves die, after which no further growth can be induced in them. Once the tops die, the potato skins start to toughen up. If you are growing storage potatoes and are impatient for the end to come, you can mow off the tops or flame them, to start the skin-thickening process, which takes around two weeks. The potatoes are ready when you can rub two together without any obvious damage to the skins. "Early" potatoes are dug before they are fully mature, and their skins will not thicken enough for storage. Yields will be lower but the advantages of an early crop may outweigh these disadvantages. See the Potato chapter for more details.

Curing and storing potatoes

After harvest, potatoes need to be cured for two weeks at 60°F–75°F (15°C–25°C) in the dark with moist air and good ventilation. This allows the skins to further toughen up, cut surfaces to heal over, and some of the sugars to convert to the more storable starches. Wounds do not heal below 50°F (10°C). Like most root vegetables, potatoes store better if they are not washed before storage. The newly harvested potato is still respiring and needs fresh air. Lack of sufficient oxygen during curing results in black heart, a condition where the tubers develop nasty black lumps of dead tissue in the centers, so be sure to provide good ventilation during curing. After two weeks, the temperature should be reduced to the winter storage temperature of around 40°F (5°C). (In summer, potatoes can be stored at 60°F–75°F/15°C–25°C for up to six weeks—at higher temperatures they will sprout.) The two-week point is a good time to sort through the potatoes and remove any rotting ones. We find that if we do one sorting after two weeks, we don't need to check them any more after that—pretty much anything that was going to rot has already done so.

Store in a moist, completely dark cellar, ideally at 40°F (5°C), up to 50°F (10°C). Ventilate as needed during times of cool temperatures, to maintain the cellar in the ideal range. We store our potatoes in open plastic crates, which do not rot or hold fungal spores, and allow ventilation.

Potato sprouting conditions

The rate of growth of sprouts on potatoes is directly related to the degree-days above 40°F (5°C), so storing potatoes above 40°F (5°C) (for best flavor), clearly runs the risk that at some point they will start sprouting. If eating potatoes do start

The shaded entrance to our root cellar. Credit: McCune Porter.

to develop sprouts, it's a good idea to rub off the sprouts as soon as possible, because the sprouting process affects the flavor, making them sweet in the same way that low temperatures do.

If you are storing potatoes for seed, and have good control over the temperature of your cellar, you can manipulate the conditions somewhat to help get the best yields. The "physiological age" of the seed tubers affects both the early yield and the final yield. See the Potato chapter for more on this.

Other vegetables in cellars

Most other root crops can also be stored in a root cellar. Some people pack the unwashed vegetables in boxes of sand, wood ash, sawdust or wood chips. Perforated plastic bags are a modern alternative. Whole pepper plants with unripe peppers can be hung upside down in the cellar to ripen, or simply to store. Headed greens like cabbage can also be hung upside down, or be replanted side by side in boxes or tubs of soil. Celery and leeks can also be stored by replanting in the same way. I'm more of a fan of choosing hardy varieties of leeks and leaving them out in the garden for the winter, but people with really deep snow or very cold winters might laugh ruefully at that suggestion.

Our root cellar renovation

Our root cellar is a simple inground cinder-block room. It received no maintenance worth speaking of for at least fifteen years, when we gave it some attention to reduce the dampness. We thought the roof was leaking, so we uncovered it, removing the pile of soil on top using a tractor backhoe. To our surprise, the concrete pad of the roof was in fine condition. We returned the five-dollar, three-pound (1.4-kg) can of Thorocrete sealant we had bought to the store. We took advantage of our inspection, though, and added two inches (5 cm) of Styrofoam blueboard insulation and an EPDM sheet (ethylene propylene diene monomer rubber, sold for pond liners and flat roofs) on top of the roof before replacing the soil with a tractor bucket and some shoveling. Yes, I'm aware that I said root cellars were sustainable and yet I just mentioned petroleum products! We already had the EPDM stuff in the barn. The Styrofoam added lots of insulation without more bulk. A 4' × 8' (1.2 × 2.4 m) sheet cost about thirty dollars and we used most of five sheets—160 ft² (14.8 m²) for a cellar that's 115 ft² (10.6 m²) inside. A deeper layer of soil would be an option if you were building from scratch; as ours is an old cel-

lar, we didn't want to add more weight on the roof. After replacing the soil, we sowed crown vetch, a very hardy low-maintenance ground cover, with a nurse crop of winter rye. It was early summer, so we knew the winter rye would not head up—we just wanted something that would germinate easily and provide some shade till the vetch got going. We vowed not to let saplings grow on the roof again. Our cellar does not have an air lock or bulkhead doors, which I'm sure would help keep temperatures stable. One day!

Inside the cellar

The inside walls of our cellar are painted with Drylok sealant, a cement-based waterproofing product. It's a good idea to stack your crates a little away from the walls to let air and moisture circulate. If you fit shelves in your cellar, leave an air gap between the back of the shelf and the wall. Air produces condensation as it cools down—wet walls do not necessarily mean you have leaks, as we found out. We set pallets on the concrete floor, so the crates are off the ground, for better air circulation. Originally we had wooden pallets, but we have gradually been collecting plastic ones that come our way, as these have the advantage of not rotting. Likewise we use plastic crates for our potatoes. Our crates are ventilated, cubic, about 12" (30 cm) on a side, 10" (25.4 cm) deep, and hold about 30 lb (14 kg) of potatoes. They stack well, six high, on the pallets. Our 10' × 11.5' (3 × 3.5 m) cellar will hold 360 crates with an ample central path. That's 10,800 pounds (4,900 kg), or around five tons (tonnes).

Root cellar construction

The Bubels' book contains great designs and instructions for excavated root cellars, including a two-room version for keeping different crops at different temperatures and humidities. Excavated root cellars are not the only possibility, but the advantage is the earth insulates the cellar. Because soil is heavy, inground cellars must be strongly built as well as being well drained, so that water does not pool, freeze in winter and crack the walls. The book has details about laying out the site, working with concrete blocks, mixing concrete, making a supported roof, and drainage. What follows is a basic outline.

Plan the size of your concrete block cellar in multiples of 8" (20 cm)—concrete blocks are 16" (40 cm) with mortar. A site in a hill is easiest, but excavating a level site is also possible—the cellar will just have a set of steps down to the door. Don't make the hole much bigger than the finished size, because all the extra space will have to be backfilled. Make the two-feet (60-cm) -deep foundation 8"–16" (20–40 cm) below the frost line. Some people like a dirt floor, but we have a concrete slab with a floor drain. You can order ready-mix concrete for the foundations and floor. Buy plenty and have some forms set up for casting pavers or something useful in case there is extra left over. A week after the floor has set, build the walls, following a string-and-stakes layout. Make the walls 6.5'–8' (2–2.4 m) tall and install a door frame and any ventilation ducts you want in the walls. Once the walls are fully set, construct the support framework for the roof, using 4"×4" (10×10 cm) posts holding 2"×6" (5×15 cm) beams. On the beams are 2"×4" (5×10 cm) joists at 16" (40 cm) centers. On these, lay ¾" (2-cm) plywood covered in 6-mil poly (a use for your old hoophouse plastic?). Install any exhaust pipes you plan to have in the roof. Make a perimeter form of 2"×6" (5×15 cm) lumber around the edge of the roof,

Inside our root cellar, we set plastic crates of potatoes on plastic pallets. Credit: McCune Porter

supported every four feet (1.2 m). Then pour another ready-mix load of concrete, 6" (15 cm) thick, with steel reinforcing in the lower half. Use rebar, old wire fencing, or bedsprings.

Backfill around the cellar until you reach 2'–3' (60–90 cm) below the final height. Put a layer of gravel around the outside of the cellar, across the back, then around the corners and sloping down to end in daylight at the front (door side) of the cellar. Install 4" (10-cm) perforated plastic drainpipe on the gravel. Surround the drain pipe with gravel and backfill the rest of the way. Add 2" (5-cm) insulation to the roof, then 6-mil poly or heavier plastic. Top this with at least 30" (76 cm) of soil, up to 4' (1.2 m). Sow the ground cover plants of your choice, avoiding anything with big taproots.

Add an insulated door and, as needed, steps; and possibly an airlock or bulkhead door.

Indoor root cellars

These storage areas should be in the coldest part of the basement, far from all heat sources. The exterior walls do not need insulation, but the ceiling and inside partition walls need 3½" (9-cm) -thick insulation, with a vapor barrier. (Another use for retired hoophouse plastic.) An outside window is an advantage, or you can put a duct through the wall, so that temperature and ventilation can be controlled, either by thermostats opening the vents or by hand. To increase the humidity, pour water on the floor or drape wet burlap sacks over the crates of produce.

Ventilation

The Bubels recommend a 4" (10-cm) air intake at floor level, with a 2.5"–4" (6–10 cm) exhaust at roof level. Eliot Coleman's *Four Season Harvest* has good information about controlling the temperature in root cellars, using a single hole in the wall, high up. The lower half of the hole has an inside pipe going down to floor level, and the upper half has an outside pipe going up like a chimney above the roof line. Natural convection will cause the air to flow. Cover the ducts with hardware cloth to keep out rodents. Close the ducts when the outdoor temperature is higher than the desired storage temperature, and open them when you need to cool the cellar and the outdoor conditions are suitable, which is more likely at night.

Root cellar ecosystem

Our cellar does not yet have ventilation ducts. We simply leave the door open at night when we want to cool it down, or in the daytime in winter. We just choose a time when the forecast temperature is in the range we're aiming for. Yes, mice do come in through the open door! We encourage black snakes to live in our cellar, to keep the mice under control. (How do we encourage snakes? I mean we don't drive them out, and if we need to, we move one or two in there.) This can be a bit unnerving, as the cellar is dark. (We chose not to have a light, as leaving it on by accident could cause a lot of potato greening before we noticed our mistake.) We have developed a special door-opening technique so we can coexist with the snakes, who like to hang out on the top of the doorframe. We unlatch the door, open it a crack, then bang it closed before opening it fully. Any resting snakes have by then dropped to the floor where we can see them and avoid them. (No snakes have been hurt in this process!) People who don't like snakes will be really motivated to fit a rodent-proof vent system!

PART II

CROPS

Legumes
Brassicas
Other Greens
Roots
Cucurbits
Alliums
Nightshades
The Others
Seed Crop Production

Chapter 26

Green Beans

There are several species of beans, which do not crossbreed. Here I will be talking about *Phaseolus vulgaris*, generally called snap beans or simply green beans. By choosing varieties appropriate for your farm, and giving thought to planting dates, it's possible to get high yields and keep them coming till frost. Green beans for shelling (harvesting when dry or eating fresh out of the pod when the beans are plump) are an easy crop in the right climate.

Varieties

We like Provider and Bush Blue Lake snap beans for productivity and flavor. We sow reliable quick-maturing Provider (50 days) for the early and late crops and high-yielding tasty Bush Blue Lake (57 days) during the main season. Contender may have more flavor than Provider but is less productive. Jade is a delicious, very tender bean that grows well in hot weather but will be a dismal failure in cold conditions.

Those who like flat beans often choose Romano II as a reliable producer of tasty beans whether it is hot or cool, wet or dry. If you have a market for yellow (wax) beans, you can grow these for visual diversity—the flavor is not much different from green beans. Purple-podded beans look attractive while raw, although the color fades on cooking.

We have given up on pole beans as we don't like putting up trellises. If the advantage of standing to harvest beats the disadvantage of putting up a trellis, you'll prefer them over bush varieties. Pole beans take a few days longer to mature but can then be picked for a longer period. Half-runner beans can be grown with or without trellises and are capable of high yields. We haven't grown any we really liked.

Crop requirements

A soil temperature of 77°F (25°C) is best for germination, although a temperature of 55°F–60°F (13°C–16°C) *and rising* works for dark-seeded varieties. Air temperatures of 65°F–85°F (18°C–29°C) are best for growth. Like all legumes, beans produce nitrogen in their root nodules, although this may not peak until after the beans are harvested. To grow strong plants you can fertilize before sowing at about 120 lbs N/acre (134 kg/ha), then use the bean-produced fertility for the following crop. If your soil is already very fertile you may not need to fertilize before sowing; excess nitrogen will produce lots of leaves but delay flowering. 80–85 percent of the nitrogen produced ends up in the bean tops, so if possible turn the tops under before planting the next crop rather then moving them to the compost pile.

The nitrogen is produced by nitrogen-fixing bacteria in root nodules, so if you are growing beans on land that has not grown legumes before, or it's spring and you don't want to rely on the existing bacteria waking up, adding some powdered inoculant is a good idea. Many growers add this at each sowing—it is cheap and easy, especially if the beans have been soaked, as it sticks better on wet seed. Just before sowing, drain the soaked seed, scatter some of the black powder on the beans as if you were adding pepper to your dinner and stir gently. Each bean needs only a few specks of the inoculant to get started. Contrary to any rural myths, inoculant does not speed germination.

Beans tolerate a wide pH range, and like plenty of sun and well-drained soil. They definitely don't thrive if flooded! An open site with good air drainage will help minimize mold problems and other leaf diseases. Crops take 50–62 days from sowing to first harvest.

Sowing

We make our first sowing ten days before the last spring frost date, and cover the bed with rowcover. Our last sowing is in early August, with the beans maturing (unless we are very unlucky) before our first frost in mid-October. When frost threatens this planting, we cover with rowcover on the cold nights. Usually this lets us get several more pickings before any serious cold weather arrives. Beans are mostly self-pollinating, so rowcover does not stop pollination, and beans that have already set can grow to full size more quickly.

Dark-seeded varieties are more resistant to rot in cold soils, so use these at least for the first spring sowing. Ensure the soil has enough water before you sow, and if possible avoid irrigation for two weeks after sowing. This reduces the chances of chilling injury to the seed from the cold water, soil crusting, and even the seed rotting. To speed germination if the soil or weather is at all dry, soak the seed overnight (up to eight hours) in tepid water. In cool, wet conditions, it may be better not to soak, as the seed could then rot. If you have to postpone the sowing after soaking the beans, rinse them twice a day and drain. Plant within three or four days; if you wait any longer the rootlets will be too long and fragile.

Sow one inch (2.5 cm) deep, a little shallower in spring for warmth, a little deeper in hot weather for moisture. Place the seeds two to three inches (5–7 cm) apart. We use the wider spacing for new seed and the closer planting for one-year-old seed. Two-year-old beans often have a germination rate of only 50 percent and are not worthwhile.

Beans can be planted two rows to a bed or on the flat in single rows with enough space to accommodate the pickers. Oregon State University has a publication for commercial bean growers (not organic); it says that the best spacing for optimum yield for bush beans is 36 in^2 (232 cm^2) per plant, for example 2" in-row × 18" between rows (5 × 46 cm). This is less space than recommended for areas prone to fungal diseases. The advantage of closer spacings is that the plants are more upright, with the beans higher in the canopy.

Cultivation

Do not cultivate or harvest while the leaves are wet, since *Anthracnose*, bacterial blight and rust disease are more likely to spread under these conditions. Hoe or machine cultivate while the bean plants are still small, and once they grow taller and bush out, few weeds will cause trouble. We have tried sowing beans through biodegradable plastic mulch, poking holes with rods and popping the seeds in. This crop stayed very clean and withstood a drought caused by a crew member mistakenly pulling the drip tape out too soon. Sowing took a long time, and anyone tempted to try this would be advised to make a jig that punches multiple holes at once. In that way, the saved cultivation time, lack of diseases and extended life of the plants might balance out the extra sowing time.

Irrigation is most beneficial during bloom, pod set and pod enlargement. Time overhead irrigation so that the leaves dry before nightfall (to prevent disease). Beans need around one inch (2.5 cm) of water per week until the end of May, then up to double that in summer.

Rotations

Because they are not planted until spring is well underway, beds for beans can grow a good stand of winter cover crops ahead of the cash crop. Winter rye should be turned under three weeks ahead of sowing, to allow the allelopathic compounds to break down. With wheat, we have found two weeks to be enough time between tilling and sowing the next crop (wheat has less of an inhibiting effect on germination than rye). It is usual to avoid legume cover crops ahead of legume food crops, to reduce the likelihood of spreading pests or diseases. For the same reasons, it is better to grow beans where there have not been other legumes for three years.

Rather than stick with a legume-free period, we routinely choose to add soybeans to our summer cover crop mixes and Austrian winter peas or crimson clover to our winter mixes, because we value the benefits of legume cover crops. We haven't seen any problem that we can directly blame on a poor rotation, and until that happens we'll likely continue to add legumes frequently, to increase the soil organic matter, feed the soil microorganisms and support the nutrient cycle. The Cornell University *2011 Production Guide for Organic Snap Beans for Processing* includes information on soil and plant nutrition, organic IPM strategies and pesticides along with test results showing the relative effectiveness of products mentioned. This thorough, 45-page document includes links to pest ID photos, and several pages of web links to more information. Not to be missed!

Succession planting

Before we discovered using *Pedio* wasps to kill Mexican bean beetles (see below), we needed a new patch of bush beans coming into production every two weeks to keep up supplies of good quality beans. We made seven plantings at regular 15-day intervals from April 16 to Aug 3. We improved our plan, focusing on the regularity of

Sowing Date	Harvest Start
Apr 16	Jun 13
Apr 18	Jun 20
Apr 20	Jun 16
Apr 21	Jun 14
May 2	Jun 23
May 7	Jul 2
May 11	Jul 8
May 12	Jul 5
May 13	Jul 1
May 14	Jun 30
May 16	Jul 1
May 17	Jul 3
May 22	Jul 16
Jun 5	Jul 30
Jun 8	Aug 2
Jun 10	Aug 3
Jun 13	Aug 1
Jun 24	Aug 9
Jun 29	Aug 24
Jun 30	Aug 15
Jul 8	Aug 21
Jul 14	Aug 28
Jul 15	Sep 2
Jul 19	Sep 3
Jul 20	Sep 1
Jul 22	Sep 6
Jul 25	Sep 11
Aug 3	Sep 18
Aug 4	Sep 26
Aug 6	Sep 30
Aug 7	Sep 20

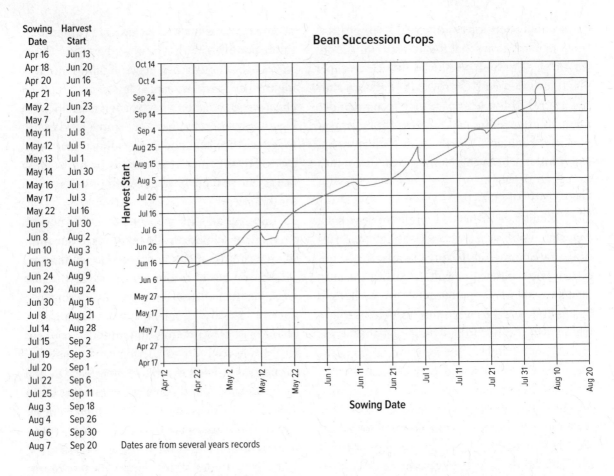

Dates are from several years records

harvest intervals (a new patch every twenty days) rather than sowing intervals. Now we only need to sow six times rather than seven. As well as saving space and sowing time, we get more beans. Much of the time we pick from two patches with an overlap period.

See the Succession Planting chapter for more on calculating dates for successions of various crops. To calculate the best planting dates for your farm, you'll need four pieces of information: the first possible planting date in spring, the first harvest date from each sowing you've done, the last planting date that will provide a crop in time before fall frost, and the number of days you want

to be picking from the same plants. For us, space is tight, and we want to pick from the same plants for three weeks, after which the yield goes down and it takes too long to "search" for the ripe beans. Some growers prefer to pick from each planting only once or twice and then move on to fresh plants.

See the table opposite from our multi-year records of various sowing dates of one variety, with corresponding first harvest dates, and then see the graph we made from this data. You'll need to draw a "best smooth line" on your graph to even out the wiggles caused by the particular weather each year. The line should have the same number of points above as below. It won't necessarily be straight—it is more likely to curve, as crops mature faster in warmer weather.

To ensure a continuous supply with a new patch coming into harvest regularly, we first determined the earliest and latest possible harvest start dates: June 16 and Sept 24 for us. The period in between is 100 days. We divided that into a whole number of intervals; in our case we went for five intervals of 20 days. If we want fresh beans every 20 days, we'll need five intervals between plantings, which is six sowings.

To determine the sowing dates for the six evenly spaced harvests, calculate the harvest dates (one every 20 days, starting June 16), mark them on the graph, and then read across to see which sowing dates will give the desired result. Harvest dates of June 16, July 6, July 26, Aug 15, Sept 4 and Sept 24 imply sowing dates for us of April 20, May 14, June 4, June 26, July 19 and Aug 7. In practice we bring the first sowing forward to April 16 if the weather is decent, and also bring the last one forward to ensure a good length of harvest if the weather starts to cool down quickly. This gives us sowing dates of April 16, May 14, June 7, June 29, July 19 and Aug 3. These sowing intervals are 28, 24, 22, 20 and 15 days.

Drying beans (soup beans)

Some varieties of beans are primarily grown for harvest as dry beans. For a commercial crop to be worthwhile, these would probably need to be grown on a field scale and harvested with a combine. As more people become interested in locally grown food, the market for sustainably produced dry beans will grow, as the market for small grains has done. Small-scale trials of various varieties can be processed by hand. There are many heirloom varieties available, as well as the standard black turtle and pinto types.

Pests

By far the worst pest of beans we ever deal with is the Mexican bean beetle (MBB), a yellow-bronze beetle with eight black spots. Adults overwinter in surface plant debris, so clean up well in the fall, and try not to plant your first bean crop near the site of your last one the previous year. Here, the MBB emerge on the first cloudy day in early June.

We buy the parasitic wasps *Pediobius foveolatus* from BioControl Network in Tennessee. When we see the first fuzzy yellow larvae (the stage the wasps attack) early in June, we order 1 unit/400 ft² (37.2 m²) of the first planting. The parasites arrive as mummified bean beetle larvae on the point of hatching. We set the open container under the bean canopy close to a supply of fresh MBB eggs and larvae. Sometimes I also shake out a few mummies here and there along the row. In a week or so we start seeing brown shriveled larvae on the leaves. Until then, we handpick the adult beetles to get the level down to one beetle per three feet (one meter) of row, so that the larvae numbers will be manageable for the parasites.

Before discovering *Pedio* wasps, we used to flame each bean planting when numbers of MBB and damage became intolerable, and move on to the next planting. A smarter move would be to plant a small early trap crop of beans deliberately at or near the site of the late beans the previous year, and a week later sow a larger production planting as far distant as possible from the first. Then flame the trap crop when there are larvae, but not yet any pupae. Later the same day, release the parasites in the production row.

The challenge for us is to get the wasps to our farm while they are still alive. Our UPS deliveries come in the afternoon, at the end of a day of touring the region. Regardless of how quickly the beasties have been transported from their "ranch" to our UPS depot, a whole day on the van is going to do them in! We solved this problem by co-ordering (early in the week) with our neighbors at Southern Exposure Seed Exchange, whose level of business requires a morning UPS delivery. Then we cover the last seven miles (11 km) ourselves.

Adjusting our planting plans has helped maximize success. We make a smaller early planting and buy just one or two units of *Pedio* wasps, which can quickly get to work for us. We plan larger plantings from the second sowing onwards.

Once hatched, the wasps are very mobile, and since the adults can fly up to twenty miles (32 km), they will likely take care of your neighbors' MBB as well. The *Pediobius* do not overwinter in our climate, but we have found that good control for a few years reduces MBB numbers so that we can sometimes take a year off from buying the parasites.

Bean leaf beetles sometimes attack our beans, but at a very low level, and the plants don't seem much affected. Other insects that can damage bean crops include potato leafhopper, seedcorn maggot, European corn borer and tarnished plant bug. Mites and slugs can also cause trouble.

Root lesion nematodes, *Pratylenchus penetrans*, can be hosted by snap beans, although they may produce no symptoms. A severe infection can cause stunted plants. Long-term planning is needed to deal with nematodes, as they are hard to destroy.

Harvest beans several times a week so they have no chance to get tough. Credit: Kathryn Simmons.

Rabbits and deer can also cause trouble. Dogs, fences and guns are the most effective ways to control mammal pests.

Diseases

The main defense against disease is to keep the leaves as dry as possible. Consider orienting the rows to take advantage of the prevailing winds, planting on raised beds, using wide row spacing or in-row plant spacing.

Fungal diseases tend to be furry. *Botrytis* grey mold can be tackled with neem extract. *Sclerotinia* white mold can be avoided by rotating susceptible crops with sweet corn or other grass crops, and avoiding nightshades, brassicas, lettuce or other legumes preceding your bean crops. Serenade (*Bacillus subtilis*) may help in dealing with an outbreak (so far, I haven't needed to find out).

Bacterial diseases of beans cause brown spots, which may have paler margins, on leaves and pods. The lesions may appear sunken or water-soaked. Virus diseases cause leaves that are typically mottled, blistered or curled.

Scout the fields deliberately once a week or more, and keep a weather eye open while harvesting or cultivating, to spot small problems before they become large ones. See the Cornell publication for detailed organic disease and pest control information.

Harvest

We pick beans three times a week, going for thinner-than-pencil-sized pods. In our climate, beans can size up in two days. It's important to train crew to nip through the stem of the pod and not leave the cap part on the plant, as removing the cap signals to the plant to continue cranking out the beans (yes please!), not to stop and focus on ripening seed. For the same reason, it is also important to pick and discard any oversize pods.

Post-harvest

Like many warm-weather crops, beans are liable to chilling injury if over-enthusiastically refrigerated. Keep the temperature above 40°F (4.5°C) or you risk the beans becoming slimy and pitted. Use hydro-cooling or forced air cooling rather than top icing. Store for 7–10 days at 95% humidity and 40°F–45°F (4.5°C–7°C). Temperatures above 45°F (7°C) are likely to lead to yellowing and development of fiber.

Chapter 27

Southern Peas, Asparagus Beans and Limas

Southern peas, also known as cowpeas and field peas, were brought to the US by enslaved Africans four hundred years ago, and originally came from India. They are in the species *Vigna unguiculata*. Asparagus beans, also known as yard-long beans, are in the species *Vigna sinesis* var. *unguiculata*, and can cross-pollinate with southern peas, but not beans of other species. Limas (or butterbeans, but not the edamame variety also known as butterbeans) are *Phaseolus lunatus*, and will not cross-pollinate any other kind of bean. All are frost-tender, warm-weather crops with similar needs and habits. Some varieties can be grown surprisingly far north.

Southern peas

Southern peas have many names. There are three basic types: strong-flavored crowder or conch peas (Colossus, Hercules, Mississippi Silver, Mississippi Purple, Zipper Cream, Dixie Lee), mild-flavored black-eyed peas (Queen Anne, Pink Eye Purple Hull, Bettergro bush, Princess Anne, all bush or semi-vine types), and the mildest-flavored cream peas (Bettersnap, Lady, Tender Cream and Texas Cream are all bush types). We really like the flavor and productivity of Mississippi Silver, a 64-day good performer with easy shelling pods. We grow this one outdoors and also in the height of summer in our hoophouse (see the Summer Hoophouse chapter). We also like Pink Eye Purple Hull, a short-bush black-eyed type, with pods held up above the foliage, easy to see and harvest. We also grow Queen Anne, a standard black-eyed type, and Whippoorwill, with five-foot (1.5-m) vines and a need for 85 days to harvest (not for the North, this one, but very drought-tolerant). Southern peas also make a good warm-weather cover crop in the South. Iron and Clay is one variety that does well for this.

All need 60–90 days of warm weather, with full sun, and preferably well-drained soil. Southern peas are a good crop for poor soil, and have even been reported to grow in Texas without irrigation, though they certainly give better results if you do have irrigation, and temperatures between the extremes. 1 oz (28 g) contains 140–250 seeds and sows 30'–50' (9.1–15.2 m). Maincrop southern peas are usually sown about two weeks after the first corn. Once the soil has reached at least 60°F (15.5°C) for 5–7 days, sow 1" (2.5 cm) deep, 3"–6" (8–15 cm) apart in rows 18"–42" (45–107 cm) apart, or do as we do and sow two rows 12" (30 cm) apart in a 4' (1.2-m) bed. Do not over-fertilize the soil before planting beans or you will get lots of foliage and not so many beans. We sow outdoors between April 25 (a bit early, really) and June 15. The faster-maturing kinds can be sown up to about 80 days before the first fall frost date.

Southern peas have been grown successfully in New York, Minnesota and Ontario, by choosing varieties with fast maturity and good cold soil emergence, such as Minnesota 13 or Minnesota 150, which sprout at 55°F–60°F (13°C–15.5°C). Using south-facing raised beds and rowcover will help northerners get their southern peas up and growing. There is a short-season heirloom variety of dwarf black-eyed peas from Italy, Fagiolino Dolico di Veneto, that is said to be similar to US "rice" cowpeas, very small, mild-flavored peas (usually white or red) which are good producers in hot climates. When grown in Pennsylvania, transplanted from 4" (10-cm) pots around the usual tomato planting date, this variety matured at the end of July. This works in Zone 6 but would be pushing it in Zone 5. Carol Deppe has selected a variety called Fast Lady Northern Southern Pea, a small, white lady pea, to mature fast in Oregon's

cool summers. Southern Exposure Seed Exchange has this and many other varieties.

If you can plant a cover crop before your southern peas, a mixture of winter wheat or rye with mustard or rape (if you're not plagued by harlequin bugs) works well. Avoid using legume cover crops if you have trouble with root or seed weevils. While the beans are growing, you could undersow with buckwheat between the rows. It keeps the soil cool and protected from compaction during picking, as well as improving soil quality and suppressing weeds. Mow or pull the buckwheat if it gets in your way.

Southern peas resist the Mexican bean beetle and most bean diseases. Some growers have had trouble with pea curculio, aphids, thrips or even stink bugs, loopers, corn borers or nematodes. Most plantings have few of these pests. Possible diseases include mosaic viruses, powdery mildew, leaf spot diseases and root rots.

The yield of cowpeas is similar to snap beans and significantly higher than limas or English peas. Pods are 7"–13" (18–33 cm) long, with up to 20 peas. The plants bear well for 3–4 weeks. 100 lbs (45 kg) of pods will yield 45–55 lbs (20–25 kg) of shelled beans, depending on variety and maturity.

Harvest as green shelling beans or as dry beans. For dry beans just wait until the pods are papery and about to split. For fresh eating, don't harvest too soon or the pods will be difficult to open. Get each crew person to test a few before picking lots, as each variety is different. Color can be a guide: the pods may have red streaks or may start turning tan. They become more flexible as the pod walls thin out. The beans can be moved slightly by rubbing the pod back and forth. In some varieties, you'll be able to feel an air-gap at the plant end of the pod. The real test is to shell

Mississippi Silver cowpeas (southern peas) growing in our hoophouse. Credit: Twin Oaks Community.

out a few pods—if that's difficult, you're really too early. Harvesting too early naturally reduces the yield (the beans are not full size yet). Harvesting "demi-sec" or half-dry is another possibility; a few demi-sec beans in with fresh ones add to the flavor without adding noticeably to the cooking time.

Southern peas can be sold shelled or still in the pods. Shelled beans can be sold in plastic bags, but do need ventilation. For shelling large quantities you will need a machine. The Roto-Fingers commercial bean and pea sheller with enhanced cleaning system is manufactured by Welborn Devices. In four seasons of use, Anthony and Carol Boutard reported it worked flawlessly: "When shelling beans, it is important to have clean, dry pods. We load the drum with approximately 40 lbs (18 kg) of beans and start it. Five to ten minutes later the beans are shelled. The machine performs well on everything from lady peas and garden peas to large horticultural beans, and threshes dry beans as well. It will not shell favas or large limas such as Fordhook 242. If you use the Roto-Fingers for dry beans, it is a good idea to drag the machine outdoors because the thin membrane inside the pod detaches and covers everything as in a snow globe." Lehman's sells a Texas Pea Sheller, which is manual but could have a motor added.

To cook, shell out and boil until tender, 15–25 minutes, covered in salted water. To freeze, shell, clean and blanch for 2 minutes in boiling water, then cool rapidly in ice water, drain and freeze.

NEVER eat raw shelled beans—they have an enzyme that makes them toxic, and leads to a very bad bellyache. The enzyme is deactivated by heating.

Asparagus beans

Asparagus beans are widely grown in Asia. They are called *dow gauk* in China and *sasage* in Japan. Although related to Southern peas, they differ in growing extensive, long thin pods. Once the weather is warm enough, asparagus beans grow quickly, produce heavily all season, and are somewhat resistant to Mexican bean beetle and drought. They thrive in hot weather and tolerate a wide variety of soils. They are a tall vining crop that needs trellising of some sort, either netting or stakes and twine, preferably eight to nine feet (2.5m) tall. With careful planning you can use the shade on the north side of a row of tall beans to grow lettuce or greens that prefer cool weather.

Although these plants will grow pods almost a yard (1 m) long if you let them, those puffy giants are not good for eating! Harvest at 12"–15" (30–38 cm), before the bean seeds are much in evidence. Chop the pods into short pieces for stir-frying or steaming. The Southern Exposure Seed Exchange catalog suggests braiding the pods or tying them into knots before cooking. In times of desperation, or for the curious, the young leaves and stems are also edible.

A quarter of an ounce (7 g) of seed (45–55 seeds) sows 20' (6 m) with seeds 8" (20 cm) apart.

We often sow closer than this, about 3" (8 cm) apart. Sow one inch (2.5 cm) deep. Our favorites are Purple Podded and Red Noodle, both a beautiful dark color that does not fade when cooked. They make an eye-catching addition to dishes of green vegetables. We have also grown the reliable Green Pod Red Seed. Green varieties need about 75 days to harvest, while red and purple ones take about 90 days. If you have a short frost-free season, you could transplant. Each plant will provide a large harvest.

Lima beans

Lima beans come in both 1'–3' (30–90 cm) bush and 6'–12' (1.8–3.6 m) pole types, with a wide range of seed sizes and colors. The plants are quite heat-resistant and drought-tolerant. Time to maturity is 66 days for our favorite, the bush Jackson Wonder Butterbean, and up to 90 days for the pole varieties, including the beautiful red-and-white-speckled Christmas lima bean. We are turning our backs on tall vines, as putting up trellises is not our favorite thing. The beans of Christmas are huge, almost as big as a quarter, but we love the flavor and ease of growing Jackson Wonder. It has smaller, buff-colored beans mottled with purple, with about four seeds per pod. We plant two sowings, on May 9 and June 16. Northern growers will want to choose a sunny spot and stick with bush varieties, such as Jackson Wonder, Fordhook, Dixie Butter Peas and Henderson Baby limas (which have pods above the leaves, an advantage at harvest time). A technique for cooler climate zones is to make a 4"–5" (10–12.7 cm) high ridge of soil running east to west and plant the seeds on the south side below the ridge. The heat will reflect from the ridge, which will protect from north winds. Transplanting is also possible.

Two ounces (57 g) of seed may contain 35–250 seeds, as size varies a lot. This could plant 14–28 poles, or a 12' (3.6 m) long row of Jackson Wonder. Seeds may be soaked overnight before planting, but definitely no longer than that or they will rot. Seeds are planted 4"–6" (10–15 cm) apart about two weeks after the average last spring frost date, if the soil is close to 70°F (21°C). Sow ½"–1½" (1–3 cm) deep, shallower if the soil is cool and crusty, deeper if hot and dusty. We sow two rows in a 4' (1.2 m) -wide bed and stake every ten feet (3 m) with a five-foot (1.5-m) stake. We then loop around the stakes with twine every foot up the stakes. This holds the beans in and gives easy access for picking. Do not work in the plants when the foliage is wet as this will encourage the spread of fungal diseases. Weed initially, then the canopy will close over and you will only need to pull a few weeds while harvesting.

Harvest for fresh shelling beans when the walls of the pods are thinning and the pods are turning a yellow-green or tan color and open easily when squeezed at the seams. Harvesting too early makes for difficult shelling. I've read that an old-fashioned laundry wringer can be used to shell out the beans once the ends of the pods have been cut off, but I can't vouch for that. The smaller limas can be shelled with the machines mentioned earlier.

Cook by barely covering the beans with salted water and simmering till tender. Freeze by blanching in a large amount of boiling water for 2½ minutes, then plunging into ice water for 3–4 minutes, draining and freezing.

Seed saving

Isolate each variety of beans from others of the same species by a minimum of 25' (7.5 m) for home use. For pure seed, isolate varieties by a minimum of 125' (38 m).

Chapter 28

Fava Beans

Credit: Lisa Dermer

Fava beans (known as broad beans in the UK) are easy to grow and have an unusual earthy, nutty flavor. Botanically, they are a kind of vetch. Like all legumes, they can fix nitrogen in nodules on their roots, leaving a fertilizer boost in the soil for the next crop. Their flowers are surprisingly fragrant and are very attractive to pollinators and other beneficial insects. They are usually eaten fresh and shelled, but they can also be dried, then cooked. In the UK, there has been some research into making foods like hummus, falafel, guacamole and tempeh from locally grown fava beans, to reduce dependence on imported garbanzos or soybeans. There are also small-seeded varieties that are mostly grown for cover crops or to feed livestock.

How and when you grow fava beans will depend on both your winter and summer climate. According to Frank Tozer in *The Vegetable Growers Handbook*, in the beans' ideal habitat they are indeterminate and so can produce over a long season, if not attacked by aphids (black fly).

Above 70°F (21°C), flowers may drop, causing the plant to act like a determinate. This means that in climates with hot summers (such as central Virginia) the plants produce a single harvest. In zones 7 and warmer they can be sown outdoors in the fall, for a spring or early summer harvest. In colder zones they are spring sown, and can be succession planted to last all summer.

Fava beans have a high protein content (26%), similar to kidney beans (25.3%) or peanuts (26%), although not as high as soybeans (36.5%). They contain levodopa, a compound humans normally make in the body, which is also synthesized industrially as a medication, especially for people with Parkinson's disease. Eating favas may help people with Parkinson's, and the effect lasts longer than synthetic L-dopa. Velvet bean, used in the South as a cover crop, is a source of the botanical L-dopa supplement.

A small percentage of people of Mediterranean, African or South Asian descent (a total of four hundred million people) have an inherited genetic sensitivity (favism) that causes illness when they eat fava beans in quantity. Favism is often erroneously called an allergy, and I am grateful to Richard Moyer for educating me on this. In people with favism, an enzyme in red blood cells functions poorly. Fava beans (and certain drugs) contain a compound that oxidizes red blood cells, and in affected people, favas overwhelm the ability of the weakened enzyme to reverse the oxidation quickly enough to keep the red blood cells alive. The dead red blood cells can cause swelling and inflammation which may look like an allergic reaction. But those are secondary symptoms, and treatment for favism is not the same as treatment for allergies. Favism is more severe in men than women, and in children than in adults. I've never met anyone with the condition, but it may be wise to mention this to your customers.

Varieties

- Broad Windsor (80 days spring sown, 240 days fall sown). A 4' (1.2-m) -tall variety with large beans, up to six of them, in 6"–8" (15–20 cm) pods. Windsor fava beans are the main ingredient in Brown Windsor soup. Windsors are not generally reckoned to have the best flavor, but are the easiest variety to find in the US. Buff-colored seeds. 17 seeds/oz (61/100 g). Reliably hardy to 12°F (−11°C).
- Aquadulce and other Longpods (80–90 days) are hardier and tastier than the Windsors. Well-filled pods to 8" (15 cm). Big reddish-brown seeds (green before they dry). Peace Seeds carries small quantities of Longpod Major. 5–6 large seeds per pod on 3' (96-cm) plants.
- Express (95 days), 3' (1 m). Up to 3 dozen 8"

(15-cm) pods per plant. The beans do not discolor when frozen. White seeded. Thompson and Morgan.
- Fava Cascine. Slightly earlier than other varieties. Long, slim pods with 6–7 beans each. Beans are a bit smaller than most favas and very tender. Seeds from Italy.
- Jubilee Hysor (95 days). A high yield of broad, well-filled pods, each containing 6–8 beans. Thompson and Morgan.
- Sweet Lorane, very hardy, good tasting. Southern Exposure Seed Exchange.
- Statissa (75–95 days spring sown). Dwarf plants, 24"–28" (60–70 cm) tall. Early maturing, short, slender pods, small beans. Buff-colored seed. 50 seeds/oz (178/100 g) Thompson and Morgan, Territorial Seeds.
- The Sutton is a fast-maturing dwarf variety, 12" (30 cm) high, that produces lots of 5"–6" (13–15 cm) pods each bearing five small, tender beans. Hard to find in the US.
- Seville Dwarf is the hardiest variety grown for food. Hard to find in the US.
- Negreta (70 days spring sown; 240 days fall sown). Height: 3' (91 cm). A fast-maturing variety with large purple seeds. When overwintered (only possible in mild areas), this Italian variety can mature almost a month before others, allowing other spring crops to be planted after the fava harvest. The 9"–10" (23–25 cm) pods are filled with 6–7 bright green seeds. Territorial Seeds.

Thompson and Morgan has a good selection of twelve varieties available on their UK website, but only three in the US.

As a cover crop, favas are a vigorous and adaptable legume, often mixed with peas, vetch,

A bed of February-sown fava beans, with supporting stakes and strings in place. Credit: Kathryn Simmons.

radish or oats. They produce lots of organic matter, and fix nitrogen if inoculant is used. In the UK "tic beans" are small-seeded cover crop or animal feed varieties sown in spring, and "winter beans" are varieties sown in the fall. In the US, I believe both are known as "bell beans." Bountiful Gardens sells them.

Banner (60–75 days) is one variety usually grown as a cover crop. Hardy down to 7°F (−14°C). Sow in the fall, or early spring in cool weather areas, the same time as peas. High yielding. Grows very thick stands 6' (1.8 m) tall by late spring. Tolerant of some waterlogging in winter. Besides producing silage or green manure, these small favas make good eating too. Sow up to 3" (8 cm) deep for the best stand, but they can also

be broadcast and tilled or raked in. Territorial Seeds, Bountiful Gardens.

Crop requirements, sowing, yield

Fava beans thrive in well-drained, fertile soil with a pH of 5.5–7.0. Deeply cultivated soils can yield almost twice as much as poorer soils. Sow in a sunny location, or partially shaded if you are pushing them into the hot end of their preferred temperature range.

For growing as a cover crop broadcast at 150–250 lbs/ac (168–280 kg/ha) or 5–10 lbs/1,000 ft² (25–50 kg/100 m²). For food crops, sow staggered double rows 12"–18" (30–45 cm) apart, planting seeds 1.5"–2" (4–5 cm) deep every 6"–8" (15–20 cm). For multiple pickings over a longer harvest period, try seeds closer, about 4"–5" (10–13 cm) apart. For maximum yield for a single harvest aim to have 2–3 plants per square foot (22–33 per square meter). Dwarf varieties can have closer rows, 9" (23 cm) apart. It's probably better not to sow pre-sprouted seeds if the conditions are cold, as the seed may rot. On the other hand, if you are sowing a bit late in the spring, pre-sprouting might help you get a good harvest before the summer gets too hot.

Twenty plants per person will be enough. The seeds are large and heavy, and seed counts vary a lot by variety: 17–50 seeds per oz (61–178/100 g) or about 300–1,000/lb, 600–1,800/kg.

The usual storage life of the seed under refrigeration is two to three years.

When to sow

Fava beans like mild, damp weather best and dislike heat. When temperatures are 40°F–75°F (4°C–24°C) they germinate in 7–14 days. They are cold tolerant down to at least 15°F (−9°C), some varieties 7°F (−14°C). Because they are so cold-tolerant, favas are usually direct sown, but like almost all plants, they can be successfully transplanted with care, if your situation justifies the extra time and attention. William Woys Weaver in Zone 7a sows favas in small pots in the greenhouse about Jan 25. He has a short, hot spring and transplants about April 10 in order to have them in full bloom by May 15 and harvest by June 1. He induces tillering (bushing out, producing side shoots), by pinching out the tops when the plants are 6"–8" (15–20 cm) tall. This increases the yield.

If your Winter Hardiness Zone is 7 or warmer, you can overwinter for an earlier harvest (using one of the hardier varieties if your climate is borderline). Rowcover can be used to protect fava beans from temperatures between 10°F–15°F (–12°C to –9°C). Or you can overwinter in a hoophouse and provide a crop before the summer gets too hot.

Don't sow too early: for best winter survival, aim to have short plants with about six leaves before you get to really cold weather. Overwintered plants will tiller in the spring (grow multiple stems). Spring sown plants do not tiller, so the yield from overwintered plants can be higher.

According to Southern Exposure Seed Exchange, in warmer zones such as Gulf Coast states and Southern coastal areas, favas are sown from October to December for harvest in March. They will still germinate in the fall at temperatures up to 90°F (32°C), but do need average daytime temperatures below 70°F (21°C) during flowering.

In the Pacific Northwest favas are sown Oct 1–Nov 15 or in late February. Temperate maritime climates offer the best conditions for fava beans for both fall and spring sowings. Carol Deppe recommends sowing after the fall frost has killed the aphids. When I lived in England we would sow hardy varieties in October or early November and again in late February/early March, then switch over to the less hardy varieties for a later spring sowing.

In zones with cold winters and summers below 70°F (21°C), favas are spring sown only, and can be succession planted to last all summer. Sow as early as the soil can be worked in spring and then sow a second and possibly a third crop later in spring to harvest throughout summer.

Here in central Virginia we have sown them in November, but our rate of winter-kill was too high for it to be worthwhile. Last year we sowed on Feb 15 and they popped up March 15, a whole month later, so even if it is too cold for them to germinate they will just sit there in the soil until conditions are right. We need to get an early spring start with ours, as summer heats up quickly. Our single harvest happens in mid-June, a good time of year to have an unusual crop come in.

Cultivation

The best conditions for growing fava beans are temperatures around 65°F (18°C), but anywhere in the range already given for sowing will work. Grow fava beans like peas, except without trellising. They have sturdy, non-branching stems 1'–4' (30–120 cm) tall, depending on the variety. They may be OK with no support at all, or you can "corral" them as we do: put a stake or short T-post in the ground every 8'–10' (2.5–3 m) along each row just outside the plants and then string-weave around the double row, boxing the plants in. This will keep them upright in windy locations, and when the heavy load of beans threatens to pull the plants down.

Don't overwater early in their life or you will get too much soft vegetative growth, and may have problems with aphids. During dry spells water enough to keep the plants actively growing.

Fava bean flowers have a delightful smell, and attract bees. Credit: Kathryn Simmons.

When plants are in full flower and pods are beginning to form, water more generously. Watering during flowering will increase the number of pods set; watering as the pods swell increases the size of the beans.

Once the plants have reached close to their full height, you can pinch out the top 4" (10 cm) of each stem to limit vegetative growth and hasten ripening. This will condense the harvest period into two to three weeks, even in a mild climate. It also discourages black aphids, which may attack plants in late spring (at least, they do in the UK and the Pacific Northwest). The pinched-out shoots can be eaten, lightly steamed or sautéed.

Season extension and rotations

For a continuous supply, sow at 2–3 week intervals, as long as the weather is suitable and will remain in the 40°F–75°F (4°C–24°C) range for at least the first 6–8 weeks of growth.

We've sown fava beans in our hoophouse on Nov 15 and harvested them in mid-May, a full month sooner than our outdoor spring sowing. This worked well, and legumes add diversity to the rotation.

In most climates, fava beans finish up in summer, and the space can then be used for sowings of salad greens or fall brassicas.

Harvest and post-harvest

From a spring sowing, most varieties take 90–120 days until harvest. The pods stand upright in pairs or fours from the leaf axils, ripening up from the bottom of the plant. Harvest by bending the pod down and pulling it downwards from the plant. Some people harvest the pods young and cook them whole, although this does, of course, reduce the yield from the plants. Generally, people let the pods get big, when they can feel the beans inside. The beans are then shelled and cooked. There is quite a bit of leeway in when you can harvest them for eating as shelled beans. The pods start to get thinner-skinned and spotty, then leathery and dark. If you wait too long, the beans will be demi-sec, and need longer to cook.

Favas can be refrigerated and stored in the shell for three to four days, or shelled and used within two days. The beans tend to get unpleasant-looking brown spots if kept for too long after shelling. Some growers sell them ready-shelled, in plastic bags.

Some people squeeze the beans out of the skins after cooking, but this is not necessary unless you want to make a dish from mashed beans without using an electric gadget.

Seed saving

Isolate fava bean varieties by at least 100' (30 m) for home use, and a mile (1.6 km) for commercial sale. They do not cross with other types of bean. Because the pods dry to a leathery wrinkled state, they do not shatter and so are forgiving about harvest date.

Chapter 29

Edamame

Edamame is an interesting, high-protein, fast-growing, easy-care, warm-weather crop. Edamame is the name for varieties of soybean that are eaten as a fresh vegetable or as a snack food like roasted peanuts. There is a huge potential market for edamame, which appeal both to those seeking local fresh produce and to those appreciating the benefits of soy. Most small-scale growers sell them as pods or as bunched plants. For certain metropolitan markets, edamame may be processed into green soybean pudding, ice cream or green soymilk. Edamame varieties differ from field soybeans in being larger, milder, sweeter, tenderer and more easily digested—they contain a smaller percentage of oligosaccharides, the gas-producing starches.

Edamame has been popular in Japan and China for centuries; the word means "beans on branches" in Japanese. The Kitazawa Seed Company lists the many names this vegetable is known by in Asia.

Varieties

Research is underway testing and breeding new varieties for various locations. Extension Services in West Coast states have a lot of information, and the University of Kentucky also has some publications. In trials in Pennsylvania, Envy gave the best yield of nine varieties tested, but Shironomai gave the highest marketable yield.

Research at Virginia State University by Tadesse Mebrahtu and others aims to find varieties with high yields, large seeds, sweet flavor, high protein and suitability for mechanical harvesting and shelling without shattering. Their research involves Asmara (medium size, high-yielding, good flavor and appearance), Kanrich (large and productive, but less good on flavor and appearance), Kahala (a tall, high-yielding plant), Owens (a short plant that matures early, with large, nutty, dark beans), Randolph (large and productive, but not so good on flavor and appearance) and several varieties which still have numbers rather than names. All these yield higher and later than the earlier and more compact Envy. Butterbean scores well in taste tests but is considerably lower-yielding.

Patricia Stansbury at Bon Air, Virginia, tested the VSU varieties with a producer grant from SARE (Sustainable Agriculture Research and Education), and now grows seed for sale.

Catalogs do not always make clear whether a variety is open-pollinated (OP) or a hybrid. Varieties available commercially include the following:

- Beer Friend (OP), 85 days
- Besweet 292, 85 days
- Black Jet, 104 days
- Black Pearl (OP), 85 days
- Butterbeans (OP), 90 days
- Davis Family Blend, 75 days
- Early Hakucho
- Envy, 75 days
- Kouri, 85 days
- Lucky Lion, 75 days
- Misono Green (OP), 85 days
- Sayamusume (OP), 85 days
- Shirofumi (OP), 90 days
- Tamba Kuro Otsubu, 150 days
- Tankuro, 85 days
- White Lion, 75 days

Crop requirements

Edamame will grow in wet or dry weather and in varying soil conditions, so long as it has warm weather. The pods of most varieties are ready for harvest 75–90 days after sowing, though some black-seeded ones take up to 150 days. All the pods are usually picked in a single harvest, making this an ideal crop to grow in succession, sowing several times during the growing season. In cooler climates, edamame can take considerably longer to reach maturity than the catalogs suggest—in southwest Washington, with only 70 percent the heat unit accumulation of the Midwest, crops take 40 days longer than advertized.

All soybeans, including edamame, are ranked in maturity groups with 0 being the fastest maturing (70–90 days) and 8 the slowest of the groups suitable for the mainland USA. All varieties except the earliest-maturing ones are also sensitive to day length, meaning that they will not shift from the vegetative (leafy) stage to the reproductive (flowering) phase until the length of the daylight falls below a critical level. Northern growers with long summer days should probably stick to the groups 0 and 1, even with a hoophouse to boost production by raising temperatures, and add 25–40 days to the "days to maturity" listed in seed catalogs. Virginia is in the Group 4 belt, while Groups 6–8 are suited to Florida and south Texas. If you plant varieties more suited to areas north of your farm, they will flower earlier, while plants are small, and so have lower yields. Varieties more suited to areas south of you will grow longer before being triggered to flower. The danger then is that pods will not mature before cold weather arrives.

Where the seasons are long enough, edamame can also be grown as a summer "catch crop" on a bed following a spring crop, and before a fall crop. Maintain sufficient organic matter in the soil, supply ample moisture and don't miss the harvest window. A benefit of growing edamame is the nitrogen that the root nodule bacteria adds to the soil. A following crop of greens will probably do well. There may be a soil gain of one pound of nitrogen per bushel (60 lbs) of harvested pods (1 kg/60 kg). Soybeans are only a moderate nitrogen-fixer compared to alfalfa.

Edamame pods stay very clean and attractive when the crop is grown in a hoophouse. Credit: Twin Oaks Community.

Sowing

Make a first sowing, in a sunny location, just after the average last frost date. Ideal soil temperature is above 55°F (13°C), with air temperatures around 68°F (20°C). Soil pH of 6.0–6.5 is best. Sow 1"–1.5" (2.5–3.8 cm) deep (less in cooler or wetter soils), with about 2"–3" (5–7.5 cm) between seeds and 6"–24" (15–60 cm) between rows. If you have not had many soybeans growing on your land in the past, use soybean inoculant, which is not the same as general-purpose pea and bean inoculant. Germination is usually rapid (within two weeks) and the bushy plants will outcompete weeds after an initial cultivation.

Outdoor sowings can continue every two weeks until about 90 days before the average first frost date, depending on the variety chosen. For us, outdoor sowings are possible from April 20 (with rowcover) to mid-July (fast-maturing varieties). Sowing varieties with different growth periods on a single sowing date can also provide a long harvest season.

Edamame may also be transplanted, which makes growing the crop possible in areas with a shorter season. Tobacco transplanting machines were used successfully by growers in Kentucky, in a 2001 trial using a SARE Producer grant.

Pests and diseases

Edamame are relatively pest free. Japanese beetles can be a problem. The best hope is the parasitic spring *Typhia* wasp. Grow forsythia to entice the *Typhia* wasps down from the tree canopy to bean level, especially if you have tulip poplar trees, a favorite of the wasps. Some edamame varieties are relatively resistant to Mexican bean beetles, so if you have this pest, grow several varieties to compare. Be ready to buy the beneficial wasp *Pediobius foveolatus*, which parasitizes the MBB larvae. Cucumber beetles can also be vectors of diseases.

Rabbits, groundhogs (woodchucks), deer and (I've read) elk can also be trouble. Bird netting and rowcover may protect young plants against the smaller mammals.

Harvest

Pods are ready when the beans are big and lumpy, almost touching within the pod, and the hairy pods are just changing from bright green to golden. The exact color does seem to vary between varieties, and is somewhat a matter of personal taste. As the beans age the flavor becomes more starchy, less sweet and less nutty, so don't wait too long!

The pods pretty much all mature at the same time, so usually a single harvest is the most efficient method. When we've had too many to pick at once, we harvested the bushes two or three times, a few days apart. That planting was an open-pollinated variety—hybrids don't offer a second chance, having a very short harvest window (one week or less). Just remove the whole plant. Commercial growers have used the same harvesting equipment used for green beans, but losses were as high as 24 percent. For manual harvesting, one pound (0.5 kg) of marketable pods per three feet (meter) of row is a reasonable expectation.

Because edamame is a legume, you may want to clip the stems above soil level and leave the roots with their nitrogen-fixing bacteria in the soil. It is quicker to pull up the plants, but less nitrogen will remain in the soil.

According to an MSU Strategic Marketing Institute paper by William A. Knudson, pod appearance is very important for Japanese markets — the pods need to be bright green and blemish-free (less mature than the buttery beans preferred by US markets). For Japanese markets, clip the leafy top part of the plant off and sell the plants with the pods attached, in bunches of four to six plants. Chinese consumers prefer fresh-shelled beans. For western markets, which accept a wider range of appearances, sell the pods, after removing them from the stems.

Post-harvest storage

Edamame are not easily stored in pods for long. They need cool conditions with some airflow to prevent molds growing. A University of Kentucky study recommended washing, draining and icing the beans before refrigerated storage in mesh bags. In Asia, the shelled beans are available as a frozen vegetable year-round.

When pods are left on the pulled plants, the flavor can be retained for several days.

Nutrition

As already noted, edamame are more digestible than regular soybeans. The nutritional information available sometimes appears contradictory. They have 16 percent or more protein, twice the amount of lima or green beans, and are rich in fiber, essential fatty acids, calcium and vitamins A, B, C and E. They are also high in isoflavones, which may help prevent cancer, osteoporosis and cardiovascular disease. Some isoflavones are phyto-estrogens (plant-produced estrogens), which may help reduce some symptoms of menopause, including hot flushes, weight gain and depression.

Recent research shows that larger seeds have a higher percentage of oil and less protein than smaller seeds. Asmara has 43 percent of its oil content as oleic acid, which reduces serum cholesterol levels; Randolph has 39 percent. I haven't seen figures for other varieties. High protein usually means less sweetness. On the other hand, oil content and total sugars can both be high in the same variety, but an excessively oily taste is unpleasant.

A nutritional downside of soy is the presence of phytic acid in unfermented soy. This binds minerals, proteins and starch, making them less available. Edamame has a lower phytic acid content than mature dried soybeans. Trypsin inhibitors in soy interfere with the digestion of protein, and too much soy can suppress thyroid function. So, as usual, the advice is to eat a varied diet!

Cooking

As edamame is an unusual vegetable, it's good to provide your customers with some instructions. Boil the pods in salted water for 6–8 minutes, then drain and serve. Diners pop the beans out of the shell with their mouths or fingers. To microwave, put 2 cups of washed pods in a suitable dish and cook for 4–5 minutes on full power. For a small number of diners, it is also possible to shell out the cooked beans. It is always easier to cook the pods before shelling.

If adding edamame to a dish, cook them separately and shell them first. They may then be substituted for peas or lima beans.

Season extension

Edamame can be grown in a hoophouse later than outdoors. This can be a good way to use the covered space in late summer, before fall greens are planted, and it also helps by providing a rotation away from nightshades, lettuce, spinach and brassicas. We sow July 27 after early summer squash and harvest in late October or early November (ripening slows down in cooler weather). We have followed edamame with a sowing of bulb onions for outdoor transplanting the following spring; or lettuce, spinach, arugula or mizuna; or kale, chard, senposai and komatsuna transplanted on Oct 30.

A hoophouse or other protected space would also make an early season crop possible. In Kentucky, researchers are working to start harvesting fresh edamame from July 1. They start the crop in greenhouses for field transplanting. Patricia Stansbury in Virginia is also working with researchers to provide fresh edamame through a very long season.

Seed saving

Using open-pollinated varieties, let the pods go brown on the plant, but be sure they don't get moldy. Bring the pods under cover to finish drying, then shell out. Don't keep seed beyond the season after the one the seed was produced in, as the germination rate drops rapidly. Soybean seed is easy to save; expect about 1 lb/10' (1 kg/6 m).

Chapter 30

Snap Peas and Snow Peas
(and a few words about shelling/English peas)

Credit: Hildegard Ott

Shelling peas (also known as podding, garden or English peas) and their relatives snap peas and snow peas are cool-weather crops, presenting a challenge to those of us in climate zones where spring heats up quickly. We have found dwarf, fast-maturing edible-podded peas to be the answer to this challenge.

Shelling peas are more of a backyard crop, as the yield of the edible crop is small for the time and space involved. Some CSA growers in cooler parts of the country, with a worthwhile season-length, can make a go of shelling peas by offering them as a U-pick crop. Snap peas and snow peas can be worthwhile commercial crops across a much broader range of climates—because the pods are eaten, the food production from a given area is higher. As an early crop, peas offer lots of eye appeal among the spring offerings—a crop that is neither greens nor stored roots! And the

flavor is the very essence of spring. Here in Zone 7 we harvest peas from mid-May to late mid-June, by which time the plants have given up and, conveniently, our bush green beans start to come in.

Varieties

Because our spring is short, we grow fast-maturing dwarf pea varieties only. We used to grow taller varieties, but we got tired of putting up big trellises for what is here a short-season crop. We do provide our peas with a little support, because we interplant them in beds of spinach (more on that later). With raised beds and dryish spring weather, it is possible to grow dwarf varieties without support at all.

Our far and away favorite snap pea, Sugar Ann, matures in 52–58 days, on short 2' (60-cm) vines. The 2.5" (6–cm) pods are deliciously sweet and crisp. Wait until they are completely round in cross-section to pick them, as the sweetness in the flavor develops at the last minute. (To see a cross-section accurately you will be forced to bite the pea in half!) Sugar Ann is also fairly disease-free in our experience, although catalogs make no such claims.

We have also grown and liked Cascadia, a 65-day variety, with 2.5' (75-cm) vines and 3" (7.5-cm) pods. It is tolerant to powdery mildew and resistant to enation virus. Sugar Sprint, 58 days, 2' (60-cm) vines, 3" (7.5-cm) pods, has also worked for us. It is resistant to powdery mildew and enation virus. These varieties have good flavor, and I learned from an observant colleague that they are fully ripe at a more oval pod shape than Sugar Ann.

After trying a succession of varieties with different maturity dates, we have switched to sowing only the fast-maturing Sugar Ann and enjoying a bountiful, if short, pea season. Those with longer springs can extend the harvest by sowing some of the slower-maturing varieties at the same time as the early ones.

A snap pea variety that has unfortunately given the whole family a bad name is the one called Sugar Snap. It has no mildew resistance at all, and its 6' (120 cm) vines need a lot of support. I've heard tales from those outside the powdery mildew zone, that this variety is the best of all for flavor, yield and long harvest period. I haven't tried Super Sugar Snap, which claims to be resistant to PM, but I do appreciate the plant breeders who worked on producing that variety.

Sugar Daddy was a big disappointment and I've heard others say so too: Poor flavor; feeble plants.

As for snow peas, we fell in love with Fedco's Sumo and have been hoping to see it return one

Our earliest crop of Sugar Ann snap peas is grown in our hoophouse. The young squash plants will take over the bed when the peas are finished. Credit: Kathryn Simmons.

year soon. Although the pods of Sumo can look a bit imposing, being large and lumpy, the tenderness, flavor and yields are wonderful. Meanwhile we like both Dwarf White Sugar and Dwarf Grey Sugar. Both are listed as 59 days to maturity, but we have found the White several days earlier than the Grey; both varieties mature almost a week later than Sugar Ann. Dwarf White Sugar has 3' (90-cm) vines and 2"–2.5" (5–6-cm) pods. Dwarf Grey Sugar has 2.5' (75-cm) vines and 2.5"–3" (6–7.5-cm) pods. The purple flowers are beautiful and the seed has great vigor in cool soils. This variety is the one most often suggested to people wanting to grow pea shoots, so you can get double value from it by taking thinnings and leaving the rest to bear pods.

Here's the few words about shelling peas. If your climate is like England—mild and damp for a long time—then they are a lovely option. Just don't brag to me—I miss them. When I lived in England we grew Little Marvel for an early crop (63 days), Lincoln for second earlies (70 days) and Alderman (75 days) for the maincrop. Fedco has them all.

Crop requirements

Peas like a pH of 6 or higher, and abundant potassium and phosphorus. Being a legume, they do not need high levels of nitrogen in the soil, so interplanting in a standing crop, or doing a fast turnaround without adding any more compost, will work fine in reasonable soils.

Pea plants dislike heat and have shallow roots, so if your climate tends to get a bit hot, plan to plant peas to the north of a tall crop, mulch the soil and water frequently. A soil with plenty of humus will help hold moisture, as will organic mulches. Of the shelling peas, smooth-seeded varieties germinate better in colder soils than wrinkle-seeded ones. The sweeter the pea, the easier it rots in cold soils.

Soil temperatures should be at least 40°F (4°C) for sowing English (shelling) peas and warmer than that for edible-pod peas, which rot more easily in cold soil. The best temperatures are 50°F–75°F (10°C–24°C). Outside this range, the percentage of normal seedlings that germinate is less than 90. Expect emergence to take 14 days at 50°F (10°C), 9 days at 59°F (15°C), 7.5 days at 68°F (20°C) and 6 days at 77°F (25°C). If you don't have a soil thermometer, I recommend getting one—it's sure to pay for itself in reduced wastage of seeds sown in the wrong conditions. It can take over a month for peas to emerge at 41°F (5°C) and

while you wait you'll worry the seed has rotted. So use the thermometer.

Alternatively, the flowering of forsythia is said to be a good phenological sign for sowing snap and snow peas. Here, that's March 10–28. Since we sow into beds already covered by rowcover, and we reckon the forsythia would have flowered sooner if it had been under the rowcover too, we aim for March 10, which is about six weeks before our last frost. Southern Exposure Seed Exchange recommends March 1–April 15 for the mid-Atlantic Piedmont.

There are now National Organic Program-compliant organic seed coatings which help reduce rotting in cold soils. Natural II is an example. I haven't tried them, so I can't vouch for them.

Sowing

Pre-sprouting, or at least pre-soaking, can help speed good germination. Soak the peas overnight in lukewarm water. They absorb a lot of water, given the chance. I've heard it isn't good to soak seeds for more than about twelve hours, as they can expire from lack of air. Our method is to soak overnight, then drain, inoculate and plant. Any leftover seeds can be kept in the fridge for a while, where they will slowly sprout. They can be sown (sprouted) up to a week later to fill gaps in the row, or elsewhere for pea-shoots.

We sow 8–10 seeds/ft, (3 seeds/10 cm), about 0.75" (2 cm) deep, or even shallower. Sowing too deeply results in a high rate of rotted seed. Sow as soon as the soil is warm enough in spring, choosing a well-drained site. Deeper sowing can be tried in summer for fall crops, in order to keep the seed cooler and damper. An alternative method is to sow in a 3" (7.5-cm) -wide band at 25 seeds per foot (about 1 cm apart).

There are about two thousand seeds per pound (four or five per gram). The count varies, so once you have settled on a particular variety, do your own calculations to make sure you buy the right amount of seed. To sow a hundred feet of peas takes half a pound of seed (744 g/100 m). We sow double rows in the centers of our 90' (27.5 m) beds and use almost a pound of seed per bed. Having two rows of peas rather than one means that the plants are slightly offset from each other and can support themselves better as they grow.

Rows should be spaced according to height. Dwarf varieties can be three feet (1 m) apart or even half that, but tall varieties need five feet (1.5 m).

I learned of the "Minnesota Method" of pea planting from a crew member: if the soil is wet (even muddy), push each seed into the mud to the right depth, then push soil to cover. This only works if the mud is warm enough, but to my surprise it did work for us when we had a very wet spring.

Young pea plants are quite frost-hardy, so spring frosts won't harm them. But if you have pea plants growing in the fall, you'll find that frost will kill the blossoms, and stop pod production.

Inoculant helps increase yields (it doesn't improve the germination rate). The nitrogen-fixing bacteria in the inoculant powder will stick to the roots emerging from the seed and colonize the plant as it grows. These bacteria may well be present in your soil, assuming you regularly grow peas and beans, but it doesn't hurt to add some to the seed, especially when the soil bacteria are still sluggish from winter inactivity. The powder is best dusted on damp seed just before planting. I tell our crew to think of adding a generous amount of pepper to a meal, not more. Stir or roll the seed with the inoculant, so that each seed has some powder stuck to it. Inoculant is not expensive; it's very worthwhile. Check that the kind you are applying is suitable for garden peas, as most combination inoculants are.

In cool climates succession plantings of peas can be made, but peas sown two weeks after the first sowing will usually mature only one week later than the first sowing, as the spring warms up. Sowing several different varieties on the same (first possible) date usually works better to achieve an extended harvest period.

Transplanting

Peas can be transplanted, but not easily. Choose a method that minimizes root damage, such as soil blocks or plug flats. People in cold climates certainly get creative in their efforts to grow peas. I have heard of people sowing peas in soil in a length of plastic guttering, then sliding the whole length out to make an instant (short) row. Some use ten-feet (3-m) lengths of plastic guttering, others use a reciprocating saw to rip a length of plastic waste pipe into two pieces, narrower than guttering.

The old method was to make a stack of upside-down cut turfs and cut them up into blocks, sowing one pea in each. The number of people using such methods is a testament to the popularity of peas.

Relay planting (interplanting)

Interplanting peas into spinach beds saves time and space and makes good use of resources. In the fall we plant beds of spinach with four rows in a four-foot (120-cm) bed, leaving an empty imaginary fifth row down the center. When winter arrives, we cover the beds with sturdy double hoops

Interplanting snap peas and spinach makes good use of time and space. Credit: Kathryn Simmons.

and thick rowcover, to save the spinach from getting weather-beaten and to help it grow fast.

In early spring, we hoe or weed these beds, harvest the leaves from the inner rows of spinach, then make two shallow drills (furrows) next to each other down the middle and sow peas. Because the bed is already warm under the rowcover, we can sow earlier than otherwise. The peas get warmer conditions and we get double value from the rowcover, and can also move on to the next crop without needing to do another round of rototilling. Having two crops together keeps our attention on the need for weeding and harvesting.

By the time we want to stake the peas, we need the rowcover for warm-weather crops and need to uncover the spinach to slow down the bolting. We clear the spinach in April and May and let the peas grow. Eventually we are left with a bed of peas only.

Cultivation

We find we can get earlier crops by planting on raised beds, which dry out earlier in the spring. We hoe or hand weed as needed. Back when we sowed rows of peas on flat ground, we sowed oats between the rows, in the walking paths, and used a lawn mower to keep the "grass" short. When the crop was finished we tilled the oats along with the peas. This helped suppress weeds and gave us nice conditions underfoot.

To support our peas, we use wood stakes every ten feet (3 m), and a version of Florida string-weaving (more commonly used for tomatoes) rather than putting up any wire. We do several rounds of string-weaving up the stakes when we install them, for the peas to twine onto, rather than waiting and then enclosing the plants with the weaving. When we harvest we "pick and tuck" errant vines.

Rotations

A benefit of growing peas is that it helps with crop rotation (as well as providing nitrogen in the soil). Our main crops are heavily in the nightshade, brassica, cucurbit and sweet corn groups, so legumes are always welcome, wherever and whenever we can fit them in.

Pests and diseases

Because peas are an early crop, they reach harvest before many potential diseases develop into a problem. However, here are a few that we have encountered:

- Aphids can be sprayed off with water from a hose. Early morning is the most effective time;
- ATTRA has a very useful publication on the use of baking soda (sodium bicarbonate, $NaHCO_3$) and potassium bicarbonate ($KHCO_3$) to control powdery mildew (PM) and other fungal diseases of plants. Other possibilities for PM include milk (diluted with water, sprayed twice weekly) or Actinovate;
- Fusarium root rot causes browning and drying up of foliage from the ground upwards. Fusarium wilt causes downward curling leaves and a stem that splits lengthwise near the base. If there is orange or red discoloration, you have fusarium;
- Bean yellow mosaic virus and pea leaf roll virus are other troubles you could look into if you have blotchy yellow leaves or rolled leaves.

Harvest

Snap peas are best when completely filled out. Train your crew to test a few each morning before picking, to be sure they get the fully flavored ones. Under-ripe snap peas are a sad waste of plant potential. Snow peas, on the other hand, should generally have flat pods with discernable bumps where the peas are.

Early morning picking is said to slow the spread of PM. On the other hand, it's usually considered best to avoid handling pea plants while they still have dew on them, to reduce the spread of other fungal diseases. We pick peas three days a week, following our phonetic system outlined in the Harvest chapter.

It's important not to pull up the plant when harvesting (pea roots are shallow), so grasp the stem just behind the pod and pinch off the pod with your thumbnail. Alternatively use two hands. We often harvest into "berry buckets" on long ropes around our necks—see the Harvest chapter for details. Having the bucket around the neck leaves both hands free for harvesting. Look backward in the direction you came from, to check for missed peas. Pea harvesting is one of those tasks some people are good at, others not at all, so watch and allocate your crew accordingly.

Yield per hundred feet is about twenty pounds of pods (9 kg/30 m). Another source suggests 2,500–3,500 lbs/ac (2600 kg/ha).

Post-harvest

The sugars in edible pod peas quickly turn to starches, so chill the peas as soon as possible, and get them to the eaters quickly. The strings along the "top lip of the smile" of pea pods should be removed before cooking. We pull ours as we pick, but most markets would expect to see the stem of the pods intact. The flavor of edible-podded peas is spoiled by over-cooking, so educate your customers to steam or stir-fry for no more than three minutes. And, of course, they are delicious raw. All peas freeze well, if you have more than you can find immediate markets for.

Fall sowing

Sowing for a fall harvest is possible in zones 5b–8a. Timing is critical, as is luck. It's a gamble, unless your climate provides a long, somewhat cool frost-free autumn. If the weather is too cool, growth will slow down too much and the plants won't ever produce peas. (Remember—flowers and pods are less frost-hardy than vines.) The seed will germinate quickly in warm soils. Sow one to two inches (3–5 cm) deep. Air temperatures of 75°F (24°C) are ideal. Shading the rows is a good plan—perhaps plant to the north of sweet corn or sunflowers, pole beans or trellised tomatoes.

Sow two months or more (70–90 days) before frost, to allow enough time to get a crop. Fedco recommends the first week of July in central Maine. Here in Virginia, Southern Exposure Seed Exchange recommends July 7–Aug 15 for the mountains, July 15–Aug 1 for the inland plains and Aug 7–Aug 21 for the coastal parts of our region. Aug 1–15 is too late in southern Missouri and similar zones.

Choose varieties with good disease tolerance or resistance, and add an organic mulch once the seedlings have emerged. If necessary, the plants can be protected with rowcover as the cold weather arrives.

Pea shoots

Pea shoots are short lengths of young stem with leaves and tendrils. They can be grown specially or cut from vigorous plants grown for pods, if you are willing to risk a small reduction in pod production. We often sow leftover soaked pea seed closely in another bed, to use for shoots. It's also possible to cut pea shoots in early spring from overwintered Austrian winter peas grown as a cover crop.

If sowing in spring deliberately for pea shoots, choose a well-drained site and prepare a good seedbed, as the plants will be very close and avoiding disease is important. Sow the peas about two inches (5 cm) apart in all directions, or in close rows. A first cutting can be taken about 32 days after sowing. Snip the top two to six inches (5–15 cm) of the plant. The remaining shoot will then branch, and after twenty to thirty days, another cutting can be taken. There will be twice as many pea shoots, due to the branching. Monitor the quality, which is a pompous way of saying "taste them," and stop cutting when they start to get bitter. In cooler parts of the country, it might be possible to mow off the plants in July and start harvesting again in the fall.

Pea shoots are eaten in salads or very briefly steamed or added to stir-fry. They are high in carotenes (antioxidants), vitamin C, vitamin K and folate.

Seed saving

Peas can cross-pollinate (which is why odd plants sometimes show up in rows sown from commercial seed). Mostly they self-pollinate. The isolation distance is 50' (15 m) for home use. For pure seed for sale, isolate by 150' (46 m).

Chapter 31

Peanuts

Peanuts are an interesting and rewarding crop to grow on a small scale. Farmers who offer farm tours will enjoy showing how the flowers "peg down" to develop the nuts under the soil. Those concerned about local food security after peak oil will appreciate peanuts as a source of locally grown protein. Everyone can enjoy the challenge of trying a new crop. Peanuts have been grown in the US since the 1700s. The first commercial planting was around 1800, in North Carolina.

Clemson Extension Service has this description: "The peanut (*Arachis hypogaea*) is a tropical plant that originated in South America. Peanut is a self-pollinating plant that looks like a yellow-flowered, low sweet pea bush growing slightly higher than 1 foot [30 cm] with a 3-foot [1-m] spread. After the flowers wither, a flower stalk called a peg elongates and pushes the ovary or pistil of the flower into the soil to a depth of 1 or 2 inches [2.5–5 cm]. The pistil then develops into the pod containing the peanuts." This useful publication is listed in the Resources section.

Crop requirements

Peanuts need a frost-free period of at least 110 days. 2520–2770 growing-degree-days above a base of 57°F (14°C) are needed for Virginia type peanuts. An easy way of tracking that is at weather.com/outdoors/agriculture/growing-degree-days. Peanuts like warm or hot conditions, with adequate but not excessive water. If you can provide these conditions, as well as relatively light, loose, free-draining soil, and attention to controlling weeds, then I encourage you to give peanuts a try. Buy some seed peanuts (we get ours from Southern Exposure Seed Exchange). Twenty-eight grams will plant about 25 feet (7.6 m). Prepare a bed or row ahead of time. You can warm the soil by covering with rowcover or plastic for three to seven days before planting. The pH should be 5–6, and if calcium levels are at all low, add gypsum—the start of blooming is the latest time you can effectively do this. Calcium deficiency results in unfilled pods and low yields (see the NCSU website, listed in the Resources section, for more on soil nutrient requirements). We plant our peanuts in a permanent raised bed, which helps provide looser soil.

At the Coastal Plain Experiment Station in Tifton, Georgia, they are trying peanuts interplanted with crimson clover. See the New Farm article in the Resources section for lots about no-till and reduced-till options, as well as flame-weeding. The NCSU website has an interesting "risk-assessment calculator for reduced tillage systems."

Those thinking of pre-sprouting or transplanting peanuts should be alerted to a peculiarity of the crop: they do not germinate well without soil. Exactly what peanuts need from the soil remains a bit of a mystery. So if you want your peanuts to get an early start, don't try to start them in a soil-less mix, or in paper towels or a glass jar. Use actual soil, or sand/compost/soil mix. They can then be successfully transplanted.

Varieties

If you have a relatively short growing season, I recommend starting with the 110-day variety Tennessee Red Valencia. It's easy to grow, very productive and has large, sweet kernels, two to five per shell. If you don't want to hill, this variety is said to be most accepting of this. Another 110-day variety is the unusual Carolina Black, an heirloom with a black skin. It's sweet-tasting and slightly larger than the Spanish kinds. Virginia Jumbo needs 120 days, produces large rich-flavored kernels, two per shell, and prefers loose soil. Whopper also needs 120 days, and its claim to fame is obvious. Carwile's Virginia is a 130-day heirloom variety with superior flavor and good drought resistance (though only average disease resistance). So far, I'm blissfully ignorant about peanut diseases. See the Resources section if you need to

know. Peanuts are prone to the same diseases as beans, with bacterial wilt being one to especially watch out for. *Sclerotinia* blight is more prevalent at high pH.

Sowing

Sow around the last frost date. The soil temperature should be 65°F (18°C) at noon, at a 4" (10 cm) depth, for three consecutive days. For us, that's late April. The beginning of June is about the latest I'd want to try here. Shell the nuts (don't split the kernels into halves) and sow 2" (5 cm) deep, 12" (30 cm) apart, in rows 30"–36" (76–90 cm) apart. Inoculating the seeds with *Rhizobium* bacteria can produce enough nitrogen in root nodules to sustain the crop, as with other legumes. Active nodules are pink or red inside. The seedlings may be slow to emerge, so preemptive weeding may be needed. If you have not interplanted, you could try pre-emergence flame-weeding (this is also being tried at Tifton). The seedlings look somewhat like peas or clover. Because they grow slowly for the first 40 days, they will not thrive if you lose them in weeds (guess how I know?!).

Interplanting lettuce and peanuts

Relay planting is a version of undersowing or interplanting, where the second crop is planted while the first is still growing. We sow a row of peanuts in the middle of a lettuce bed and get great results. The timing is a little tricky, so try it at least twice before deciding whether it suits you. We are still fine-tuning this one! We sow the peanuts April 29–May 12 (around our average last frost date) into the middle of the bed with lettuce transplanted on April 22–May 15. The ideal seems to be to plant regular size lettuce transplants (not overgrown ones!) on the same day you sow the peanuts, or up to two weeks later. We use romaine lettuces and small Bibbs for these plantings, not large spreading leaf lettuces.

When we sowed peanuts in an empty bed, the slowly emerging peanuts got lost in weeds and the slow-growing unusual seedlings were hard for some of our newer crew to distinguish from the weeds. We frequently hoe in our lettuce beds, so as long as we remember that the peanuts are there and don't hoe them off, they do well. In hot springs we have had shadecloth over the whole bed for the lettuce, and the peanuts come up very nicely. In cooler springs we use rowcover.

The lettuce grows faster in cooler, wetter springs than peanuts do, so if necessary, we harvest the inner rows of lettuce a bit earlier than we might have expected, before the peanuts get swamped. All the lettuces are harvested before the peanuts grow large, leaving the peanut canopy to fill out the space.

Cultivation

Once the seedlings are 12" (30 cm) tall, you can hill them up, as you would potatoes, to increase yields. We skip this step—the peanuts have no difficulty pegging in the loose soil of our raised beds. We've grown Virginia Jumbo and Carwiles without hilling and got plenty of peanuts. But if you are not interplanting and have a good hiller attachment on a tiller or wheel hoe (or a visiting school group, perhaps), then hilling can deal with weeds and provide more loose soil for the nuts to develop in. Avoid disturbing the soil after pegging has begun.

Watering

There are two periods when adequate watering is vital. The first is at planting, to ensure germina-tion. The second is from fifty days after planting, when pegs start to enter the soil, until two weeks before harvest. This is the time when the pods are filling. Stop watering ten to fourteen days before harvest, to help the peanuts dry for storage.

Harvesting and drying

You can wait for a light frost to kill the tops, or you can simply harvest when enough peanuts are mature. If the fall is wet, you might be advised to pull or dig the plants early, when the seed is almost ripe, and dry them indoors. This will minimize losses from rotting.

In dry, light soil, you can just pull the plant and the peanuts will come up attached to the stems. Then you can bunch and hang them up-side down in an airy barn where mice are not a problem. Mice really love peanuts and will run off with them and stash them at a phenomenal rate. If the soil was dry and loose enough for you to pull the plant, the nuts will likely be relatively clean, so you can dry (and cure) them on the vine (it takes up to three weeks).

In heavier soils, dig the plants up lightly—no need to go deep. You can pull the nuts off the stems into a bucket and collect the loose ones from the soil. If needed, wash the peanuts, spreading the loose ones out on trays to dry. Mold is an enemy of efforts to dry peanuts. Moldy peanuts contain toxic aflatoxin, so get them dried quickly. Set up fans indoors, or dry the peanuts outdoors in the sun: three to seven days is a likely amount of time.

See below for information on a wonderful solar dryer. If you don't have one you could use our previous method, which involved spreading the peanuts on clean compost-screening trays on wheelbarrows. Just wheel the barrow from place to place to catch the sun, and under cover if rain threatens. A second compost screen inverted over the first keeps mice away. This system works for very small-scale production. You could also make mouse-proof lidded trays from rat wire and dry the peanuts on greenhouse benches, with fans.

Boiled peanuts

Immediately after harvest you can eat all or some of your bounty, without drying or curing. These are a special treat. Here's how we cook a big pot of them for a party: fill a large pot two-thirds full of water, add two handfuls of salt and enough peanuts (in their shells) to fill the pot, but still have enough room to move. Cook for five whole hours, strain and serve. They are a big hit!

Curing and storing

If vine-drying, you can also cure the peanuts on the vine, as described above. Or else pick the good nuts off the plants once they have dried and cure them in the sun for a few days. Peanuts will cure to a storable state in two to three weeks indoors at ambient temperatures or six days in a solar dryer of the type described below, kept facing east. It's important not to heat the nuts higher than 85°F (29°C). Drying too fast causes skin slippage and kernel splitting. To find out when peanuts have finished curing, taste some and see. If they still have the watery crunch of water chestnuts, they are not ready; once they taste good, they're cured. For long-term storage, a sealed container in a freezer works well. They need to be kept very dry. Like most seeds, peanuts are very good at drawing moisture from the air.

Solar drying and curing

We built an excellent solar food dryer, following plans from *Home Power* magazine. Detailed plans are on their website. It can be built for around $150, and operates by natural convection. The food is placed on racks in the back portion, out of direct sunlight. Solar energy passes through a sloping box topped by a glazing material. Layers of black metal window screening absorb the heat, and the hot air rises into the drying chamber. This food dryer is actually designed to reach temperatures of 110°F–180°F (43°C–82°C) in order to dry food in two days or less. Our use for curing peanuts could be called "off-label," as we are trying to avoid such high temperatures. Instead of turning the dryer on its wheels to track the sun and get maximum solar gain, we keep it facing east, so it gets early daytime warming but does not overheat. Proper attention to opening vents in the morning and closing them later in the day to prevent rehydration is important.

Roasting

To roast peanuts, spread them shelled or unshelled in a single layer and cook them at 350°F (175°C) for 20–25 minutes in the shell, or 15–20 minutes shelled.

Seed saving

After your first year, you can save your own seed and grow as much as you care to. Seed saving on this scale couldn't be easier! Peanuts are largely self-pollinating but are visited by various bees, so cross-pollination can occur. Grow only one variety (or if you grow more, isolate the varieties from one another). Commercially, isolation distances of a thousand feet (300 m) are used. Seed quality is easier to maintain in unshelled peanuts, so store them this way until planting time. After drying and curing, set aside sufficient quantity for next year's seed, in a very mouse-proof place. You can select for size and number per shell at the storage stage. To select for flavor, save at least twice as much as you need. At planting time, open the shell, eat one kernel, then decide whether or not to plant or eat the rest in that shell. Don't eat too many, of course! To select for disease resistance or per plant yield, you will need to make the assessment at harvest time.

Peanuts in the hoophouse

Theoretically, peanuts could be a good summer crop, being heat-loving and a legume (most hoophouse crops are brassicas, lettuce, cucurbits or nightshades, so legumes help with the rotation). We tried growing them in our double-poly hoophouse, but didn't rate it as a success: We got spindly plants and low yields compared to our outdoor planting, so we're in no hurry to repeat the experiment. Possibly we didn't water enough. Or possibly mulch would have helped keep soil moisture in. Perhaps others will have more luck. I'd love to hear about it.

The solar drier at Twin Oaks was built following the design in Home Power magazine. Credit: Twin Oaks.

Chapter 32

Broccoli, Cabbage, Kale and Collards in Spring

Kale can make a higher net profit than the same area of outdoor tomatoes! Brassicas (cole crops, also previously known as crucifers) can be reliable and valuable workhorses, providing large harvests over long periods. I have grouped broccoli, cabbage, kale and collards together because their requirements, and the ways we grow and harvest them, are very similar. Some other brassicas (Asian greens, turnips and rutabagas) have their own chapters. The Winter Hoophouse chapter covers many brassicas too. I have written separate chapters on growing brassicas in the spring and the fall because there are significant differences in raising the transplants, fitting the planting into your rotation and dealing with pests and diseases. In northern regions, these brassicas are planted only in the spring and stand all summer, through

the fall and into the winter, until cold weather kills them. In the South, they are a spring/early summer and a fall/overwintered crop and summer is the impossible time for them. In parts of California and the Pacific Northwest, they'll grow year round.

Growers scaling up their production can do so by introducing different brassicas, extending the season for favorite greens, introducing new mini sizes of broccoli or cabbage by reducing the spacing, and of course by planting more land.

Crop requirements

Brassicas do best on fertile, well-drained soil with adequate moisture. They require a lot of potassium, so a sprinkling of wood ash, kelp meal or granite dust could be helpful. Calcium, boron, iron, manganese and molybdenum are also important for good brassica crops—healthy, biologically active soils can usually supply enough of all these. Optimal growing temperatures for most brassicas are 60°F–65°F (15°C–18°C).

Broccoli varieties

Have a clear mind about your goals. Do you want crown-cut broccoli (big heads) or bunches? Are side shoots valuable or a hassle? How long of a harvest season do you want? How much do you want at once? Our goal for broccoli is enough for a hundred people to eat it fresh every other day, plus some left over to freeze for the winter. We also like to have some good side-shoot production. To extend the season as long as possible, we grow several varieties with different numbers of days to maturity, and have two sowing dates for each. This gives us the maximum harvest period from just two rounds of transplanting.

Days-to-maturity listed are often days from transplanting, so add the time to grow the transplants, or 21 days if direct sowing, to determine when to sow. If the days-to-maturity listed are days from direct sowing, subtract 21 and add back in the 35–56 days to grow the transplants. We have recently grown Tendergreen (47d from transplanting) as our early variety, replacing Packman. Windsor (56d), Gypsy (58d) and Premium Crop (61d) fill out our mid-season. Diplomat (68d) and Arcadia (63d) are our slow-maturing varieties.

Johnny's rates Windsor, Gypsy, Bay Meadow and Green Magic as their best at tolerating hot weather (for spring sowing) and Arcadia, Diplomat, Marathon and Belstar as having the best cold tolerance (for fall crops). Our climate is different from Maine and doesn't divide so simply into hot and cold. Our broccoli gets either a cold start and a hot finish (spring sown) or a hot start and a cold finish (fall crops). We do plant Arcadia in spring and harvest successfully in quite hot weather, although it's not rated as having good heat tolerance. Fedco finds Windsor good on both cold and heat tolerance, as they do Fiesta (86 days from direct sowing).

We have not had success with open-pollinated broccoli, or with broccoli raab, which are waving yellow flowers at us before we've realized they are ready. I think small vegetables are better suited to those cooking for less than eight, rather than catering-scale cooking, as at Twin Oaks.

Cabbage varieties

We aim for some quick-maturing cabbage varieties, choosing Early Jersey Wakefield and Faroa (64d). We also like some large solid cabbage for making sauerkraut: Early Flat Dutch fits the bill. And we look for a green cabbage that holds well in hot weather, to extend the season as long as possible: Tendersweet (71d) is a tasty variety that works well for us. We like a red that's fairly large and yet quick-maturing (many reds are too slow for us, taking us deep into hot weather). We used to favor Super Red 80 (73d), but gave it up after two poor years.

Varieties of kale and collards

Our kale, for both fall and spring planting, is Vates Dwarf Blue Scotch Curled, the most cold-tolerant we have found. There are special spring kale varieties such as Hanover, but we prefer the curly-leafed kind over the flat-leafed as it holds up better against wilting and wind damage. We have not been successful with Nero di Toscana, the dinosaur kale, as it harbors too many aphids in the curled-back edges of the leaves. White Russian

Vates is our all-time favorite kale variety for outdoor growing. Credit: Kathryn Simmons.

and Red Russian, which do very well in our winter hoophouse, do not do well outdoors for us, as they get weather-beaten.

Our collards are Morris Heading, a slightly savoyed leaf, finishing with a head later in spring. We buy it from Southern Exposure Seed Exchange. We've also grown Georgia. Northerners seem to prefer smaller-leafed collard varieties such as Champion. Green Glaze has shiny leaves more resistant to caterpillar attacks.

In our climate, there are no kale and collards varieties that will grow through the summer; ours have all bolted by mid-May. We find that by doing some spring planting, we can boost our total production (those overwintered plants get a bit tired!). In addition, the spring plantings will bolt a little later than the overwintered ones, so we extend our season.

Kohlrabi

Kohlrabi is an unusual crop that I'm squeezing into this chapter. It's sure to attract attention at your market booth, or be a discussion piece in CSA bags. It is tender and crunchy with a flavor between a cabbage and a turnip. It's easy to grow and doesn't wilt as soon as you harvest it. It can be eaten raw (sliced or grated) or cooked, and the leaves can also be cooked and eaten. It also stores well, offering flexibility about when it is sold, which is always an advantage for growers.

We like the open-pollinated Early White Vienna and Early Purple Vienna (58 days from sowing) and the purple hybrid Kolibri (45 days), which is more expensive but beautiful, quick-growing and uniform. You'll notice that the purple ones grow slower than the white (green) ones. A mixture of the two colors looks very attractive, and they're both white under the skin, so our strategy is to grow 25%–35% purple and mix them in. 4 g = about 1,275 seeds, sows 135' (41 m). 5,500–9,000 (avg. 6,500) seeds/oz, (200–320 or an average 230 seeds).

The other type of kohlrabi is the lumpen giant winter storage variety, such as Gigante (130 days from sowing in spring for late fall harvest). It regularly exceeds 10" (25 cm) in diameter and 10 lb (4.5 kg) in weight. The world record is 62 lb (28 kg), leaves included. Gigante's crisp, white, mildly tangy flesh doesn't get woody even when it grows enormous.

Brassica pests seem less interested in kohlrabi than in other brassicas, and as the leaves are not generally eaten by humans, there is scope for allowing a certain number of leaf-eating pests, while still having an attractive crop and a high enough yield.

Purple varieties seem slightly more resistant to cabbage worm and also less susceptible to cracking.

Amounts to grow

We plant cabbage, broccoli and collards at 18" (46 cm) and kale at 12" (30 cm). Yield can be 1 pound (450 g) per plant for broccoli. On a larger scale, you can expect five tons per acre (11 t/ha) of broccoli, 15 tons per acre (34 t/ha) of cabbage.

We like large broccoli heads. For a 25 percent higher total yield, but smaller heads, you can produce three-packs of 1½ lb (0.7 kg) heads. Plant double or triple rows with a 24" (60 cm) aisle between outer rows of one set and the next. The in-row spacing is only 6"–8" (15–20 cm).

For markets supplying small households, mini-cabbages spaced at 8"–12" (20–30 cm) in the row, with rows 12"–18" (30–45 cm) apart, may give higher total yields as well as heads of a more useful size.

For a hundred enthusiastic vegetable eaters, in spring we grow up to 2,000' (600 m) of broccoli, 600'–700' (180–200 m) cabbage, 1,080' (330 m) of kale and 540' (165 m) of collards. We harvest kale and collards by the leaf, for multiple harvests.

Rotations and cover crops

In our rotation, early food crops are preceded by easy-to-incorporate cover crops that winter-kill and break down easily when disked in February. For us, that's oats and soy sown in August. They, in turn, require a preceding food crop that is finished in August. Our main spring broccoli and

Kohlrabi is an eye-catching crop. Credit: McCune Porter.

cabbage patch follows an early sweet corn crop the previous year.

When our spring brassicas become bitter in early July, we bush hog to destroy harlequin bug habitat. We may simply mow until early September, then put in the winter cover crops. If conditions allow, we disk and sow a round of buckwheat, perhaps with soy. We need six weeks or more for a good growth of soy, but only four weeks for buckwheat. Our goal is to disk this patch and sow winter rye, hairy vetch and Austrian winter peas Sept 7–14, to make good growth for no-till tomatoes the following year. None of our other food crops are finished early enough for the plot to be available in time to sow the vetch mix. Spring cabbage and broccoli provide this important niche in our rotation. The combination of spring cultivations and summer cultivation with early fall cover crops can provide opportunities to deal with both winter and summer annual weeds.

A rotation of one year in four is ideal for brassicas to avoid disease and pest buildup. When possible, plant one year's brassicas at some distance from the previous year's, rather than in adjacent plots. The spread of disease can be reduced by avoiding walking through wet crops and by controlling weeds in the brassica family, such as rock cress, pepperweed, garlic mustard and wild rocket.

We often plant our earliest cabbage in permanent raised beds, as we cultivate these with the walk-behind tiller and can work them a bit sooner than the flat fields, where we need to wait for the soil to dry enough to use the tractor and disks. We always grow collards and kale in the raised beds.

Sowing

In early spring, transplants have the advantage over direct-seeded crops—they grow faster under protected conditions and bring earlier harvests. At other times of year, you may prefer to direct seed.

Work back from the desired harvest date to calculate the sowing date. Naturally you'll need to allow for your climate and choose a realistic transplanting date. We start our very first cabbages in mid–late January, as soon as I'm psychologically prepared to start a whole new year. Kale and collards follow at the beginning of February, broccoli and maincrop cabbage a few days later. A second broccoli sowing happens fourteen days after the first (to extend the harvest season) and a third (to fill gaps) ten days after the second. We direct sow kohlrabi four to six weeks before the last frost, 0.25" (0.7 cm) deep, 1" (2.5 cm) apart. For an earlier crop, we start transplants eleven weeks before the last frost date, and set them out at six to eight weeks old.

Our general formula is to produce 20 percent more starts than we intend to grow, so we can pick the best and sail through 20 percent of a disaster. Despite being a cool-weather crop, brassicas actually germinate very well at high temperatures: the ideal is 77°F–85°F (25°C–30°C), and 95°F (35°C) is still OK. Given enough water, the seedlings will emerge in four and a half days at the low end of this range, and at the top in only three days. The minimum temperature for good germination is 40°F (4.5°C), but you'll need to wait more than two weeks for emergence if it's that chilly.

There are several systems for growing transplants: open flats, cell-packs or plug-flats and outdoor seedbeds. People who use cell-packs for brassicas often use the 48-cell inserts in a standard germination tray, or the 72-cell size with 1.5" (4 cm) cells.

We start most of our spring brassicas in open flats, sowing 3–4 seeds per inch (5–10 mm apart). We press a plastic ruler 0.25" (6 mm) deep into the seed compost to make a small furrow, spacing the rows 2"–3" (5 cm) apart. As soon as the seed leaves are fully open, we spot (prick) them out to 4" (10-cm) -deep flats, with 40 plants per 12" × 24" (30 × 60 cm) flat. The plants grow to transplanting size in these flats.

Once they have emerged, the seedlings need good light, nutrients, airflow and protection from bugs. 60°F–70°F (15°C–21°C) is a good temperature range for growing them. For the last two weeks before transplanting, harden off the plants by moving them into cooler, breezier, brighter conditions.

For kale we have had good success sowing in a hoophouse bed and growing the young plants there to transplanting size (about five weeks old).

Rolling out re-used drip tape for new strawberry beds.

Pam shows one of their homemade drip tape shuttles.

Credit: Luke J. Stovall.

Credit: Luke J. Stovall.

Credit: Luke J. Stovall.

Mulching new strawberry beds with newspaper and dried sorghum-sudan grass cut from the patch behind a few weeks earlier.

Credit: Luke J. Stovall.

Setting out double layers of overlapped newspaper before topping with organic mulch doubles the effectiveness of the mulch without making it any deeper.

The Colored Spots Plan

Harvesting Roma paste tomatoes.

Hanging our garlic in netting to cure gives the crew a pleasant break from hot outdoor work.

Harvesting Jalapeno hot peppers.

A broccoli plant one week after transplanting into pre-rolled hay.

Vates is our all-time favorite kale variety for outdoor growing.

Unless precision-sowed, carrots need thinning as soon as possible to 1" (2.5 cm) apart.

A healthy young zucchini plant.

Credit: Kathryn Simmons

Credit: Kathryn Simmons

Winter squashes need a big space, but are easy-care. This is a Cheese Pumpkin squash.

Credit: Kathryn Simmons

Onions are a top priority crop for keeping weed-free, as the vertical leaves offer no competition to weeds.

Yellow potato onions in early April.

General Lee is our regular slicing cucumber variety.

Corona peppers are one of our favorite orange varieties.

Harvesting and boxing up our four varieties of sweet potatoes.

Credit: Wren Vile.

Credit: Kathryn Simmons

Once the vines have run, sweet potatoes cover the soil and leave no room for weeds.

Credit: Kathryn Simmons

We often use soybeans as a traffic-tolerant, nitrogen-producing undersown cover crop that also deters weeds.

Garlic emerging through the hay mulch in mid-November.

Garlic harvest.

Credit: Pam Dawling

Credit: Twin Oaks

Credit: Kathryn Simmons

Garlic plants in spring.

Credit: Twin Oaks

Garlic hanging in netting to cure.

The bare-root transplants adapt well to outdoor conditions.

Stressed plants are liable to bolt or, in the case of broccoli, "button" (produce tiny heads). The plants are more likely to button if the transplants are over-large (stems more than 0.25"/6 mm in diameter), so it is better to make several sowings and only set out plants of a moderate size.

Transplanting

We aim to transplant at four or five true leaves (5–8 weeks after sowing). This is early to mid-March for kale and collards (usually five weeks old) and the earliest cabbage (7 weeks old), very early April for maincrop cabbage and the first broccoli (7 weeks old), a week later for the second sowing of broccoli (it is sown two weeks later, but catches up one week). Soil temperatures of 65°F–75°F (18°C–24°C) are ideal. We transplant every afternoon until done. Water the seedlings well before transplanting, and plant up to the base of the first true leaves to give the stem good support. Press the soil very firmly around the plants so that the roots have good soil contact and won't die in an air pocket. Water within half an hour of planting and again on the third and seventh days, and then once a week.

We grow spring broccoli and cabbage in beds on 4.5'–5' (1.3–1.5 m) centers, at two rows per bed, with plants 18" (50 cm) apart. After spreading compost and incorporating it, we mulch the beds with rolled-out, spoiled hay bales, then make "nests" (holes down to the soil) in the hay by hand, using a stick of the right length to measure spacing along one row. The person working the opposite side of the bed does not measure but eyeballs the spacing, to stagger the two rows. After planting and watering we pull the mulch in around the base of the stems. We do not mulch kale or collards, because they will not be there so long.

When broccoli plants have five to eight true leaves, they are susceptible to cold stress if subjected to continuous temperatures below 40°F (4.5°C) for a few days, or a longer spell at 50°F (10°C). For cabbage, the sixth and seventh leaf stage is the most vulnerable. So if the weather deteriorates at the critical time, give your plants extra protection. High daytime temperatures can to some extent compensate for low nighttime temperatures—it's cold days and cold nights together that do the damage.

Our spring broccoli and cabbage planting is traditionally marked by cracking open a brand new roll of rowcover, as this crop uses lots, and we often need every piece we've got. We also take rowcover directly from the early turnips, leek seedlings, oriental greens and even the kale and collards (only three weeks after transplanting) to the tenderer broccoli. It's more efficient to take rowcover and sticks directly from one crop to another, rather than packing them away in the shed just to haul them out again a few days later.

Two weeks later, we check the plants and fill any gaps. Plants from the first sowing are usually too big by then. We use the younger plants (fill gaps in the first sowing with second sowing plants, gaps in the second sowing with plants from the "filler" third sowing).

Farmscaping for brassicas

Putting 5 percent of the crop area in plants that attract beneficial insects can seriously reduce pest numbers. Sweet alyssum, yarrow, dill, coriander (cilantro), buckwheat, mung beans, black oil-seed sunflower, calendula and cleome all work well to attract a range of insects that eat or parasitize

Spring cabbage and broccoli beds, with some brussels sprouts in the distance. Credit: McCune Porter.

aphids. Pans of water and gravel will help attract aphid midges and lacewings. The gravel provides surfaces for the insects to land on while drinking. Mix flowers to have something blooming all the time. Beneficials will generally move up to 250 feet (75 m) into adjacent crops. Cleome is a good trap crop for harlequin bugs, which can be captured in buckets of soapy water by holding the tall open flower heads over the bucket and shaking.

At the "gap-filling" stage, around April 25, we also plant out alyssum (sown March 3) in the center of the beds, one plug every five to six feet (1.5–2 m), to attract beneficial insects. We also plant "insectaries" around the garden, usually at the ends of beds. These flowers are planted inside rings sawn from a plastic bucket. The rings alert the crew that something special is there, not just a clump of weeds.

Caring for spring brassicas

About a month after transplanting broccoli and maincrop cabbage (mid-May), we remove the rowcovers and the sticks we use to hold them down, taking them directly to cover newly planted lettuce, eggplant, watermelons, muskmelons and okra.

Our broccoli and cabbage patch is next to our garlic patch, which works well. In the middle of May, we weed the brassicas. Then we weed the garlic, gather the weeds and mulch from the garlic and use it to top up the mulch around the broccoli and cabbage. This helps the garlic dry down for the last three weeks before harvest and helps maintain cool, damp soil to maximize the length of the broccoli harvest season, and the size of the cabbages.

We use only overhead sprinkler irrigation (not drip tape) for spring brassicas, which helps cool the leaves and can wash off aphids. Organic mulches help keep the soil cool, as well as adding lots of organic matter to the soil. Brassica roots are relatively shallow, so long droughty spells without irrigation can cause problems. One inch (2.5 cm) of water per week is about right. If you are using plastic mulch, you'll need to use drip irrigation under it. The black plastic can get too hot later in the spring, but luckily the plant canopy will cover some of it.

With broccoli, the first weeks after transplanting are used for vegetative growth, adding leaves until there are about twenty, when "cupping" starts—the leaves start to curl up, forming a convex shape rather than growing straight out. The cupping stage is usually ten to fourteen days before harvest starts, depending on temperature. If the weather gets too hot—more than 80°F (27°C)—too soon, broccoli may grow only leaves, and not head up.

To reduce the splitting of cabbages that can follow a sudden big rainfall, it is possible to either hoe energetically, destroying some of the roots, or give the cabbages a sharp quarter-turn twist. Either action will slow the water uptake into the leaves.

Kale and collards are easy to care for—hoe the weeds and water enough to keep the crops growing vigorously.

Pest control

Using rowcovers keeps many pests off the plants while they are small. We have not had much trouble with aphids, perhaps partly because of our overhead sprinklers; insecticidal soap sprayed three times, once every five days, can usually deal with them. Our worst pest is the harlequin bug. For lack of a better organic solution, we handpick them. Ladybugs are reputed to eat harlequin bug eggs.

Sometimes we have had enough cabbage worms to make Bt (*Bacillus thuringiensis*) necessary, but usually paper wasps eat the caterpillars. The action level threshold is an average of 1 cabbage looper, 1.5 imported cabbageworms, 3.3 armyworms or 5 diamondback moth larvae per 10 plants. Below this level you can do watchful waiting rather than spraying with Bt or spinosad. We are lucky enough to have the naturally occurring wasp parasite of cabbage worms, the Braconid wasp *Cotesia* species, which are found as small cottony white or yellowish oval cocoons in groups on brassica leaves. The *Cotesia* wasps like umbel-

liferous flowers, and overwinter on yarrow as well as brassicas. If you find *Cotesia* cocoons and your brassicas aren't diseased, you can leave plants in the field over winter. Or you could collect up leaves with cocoons in late fall and store them at 32°F–34°F (0°C–1°C) until spring. Hopefully no one will clean out your fridge without asking.

Richard McDonald has good information in his *Introduction to Organic Brassica Production*. He reports that broccoli plants with six to sixteen leaves (just before cupping) can lose up to 50 percent of their leaf area without reducing yield. Moderate defoliation (20–30 percent) causes the plant to exude chemicals that attract parasitic wasps and predatory insects. If you relax and allow this amount of defoliation early on, you can encourage these beneficial insects to move in and begin foraging in the area.

Once the plants cup, you want to prevent further defoliation by having the most beneficials and the fewest pests on site. If pest levels are above the action threshold, cupping is the stage to take action, and probably not earlier.

To float out worms and aphids after harvest, use warm water with a little vinegar and soak for up to fifteen minutes, then rinse.

Trap crops

A row of mustard greens can be used to lure flea beetles. They like the pungent compounds in brassicas. Once you have lured the flea beetles you need to deal with them before you create a flea beetle breeding ground. Flaming the mustard plants is one possibility. If you have poultry that likes eating flea beetles, you could cut off some of the leaves and carry them to the chicken run. Bug vacuums are also a possibility. Another approach is to hold an inverted bucket lined with sticky trap

We plant our Morris Heading broccoli three rows to a bed, to give it ample space. Credit: Kathryn Simmons.

compound over the plants and rap the stems with a stick. If you're lucky, the pests will stick in the bucket.

Harvest

We like our broccoli heads to get as large as possible (without opening up) before we harvest. We test by pressing down on the head with our fingertips and spreading our fingers. We harvest as soon as the beads start to "spring" apart. This may be a little late for other growers. We also look at the individual beads and aim to harvest before the beads even think about opening. We cut a fairly long stem, which reduces the number of side shoots, letting the remaining ones develop more strongly. We cut diagonally to reduce the chance of dew and rain puddling, which can cause rotting of the stem.

Kale and collards overwinter here, and we also plant more in spring to boost production for harvest until the end of May. (The overwintered plants bolt earlier—mid-May.) We harvest both kale and collards by snapping off the largest leaves. We use a marker flag to show which bed to harvest next, and move it to the next bed after each harvesting.

Our cabbage heads up from May 25 and some hold until July 15. Our broccoli harvest is about 35 days long, May 20–June 30, possibly a week later. For storage cabbage, we set the cut heads upside down on the stump, in the "basket" of outer leaves, and come back an hour later to gather them into net bags. This allows the cut stem to dry out and seal over, improving storability. We collect any non-storable cabbage into "Use First" buckets.

Kohlrabi may be harvested at 2"–3" (5–8 cm) in diameter or even up to softball size, depending on your market and your growing method (direct sown and thinned, or transplanted). If left growing for too long the swollen stem becomes woody. Cut the plant from the ground with a sturdy knife. The base of the globe can be quite fibrous, so either cut the wiry root just below the soil surface or cut higher, leaving a small disk of the globe behind, attached to the root. Snip or lop off all the leaves if storing, or leave a small top knot if the kohlrabi will be sold immediately. We harvest in spring from around May 10 to June 30.

We harvest all our brassicas three times each week, and take the produce directly to our walk-in cooler. Some growers place ice on top of the broccoli or hydrocool it (submerge it in cold water or shower it from above with cold water) if they are dealing with a large volume compared to the cooling power of their refrigeration. Suitably chilled, broccoli can store for ten to fourteen days.

Chapter 33

Broccoli, Cabbage, Kale and Collards in Fall

Crop requirements

Here in central Virginia, most brassicas are planted in spring and again in fall. Although it can be hard to think about sowing seeds in midsummer, it's really worth making the effort to grow fall brassicas because as they mature in the cooler fall days they develop delicious flavor. And weeds and pests slow down then, too—once established these crops need little care.

The most challenging part of growing fall brassicas is getting the seedlings growing well while the weather is hot. Unlike some cool-weather vegetables such as spinach, brassicas actually germinate very well at high temperatures: the ideal is 77°F–85°F (25°C–29°C), but up to 95°F (35°C) works. Some people find refrigerating the damp seeds for 24 hours gives a better germination rate. Given enough water, summer seedlings will emerge in only three days. Once they have

emerged, the challenge begins. As well as temperature and moisture in the right ranges, the seedlings need light (very plentiful in midsummer!), nutrients, good airflow and protection from pesty bugs. This last is the most challenging for us. We deal organically with flea beetles, harlequin bugs and, sometimes, cabbage worms. Our main defenses are farmscaping and rowcover. See the previous chapter for more on pest control.

This chapter focuses on broccoli, cabbage, kale and collards, although I briefly mention cauliflower, kohlrabi and brussels sprouts.

Rotations

Because fall brassicas are transplanted in summer, it's possible to grow another vegetable crop, or some good cover crops, earlier in the year. An overwintered cover crop mix of winter rye and crimson clover or hairy vetch could be disked in at flowering, and be followed by a short-term warm-weather cover such as buckwheat, soy or cowpeas. Brassicas are heavy nitrogen consumers. To minimize pests and diseases, don't use brassica cover crops.

Our system for fall broccoli, cauliflower and cabbage uses a nerve-wrackingly fast turnaround:

- We plant potatoes (Irish potatoes) mid-March, harvest early–mid-July, spread compost, and disk it in;
- Then we transplant the broccoli, cauliflower and cabbage right away;
- After four weeks, we cultivate and undersow the brassicas with clovers (see the Aftercare section);
- The clovers become an all-year cover crop over the winter and the next season.

Usually this works well for us, and the cover crop is relatively weed-free due to the intensive cultivations of the first seven months of the year.

There is research at Virginia Tech into organic no-till broccoli planting combined with cover crops and farmscaping (to encourage beneficials). Broccoli is transplanted into mow-killed cover crops such as foxtail millet and soybeans. Adding a nitrogen source is important, or the yield may be reduced.

Systems for growing transplants

The same systems you use for growing transplants in spring may also work well for fall. Direct sowing, in drills or in "stations" (groups of several seeds sown at the final crop spacing), is also possible. If you use flats, it can help to have them outside on benches, above the height of flea beetles. Benches in an enclosed shade-house might be ideal too, but a shade-house is still on our wish list.

We use an outdoor "nursery" seedbed and bare root transplants, because this fits best with our facilities and our style. The nursery bed is near our daily work area, so we'll pass by and water it. (Our greenhouse is off to one side of our gardens.) We're always relieved to close the greenhouse transplant season in May, as we no longer need to water little plants many times each day to keep them alive. Having the seedlings directly in the soil "drought-proofs" them to some extent; they can form deep roots and don't dry out so fast. Other people manage fine with flats in June and July, but we avoid it.

For the seedbeds we use rowcover on wire hoops. Overly thick rowcover or rowcover resting directly on the plants, can make the seedlings more likely to die of fungal diseases in hot weather—good airflow is vital. I have been searching for a lightweight small mesh netting that would keep bugs off, but let in lots of air. Purple Mountain Organics sells ProtekNet, which has worked well in our trial season.

Timing

The number of days to harvest given in seed catalogs is usually that needed in spring—plants grow faster in warmer temperatures. To determine when to sow for fall plantings, start with your average first frost date, then subtract the number of days from seeding to transplant (21–28), the number of days from transplanting to harvest for that variety (given in the catalog description), the length of harvest period (we harvest broccoli for 35 days minimum) and another 14 days for the slowing rate of plant growth in fall compared to spring. For us, the average first frost is Oct 14–20, so we sow 53-day broccoli 21 + 53 + 35 + 14 days before Oct 14, which is June 13–19.

The number of days quoted for fall varieties of cauliflower already allows for the expected rate of growth at that time of year, so the 14 days for slower growth isn't a factor. However, cauliflower is tenderer, so allow for the possibility of a fall frost earlier than average.

Planning

The third week of June is a good time to start sowing fall brassicas in our region, but we do modify this slightly. I go on a two-week vacation in July, thanks to reliable crew members who will take extra care of the crops while I'm not there. I try to arrange to go away in the small window after every part of the garden has been planted once and before the second round really gets complicated. So I might shift the sowing dates a week, in order to be home to join in the time-consuming and exacting transplanting.

Our ten-year rotation plan shows us a long way ahead how many row feet of fall broccoli and cabbage we can fit in. By the time we order our seeds at the turn of the year, we know roughly what we expect to grow. In February we make a more detailed plan for the fall brassicas and draw up a spreadsheet of how much of what to sow when. See the chart on the following pages for which varieties we like and how much we grow.

The more winter-hardy brassicas (kale, collards, kohlrabi and brussels sprouts) are grown in our raised bed area, which is more accessible for winter harvesting and more suited to small quantities. Each bed is 90' × 4' (27.4 × 1.2 m) and fits four rows of most crops, but three of the bigger collards. The planning for those takes place in May. We have a general idea of how much of each crop we'd like to grow, but often space for brassicas is tight, because we love them so, and yet we don't

A healthy young Morris Heading collard plant. Credit: Kathryn Simmons.

want to throw the rotation plan out of the window. We wait till closer to planting time to figure out exactly how much kale, collards, Asian greens, kohlrabi and the root brassicas we can fit in.

Sowing

Our rough formula for all transplanted fall brassicas is to sow around a foot (30 cm) of seed row for every 12'–15' (3.6–4.6 m) of crop row, aiming for three seeds per inch (about 1 cm apart). This means sowing 36 seeds for 10 plants grown on 18" (46 cm) spacing. And we do that twice (72 seeds for 10 plants!), two sowings a week apart, to ensure we have enough plants of the right size. Each week after the first week, we weed the previously sown plants, and thin to one inch (2.5 cm) apart. Then we check the germination, record it and resow to make up the numbers. Our seedbeds have an eight-week program—see the spreadsheet for examples of our timings, quantities and varieties.

Brassica Planting

1. On the same day of each week, sow, label, water, hoop and rowcover the "Feet Plan" for that week. Allow 3 hours. Make a map.
2. Check and record the % germination of the previous two weeks' sowings (100% = one plant per inch). Weed and thin to 1".
3. In a fresh row, sow topups for varieties with a germination less than 80%. Enter the info in the column for the current week. Example: If Arcadia week 1 germ = 70% in week 3, sow 30% the length sown in week 1 (30% of 15' = 5'), in the week 3 bed, and write 5' in the Arcadia row in the week 3 col. There are no sowings in weeks 5–8 except kohlrabi, resows and kale beds.
4. Transplant Chinese Cabbage at 2 weeks, others at 3–4 weeks old:
   Week 4 (7/13–7/19): Transplant week 1 cabbage and broccoli, week 2 Chinese Cabbage (Blues).
   Week 5 (7/20–7/26): Transplant week 3 Chinese Cabbage, week 2 cabbage, broccoli, senposai, Yukina Savoy, any other week 2 sowings.
   Week 6 (7/27–8/2): Transplant week 3 collards and fall greens, and any week 3 resows.
   Week 7 (8/3–8/9): Transplant week 4 collards, kohlrabi, resows. Also fill gaps in week 4 plantings (= week 1 sowings).
   Week 8 (8/10–8/16): Transplant week 5 kohlrabi, anything you didn't keep up with, and replacements in weeks 4 and 5 plantings (weeks 1 & 2 sowings).

Vegetable	Row Ft	Plants /100ft	Plants	Sd/unit	Seed Need	Varieties	Unit	Bed Week 1: 6/22-6/28 Date: Feet Plan	Feet sown	% Germ	Bed Week 2: 6/29-7/5 Date: Feet Plan	Feet sown	% Germ
Broccoli	1,980	67	1,327	200	33.2			111			111		
	540	67	362	200	9.0	Arcadia s/f 94d	gm	30			30		
	540	67	362	200	9.0	Blue Wind 72d	gm	30			30		
	360	67	241	200	6.0	Gypsy 78d	gm	20			20		
	540	67	362	200	9.0	Premium Crop s/f 82d	gm	30			30		
Cabbage	1,080	67	724	160.71	22.5		gm	103			103		
	90	67	60	160.71	1.9	Early Jersey Wakefield	gm	9			9		
	180	67	121	160.71	3.8	Farao 60+21d	gm	17			17		
	360	67	241	160.71	7.5	Melissa 85d	gm	34			34		
	90	67	60	160.71	1.9	Ruby Perfection 85d	gm	9			9		
	360	67	241	160.71	7.5	Tendersweet 71d	gm	34			34		
		67	0	160.71	0.0		gm	0			0		
Collards	540	67	362	240	7.5		gm						
	540	67	362	240	7.5	Morris Heading 70d	gm						
Fall Greens	660	100	660								48		
	90	100	90	160.71	2.8	Blues 53d	gm				8		
	90	100	90	160.71	2.8	Green Komatsuna 35d	gm				8		
	30	100	30	160.71	0.9	Red Komatsuna 35d	gm				3		
	270	67	181	160.71	5.6	Senposai 40d	gm				15		
	90	100	90	160.71	2.8	Tokyo Bekana 45d	gm				8		
	90	100	90	160.71	2.8	Yukina Savoy 45d	gm				8		
Kohlrabi	360	100	360	100	18.0		gm						
	180	133	239	341	3.5	E White Vienna 58d	gm						
	180	133	239	342	3.5	EPV 58d	gm						
Totals								213			261		

Sowing dates for crops with various days to maturity

Days to Harvest	Harvest mid-Sept– mid Oct	Harvest late Sept– mid-Oct	Harvest from mid-Oct	Days to Harvest	Harvest mid-Sept– mid Oct	Harvest late Sept– mid-Oct	Harvest from mid-Oct
30d	Jul 27	Aug 16	Aug 31	90d	May 28	Jun 17	Jul 2
40d	Jul 17	Aug 6	Aug 21	100d	May 18	Jun 7	Jun 22
50d	Jul 7	Jul 27	Aug 11	110d	May 8	May 28	Jun 12
60d	Jun 27	Jul 17	Aug 1	120d	Apr 28	May 18	Jun 2
70d	Jun 17	Jul 7	Jul 22	130d	Apr 18	May 8	May 23
80d	Jun 7	Jun 27	Jul 12				

Bed Week 3: 7/6-7/12 Date:			Bed Week 4: 7/13-7/19 Date:			Bed Week 5: 7/20-7/26 Date:			Bed Week 6: 7/27-8/2 Date:			Bed Week 7: 8/3-8/9 Date:			Bed Week 8: 8/10-8/16 Date:		
Feet Plan	Feet sown	% Germ	Feet Plan	Feet sown	% Germ	Feet Plan	Feet sown	% Germ	Feet Plan	Feet sown	% Germ	Feet Plan	Feet sown	% Germ	Feet Plan	Feet sown	% Germ
30			30														
30			30														
48																	
8																	
8																	
3																	
15																	
8																	
8																	
			40			40											
			20			20											
			20			20											
78			70			40			0			0			0		

Transplanting

We aim to transplant most brassicas at four true leaves (three to four weeks after sowing), but it often slips to five weeks before we get finished. It is recommended to transplant crops at a younger age in hot weather than you would in spring, because larger plants can wilt from high transpiration losses. If we find ourselves transplanting older plants, we remove a couple of the older leaves to reduce these losses. It would make an interesting experiment to see which actually does best: three-week transplants, four-week transplants or five-week transplants with two leaves removed. The larger root mass of the older plants would give them an advantage. On the other hand, old, large transplants can head prematurely, giving small heads. By that point in the year, my scientific curiosity has been fried by the sheer workload of crop production!

A hardworking overwintered bed of Vates kale. Credit: Pam Dawling.

We plan to have transplanting crews six days a week for an hour and a half or two hours in late afternoon or early evening, for two to three weeks (not counting the kale transplanting, which happens later). We water the soil in the plot if it is dry. It is very important at this time of year to get adequate water to the plants undergoing the stress of being transplanted. Likewise, good transplanting technique is vital, so be sure to train the crew well. See the previous chapter for details, and water a lot more than you do in spring. If you have drip irrigation, you can more easily give a little water in the middle of each day, which will help cool the roots. With only overhead sprinklers, you may not have the option of watering every day.

We transplant the broccoli, cabbage and cauliflower in 34" (86 cm) or 36" (91 cm) rows, the row spacing we use for potatoes. We hammer three-foot (one-meter) stakes every fifty to sixty feet (fifteen to eighteen meters) along the row and attach ropes between them. These both mark the rows for transplanting and support the rowcover we use for the first few weeks after transplanting to keep the bugs off. We find that an 84" (2.1 m) width rowcover can form a square section tunnel over two crop rows, giving better airflow than if it was floated directly on the plants. Setting out drip tape, watering the soil before planting and setting the plants by an emitter all help survival during the hot summer days. See the Transplanting Tips chapter for more on successful transplanting.

We plant kale by a mixed direct-sow/transplant method that allows for the possibility of patchy germination, and requires less watering than if we direct sowed it all. We sow two beds with rows 10" (25 cm) apart and then carefully thin them, leaving one plant every foot (30 cm) (those plants grow quicker than transplants, as they have no transplant shock). Meanwhile we use the carefully dug thinnings from those beds to fill gaps and to plant other beds, at the same plant spacing. Another reason we use this system is that we want a lot of kale, and there isn't time to transplant it all. Vates kale is the hardiest variety we have found, although I'd love to find a taller Scotch curled variety that could survive our winters (Winterbor does not survive as well as Vates).

Collards are transplanted bare-root, from the nursery bed. We only plant a relatively small amount of collards, and they have a wide spacing, so it is easier to transplant them all. We plant at 18" (46 cm) within the row, with rows a foot (30 cm) apart.

Other brassica crops

Brussels sprouts are rather a challenge in our climate, and not economically viable, in my opinion. We have worked on finding the best variety (Oliver) and timing for our situation, and we transplant into a mulch of spoiled hay to help retain moisture and an even soil temperature. Despite our efforts we have lost crops to under-watering and to the depredations of sap-sucking harlequin bugs. Additionally, harvest timing is critical, as brussels sprouts will not overwinter here.

Cauliflowers can be a tricky crop for us too. They are almost impossible in spring but delicious in the fall. Be sure to check the "days to harvest" for each variety (they vary widely) and sow at a realistic date to get a crop before too many frosts endanger the curds.

We transplant Vienna kohlrabi around July 29–Aug 12 from sowings made July 9–16 at 8" (20 cm) apart in the row, with 9"–10" (23–25 cm) between rows—4 rows in a 48" (1.2-m) -wide bed. Later sowings (up until early September) would

also work for the fast-maturing varieties. Superschmelz kohlrabi (60 days from transplanting) can also be summer sown for fall harvest. It produces 8"–10" (20–25 cm) bulbs, which remain tender and an attractive globe shape.

See the Asian Greens chapter for more about napa cabbage, pak choy, Tokyo Bekana, senposai and komatsuna. We grow some winter brassicas in our hoophouse: Russian kales, turnips and Asian greens—more about those in the Winter Hoophouse chapter.

We haven't tried this, but Early Purple Sprouting Broccoli, which is one of the broccoli staples in the UK, is hardy down to –10°F (–24°C), and is grown overwinter for March harvest.

Aftercare

About a month after transplanting the broccoli, cauliflower and cabbage (late August–early September), we remove the rowcovers, stakes and ropes and the sticks we use to hold down the rowcover edges, then hoe and till between the rows. Then we use an Earthway seeder on a shallow setting, with the "light carrot" plate, to sow four rows of mammoth red clover, sometimes with white clover and crimson clover too, between each pair of brassica rows. The rows are about six inches (15 cm) apart. I have also successfully broadcast the clovers. We use overhead sprinkler irrigation to get the clover germinated, which also helps cool the brassicas. The ideal is to keep the soil surface damp for the few days it takes the clover to germinate. Usually watering every two days is enough.

If all goes well, we keep the clover growing for the next season, mowing several times to control annual weeds. This is one year in ten of our rotation, which we designed following Eliot Coleman's method in *The New Organic Grower*. If

Fall broccoli rows. Credit: Pam Dawling.

there are too many perennial weeds, or the clover did not germinate well enough, we disk the clover patch in late spring and plant a mix of warm-weather cover crops, such as buckwheat, soy and sorghum-sudan hybrid, for the rest of that season.

Aftercare for kale and collards is a simple matter of hoeing, weeding, watering as needed, and watching for pests. We do not normally use rowcover in the winter for kale and collards, as they will survive without. In harsh winters we lose the collards. Using rowcover would keep crops in better condition later into the winter. We don't like the extra work of dealing with the rowcover, nor the cost, but your situation may warrant it.

Harvest

Cabbage heads up from Sept 25 and holds in the field till late November. Cabbage is mature when the outer leaf on the head (not the outer plant leaves, which are left in the field) is curling back on itself.

Our main broccoli harvest period is Sept 10–Oct 15, with smaller amounts being picked either side of those dates. See the previous chapter for harvesting details.

Cauliflower heads need to be harvested before they get frosted. We use gaudy plastic clothes pins (easy to find) to clip the leaves over a developing curd once frosts threaten. The leaves are frost-hardy. In a back-and-forth climate like ours, it is possible to harvest some heads in between one frost and the next if we are lucky. Whether it's worth taking the chance is a valid question.

Kale and collards are harvested (by snapping off the bigger leaves) all winter in small amounts, and then in larger amounts as spring warms up, until the end of May, when they bolt.

We harvest Vienna kohlrabi in the fall from Oct 20 to Nov 15. It stores well in perforated plastic bags in a walk-in cooler. In a plentiful year we have eaten stored kohlrabi all winter into early spring.

Chapter 34

Asian Greens

There are now many varieties of greens that grow quickly, look appetizing and bring fast returns for the grower. Most of them are Asian in origin, although some of the newer ones have been developed in the US. They are mostly brassicas, and can be grown at whatever time of year you normally grow cabbage or kale. We have a shorter brassica season in the spring and a longer one in the fall. In spring, brassicas will bolt when the weather gets hot; whereas in the fall, when the weather is cooling and the days shortening, they will not. Having greens to eat in the fall depends on getting them germinated and planted while the weather is still hot in June and July. See the previous chapter for more on starting brassicas in hot weather. Some of the fastest-growing Asian greens are ready for transplanting just two weeks after sowing (or you could direct sow them). These crops are a quick way to fill out your market booth or CSA bags. We find that people are very ready for some fresh greens as the summer begins cooling down. Asian greens are faster growing than lettuce and the seeds are better able to germinate in hot weather. They can be used as a catch crop in spaces where other crops have failed or otherwise "finished early." If you keep a flat of seedlings ready, you can pop plugs into empty spaces as they occur.

Flavors vary from mild to peppery, so if you or your customers don't want hot mustards read the variety descriptions in the catalogs before growing a large area. We found ourselves left holding large bags of Red Giant and Southern Wave mustard seed after an impulse buy. Happily, we were able to use it up in baby salad mix this spring! Colors cover the spectrum from chartreuse through bright green to dark green and purple.

These greens are nutritious as well as tasty. They are high in carotenoids, vitamins A and C, calcium, iron, magnesium and fiber. They also contain sulphoraphanes (antioxidants), which fight against cancer and also protect eyes from the UV damage that can lead to macular degeneration. If you need more persuading, these vegetables help prevent high blood pressure, heart disease and stroke.

Asian greens fall into two main groups: the turnip family, *Brassica rapa*, of Asian origin; and the cabbage family, *B. oleracea*, of European origin. Some crops are *Brassica rapa* var. *pekinensis* (napa cabbage and michihili); others are *B. rapa* var. *chinensis* (bok choy) or *B. rapa* var. *japonica* (mizuna). I've given the Latin names where I can, but sometimes different sources offer different names. See *Grow Your Own Chinese Vegetables* by Geri Harrington for a glossary of alternative names for these crops. Kitazawa and Evergreen Seeds have the most choices, many more than I can mention here.

Favorite varieties

Our favorite Asian greens include:

- Napa cabbage, *Brassica rapa* var. *pekinensis*. This is one type of wong bok, which in turn is one of two kinds of pe tsai/Chinese celery cabbage. We like Blues (52 days from seed to harvest) best, though Kasumi has the best bolt tolerance and is larger: 5 lb (2.3 kg) compared to 4 lb (1.8 kg). Orange Queen is a colorful but slower-growing variety (80 days). All are hardy to about 25°F (−4°C);
- The cylindrical type of Wong Bok Chinese cabbage, also *Brassica rapa* var. *pekinensis*, produce 16" (40-cm) tall heads 6" (15 cm) across. The light green leaves are very tender, excellent for stir-fries and pickling. This type is more stress tolerant and resistant to bolting and black speck compared to napa cabbage, but cannot be stored after harvest as long.

We like Jade Pagoda (72 days) and the open-pollinated Michihili, which matures 70 days after sowing;
- Maruba Santoh, *Brassica rapa* var. *pekinensis*. This is a fast-growing chartreuse (yellow-green) tender-leafed plant that can be harvested as baby leaves. Alternatively, the leaves and wide white stems of the mature plant provide crunch for salads. It takes only 35 days to maturity, and is fairly bolt resistant;
- Tokyo Bekana, *Brassica rapa* var. *chinensis* or var. *pekinensis* (opinions vary). This is another fast-growing tender chartreuse leafy plant, similar to Maruba Santoh but more frilly. We have often used it for salad leaves to get us through late-summer lettuce shortages. It has a mild flavor but even so, I have been surprised that many people don't even notice they are not eating lettuce— I suppose enough salad dressing masks all flavors! It takes 21 days to become a baby crop and 45 days to full maturity;
- Pak choy/bok choi, *Brassica rapa* var. *chinensis*. This was previously known as Chinese mustard cabbage. The sturdy white leaf stems and big green leaves are usually harvested as a head, 12"–15" (30–38 cm) tall, but can be picked as individual leaves, for bunches of mixed braising greens or stir-fry combinations. It needs 45–55 days to maturity. We grow Prize Choy or Joy Choi. There is also red choi (a 45-day, red-veined baby leaf or maroon-leaved full-size version);
- Mizuna/kyona, *Brassica rapa* var. *japonica*. This is very easy to grow, cold and wet-soil tolerant, and also fairly heat tolerant (well, warm tolerant). It has a mild flavor and regrows vigorously after cutting. The ferny

leaves, available in green or purple, are very pretty in salad mixes and add loft. Use for baby salads after only 21 days, or thin to 8"–12" (20–30 cm) apart, to grow to maturity after only 40 days;

- Mibuna is similar to mizuna, but less ferny and more spoon-shaped;
- Tatsoi/tah tsoi, *Brassica rapa* var. *narinosa*, is a small plant, a flat rosette of shiny, dark green spoon-shaped leaves and white stems. We usually direct sow this and then thin into salad mixes, leaving some to mature at 10" (25 cm) across for cooking greens. We also transplant at 6" (15 cm) if that suits our space better. It has a mild flavor, an attractive appearance, and is easy to grow. Tatsoi is extremely cold tolerant, hardy to 22°F (–6°C). Kitazawa Seeds have a Red Violet tatsoi, with an upright habit, which sounds good. Tatsoi takes 21 days to become baby salads; 45 days to reach cooking size;
- Komatsuna, *Brassica rapa* var. *perviridis* or *Brassica rapa* var. *komatsuna*, is also known as mustard spinach and Summer Fest. It is available green or red, and grows into a large cold-tolerant plant 18" (45 cm) tall. Individual leaves can be picked and bunched, or the whole plant can be harvested. The flavor is much milder than the English name suggests, being mildly peppery. It reaches baby salad size in 21 days and full size in 35 days;
- Yukina Savoy, *Brassica juncea* or *rapa*, is like a bigger tatsoi, with blistered dark green leaves, greener stems and delicious flavor, about 12" (30 cm) tall. It is both heat and cold tolerant. We transplant this at 12" (30 cm). It needs 21 days to reach baby size, 45 days to full size;
- Senposai is the star of Asian greens as far

Senposai is a star performer among greens. Credit: Kathryn Siimmons.

as I'm concerned. It is quite heat and cold tolerant, a big plant producing large, round, mid-green leaves which are usually harvested with the cut-and-come-again method. It can be very productive and grows fast. Transplant it at 12"–18" (30–45 cm) spacing; it really will use all this space. It cooks quickly (much quicker than collards), and has a delicious sweet cabbagey flavor and tender texture. It is a cross between komatsuna and regular cabbage. It takes only 40 days to mature.

Non-stars for us

- Hon Tsai Tai, *Brassica rapa*, is like a purple broccoli raab, mostly stem with small clusters of buds (which in our case, turned to flowers while we blinked). Also known as Choy Sum. In climates cooler than Zone 7 this might be productive, at least in the fall. For spring it seems like it would be a challenge most places. It matures very fast—about 35–40 days. Hardy to 23°F (–5°C);

- Broccoli Raab, *Brassica rapa ruvo*. We had the same trouble with this as with Hon Tsai Tai;
- Mei Qing Choi, *Brassica rapa* var. *chinensis*. This is a miniature 6" (15 cm) pak choy, and we don't do well with miniature crops. I avoid varieties labeled "compact" in the catalogs. These might suit your market, but we do better with larger vegetables. It matures in less than 45 days, a definite plus;
- Bitamin-Na/Yokatta-Na/Vitamin Green, *Brassica rapa* var. *narinosa*, is a slender, white-stemmed plant, about 12" (30 cm) tall. It can be planted 4" (10 cm) apart, or direct sown and thinned. It tolerates heat and cold and is quick-growing with good flavor, not pungent: 21 days for salad mix, 45 to its full size;
- Mustards such as Red Giant and Osaka Purple, *Brassica juncea*, and American Mustards (eg Southern Green Wave) are too hot for us, even at 3" (8 cm) leaves. They are hardy to light frosts and the colors are very attractive. They take 21 days to baby leaves, 40–45 days to full size;
- Tyfon Holland Greens is an industrial-strength plant, another hybrid of komatsuna with a heading brassica. This could be good in a survival situation, or to grow for goats. It's not a gourmet green;
- Tenderleaf, *Brassica rapa*, is another big, open-pollinated sturdy plant. It is quick-cooking and mild-flavored, despite appearances. It is selected from a cross of Tendergreen and tatsoi, and is very disease-resistant

and cold tolerant. It can be sown later in the fall than other greens, so could be the perfect solution for a last-minute crop where your original plan didn't work. It can be a useful salad mix crop at the baby stage. We just let ours get too big and gnarly.

Asian greens we've yet to try

- Celery cabbage/pe tsai, *Brassica rapa* var. *pekinensis*. This is the second (non-wong bok) type of pe tsai, or pei tsai: a small loose-leaf type, fast-growing vegetable with light green leaves and white petioles, that can be ready for harvest in 3–4 weeks after sowing. It is more heat tolerant than napa. According to Evergreen Seeds, Tokyo Bekana is in this group, but in our experience, the Tokyo Bekana we have grown is far from small!
- Hong Vit, *Raphanus sativus*, is a leaf radish with pink stems, hairless leaves and a mildly radishy flavor. It takes 21 days for a baby crop, 35 days to fully mature leaves.

Non-Asian (including American Asian-type) brassicas

- Arugula, the by-now-famous spicy little leaf. It is very cold tolerant;
- Cresses are hot and spicy, often eaten quite small;
- Mizspoona, *Brassica rapa*, is a large sturdy plant, taking 40 days to maturity. It has a sweet flavor with a good balance of mild zinginess. (From Frank Morton and Fedco);
- Pung Pop Mustard Gene Pool, *Brassica juncea* (also from Frank Morton and Fedco), is a cross of pungent Indian mustards for those who like Big Flavor. It needs 40 days to harvest;

- Pink Lettucy Mustard Mix, *Brassica rapa* (Frank Morton, Fedco), is fast-growing and adds a touch of color to the brassica portion of winter salad mixes. It is available in a varied mix of colors and shapes, is cold tolerant, and is ready in 40 days;
- Chinese Thick-Stem Mustard (Fedco, Brett Grohsgal). Multiple cuttings of this balanced flavor salad mix crop will fill your CSA bags. It is extremely cold tolerant;
- Torahziroh, *Brassica oleracea algoblabria*, is a robust producer of high yields of large leaves with a good, not overpowering flavor. It is related to crops known as Chinese kale or Chinese broccoli. Relatively slow to bolt, it is ready in 45 days;
- Ornamental and garnish kales and cabbages add color and texture. We like Nagoya Red and White and Red Chidori;
- Every year there are new salad mix mustards, such as Johnny's Ruby Streaks and Golden Frills;
- Abyssinian Cabbage, *Brassica carinata*, including Tex-sel Greens, is winter hardy in Zone 7. It is a non-heading, fast-growing, tasty cut-and-come-again crop. Because it bolts readily in spring, it is best used in the fall;
- Portugese Cabbage, *Brassica oleracea costata*, has recently become more available in the US.

Asian non-brassicas:

Chrysanthemum greens/shungiku, *Chrysanthemum coronarium*. These have a very distinctive aromatic flavor, which you may or may not love. The flowers are very pretty, if you give up harvesting the plants. Allow 21 days for baby greens, 45 days to reach full size.

Crop requirements

These crops have similar care requirements to other brassicas, with the addition of extra attention to providing enough water during hot weather to prevent bitter flavors and excess pungency, and closer monitoring of pests, which may have built up large populations during the summer. Very fertile soils grow the best Asian greens, so turn in leguminous cover crops or compost to provide adequate nutrition.

Sowing

We almost always transplant our brassicas because we use our growing spaces very intensively and transplanting gives the previous crop some extra weeks to finish up. If we have four weeks between the end of one crop and the brassica transplants going in, we sow a buckwheat cover crop to add organic matter and smother weeds. We'd usually rather make use of this cover crop opportunity than direct sow greens. We grow a lot of brassicas and our crop rotation is always pushed and stretched by the amount of brassicas we'd like to plant.

In spring we sow in flats in a greenhouse, to get an early start. In summer we make an outdoor nursery bed, sow at about three or four seeds per inch (5–10 mm apart), and cover with rowcover. The seedlings emerge in as little as three days in summer temperatures. I prefer outdoor seedbeds for summer sowings, because it is easier to keep the plants watered. The third option, direct sowing, has the advantage that thinnings can be used for salads.

We start sowing our fall Asian greens for outdoor planting around June 26 and repeat a week later for insurance (July 3), the same dates we sow fall broccoli and cabbage. The last date for sowing these crops is about three months before the first fall frost date. In our case that means July 14–20.

Transplanting

In spring, we transplant Asian greens at 4–5 weeks of age, about a month before our last frost date. In summer, the faster growing types are ready to transplant two weeks after sowing. Napa cabbage, Tokyo Bekana and Maruba Santoh are in this category. Most others transplant best at 3–4 weeks of age (less time than needed in spring). Adding this up, you'll see we are transplanting them outdoors from July 10 to July 31.

To minimize transplant shock, water the plants well an hour before transplanting, get them in the ground as quickly as possible and water again. Shadecloth or rowcover will help keep the breezes (if any!) and strong sun off the plants.

Irrigation

These crops are relatively shallow-rooted, and need plenty of water to grow pleasant-tasting leaves. One inch (2.5 cm) of water per week is enough, except during very hot weather, when two inches (5 cm) will work better. Drip irrigation saves water and reduces disease and weed pressure. On the other hand, overhead irrigation can be cheaper and easier to set up for crops that will be harvested before much time has passed. And overhead sprinklers can wash off aphids, and could be all the control measure you need for that pest.

Pests and diseases

Our worst brassica pests are harlequin bugs. We usually try to pick and kill as many as possible in the spring, and again when we plant out fall brassicas, in hopes of keeping the population under control. If you aren't in the harlequin zone, read on for other pests.

Some years we have flea beetles, which we have "done in" with Spinosad, an enzyme produced by a natural, although rare, soil organism. Hb nematodes will also control them, as will neem oil or the braconid wasp *Microtconus vittatoe Muesebeck*. Garlic spray, Miller's Hot Sauce, kaolin and white sticky traps have also been suggested. You can also catch them with a vacuum cleaner, or inside a bucket coated with Tanglefoot paste (hold the inverted bucket over the plant, shake it and catch the jumping beetles in the goo). Useful (reassuring?) information is that brassica flea beetles are a different species from the ones that plague eggplant, and they can only fly a few hundred yards (meters).

Aphids generally are more of a problem in the cooler weather of early spring, before their predators have arrived in high enough numbers. If they get out of hand insecticidal soaps can be used.

Rowcover will keep caterpillars off the plants, and Bt (*Bacillus thuringiensis*) will kill them if rowcovers fail. Bt degrades rapidly in sunlight so is best applied early evening or early morning, whichever seems likely to catch most caterpillars. The beneficial fungus *Beauvaria bassiana* infects caterpillars, but can get costly. Caterpillars have many natural enemies. In our garden the paper wasps eat caterpillars, and we also have the parasite *Cotesia glomerata*.

Coming from the moist and verdant islands of Britain, I used to think slugs were an endangered species in Virginia, as I rarely saw any. When we put up our hoophouse, though, I found we were farming them! Slugs can best be caught at night with a flashlight. (Well, actually with scissors, by flashlight!)

Grasshoppers can be a problem. We are trying to determine when the young hatch in July, so we know when we need to be most alert and attentive to keeping them off our plants.

In our hoophouse we have been troubled in January by vegetable weevil larvae, which come out of the soil at night and make holes in the leaves. We have used Spinosad against them with some success.

A net fabric with small holes is better than rowcover in hot weather, as airflow is better and it heats less. Some people use nylon bridal netting from a fabric store. ProtekNet Pest Control Netting is made of clear high-density polyethylene with UV resistance and a lifespan of eight to ten years. Its light transmission is 90 percent. It is available from Purple Mountain Organics in Maryland. The 1.35×1.35 mm (60 gm/m^2) mesh is one-sixth the length of a cucumber beetle. It also protects crops against weather damage. Enviromesh from Agralan is another promising-sounding product to keep insect pests from crops.

Diseases are not much of a problem for us, though if they are for you, you may be justified in feeling grumpy with me after reading that! Most of these greens are fast-turnaround crops, so if some get sick, pull them out and move on in life. If it's fall you can probably sow some spinach to provide greens without antagonizing the brassica disease gods. Clubroot is perhaps the longest-lasting disease, requiring land to be taken out of brassica production for ten years. Other diseases include various molds and wilts. See ATTRA's *Cole Crops and Other Brassicas: Organic Production*.

Pests and weeds will slow down as the winter approaches, so we can too!

Harvest

Some of these greens are harvested as whole heads; others can be harvested by the leaf and bunched or bagged. The open rosette types, such as tatsoi or the bigger Yukina Savoy, are usually gathered closed and banded with plant ties or rubber bands. Most can be grown for baby salad mix. With mizuna we do a "half buzz-cut," snipping off leaves on one half of the plant an inch (25 mm) above the ground each time we come by.

After harvest, get the crops into shade and a cooler as soon as you can. Some of the heading types can be stored in a walk-in cooler for quite a while, almost as long as regular cabbage. I've heard that pak choy can be preserved by drying the separated parboiled leaves on a laundry line! Sounds adventurous and committed!

Season extension/overwintering

The fastest-growing varieties can be succession sowed for a continuous supply. Some of these greens are very cold hardy, and can be harvested all winter in milder climates or kept alive until they revive in the spring to provide earlier harvests than spring-sown crops. Rowcovers on hoops will help keep these crops in marketable condition, and ameliorate the microclimate, for better growth rate. Seed companies such as Wild Garden Seeds and Gathering Together Farm specialize in producing very cold-tolerant varieties.

Hoophouses are the place to be in winter, if you are an Asian green. With the nighttime protection of two layers of plastic and an air gap, September sowings of these crops can thrive on the sunny days and grow at a surprisingly fast rate.

For information on crops to grow in the winter hoophouse see the relevant chapter. There is a date when the daylight falls below ten hours, after which little growth will happen till spring. The dates depend on your latitude. Here at 38° N, that's November 20 to January 20. My neighboring grower Ken Bezilla points out that these dates are modified somewhat by the time it takes to cool the soil and the air. In practice, the effective dates for us may be closer to December 15–February 15. Brassicas are the most productive crops in these conditions.

Seed saving

If you want to save seed from a brassica, you need at least 600 feet (200 m) isolation (seed for home use only). For commercial seed you need a quarter of a mile (400 m) with barriers or half a mile (800 m) without. If you plan to grow seed of more than one brassica, carefully choose ones that won't cross. Beware of the possibility of brassica crops being wrongly classified, as the classification is in flux. Also beware of brassica weeds. Brassicas are outbreeding plants and, as such, are in danger of inbreeding depression if a large enough population of plants is not grown to ensure genetic diversity. Save seed from at least 60 to 75 plants, and preferably 125 to 150. Plant twice this number in fall, pull out any atypical plants and leave the best over the winter. In the spring, let them bolt, and as the seedpods dry, pull up the plants, and perhaps hang them up to finish drying under cover if your weather is damp. If you have high humidity, use a fan. Hanging the plants inside paper sacks will help reduce loss of seeds once the pods start to shatter. You can stomp on the bags to shatter the pods, and then winnow and screen the seeds. See the excellent 24-page guide on brassica seed production from the Saving Our Seed Project.

Chapter 35

Spinach

Spinach is a wonderful crop, which grows under cool, and even cold, conditions. Here spinach is a fall and spring crop, with the option of overwintering the fall-planted crop under rowcover for winter and early spring harvests. Spinach grows very fast, can be ready for harvest in less than 50 days, and can be a welcome change after summer crops. Its flavor is sweetest and nuttiest in cold weather. (Summer can be short of leafy greens, but there are hot-weather spinach substitutes such as chard and New Zealand spinach; see the next chapter.)

Day length of more than fourteen hours triggers bolting in spinach. All of us, wherever we are, have twelve hours of daylight at the spring equinox and the fall equinox, and less than that from fall to spring. So, provided temperatures are in the right range, we have over six months of suitable spinach growing conditions. Hot weather will accelerate bolting once the daylight trigger has been reached, as will overcrowding (with other spinach or with weeds) and under-watering. The exact temperature that triggers bolting varies between varieties. Here we reach fourteen hours of daylight on May 8, and spinach is definitely a lost cause after that date.

Varieties

For people who love heirloom vegetables, the Bloomsdale varieties—Winter and Long-Standing—are reliable. For bolt-resistance, though, we go with the hybrids. Here is something that was news to me: apparently on the east coast, bubbly savoyed spinach varieties are preferred, while on the west coast smooth-leaved ones are popular. (When fact-checking the only reference I could find was Bobby Flay's cooking show on TV, but I'm sure I learned it elsewhere!) So if you have recently moved a long distance and intend to sell spinach at the farmers' market, find out which kinds are more likely to sell. Tyee is the variety we like best, and most years it's now the only one we grow. I order a big bag of semi-savoyed Tyee each spring and I'm all set. It is worth noting that seed viability declines to less than 80 percent after a year.

One spring I was entranced by the catalog description of Giant Viroflay. It certainly grew big leaves, but not for long, and then it was all over. We experimented with smooth-leaved spinach, thinking it would be easier for cooks to clean, and it probably is. The disadvantage is that it wilts very quickly after harvest, and needs to be picked and chilled promptly.

Space is a popular smooth-leaved type. Paul and Sandy Arnold in Argyle, NY, use Tyee for spring plantings from transplants, because it is fast-growing. In summer they switch to direct seeding the more bolt-resistant, smooth-leaved Space, sowing in the evening and watering frequently until the seeds germinate. Space can grow all summer there without bolting. Sowings are made once a week, unless 90°F (32°C) weather is forecast, in which case they wait for it to cool off before planting.

Other varieties that are recommended for bolt tolerance include Indian Summer and Olympia. Samish, Olympia and Avon are other cold-hardy varieties.

John Navazio has done some research into disease resistance and found that the glossy-leaved Wintergreen has some *Fusarium* resistance. So does Ozark II, but it is a research variety not commercially available. Tyee has some resistance to Downy mildew, but Samish and Unipack 144 have better resistance. Cascade is prone to many diseases.

Crop requirements

Spinach is a cool-weather crop and is hard to germinate in warm conditions. Like most leafy greens, it appreciates a fairly rich soil, plenty of water, good drainage, pH of 6.0–7.0, and moderately good light. Partial shade can be an advantage once the weather starts to warm, so take this into account when planning crop layouts.

Sowing

The number of days to emergence is 62.6 at 32°F (0°C). I hate the thought of 62 days! Surely it will warm up before then? At 41°F (5°C) seed takes 22.5 days to emerge; at 50°F (10°C), 11.7 days; at 59°F (15°C), 6.9 days; at 68°F (20C), 5.7 days; at 77°F (25°C), 5.1 days; at 86°F (30°C), 6.4 days; and don't waste time trying at higher temperatures than that. So, in terms of days to emergence, spinach comes up quickest at 59°F–86°F (15°C–30°C).

But days to emergence is only half the story with spinach! (I found this out the hard way.) The percentage of normal seedlings that do emerge is also vitally important. At 41°F (5°C), 96% of the seedlings will be normal; at 50°F (10°C), 91%; at 59°F (15°C), 82%. Then there is a rapid drop-off: only 52% normal at 68°F (20°C) and just 28% normal at 77°F (25°C). So, combining these two factors, temperatures of 50°F–59°F (10°C–15°C) should produce plenty of normal seedlings in a reasonable 7–12 days.

Sow seed half an inch (1.3 cm) deep. Rows can be anywhere from 3"–12" (8–30 cm) apart, depending on your equipment and the space available. Sowings are usually thinned to 4"–6" (10–15 cm).

In spring we direct sow spinach or transplant from Speedling plug flats (200-cell size) or barefoot root from a hoophouse nursery bed as early as possible. We typically plant four rows of a smallish crop like spinach on our 4' (1.2-m) -wide raised beds. We use rowcover until the plants are established. See the Season Extension section in this chapter for tips on establishing spinach in both cold and warm weather.

Eight weeks before the first fall frost date is a good time to start planting spinach again, if it's not too hot. Our average first frost date is October 14, which would indicate mid-August as a starting date, but it's too hot here then, so we wait until September 1–5. We might risk an earlier sowing if the season was a cool one. This big planting feeds us from the middle of October until late April. We either direct sow (when the weather cools enough for fall deadnettle and chickweed to germinate) or transplant (after other crops finish in September, October or even early November). We use raised beds and often transplant some of the extra plants moved from the direct-sown beds.

As well as being in USDA Hardiness Zone 7, we are in Zone 7 for summer temperatures—right in the middle on both scores. The American Horticulture Society publishes the heat zone map and the USDA publishes the plant hardiness map.

Our second fall sowing is usually September 20–30, and this one overwinters as small plants and does not reach harvestable size until early spring. This planting will bolt in the spring at a date between the first fall sowing and the spring sowings. In some climates, fall and winter harvesting reduces the survivability of spinach plants; small plants, however, overwinter well—the ideal size would fit under a small teacup. Smaller or larger plants are not as cold tolerant.

In climates colder than Zone 7, make the fall sowings earlier in order to have adequate growth before hard freezes, sowing as soon as temperatures are cool enough for germination. Sowings after the end of September can be made in warmer climates: until Oct 15 in South Carolina and Nov 15 in Louisiana.

We use double hoops and rowcovers and pick spinach throughout the winter, whenever leaves are big enough. We usually have about seven beds and can pick one each day in October, November, February and March, when the weather is not too awful to go out. Spinach will make some growth whenever the temperature is above about 40°F (5°C), so we can also make occasional harvests in December and January.

Transplanting

Spinach transplants very easily bare-root or from plug flats, so seedlings can be started in more ideal conditions than you might have outside. Just be sure to water the transplants daily until you are sure they have taken. Setting a plant beside every drip emitter (while the irrigation is running) is a wonderful method for those using drip tape.

We plant on a 6" (15 cm) spacing in the row, with rows about 9" (22 cm) apart. Closer row spacing is possible if you have raised beds.

Relay planting

We sow snap peas or snow peas in the center of beds of overwintered or spring-planted spinach. This is a version of undersowing, where the second crop is planted while the first is still growing. The trick is to plan ahead and leave room in the center of the bed when planting the spinach. We leave an extra wide central space between the inner rows of spinach (less than a whole extra

In spring we sow peas in our spinach beds and they take over as the spinach passes its peak. Credit: Kathryn Simmons.

row's worth, as peas are a vertical crop). We've found the Row Marker Rake from Johnny's to be a very worthwhile investment for making consistently parallel rows, and making faster hoeing possible.

Because spring heats up quickly here, we have a short season for peas. We plant a double row of peas in the middle of each spinach bed and take care of the two crops together, with the spinach gradually giving way to the peas in April and May. The crops share the rowcover, the warmer soil, the cultivations, the compost and, above all, the space. One tilling is eliminated and the bed is doubly productive.

We aim to sow our peas on March 1st, or whenever the forsythia blooms. Snap pea seed is more vulnerable to rotting in cold soil than shelling peas, probably because the seed is higher in sugars as opposed to starches. If you're growing shelling peas, you can sow earlier. In preparation,

we hoe and weed the spinach and soak the pea seed overnight. Before making the furrows for planting the peas, it helps to harvest the leaves from the inner rows of spinach. Because the bed is already warm under the rowcover, you can sow the peas earlier than in uncovered soil. See the chapter on Peas for more from their perspective.

By the time we need to take off the rowcover and use it for other crops, the peas are well up. We do another round of weeding and install the pea stakes.

We harvest spinach leaves throughout the spring, about once a week for each bed. As they start to bolt, we harvest whole plants. Eventually we are left with a bed of peas only. Having two crops together keeps our attention on the need for weeding and harvesting.

One year we discovered the built-in, fail-safe feature of this method: if you fall behind with the string-weaving and the cultivation, then the bolting spinach will support the peas!

Growing in the hoophouse

It is quite phenomenal how much beautiful growth spinach can make in our winter hoophouse! We just harvest whenever it looks big enough, working our way up and down the beds and the various plantings. See the Hoophouse in Winter and Spring chapter for more details.

We make six sowings of spinach in our hoophouse between Sept 6 and Jan 24. This last sowing is for transplants to be moved outdoors. We've tested this method against plug-flat grown plants started on the same date, and the transplants do equally well. The advantage of the hoophouse open-ground sowing is that it's quicker and also less work from then on—the crop gets watered along with everything else and dries out less quickly than flats do. We transplant the Jan 24 sowing outdoors around Feb 21, at four weeks of age. At that time of year here, bare root transplants don't get hot enough or dry enough to suffer.

Pests and diseases

Voles can be a problem for us, as they eat the roots. We use plastic "Intruder" mousetraps from the hardware store, baited with peanut butter or fruity bubble gum. Eliot Coleman in *The Winter Harvest Handbook* shows an interesting box for holding mousetraps, which voles find inviting. We plan to try that.

We have not been troubled much by insect pests, but others have to deal with webworms and leaf miners. Cornell's *IPM Production Guide for Organic Spinach* and the Seed Alliance's *Spinach Diseases Field Identification* are good sources of information on pests and diseases.

Diseases have not been a problem for us either. The main troubles we have are yellowing leaves from near-drowning experiences, frost-damaged patches and abrasion damage from rowcovers.

Harvest

There are three main ways to harvest spinach; four if you count using the thinnings from direct sowings.

We generally harvest by the leaf, with scissors. Harvesting one leaf at a time has the advantage of producing an attractive product—all whole leaves. I believe it also is least damaging to the plant and likely to give the highest total yield over the life of the plant. It can be rather slow, depending on the skill of the harvester (and the outdoor temperature!).

The "buzz-cut" method is much faster, and the plants live to carry on producing more. Gather the leaves and cut them all, one inch (25 mm) above the crown. The plants do take longer to recover from buzz-cutting than from leaf-harvesting. If you are making a salad mix, the fact that the leaves are cut might not matter. This is a good method when time is at a premium. We use it in the spring, when the plants start to grow tall in preparation for bolting. If your plants are frost damaged, with yellow-white patches, this method might not work as well as the leaf-harvesting method, as you might have a lot of sorting to do.

The final method is the single harvest: cut the plants, then till in the crop residue and start again. This is the quickest method and produces the most beautiful spinach, but obviously it isn't suitable in late fall if you hope to overwinter your spinach, as it kills the goose that lays the golden eggs, and it will be too late in the year to resow. (Season extension into winter relies on extending the harvest period by keeping plants alive and productive for multiple harvests.) In early fall or in spring this method has its place as a quick catch-crop before a longer-term food crop or winter cover crop is sown or transplanted.

Season extension

Unlike spring sowings, plants started in the fall won't bolt, as the days are getting cooler and shorter as the plants mature. If your climate is suitable for overwintering spinach, or you can cover the crop, you can harvest throughout the winter or keep the plants alive until they begin growing quickly again in early spring. Overwintered plants will bolt a few weeks ahead of spring-started ones.

Spinach is cold tolerant down to 20°F (−7°C). In areas where continuous snow cover is not to be

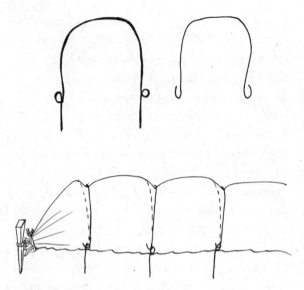

Double wire hoops and rowcover protect spinach from winter temperatures and abrasion by the rowcover. Credit: Pam Dawling.

had, rowcover on hoops will protect the plants. If you use rowcover without hoops, the abrasion of the fabric on the leaves causes unsightly damage, and the leaves are more likely to freeze to the fabric if you have wet weather followed by freezing nights. Hoops also create a nice microclimate, noticeable when you stretch your arm in to harvest. We use double hoops every six feet (1.8 m) along a bed.

We make the hoops ourselves. The inner hoops are #9 galvanized wire, cut to eight feet (2.4 m) using large bolt-cutters. We bend the wire into a hoop with 12" (30-cm) "legs" and round "eyes" where the hoop meets the soil. The rowcover goes over the top of these hoops, then the thinner hoops, made of #14 wire cut six feet (1.8 m) long, hook into each eye of the bigger hoops, preventing the rowcover blowing away. This works best when the rowcover is pulled taut along the length of the row and the ends are fastened down with heavy weights (or perhaps tied with twine to stakes), and when the thinner hoops are a little smaller in radius then the thicker ones, so that the rowcover is actively held down. We hold our rowcover edges down by rolling them under around wood sticks. Rolling under, rather than over, means that when the stick starts to roll off the edge of the bed, it tightens itself into the rowcover instead of unrolling. (These little things can make so much difference!) For harvest, the sticks on one side can be extricated and the rowcover pushed up to lodge at the top of the hoop between the pairs of wires. If the wind does pick up an edge, the cover does not blow away because the outer hoops hold it down.

Another way to protect spinach over the winter is to broadcast oats into it at planting time. The oats grow up with the spinach and protect it from the worst of the early winter weather before the frost kills them. In climates too warm to frost-kill oats, you don't need them to protect spinach!

Spinach that has been overwintered will benefit from a nitrogen boost in the early spring, either side dressing or foliar feeding.

For successful warm-weather sowing, sprouting the seeds is a good method. We soak the seeds in water overnight, then drain and put the jar in the refrigerator for a week. Turning or rolling the jar daily helps get even germination, but it's also possible to totally ignore it for the week and still have very useable seeds. I speak from experience!

Another method we have used to get spinach sprouted in hot weather is to use a float tank and Speedling trays. We made a float tank from cinder blocks lined with old carpet (to protect the plastic from abrasion) and black plastic. Speedling flats will float on water in the tank. Spinach plants will drown if left floating continuously, so we drain the flats overnight (when it is a bit cooler) and float them all day.

Laying shadecloth over the spinach beds a week before sowing cools the soil and helps germination. Shadecloth on hoops can help spinach transplant establishment. Noontime watering can help cool the soil and give the seed the best chance.

Seed storage

We usually put our spinach seed in double zip-lock bags in the freezer from April till we need it at the beginning of September. During the winter our seeds are stored in plastic food containers in a cool basement or shed. If kept at or below 8 percent moisture, spinach seed can keep for three years.

Seed saving

It was quite a revelation to me when I found out that spinach has male plants and female plants. The isolation distance for spinach is a quarter to a half mile (400–800 m) for home use seed and one to two miles (1.5–3 km) for pure commercial seed. Grow at least 120–200 plants for seed to avoid inbreeding depression, which comes from selecting too few of an out-crossing (not self-pollinating) crop. When the plants start to bolt, thin them to 2' (60 cm) apart. When they reach 18" (45 cm) tall, cut them back to 12" (30 cm) to encourage shorter, branched plants and multiple seed heads. Pull them up when the lower parts of the branches have ripe seed. To prevent seed shattering, pull the plants in the early morning when they are still covered in dew. Cut off the roots and hang the plants under cover upside down until they are fully dried out. Strip the seed by hand and clean by threshing over an inclined canvas that allows seed to roll to the bottom while trash is left behind. The Organic Seed Alliance publishes an excellent reference on this subject, the *Spinach Seed Manual*.

Chapter 36

Chard and Other Summer Cooking Greens

Swiss chard

Spinach bolts in the spring, brassicas (crucifers) become bitter when hot weather arrives, but Swiss chard and a few other vegetables can provide fresh, tasty, succulent greens through the summer with little effort and very few troubles. Chard (*Beta vulgaris* var. *cicla*) is the same species as beetroot and, like beets, is a biennial. Hence it will not flower until the second year after planting, and can provide fresh greens all summer and fall, until halted by hard frosts. Even then, the root may survive and regrow the next spring.

Varieties

Swiss chard is available in several shades of green, a mix of rainbow shades, and various selections of single colors.

The white-stemmed, green-leafed varieties are more productive, as they have more chlorophyll than the colored leaf kinds. They are also less likely to bolt under stress. Fordhook Giant, the most easily found variety, is the workhorse. It has thick savoyed (crinkled) leaves and is very bolt-resistant. Some growers have other favorites. Lucullus has a more tender leaf and is a lighter shade of green. Monstruoso has very broad stalks. Large White Ribbed is said to be more productive but not quite as tender.

Red kinds seem to have a stronger flavor, closer to beet greens, and perhaps because of that are less troubled by leaf miners. Red varieties, however, are more likely to bolt if subjected to prolonged periods below 50°F (10°C) followed by extended dry conditions. Charlotte resists bolting better than Rhubarb/Ruby Red and is more productive, with broader stems.

Bright Lights and Rainbow are two varieties of multicolored chard. Bright Lights has thinner stems and brighter colors than Rainbow.

Crop requirements

Although it can be sown earlier and used as a spring crop, chard really comes into its own if sown later in spring, to mature as spinach finishes up. It only takes about 50 days from sowing to maturity.

Chard needs a soil rich in organic matter, with a pH over 6.0, if it is to crank out large quantities of healthy leaves. It is an easy crop that will grow in partial shade or full sun and in all soil types. It does need regular watering throughout the season, however: water stress can cause bolting if the weather has been cool. If yield drops, consider side dressing with compost, soybean meal, cottonseed meal or alfalfa meal, or foliar feed with kelp or fish emulsion. Nitrogen is the vital ingredient for growing luxuriant tender chard. Be careful (as always) not to apply too much, or the colors of the stems will fade.

Chard plants continue producing leaves into the winter, so for cooking greens there is no need to sow succession crops. If you want baby chard for salads, see the Winter Hoophouse chapter.

Sowing

If you want chard in spring, you can direct seed two or three weeks before the last frost or start flats earlier. If you prefer spinach in spring, and switch to chard for summer, sow in plug flats or soil blocks three weeks before your last frost date (we sow around April 6). We like the Winstrip 50-cell open-bottomed plug flats for this crop. (Un-fortunately these are now hard to find in small quantities). Using transplants enables us to get an earlier crop from the future chard bed, or finish growing a good winter cover crop, before we put the bed into production.

Each "seed" is actually a dried fruit containing several seeds, just like a beet seed, so you'll likely get several seedlings in each cell. When the seedlings have a leaf or two, thin them to one per cell by snipping off the extras.

The temperature range for germination is 41°F–95°F (5°C–35°C), with optimum germination temperatures of 59°F–77°F (15°C–25°C). Germination time is 4–20 days. There are 70 seeds/g, 2000 seeds/oz. For salad mix, you will need 2 oz (60 g) per 100' (30 m).

For direct seeding, sow about six seeds/ft (5 cm apart), 0.5" (1 cm) deep, in rows 12"–24" (30–60 cm) apart. This will use ⅓ oz (9 g) per 100' (30 m). As they grow, the plants can be thinned gradually until 12" (30 cm) apart. Sowing at closer spacing provides smaller leaves, and so is better for salads. One ounce (28 g) of seed will sow 100' (30 m) at close spacing. The thinnings are an excellent salad crop.

It's also possible in warmer areas to sow chard for a fall crop. This might be a useful Plan B if some other crops have failed. The last planting date is ten weeks before frost. The winter-kill temperature is 15°F (–10°C).

Transplanting

We plant our chard out around April 29–May 6, at 3–4 weeks of age. We transplant into beds already mulched with rolled out bales of spoiled hay, making "nests" through the hay down to soil level, at 12" (30 cm) spacing. The plants will grow large, so we put only two rows in a 4' (120 cm) bed with 1' (30–cm) paths. The mulch controls weeds and keeps the soil cooler and damper through the summer.

Interplanting with a fast-growing crop such as lettuce or scallions (green onions) might be a possibility, as an alternative to mulching before transplanting. Transplant the chard at the usual in-row spacing, with about 16" (37 cm) between rows. Transplant lettuces or clumps of scallion plants between the chard rows. Keep these crops growing fast and harvest the scallions and lettuce after about five weeks, before the chard gets too big.

Pests and diseases

Chard generally has few pests and diseases. One disease we have had some trouble with is *Cercospora*, a fungus that grows as small round tan or brown spots on the leaves. Look for a purplish halo, or take a hand lens and look for black fruiting spores in the centers of the spots. We remove the affected leaves to a hot compost pile and hope for the best. Fusarium wilt can cause seedlings to wilt and shrivel, or older plants to wilt and turn yellow.

Slugs can be a bother in cool climates, but less so in summer. Root-knot nematodes have caused trouble in Florida; crop rotation helps avoid them. Flea beetles have been reported on chard in Pennsylvania. Leaf miners may be seen in northern areas. Use rowcovers in spring when the miner flies are most active, or against flea beetles. Blister beetles have also been known to cause problems in the South. Adult beetles emerge from the soil in areas where an infestation occurred the previous year. Their preferred crops are tomatoes, potatoes and eggplant, with beets and chard as second choice. If numbers justify action, handpick with gloves on (to avoid getting blistered by crushed beetles), or brush into a scoop improvised from a plastic bottle. Drop the beetles into soapy water. If numbers of blister beetles are low, live with them and know that their larvae eat grasshopper eggs, and will save you from marauding grasshopper plagues later in summer.

Harvest

Chard will reach maturity fifty to sixty after days from sowing. That's late May for us, right when the last spinach is bolting. Chard is picked as individual leaves and sold as bunches. We pick outer leaves (discarding the damaged or tough ones) and stand them in tall buckets, adding an inch (25 mm) of cold water to help keep the leaves from wilting. Pull the stems outward and twist low down (or cut with a knife), to avoid leaving stubs, which will "cage in" the developing stems and take energy for the plant to maintain. Remove any inedible tough leaves as you harvest and place them around the plants to top up the mulch, unless the leaves are diseased.

Towards fall it is possible to extend the growing season by allowing the outer leaves to get large, protecting the heart of the plant while you harvest the younger inner leaves (making sure not to damage the growing point). If grown for baby salad mix, chard will be ready to harvest 35 days after sowing.

The yield from 100' (30 m) is about 50 lbs (23 kg).

Post-harvest

Chard needs quick cooling and refrigerated storage after harvest. It stores well frozen if carefully blanched for three minutes, chilled, drained and frozen in airtight containers. Depending who your customers are, you might need to label this crop clearly, and describe its wonderful features. I have seen it marketed as "The Big Spinach." Here we had an amusing case of mistaken identity, when cooks took a bucket of ruby chard, threw away the leaves and made an "interesting" faux rhubarb pie for dessert, from the stems.

Overwintering

To keep chard in good condition overwinter, either cover with hoops and rowcover (in milder areas, Zone 6 or warmer), or else mulch heavily right over the top of the plant, after cutting off the leaves in early winter. Chard is hardy to 15°F (–10°C) without rowcover.

Seed saving

If you want to save seed, you'll have a long wait—chard is biennial. It crosses with beets, so don't try to grow seed of both. The isolation distance for chard is 600' (180 m) for seed for home use, half to one miles (0.8–1.6 km) for commercial seed. Chard color mixes are not easy for small-scale seed saving as they are actually a combination of varieties, and each color needs to be saved as a separate variety.

Population size and seed production details are the same as for spinach—see the previous chapter. Seed longevity is three years when stored well.

Perpetual spinach/spinach beet

Spinach beet, also known as perpetual spinach, is by far the closest to real spinach in appearance and flavor. It is a kind of chard with narrow green stems and plentiful glossy green leaves, which are generally smaller than other chard leaves. It is a trouble-free, adaptable crop, and deserves to be much better known.

Ruby chard is an attractive plant. Credit: Kathryn Simmons.

Beet greens

Beets grown for the roots also provide leaves that may be used for greens, or beets may be sown solely to supply leaves in warmer weather, when the roots would be too tough to use. Beet greens are a tasty and colorful vegetable for cooking. Beet seed may be pre-sprouted in the fridge, to get around the difficulty of outdoor sowing when hot. In the American South, there is a hot-weather fungal disease that mimics the cold-weather damping off disease and can cause beet seedlings to disappear almost overnight. Catalog descriptions can help you decide which beet varieties offer the most promise as greens.

Unlike the other spinach substitutes listed below, beets are biennial and will not bolt in the summer. The crops below will resist heat to quite an extent, but may eventually flower and set seed. Planting a succession is a way to get around this problem.

Other hot-weather spinach substitutes

Some of these spinach substitutes look a bit like real spinach, and some taste similar. Others are fairly different, but are cooked and eaten in the same ways as spinach. All are warm-weather crops, so wait till the soil temperature is at least 60°F (16°C) before direct sowing. If sown in mid-June, they can follow an earlier crop such as lettuce or peas.

Aztec red spinach (*Chenopodium berlandieri*) is closely related to true spinach. (The chenopods are now considered a subfamily of the Amaranth family.) An attractive red and green plant that is known in Mexico as *huauzontle*, this crop can make a dramatic statement in the vegetable garden. It can grow to 8'–12' (2.4–3.7 m) tall, although it is a skinny plant, not bulky. The Aztecs grew it between rows of corn. Thin to 15" (38 cm) apart. Each plant can produce a pound (0.5 kg) of colorful leaves, which steam in just one minute and keep their color when cooked. Hot weather increases productivity, while cooler fall weather increases the color intensity of the red leaves.

Orach is another member of the Chenopodiaceae family, and comes in several green, red and purple color schemes. Botanically it is *Atriplex hortensis*. It can be hard to transplant and likes plenty of water. This ornamental is also salt tolerant. The plants produce small leaves and set seed liberally, although orach is not usually invasive. Thin to six inches (15 cm) apart. Nowadays orach

has a new role as baby leaves in elegant salad mixes, but it can also be grown to full size and eaten steamed. The flavor is good, and the color is retained after cooking.

Good King Henry, *Chenopodium bonus-henricus*, also known as Mercury or Lincolnshire spinach, has thick, long-stemmed, arrow-shaped leaves. It is a hardy perennial that vigorously self-seeds. It is rich in vitamins A and C and calcium. This is a fairly untamed plant which bolts easily, so don't expect a long picking season. Early in the year the emerging shoots may be picked and eaten like asparagus.

Magenta lamb's quarters, *Chenopodium album*, has leaves of a lovely magenta color. It has a mild flavor raw or cooked. This is basically a giant weed, which grows to 6' (1.8 m) and re-seeds readily, so keep it from seeding if you don't want an invasion.

Strawberry spinach or **beetberry greens**, *Chenopodium capitatum*, is an ancient plant from Europe. It is similar to lamb's quarters in habit, but only eighteen inches (45 cm) tall. The triangular, toothed leaves are thinner than spinach leaves, very nutritious and high in vitamins. This plant is also grown for the small, mildly sweet, strawberry-like fruits at each leaf axil. It may re-seed vigorously and/or become invasive.

Amaranth is a genus of food plants (*Amaranthus*) found across the globe. There are two basic types: seed amaranths, used as a grain, and leaf amaranths, of which several varieties are sold. Leaf amaranth is also known as callaloo. Some are very attractive, looking like coleus. Thin to at least six inches (15 cm) apart and when the plants reach six to eight inches (15–20 cm), pinch out the tops to get bushier plants. The crop is ready fifty days after sowing. Some people say that ama-

Malabar spinach thrives in hot weather and needs a tall trellis. Credit: Southern Exposure Seed Exchange.

Rainbow chard is a treat for the eyes. Credit: Wren Vile.

ranth should not be eaten raw, but I have failed to discover why, and others recommend it as salad. It is tasty steamed or stir-fried. The leaves are tender with a somewhat nutty sweet flavor. Red-root pigweed is an amaranth. If you have this weed and its particular striped flea beetle, you will also find your edible crop full of holes and not saleable. For this reason, we don't grow amaranth crops.

New Zealand spinach (*Tetragonia expansa, Tetragonia tetragonioides*) is salt tolerant and will even grow in sand. It is a sprawling bushy plant with small, fleshy, triangular leaves. Thin to at least six inches (15 cm) apart. It is very slow to germinate and needs hot weather to really get going. Regular trimming encourages lush growth. Scissors can be used to harvest the shoot tips. If it seeds, you'll get lots of plants the following year. The flavor is very mild—I rate this one as not particularly like spinach.

Malabar spinach (*Basella alba, Basella rubra*) is a vining plant with crinkled heart-shaped leaves on green or red vines. A tropical plant from Asia and Africa, it needs tall trellising and will reward you with its attractive appearance. Germination can be erratic, so don't give up too soon. Soaking the seed in warm water before sowing may help. Thin to at least six inches (15 cm) apart and, to promote a more branched plant, pinch out the central shoot after the second set of leaves. It is little troubled by pests and will produce an abundance of moderately small leaves, looking like real spinach, two months from sowing. Individual leaves may be harvested as needed. The taste is slightly seaweedy (it's also known as "land kelp") and the texture is somewhat mucilaginous in the way that okra is. It can be eaten raw if you like the chewy texture.

Melokhia (*Corchorus olitorius*) is an Arabic summer cooking green which grows quickly to a height of three feet (one meter) in hot weather. Only the small leaves are cooked and eaten. Jute fiber is extracted from the mature plants. Seed is available from Sandhill Preservation.

Chapter 37

Lettuce
All Year Round

Overview

We grow lettuce outside from transplants from February to December, in coldframes starting in September for harvest in November and December, in a greenhouse for leaf harvest until spring and in a solar-heated hoophouse from October to April. We have in the past overwintered lettuce with hoops and rowcover outside, and harvested it as leaves all winter. We have also overwintered in coldframes, for harvest until March. Notes are also included for options other than the ones we choose.

Lettuce types

There are five different general types of lettuces, suited for various tastes and situations. The bolt-resistance generally goes from leaf types (first to bolt) through romaines and butterheads, to icebergs, with Batavians as the best, apart from Jericho and Kalura romaines.

Leaf lettuces include the more familiar oakleaf types and very frilly ones that add interesting texture and appearance to mixes as well as important "loft," which prevents mixtures from falling into unpleasant flat heaps. Leaf lettuces are usually the quickest to produce harvestable-sized leaves and in general have more heat-tolerance (but not bolt-resistance) than romaines. Despite the name, leaf lettuce can be harvested as whole plants, not only leaf-by-leaf for mixes.

Romaines (cos) lettuces are upright, usually green, often very crisp and flavorful. They have double the vitamin A and C of other kinds of lettuce.

Butterhead (Bibb, Boston) types are also usually grown as full heads. They are high in vitamin A, folacin and fiber. The Bibbs, which have soft, sweet tender leaves, can be used at the baby stage in mixes.

Iceberg (crisphead) types are usually grown as full heads, and are less useful for leaf lettuce mixes. They have high water content and are less nutritionally dense.

Batavian lettuces (also called summer crisp or French crisp) are tasty, thick-leafed varieties which have excellent heat and cold tolerance. Batavian varieties also have particularly good hot-weather germination. These are sometimes classified with icebergs as crisphead types, but I think they are very different.

Varieties for spring and summer

Some standards like the Salad Bowls are available year after year. Others can be in fashion for a few years and then disappear. We like Pirat, Panisse, Oscarde, Galactic/Merlot, Green Forest, Buttercrunch and Nancy for spring use.

For summer, Jericho and Kalura are outstanding romaines. Batavians include Cherokee, Sierra, Nevada, Concept, Pablo, Cardinale, Magenta and Loma, a smaller, frilly-edged one. Anuenue is the only iceberg we grow in summer. Other growers recommend the icebergs Ben Shemen and Queensland. We have also had luck with the De Morges Braun leaf type in fairly hot weather.

Varieties for fall and winter

Varieties we have found particularly cold-hardy for outdoors include Brune D'Hiver, Cocarde, Esmeralda, Galactic, Hyper Red Wave, Integrata, Kalura, Lollo Rossa, North Pole, Outredgeous, Rossimo, Rouge d'Hiver, Sunfire, Tango, Vulcan and Winter Marvel. The Salad Bowls are not so good outdoors in cold weather but do very well under cover. Icebergs do not survive frost.

Crop requirements

Lettuce seed remains dormant unless triggered by adequate light and warmth. It needs light to germinate, so don't sow too deep: 0.25"–0.4" (6–10 mm) is enough. Some sources recommend not covering the seed at all, but this can make it hard to keep the seed damp. Dormancy is more pronounced in fresh seed, which has higher levels of the hormone that controls germination. The optimum temperature range for germination is 68°F–80°F (20°C–27°C). Germination takes 7 days at 50°F (10°C), 4 at 59°F (15°C), 3 at 68°F (20°C), only 2 days at 77°F (25°C), back up to 3 at 86°F (30°C) and will not occur reliably at hotter temperatures. Even a few hours at temperatures higher than the optimum can induce dormancy, so store seed in a dark, cool place—in summer, refrigerate seeds between sowings.

Lettuce prefers a well-draining soil high in organic matter, with a pH of 6.0–7.0, not lower. Fertile soil with good tilth is essential, to allow a good root system to develop. The roots also need air exchange, so do not let the soil get crusted. Lettuces need to keep growing fast to taste good. On the other hand, don't overdo the nitrogen as it seems this can boost the growth of E. coli bacteria.

Optimum growing temperatures are 60°F–65°F (15°C–18°C), with a minimum of 40°F (4.5°C) for growth to occur. If nights are cool, 80°F–85°F (27°C–29°C) days can be tolerated. If temperatures are too hot, lettuce will bolt, although some varieties are a lot more heat tolerant than others. Lettuce is more cold tolerant than many people realize. If the plants are sufficiently hardened (prepared by growing in gradually lower temperatures) they can withstand freezing. In a hoophouse or unheated greenhouse they may freeze every night and thaw every morning with no ill effects. Outdoors, a combination of cold nights, chilly days and wind damage will eventually kill them.

Lettuce requires a relatively large amount of water throughout its growth. Insufficient water is the main cause of bolting and/or bitterness. Bolting is also more likely with long days, mature plants, poor soil, crowding and hot weather. Another cause of bolting is vernalization—the chilling of plants with stems thicker than a quarter of an inch (6 mm) at sustained temperatures below 50°F (10°C) for two weeks or more, followed by warmer weather.

Lettuce seedlings in open flats. Credit: Kathryn Simmons.

Sowing

Lettuce for mature heads can be direct seeded whenever the weather is suitable, or it can be transplanted. Some growers like to use pelleted seed for direct sowing as it is easier to space the seeds as needed. For raw seed, sprinkle thinly in a shallow drill. Minimum soil temperature for germination is 35°F (1.6°C). Emergence takes 15 days at 41°F (5°C), 7 days at 50°F (10°C) and 3 days at 68°F (20°C).

Direct seeding is used for growing baby lettuce mix, which is cut when small and then allowed to regrow for further cuttings. Mesclun, salad mix, spring mix and misticanza are all names for mixtures of baby lettuces, sometimes with other greens.

Seeds for transplants can be sown in cell-packs or plug flats, putting three seeds in each cell, and later reducing to one seedling with scissors. Cells with diameters from 1"–2½" (2.5–6 cm) can be used. The 96-cell size (1" × 1½", 2.5 × 3 cm) works well, as does the 200-cell size (1" × 1", 2.5 × 2.5 cm), if you can be sure to get the transplants out before they get root-bound. If warm germination space is limited in early spring, sow seed in a small flat, then spot the tiny seedlings into bigger flats or 606-cell packs (2" × 2¼", 5 × 5.6 cm) to grow on in cooler conditions before planting out. Soil blocks are also possible, but take more time. I recommend sowing several different varieties each time—not only for the beautiful effect, but also to spread your risks in case one kind bolts or suffers disease.

Another option, from mid-April to October, is to use an outdoor nursery bed rather than sowing in flats. For us, this is less work. We simply sow four three-foot (one-meter) rows for each final planting of 120 lettuce, water, weed, then transplant the bare-root plants directly from the seedbed. In very hot weather, indoor sowings might give more reliable germination.

Fall crops to be finished in a hoophouse or other protected structure can either be direct sown or transplanted as bare-root seedlings from an outdoor nursery bed or from plug flats. Because conditions in a hoophouse can be warmer than ideal for lettuce germination until well into fall, it often works better to start plants in a cooler location, then move the plants.

Transplanting

One advantage of transplanting is the ability to grow lettuce when outdoor temperatures don't favor germination. Another is getting a jump on the weeds by eliminating the need to weed around seedlings. A third is that there is time for another crop to mature in the space while transplants are growing, increasing overall yield. In early spring, an earlier harvest is more possible from transplants than from a direct-seeded crop, since the transplants grow indoors while it is still too cold to direct seed. Transplants are tougher than seedlings in withstanding some pests and fungal diseases.

Transplant the seedlings at three to six weeks of age (four to six true leaves) depending on the time of year and how fast they are growing. See the Lettuce Logbook below for our sequence of sowing and transplanting dates. Early in spring, plants will take a long time to size up. As the weather warms they will grow more quickly, until it gets hotter than ideal for them. In September in our climate, four-week old plants will be a good size; as temperatures cool down in the fall, we need more time to grow the plants to transplanting size. Older transplants generally are slower to head up and do not produce good heads. If plants have become a little too large, remove the outer leaves to reduce transpiration losses. Harden off and then transplant. Water well the day before, and again one hour before transplanting.

Handle transplants only by their leaves or the root ball—try not to damage the roots or stem. Transplant seedlings 8"–12" (20–30 cm) apart, firm them in and water. Use the 12" (30 cm) spacing outdoors for growing full-sized heads, and the closer spacing if you will be harvesting individual leaves. Watering in with seaweed solution helps plants recover from transplant shock. Tools to speed transplanting and increase spacing accuracy include measuring sticks or plywood triangles, rollers with accurately spaced blocks attached, transplanting wheels, row-marker rakes and pre-set drip irrigation tape run for 15–20 minutes before planting, to make wet spots to plant into (accurate spacing saves a lot of time at cultivation). In warm weather, lettuce can be ready to harvest four weeks after transplanting, as baby heads or as individual leaves and six weeks after transplanting as full-sized heads. In summer, heads can be ready in as little as three weeks from transplanting.

Water new transplants daily for the first three days, then once or twice a week after that. Deeper weekly waterings equivalent to an inch (25 mm) of rain are better than frequent superficial irrigation, as roots will grow deeper, giving the plant greater resistance to drying out.

Caring for the crop

Hoeing or tractor cultivation will likely be needed to remove weeds, as these compete with the crop

A young lettuce plant in spring. We call this variety Freckles, although it is actually sold as Forellenschluss. Credit: Kathryn Simmons.

and could get mixed in at harvest. All cultivation should be shallow, as lettuce roots are near the surface. Crop rotations including cover crops can do a lot to reduce the weed problem.

Lettuce requires a relatively large amount of water throughout its growth. In cooler weather, water late morning or early afternoon, to give the leaves time to dry before sunset. This reduces the chance of fungal diseases. Spraying with seaweed extract can double the size of lettuce in one to two weeks, enabling harvest up to three weeks earlier.

Reducing nitrate accumulation in the winter

During periods of short daylight length, there is a health risk associated with nitrate accumulation in leafy greens. Nitrates are converted in the body into toxic nitrites, which reduce the blood's capacity to carry oxygen. Additionally, nitrites can form carcinogenic nitrosamines. Plants make nitrates during the night, and when the day length is short, the nitrates do not all get converted into leaf material. It takes about six hours of sunlight to use up a night's worth of nitrates. In winter, leafy vegetables can easily contain the acceptable daily intake level of nitrate for an adult in a small handful of leaves, unless special efforts have been made to reduce the nitrate levels. Spinach, mustard greens and collards contain about twice as much as lettuce; radishes, kale, and beetroots often have two and a half times as much. Turnip greens are especially high, at three times lettuce levels.

To reduce levels to the minimum possible during the high-nitrate-accumulating winter period, harvest only after at least four (preferably six) hours of bright sunlight. If possible, avoid harvesting on overcast days. Keep soil moisture adequate and ensure soil has sufficient P, K, Mg and Mo. Use organic compost. Keep the crops as warm as practicable. Avoid over-mature crops and discard the outer leaves. If your crops are in a greenhouse, ensure that CO_2 levels don't get too low by adequately ventilating as soon as temperatures reach 68°F (20°C). Once lettuce is harvested, their nitrates will convert into nitrites as long as temperatures are merely cool, rather than cold. So for healthy lettuce, refrigerate immediately after harvest.

Intercropping

Lettuce can be interplanted to increase the productivity of an area and provide better habitat for one or both crops. Cultivation is reduced, and the relay planting allows maximum use of the space. Examples include sowing or transplanting warm-weather crops such as peanuts, tomatoes or peppers into the center of beds of lettuce at the transplanting stage, or one month or more after direct seeding. We sow a row of peanuts in the middle of a lettuce bed and get great results. The timing is a little tricky, so try it more than once before deciding whether it suits you. We are still fine-tuning this one! We sow the peanuts April 29–May 12 (around our average last frost date) in the middle of the bed with lettuce transplanted on April 22–May 15. The ideal seems to be to plant regular-size lettuce transplants (not overgrown ones!) on the same day the peanuts are sown, or up to two weeks later. Both crops share the row-cover while the weather is still chilly. The lettuces are harvested before the peanuts grow large. See the Peanuts chapter for more on their side of this scheme.

Research has shown that interplanting of transplanted lettuces and tomatoes does not delay the date of first tomato harvest, or reduce lettuce yields. But lettuce sown immediately before tomatoes are transplanted will have a significantly lower yield, as the tiny lettuce seedlings cannot compete with the fast-growing tomatoes.

Pests and diseases

Organic growing involves using the least invasive methods of pest control. Grow the crop in ways that are least likely to encourage pest outbreaks. Maintain balanced soil fertility, irrigation, airflow, daylight, temperature and growing space (including removing competing weeds). Farmscape to encourage beneficial insects, including predators of pest bugs. Use foliar feeding to boost the plants' growth and health. Rowcover or netting can keep insects from eating your crop.

Monitor crops for pests regularly: look closely and record any outbreaks, the level of infestation and controls used. This will provide useful information for future years. Small numbers of pests may not be doing enough damage to worry about. Once the number of pests exceeds the "Action Level," start with the method that will cause the least damage to other life forms. Often this is physically killing or removing the pest: handpicking, vacuuming or forcing them off with a jet of water.

If the scale of the outbreak is too large for physical controls to work, the next stage can be to introduce beneficial insects (either relocated from elsewhere on your property or purchased). If the infestation is large and the rate of increase is too fast for predatory species to manage, it becomes necessary to use some kind of insecticide to save the crop. Some botanicals also kill beneficial species, so if you plan to use those, do not introduce the beneficials until after the pesticide has had time to reduce the numbers of the infestation, and to degenerate. Here are some specific management strategies for pests and diseases you may encounter. For more information on how to approach crop afflictions, consult the Sustainable Pest Management Chapter.

- **Aphids:** Farmscape with alyssum, clovers, dill, yarrow; use rowcover (but check aphids have not got in—you may only be keeping the predators out!), water jets, insecticidal soap, horticultural oil, hot pepper wax, flour, diatomaceous earth or other desiccants. Try *Beauvaria bassiana* fungus (kills ladybugs too), sugar esters, syrphid flies, aphid midges, parasitic wasps, ladybugs, lacewings, etc. Aphids thrive at lower temperatures than most of their predators, so they are a special problem in early spring, before temperatures reach 45°F (7°C).

- **Slugs:** Handpick, trap in sunken plastic dishes of beery sugar water, or use Sluggo bait.
- **Grasshoppers and crickets:** Use bait containing the parasitic *Nosema locustae*; praying mantids.
- **Cutworms:** Dig gently near plant stem, catch and kill. Use sulfur, or collars around each plant.
- **Thrips:** Try rowcover, sugar esters, insecticidal soap, predators (pirate bugs, lacewings, ladybugs), *Beauvaria bassiana* (Naturalis) or neem (although the last two kill other species too).
- **Groundhogs, rabbits, deer:** Take your pick of fencing, deterrents (hair trimmings, carnivore urine, dogs), trapping or shooting.
- **Damping off** affects young seedlings in cool grey wet conditions. To avoid: reduce watering in chilly weather, increase airflow, foliar feed with seaweed spray and compost tea. Combine one part compost with six parts water, leave for one week, then filter and spray. Use every five to ten days to prevent damping off, powdery mildew, downy mildew, *Anthracnose*, botrytis and late blight. There are also commercially available organic fungicides that use beneficial fungi.
- **Sclerotinia** (lettuce drop) fungi attack lower leaves at soil level and produce a cottony growth. The whole plant then collapses flat and limp, with leaves spread out around the collapsed stem. This disease can be difficult in hoophouses where lots of lettuce is grown. Soil solarization every fourth year, preferably including "biofumigation" with a mustard cover crop, is the cure. Sow the mustard when you transplant the last lettuce crop for that bed in spring. Cut the lettuce, let the mustard grow a couple more weeks, then turn it under. Late June is ideal. Irrigate, cover the soil tightly with old hoophouse plastic and cook for two months. T-22 Plantshield or SoilGard may prevent this disease if shaken onto the seed before planting.
- **Bottom rot**, caused by *Rhizoctinia* fungus, is another soil-borne cool-season problem. It affects fairly full-grown plants, appearing initially as rusty, slightly sunken lesions, perhaps with amber ooze. The whole plant may rot into a slimy black mess. Use solarization.
- **Tip burn** is actually a physiological disorder rather than a disease. It occurs when a sudden change to warmer breezy weather (including human-made "weather" resulting from mismanaged irrigation or greenhouse ventilation!) causes rapid transpiration. If the transpiration rate is much higher than the water uptake rate, the plant cannot get water to the outer edges of the inner leaves, which then brown and die. It is related to soil calcium deficiency, is worse in very fertile soils and is a particular problem on those lovely sunny early spring days. Reduce transpiration by shading and/or shielding from the wind. Consider misting or spraying if this doesn't seem likely to increase the chance of fungal disease. If tip burn seems to be a big problem in your location, look for varieties with resistance.

Harvest

In cool weather, leaf types are ready for harvest 50–60 days from direct seeding, 30–45 days from

A bed of fall lettuce ready for harvest. Credit: Pam Dawling.

transplanting. Head lettuce needs up to 80 days from seeding, or 60–70 days from transplanting in spring. In July head lettuce can take as little as 50 days from seed to harvest. Baby lettuce can be cut 21 days from spring or summer seeding, but may take two or three times as long from November to mid-February. Cool season lettuce mix may provide up to four cuttings, but in warm weather it will only provide a single harvest.

In warm and hot weather, be sure to harvest head lettuces every one to three days to get them before they turn bitter. In summer be prepared to harvest the heads smaller. Excessive milkiness from the cut stem is a sign of bitterness. You can also test by nibbling a piece of leaf.

Harvest methods depend on the size of the crop and the quantity cut each time. Whole heads may be cut with a knife; individual full-size or half-size leaves may be cut with knife, scissors or thumbnails. To harvest baby lettuce, use scissors, shears or a serrated knife, cutting an inch or so (a few centimeters) above the soil to preserve the growing point of the plants for regrowth. For large quantities, there is now a specialized tool consisting of a long knife with an attached fabric catching-box; see Johnny's Seeds. There are also power mowers.

Post-harvest

Lettuce should be cooled as soon as possible, as it rapidly wilts. Refrigerate or immerse in ice-cold water immediately. Forced air cooling in a cold room with high humidity is a good method for large quantities. Nylon mesh bags are useful for washing loose-leaf crops. Thorough washing before sale or serving is most important. If aphids are a problem, cover with water and wait a few minutes until they sink. A salad spinner is the ideal way to dry washed lettuce. An old washing machine, with the agitator removed, works for large quantities. If you wash lettuce in a mesh bag, you can swing it around your head; or set up a plastic laundry basket hanging by ropes from a beam or branch, "wind up" the basket with the bag of lettuce inside, then let the unwinding spin out the water. Sort the crop and return it to refrigeration.

Head lettuce to be sold wholesale is packed in the field unwashed, 20–24 heads in a waxed carton. Usually there are two layers: the bottom layer is packed stems downwards and the top layer stems up, which keeps the heads in best condition. Head lettuce can be held for two weeks at 95 percent humidity and 32°F (0°C) if necessary.

Succession planting

Lettuce grows faster at some times of year than others, so the times between one sowing and the next need to vary to balance this. Lettuce for harvest in February will take two to three times as long from planting to harvest as that for September harvest. December and January sowings grow very slowly, and early February sowings will almost catch up. Crop scheduling for a continuous supply of lettuce is tricky, and worthy of attention.

The short version is that to harvest every week you need to have sowing gaps of more than one week in the spring, less than one week in the summer (in our climate) and decreasing intervals down to as few as two days in the fall. Timing is especially critical in the fall, when temperatures and day length are decreasing: one day difference in sowing date can make almost a one week difference in harvest date.

The following month-by-month narrative (for Zone 7) is followed by charts we use to fine-tune our dates.

January: Make a first sowing indoors in the latter half of the month for the first outdoor transplants. If you have greenhouse or hoophouse space, transplant lettuce there until mid-February.

February: Sow fast-growing, cold-tolerant varieties every fourteen days, in flats indoors.

March: Transplant the first three sowings outdoors with rowcover, as they reach transplant size and are hardened off. You can use plastic mulches in early spring to warm the soil. Start harvesting leaves from the earliest plantings late in the month. Sow in flats every twelve days. Outdoor direct sowing is possible from late March or early April. (We transplant all our lettuce.)

April: Transplant the March sowings. Sow in flats every seven days. Whole heads should mature from mid-April.

May: Switch to heat-tolerant varieties, and an outdoor nursery bed, or carry on sowing in flats if you prefer. Sow every seven days. Transplant one week's needs each week.

June: Sow only the most heat-resistant varieties, every seven days, under shadecloth. Transplant one week's needs each week, using shadecloth to cover transplants for the first two weeks.

July: Sow only the most heat-resistant varieties, every seven days, in the evening, under shadecloth. Use burlap or boards to cool the soil for several days ahead of sowing: soil temperature must be lower than 80°F (27°C). Lay ice over the soil-covered seed rows, or switch to sowing in flats in a fridge. Transplant one week's needs each week, using shadecloth to cover transplants for the first two weeks. Harvest in September. Late July and early August sowings will provide October harvests.

August: Sow every five days early in the month, down to every three days later in the month. Use heat-tolerant varieties early in the month, then switch to cold-tolerant ones after Aug 20. Transplant one week's needs each week, with shadecloth. Harvest in late October and early November. Mid-August is our last chance for outdoor direct seeding (80 days before the first expected hard freeze). We transplant the later August sowings into coldframes for November and early December harvest as heads, although outdoor transplanting, with rowcover as cold weather arrives, is also possible. Harvesting would be somewhat later from outdoor plants.

September: Sow cold-hardy varieties every two days until Sept 21, then every three days. Sowings from the first week could provide the last outdoor planting, under rowcover, for December harvest; the second and third weeks' plants will be only for coldframes, hoophouses and solar heated greenhouses; the fourth week will be replacements for casualties. If you plan to overwinter lettuce outdoors with hoops and rowcover, aim to have plants half-grown by the time the very cold weather hits. Try a few different sowing dates, as the weather isn't very predictable. For us, Sept 10–18 are the best dates. Transplant one week's needs every five or six days until Sept 21, and every few days for the rest of the month.

October: Sow hardy lettuces every three to seven days (if you have some covered growing space to transplant into) until approximately Oct 15, then every seven days until Oct 31. Use 8" (20 cm) transplant spacing if you will be harvesting leaves rather than heads. Another option is to direct sow lettuce mix in a greenhouse on Oct 20 for winter harvest.

November: Sow once between Nov 1–15 in a hoophouse or greenhouse for January transplants, or take a break. This would be the last chance to transplant into coldframes for January and February harvests.

December: When daylight length is less than ten hours, little plant growth is happening. Harvest, write up crop records and plan for next year. If you have greenhouse or hoophouse space, transplant at the end of December for lettuce heads in February (or leaves in January and February). If needed, make a sowing between Dec 1–15 to transplant in a greenhouse in late January.

Lettuce logbook page

We use a Lettuce Log to set our sowing and transplanting dates and record actual dates of sowing, transplanting, starting and finishing harvest of each of our plantings, for head lettuce from transplants. The information helps us improve the sequence the next year and get closer to our goal of

a continuous supply. Unless you are in a similar climate zone, these exact dates won't be right for your farm, but you can see the general themes. The gap between one sowing and the next gets smaller as the year progresses; the gap between one transplanting and the next does likewise; the number of days to reach transplant size dips to 21 days in the summer, then lengthens as the weather cools and the days get shorter. Use these guidelines for your first year or two while you are collecting your own data.

Succession crops graph

This graph is based on our experience at Twin Oaks, using information in Eliot Coleman's *New Organic Grower*. See the chapter on Succession Planting for details on how to make and use these graphs. We recorded sowing and harvest dates for several years, then plotted our data on a graph. We used the graph to determine a sequence of sowing dates likely to provide a regular weekly harvest. This list of dates was used to compose the logbook on pg. 256.

Season extension options for hot weather

In warm weather there are various tricks to get lettuce to germinate. Use only the most heat-tolerant varieties, as others may not germinate at high temperatures and the plants will bolt and taste bitter. If possible, store seeds in a tightly closed container in a freezer. Another method is to freeze seed for four days before sowing. After freezing, always bring the container to ambient temperature before opening, to avoid dampening the seed with condensation from the relatively warm air. An easier compromise is to store the seed in the fridge.

Twin Oaks Lettuce Log 2012

#	Sow Date	Sowing Gap	Sown	Days to T/pl	T/plant	T/pl gap	Trans-planted	Harvest Start	Harvest Finish	Notes (Varieties, Success/Failure)
1	Jan 17			53	Mar 10			May 1	May 16	
2	Jan 31	14		49	Mar 20	10		May 6	May 20	
3	Feb 14	14		45	Mar 29	9		May 20	May 24	
4	Feb 28	14		40	Apr 8	10		May 24	May 26	
5	Mar 13	14		34	Apr 17	9		May 26	Jun 2	
6	Mar 26	13		30	Apr 26	9		May 29	Jun 3	
7	Apr 5	10		30	May 5	9		Jun 4	Jun 9	
8	Apr 14	9		29	May 13	8		Jun 9	Jun 15	
9	Apr 23	9		27	May 20	7		Jun 15	Jun 25	
10	May 1	8		25	May 26	6		Jun 25	Jul 3	
11	May 9	8		21	May 31	4		Jul 4	Jul 11	
12	May 17	8		19	Jun 6	6		Jul 9	Jul 19	
13	May 24	7		18	Jun 12	6		Jul 20	Jul 27	
14	Jun 1	7		18	Jun 19	7		Jul 28	Aug 4	
15	Jun 7	7		19	Jun 26	7		Aug 5	Aug 11	
16	Jun 13	6		19	Jul 2	6		Aug 12	Aug 17	
17	Jun 18	5		20	Jul 8	6		Aug 17	Aug 24	
18	Jun 23	5		21	Jul 14	6		Aug 25	Sep 6	
19	Jun 28	5		22	Jul 20	6		Sep 7	Sep 14	
20	Jul 3	5		22	Jul 25	5		Sep 15	Sep 20	
21	Jul 8	5		22	Jul 30	5		Sep 21	Sep 27	
22	Jul 13	5		21	Aug 4	5		Sep 28	Oct 2	
23	Jul 18	5		21	Aug 9	5		Oct 3	Oct 7	
24	Jul 23	5		22	Aug 15	6		Sep 28	Oct 13	
25	Jul 28	5		23	Aug 21	6		Oct 14	Oct 20	
26	Aug 2	5		24	Aug 26	5		Oct 21	Oct 27	
27	Aug 7	5		24	Sep 1	6		Oct 29	Nov 4	
28	Aug 12	5		25	Sep 7	6		Nov 5	Nov 11	
29	Aug 16	4		24	Sep 10	3		Nov 12	Nov 18	
30	Aug 20	4		23	Sep 13	3		Nov 19	Nov 25	
31	Aug 23	3		23	Sep 16	3		Nov 26	Dec 2	
32	Aug 26	3		23	Sep 19	3		Dec 3	Dec 9	
33	Aug 29	3		23	Sep 22	3		Dec 10	Dec 31	11 Beds Total
34	Sep 1	3		24	Sep 25	3				Frames
35	Sep 3	2		25	Sep 28	3				Frames
36	Sep 5	2		26	Oct 1	3				Frames
37	Sep 7	2		27	Oct 4	3				Frames
38	Sep 9	2		29	Oct 8	4				Greenhouse
39	Sep 11	2		31	Oct 12	4				Greenhouse
40	Sep 13	2		33	Oct 16	4				Greenhouse
41	Sep 15	2		35	Oct 20	4				Greenhouse & hoophouse
42	Sep 17	2		34	Oct 21					Hoophouse
43	Sep 19	2		33	Oct 22					Filler
44	Sep 21	2		32	Oct 23					Filler
45	Sep 24	3		30	Oct 24					Filler & hoop-house #2
46	Sep 27	3		28	Oct 25					Filler

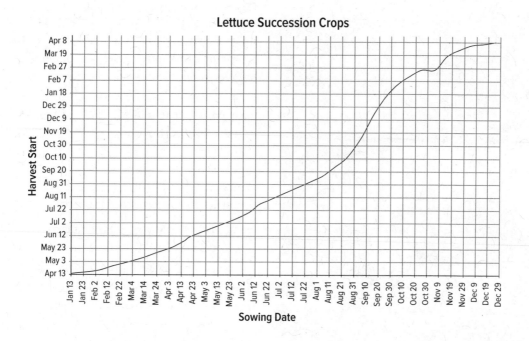

Lettuce Succession Crops

If you have fridge space, put the seeded flats in a plastic bag in the fridge for two days to break the dormancy; if you don't, a cool basement will do. If you need to do the seeding outdoors, you can improve your chances by cooling the soil for several days ahead, by watering and covering with thick organic mulch, boards or burlap bags. Soak the seed in cool water for a day, then drain and mix with a dry gritty material such as sand, corn grits or coarse bran before sowing. Optionally, store the drained seed in a jar in the fridge for two days before sowing. Sow seed in the evening. After sowing (thickly), put ice on top of the soil covering the seeds, and cover with shadecloth (50 percent shade is ideal). Substitutes for professional shadecloth (for small plantings) include tent screen windows, nylon window screen and nylon net curtains—but be sure to use something air can flow through, to prevent overheating. Water with freshly drawn cool water at midday (possibly more than once a day) until the seed germinates.

Use younger transplants (three weeks old) than you would in spring—they will recover more quickly than larger ones. Closer spacings such as 10"–12" (25–30 cm) will enable foliage to grow to completely cover the bed and keep a cooler microclimate. If possible, plant to the north of tall plants such as corn, tomatoes or pole beans. Transplant in the evening. Develop a fast and efficient technique so that you can get your crop planted and watered in the last hour before sunset. Cover the new transplants with shadecloth for at least one week, preferably until harvest (which may be as little as three weeks from transplanting). Shadecloth laid on hoops (wire, fiberglass or PVC electrical conduit) is even better than shadecloth lying on the plants, as it allows better airflow. Ideally, fasten the shadecloth to the hoops with clips or clothespins to hold the bottom edges

Weeding winter hoophouse lettuce. Credit: McCune Porter.

about a foot (30 cm) above the soil to maximize airflow. To cool lettuces growing indoors cover hoophouses with shadecloth and spray or paint glass greenhouses with shading compound.

Water much more in hot weather—bitterness before bolting is almost always a sign of water stress. Just one day of insufficient soil moisture can trigger bitterness in lettuce. Overhead watering early in the morning can be used for mature heads you want to hold in the field for a couple more days. For some crops, an organic mulch of straw or spoiled hay will moderate soil temperature and reduce irrigation needs; unfortunately this does not work so well for lettuce, as it can be hard to remove stray wisps of mulch from the harvested crop. It is also a fast-growing, short-term crop, and organic mulches take time to break down. (This would not be a problem if the following crop were to be transplanted into the leftover mulch.) Most people do not mulch lettuce, except for plastic mulches to warm soil in early spring or to control weeds. There are white and silver reflecting plastic mulches sold for summer use.

Season extension techniques for fall and winter

Heat-tolerant lettuces also tolerate cold. There are also specialized cold-hardy varieties that do not tolerate heat (because they have a relatively low water content). Sow these in fall and winter only. Choose your site with care. Protection from prevailing winds can be vital. Conserve soil warmth by using light-colored mulches to reduce radiation losses.

Rowcover will provide a temperature gain of 4–6 degrees F (2.2–3.3 degrees C), depending on the thickness. It also reduces light transmission and airflow, but the trade-off can be very worthwhile. Lettuce may survive an occasional dip to 10°F (–12°C) with good rowcover—but not 8°F (–13°C), I can tell you from experience! Adolescent lettuce are more cold-hardy than full-sized plants. 9 gauge wire, plastic or fiberglass hoops, 5'–6' (1.5–1.8 m) apart along the row to hold the rowcover above the leaves, will reduce freezing damage and provide a better microclimate. Rowcover also excludes some pests and can provide better quality produce by reducing weather damage. Make sure the edges of the rowcover are held down with sticks, rocks or lengths of rebar. There is some evidence that two layers of thinner rowcover work better than one thick layer in protecting against the cold.

It is thought that freshly watered plants are less resistant to freezing damage, so choose days with mild expected overnight lows for watering. If cold weather is expected, spray with seaweed a few days beforehand to help protect against frost. Overhead sprinklers can be used to water overnight on frosty nights and prevent frost damage to foliage, but lettuces can do surprisingly well at fairly low temperatures and "anti-freezing" night watering is better used for warm-weather crops. It is sometimes possible to save a frozen crop by spraying with water by hand early in the morning and continuing until sunlight reaches the plants.

"Low Tunnels" are hoops, about 18" (45 cm) high at the apex, covered with clear slitted polyethylene. They are more labor-intensive to use than rowcover because they are not permeable to water or air. Drip irrigation can deal with the watering. The slits in the plastic let hot air escape; at night they stay closed, trapping the heat. When warm, the edges of the slits curl, providing an escape for the hot air. Low tunnels reduce wind damage but do not provide much protection from the cold. "Quick Hoops" are bigger versions of a similar idea, using rowcover with clear polyethylene and providing more protection from bad weather. They are best suited to areas with reliably cold winters, not like Virginia's back-and-forth winters that include spells too mild to keep crops under polyethylene.

Coldframes are traditionally made from blocks, boards or straw bales, with discarded windows over the top. They are very useful on a small scale, although labor-intensive. *Solar Gardening* by Leandre and Gretchen Poisson is the best book I have found on making and using small structures. For large-scale production, the construction costs of a hoophouse are lower than for a comparable area of coldframes. Single-layer hoophouses are sometimes called coldframes.

Hoophouses (high tunnels) make winter salad growing easier and more efficient. They permit lettuce planting in October and November, providing harvests all winter and through till April. Crops continue to grow throughout the winter whenever temperatures are warm enough. Solar heating is sufficient to enable a wide range of cold-tolerant crops to be grown. Crops are grown in the ground, often in long, intensively planted beds, with drip irrigation. Heated greenhouses of a more traditional kind will also grow lettuces all winter, as will unheated greenhouses with a masonry north wall (in central Virginia at least).

The economics of season extension bear consideration. All season extension techniques require the input of more time and/or money than outdoor growing, to gain extra production. For each grower there comes a break-even point, beyond which the diminishing returns make the extra energy inputs counterproductive. At some point you might do better to turn your attentions to some other crop (kale?) and not chase after lettuce regardless of costs. Before investing a lot of money, talk with other growers. You can grow more lettuce per dollar invested in a large hoophouse than a smaller heated greenhouse constructed of expensive materials.

Chapter 38

Carrots, Beets and Parsnips

Carrots, beets and parsnips are workhorse root crops that thrive in mild weather, store well and are popular traditional foods. They can provide high yields for the time invested. Parsnips are traditional northern European root vegetables that are surprisingly easy to grow in warmer climates, while also being extremely cold tolerant. All three crops have similar requirements, so I'm writing about them together. Because these root crops all store well, they have the benefit of providing vegetables for sale after the harvest period is over, giving the farmer time to rest a little or delegate the sales to an intern while taking a vacation. CSA farmers can also use storable crops as the flexible "filler" in the box: more when other crops are in short supply, less in bountiful weeks. Beets can also provide bunches of edible greens even before broccoli and cabbage are ready in spring.

Varieties

Carrots come in several main types, and the factors influencing your choice will be soil type, climate, market and harvest time and method. All carrots sweeten up when days are warm and nights cool.

Nantes types are sweet, juicy and tender. They contain few terpenoids, the volatile flavor compounds we think of as "carroty." When poorly grown, they can be watery and bland.

Imperator types are heavier and contain more terpenoids and fewer sugars and can be prone to bitterness if something goes wrong. Many hybrid varieties are a cross between Nantes and Imperator types. Imperator types are best for storage, and Chancellor has good cold tolerance.

Chantenay types have a flavor described as "parsley-like."

Kuroda types have a drier, creamier flavor. They are tolerant to Alternaria.

Here on a sandy clay loam, we grow Danvers 126, a sturdy, open-pollinated variety suited to high production of bulk carrots. We have also grown the 75-day Bolero, a 7"–8" (18–20 cm) hybrid storage type, which looks a bit prettier, and has slightly better flavor, than Danvers; ultimately we decided the extra cost of the seed was not worth it. Resistance to Alternaria and Cercospora has been important for our fall crop. In the past I have grown Chantenay Red Core (65 days), a blocky variety with a blunt tip, five inches (13 cm) long and two inches (5 cm) at the shoulder. It resists splitting and can deal with clay.

The 56-day Nelson, a nice, slender six-inch (15-cm) cylindrical Nantes variety with a blunt root, has done well for us in summer. It is tender and crisp with deep orange color, somewhat brittle, but less so than tapered types such as Mokum. In the past, before the soil here had been improved, it was necessary to grow shorter carrots, as the long ones would break in the tight soil. Most of the varieties recommended for clay soils are both small and short. Oxheart is a large open-pollinated variety, great for clay soils. Despite being short, it has a thick 3"–4" (8–10 cm) diameter. Purple, white, yellow and red varieties are also available, if you want glamour.

Beets too, come in several types—round, top-shaped and cylindrical. The size and quality of the greens is a factor if you sell bunched beets with tops, or use the tops for greens. We like the long Cylindra/Formanova/Forono ones which are six inches (15 cm) long, very tender and easily to peel and slice for pickles or cooking (55 days, OP).

Among the round ones we like Ace (50 days, hybrid) and Detroit Dark Red (60 days), a tender open-pollinated variety. Detroit Crimson Globe is said to maintain better flavor in hot weather than most others, which can develop off-flavors. Early Wonder Tall Top (48 days) is also open-pollinated. Lutz Green Leaf (70 days) is a big long-storage variety that hit some problems with seed supply in recent years. Check that you are getting good seed before you buy.

Bull's Blood is a specialty variety grown for the dark shiny red leaves, to use in winter salad mixes. Beet leaves (and chard and spinach) grow faster in winter than lettuce does, and this beet has exceptionally dark leaves, which add strong color to the mix. Some people claim to like the roots of this one, but I find that after harvesting leaves all winter, the quality of the roots is not good. There are also golden beets, white beets and candy-striped Chioggia beets, although in my experience, what they gain in appearance they lose in flavor and tenderness. Their Italian name is pronounced *key-ojia* (or *key-oja*, if you talk fast.)

Parsnips are available in fewer varieties, and most catalogs offer only two or three. There are two main types, the "hollow crown" types, where the leaves grow from an indentation in the top of the root, and the flat crown kind. Hollow crown parsnips are harder to clean and the top of the root is not easily made usable. Most varieties are open-pollinated (Hollow Crown, Harris Model, Andover, Tender and True, Cobham Improved and Lancer) while a few are hybrids (Albion, Javelin and Gladiator). People say that hybrids are more uniform and higher yielding, but I have not tested these claims. Whiter varieties are more prone to canker, the main parsnip disease; Tender and True is resistant to it. All take 100–120 days to maturity, but there is no particular market for early parsnips, so that isn't likely to be a deciding factor. Harris Model, Andover and Tender and True have worked well for us.

Seed specifications and yield

Carrot seeds are small: 175,000–600,000 (avg. 288,000)/lb, 18,000 seeds/oz, 625 seeds/g.

Their ideal growing density is 3,000 seeds (⅜ oz, 10 g)/100' (30 m), 1 seed/cm, 625 cm/g, 600'/oz, 9,600'/lb, 2½ lb/acre (at 30 seeds/ft in rows 24" apart) or 720,000 seeds/acre.

They will yield 100 lbs/100' (1.5k/100 m) or 30,000 lbs/acre (5443 kg/ha).

Beets seeds are larger: 17,000–40,000 (avg. 35,000)/lb, 2,200 seeds/oz; 80 seeds/g.

They should be sown at 1,285 seeds (⅔ oz, 18 g)/100' (30 m); 150'/oz, 2,300'/lb, 9 lbs/acre or 315,000 seeds/acre.

They will yield: 40 lbs (18 kg) greens and 100 lbs (45 kg) of roots per 100' (30 m) or 14,000 lbs/acre (2,540 kg/ha).

Parsnip seeds come with 105,000–120,00/lb, 7,000–10,000/oz, 250 seeds/g.

They should be sown at 4,000 seeds (½ oz, 14 g)/100' (30 m) or about 3–5 lbs/acre.

They will yield 80–120 lbs (36–54 kg) per 100' (30 m).

The viability of parsnip seeds goes downhill fast, so never try to save leftover seed for another year. They are one of the few seeds I always buy fresh each year.

Crop requirements

Any decent soil will grow some carrots, beets or parsnips, but the best ones grow in deep, loose and fertile sandy loams with good moisture-holding capacity. All prefer cool temperatures, for best flavor and appearance. Old books warn against us-

A satisfying bed of young carrots. Credit: Kathryn Simmons.

ing manure before carrots as it will make them fork. This refers to uncomposted manure, not to compost. Compost will increase yields, and even reduce the culls with some varieties (research by Dan Brainard at Michigan State University). Compost not only increases the organic matter in the soil, but also suppresses some diseases and nematodes (which can cause forked carrots).

Beets need a pH of 6.0–7.0, but prefer it between 6.5–6.8. They require abundant potassium, which can be supplied by woodash. Boron deficiency can show up in beets as internal browning, or dark dead tissue, as well as distorted leaf growth. It is most likely to occur in alkaline soils after long hot, dry spells. Beets can suffer from "zoning" (white rings in the roots), if there are acute weather fluctuations.

Parsnips do not want an overly rich soil, nor one that crusts easily or is full of rocks. Their ideal pH is 6.0–6.5.

Sowing

Sow carrots and beets whenever the soil is between 50°F (10°C) and 95°F (35°C), so long as you can keep the surface damp. We start with carrots in February and sow every couple of weeks in spring. No, our soils are not 50°F (10°C) in February, but the seed comes to no harm in the ground and it's a job we can get done early. We sow once a month in summer if we need more, and then finish with a big sowing at the end of July or early August—10–12 weeks before the usual first frost—to be harvested in November. With beets we do a single sowing in mid-March and another in early August. The optimum germination temperature range for beets is 50°F–95°F (10°C–35°C). We are growing for fresh use, pickling and storage, but not bunch sales, so we don't need to do frequent sowings. Most people sow parsnips just once a year, early in spring, as they are slow-growing. But the exact date in spring doesn't seem to matter—if you have at least 110 days, they are almost guaranteed to be big enough by the end of the season. The minimum germination temperature is 35°F (2°C). Some growers like to use fluid sowing of pre-germinated parsnip seeds, to get better emergence and earlier harvests. See the Summer Germination of Seeds chapter for how to do this.

The table on this page shows days to germination as a function of soil temperature. This information (from Knott's *Vegetable Growers' Handbook* and Nancy Bubel's *New Seed Starter's Handbook*) is very useful for flame-weeding (see below).

The second table shows the success rate of germination as a function of temperature. Here we see that most viable carrot seeds can still germinate if the soil is as hot as 86°F (30°C). Also, the success rate of beet seedlings is above 100% until

Germination time and rate of success as a function of temperature

	50°F (10°C)	59°F (15°C)	68°F (20°C)	77°F (25°C)	86°F (30°C)	95°F (35°C)
Days to Germinate (d)						
Carrots	17.3	10.1	6.9	6.2	6.0	8.6
Beets	16.7	9.7	6.2	5.0	4.5	4.6
Parsnips	26.6	19.3	13.6	14.9	31.6	0

Percentage of normal seedlings (%)

	50°F (10°C)	59°F (15°C)	68°F (20°C)	77°F (25°C)	86°F (30°C)	95°F (35°C)
Carrots	93	95	96	96	95	74
Beets	156	189	193	209	192	75
Parsnips	79	85	89	77	51	1

soils reach 95°F (35°C) because each beet "seed" is actually a seed cluster. Parsnip viability starts to decrease at temperatures over 68°F (20°C)

For carrots, aim to sow 30 seeds/ft (1/cm), 0.25"–0.5" (0.6–1.2 cm) deep. Some people sow in single rows 8"–10" (20–25 cm) apart. Others sow in bands 2" (5 cm) wide, at 8" (20 cm) apart, with one length of drip tape serving two bands in one 16"–24" (40–60 cm) bed. There are precision seeders that save you from thinning, but many small growers use an Earthway type seeder, as we do, and then thin. Some people mix inert materials (such as dry sand) with the seed to help get a spaced stand. Some people bake old carrot seed to dilute the good new seed.

For beets, we often presoak the seed for a couple of hours, and in summer pre-sprout it, by draining the soaked seed and keeping it in the jar for a couple of days. (Beet seed "drowns" easily, so don't soak in water for too long.) Then we hand sow, sometimes mixing damp sprouted seed with dry bran or grits to minimize clumping. We sow an inch (2.5 cm) apart in single rows 8"–10" (20–25 cm) apart. Others sow in bands 2"–4" (5–10 cm) wide, at about 15 seeds/ft (2 cm apart), with bands 12"–18" (30–45 cm) apart. Sow 0.5" (1.2 cm) deep in spring, deeper in hot summers, but never more than 1" (2.5 cm). It is important to get good soil contact between the soil and the corky seed-balls, so tamp or roll the rows after seeding. As with carrots, avoid soil crusting. For a continuous supply of greens and baby beets, sow every two weeks until eight weeks before regular frosts usually occur, or about ten weeks before you expect a heavy freeze.

For parsnips, we sow between March and mid-April, or even late April. Further south than us, growers can sow in the late summer, in August or early September. The soil temperature needs to be below 70°F (21°C). Parsnips are slow to germinate, even under ideal conditions, so we dot radish seeds occasionally along the row to enable us to hoe before the parsnips germinate. The emerging radishes can also help prevent soil crusting. Aim for a depth of 0.25"–0.8" (0.6–2 cm), in rows 8" (20 cm) or more apart. We usually "station sow" our parsnips, putting several seeds at each spot, 1.5"–4" (3–10 cm) apart. Alternately, sow and thin to this spacing later.

Root crops do well on raised beds, because the soil stays loose and the roots can easily grow deep. Hard rain in the first 3–4 days after planting can create a crust which could stymie the emergence. To prevent this, irrigate for half an hour each day until the carrots come up. Keep the soil surface moist. Some people use shadecloth to help with this. Old books recommend covering the rows with boards, though clearly that isn't practical on a large scale.

Flame-weeding is always a two-person job here. It's safer to have someone watching for stray fires. Credit: Brittany Lewis.

I read in *Growing for Market* of a very successful innovation that solved the problem with soil crusting and poor carrot emergence for one small-scale grower. She used a soil erosion control blanket, made from excelsior wood product quilted between layers of plastic. For $44 (in 2004) she covered a 300' bed, until the carrots germinated, then she rolled the blankets for reuse. They are available from American Excelsior in Rice Lake, Wisconsin: (715) 234-6861. Call and ask for seconds.

Transplanting

Carrot and parsnip taproots get damaged by transplanting, but surprisingly, beets can be successfully moved around. For early crops in cold climates, start seed indoors in early spring and transplant it at about 5–6 weeks of age, after any real cold weather subsides. Plant out 3" (8 cm) apart in rows 12"–18" (30–45 cm) apart. We have sometimes transplanted beets within a bed to fill in gaps.

Flaming

Carrots, beets and parsnips are ideal crops for pre-emergence flame-weeding. The goal is to flame the bed the day before the expected emergence of the crop. Use a soil thermometer and the table above to figure out which day to flame. For carrots it's possible to sow a few "indicator beets" at one end of the bed; as soon as you see the red loops of the beet seedlings breaking the surface, flame the carrots. (But look for carrots too, just in case!) Beets are always a bit quicker than carrots in germinating. Another way to get an alarm call is to put a piece of glass over part of a row. The theory is that the soil under the glass will be warmer and the crop there will come up sooner than the rest. I tried this once, but the soil under the glass dried out, and those carrots came up later than the rest! Nowadays we have a "no glass in the garden" rule, for safety, so I use beets, the thermometer and the chart.

We use a handheld flamer attached to a propane cylinder that is in a wheelbarrow pushed by a second person behind the first. This person also acts as a "fire warden." Some growers mount the propane on a backpack frame. Walking along the aisle between beds and wafting the wand diagonally back and forth across the bed takes about ten minutes for a 100' (30 m) bed. Flame-weeding alone can reduce hand-weeding to one hour/100' (30 m). Hand-weeding can be reduced to 6 minutes/100' (30 m) by flame-weeding after using stale beds which have been hoed three or four times.

Caring for the crop

Carrots do very poorly with competition, so try to start early carrots in a bed that had only light weeds the year before. Later sowings can make use of the Stale Seedbed Technique, where the bed is prepared ahead of time and one or more flushes of weeds are germinated and flamed or hoed off.

Get to the initial thinning as soon as you can, spacing to about one inch (2.5 cm) apart, weeding at the same time. We usually have someone with good eyesight and hand-eye coordination take a scuffle hoe between the rows the day before the hand-weeding. If you are in an area with Carrot Rust Fly, you will want to remove all thinnings and broken foliage from the field, so you don't lure the low-flying pest with the wonderful smell of the broken leaves. We do a second thinning, to three inches (8 cm), at the stage when the baby carrots can be used for salads. If we get more weeds, we might do another round of weeding before harvesting the full-size carrots. If the shoulders of the carrots are prone to greening, you can hoe soil up over the crowns. If carrots are spaced too widely, they will be more likely to split, and the overall yield will be reduced. *The Complete Know and Grow Vegetables* by J. K. A. Bleasdale, P. J. Salter et al. has good information on researched crop spacing for maximum yields, among other gems. For carrots, they recommend 1.5 × 6" (4 × 15 cm) for maximum total yield (medium-sized carrots), and 4 × 6" (10 × 15 cm) for early carrots to minimize competition and get rapid growth. If you want to have rows more than 6" (15 cm) apart, calculate the area of these optimum spacings, then divide by your chosen row space. For example, if your rows are 12" (30 cm) apart, the carrots can be as close as three-quarters of an inch (2 cm) if total yield is more important than individual size, or two inches (5 cm) for fast early carrots.

Dry soil at the 3–4 leaf stage can cause forked or irregular carrots. Hairy carrots indicate either too little water or too much nitrogen.

Beets come up in clusters, and they too will benefit from hoeing, thinning and weeding. Beets deal with weed pressure and crowding a lot better than carrots do, so if you have to choose which to weed, the carrots win! Once again, we thin in stages, so that at the second thinning, the baby beets can be used as a crop. For mature beets, allow each a minimum of three inches (7.5 cm). Cylindra beets can be left a bit closer, and will push themselves up out of the soil as they grow. *Know and Grow Vegetables* recommends establishing five plants/ft² (54/m²) for early beets. This translates to a final spacing of 4"×7" (10×18 cm). For maincrop beets, aim for 10–15/ft² (107–161/m²), or spacing of 1"×12" (2.5×30 cm) for maximum total yields of small-sized roots.

Parsnips give their highest yield at 3 plants/ft² (32 plants/m²) for large varieties and 6–7/ft² (64–75/m²) for small varieties. The roots will be relatively slender: 1.5"–2" (3.8–5 cm). For larger roots, but lower total yield, plant large varieties at 2/ft² (21/m²). *The Complete Know and Grow Vegetables* recommends that the ratio between row spacing and in-row spacing not be greater than 2.5:1. For example, to achieve 2/ft² (21/m²), the maximum row space would be 13.4" (34 cm) and the in-row spacing 5.5" (13.5 cm). In rounder numbers, this could be 12"×6" (19×8 cm).

Pests and diseases

Carrots can be troubled by Alternaria and Cercospora, which both reduce yield and quality. Alternaria blight starts on the oldest leaves, which blacken and shrivel. We have had this in the summer, and our response is to cut our losses and harvest them right away. Cercospora leaf spot starts as small dark spots with yellow edges on younger leaves and stems. Copper fungicides (if you use them) can be employed as a preventive measure or control.

There are several insect pests of carrots. The main two are the carrot rust fly and the carrot weevil, both of which tunnel in the roots. I have been trying to figure out which we have, since if we don't have any rust fly, we can stop being so careful about removing all the thinnings from the field. Eric Day, the entomologist at Virginia Tech, tells me carrot rust fly has not been seen in Virginia for twenty years, so I'm ready to relax my vigil. I believe the rust fly usually tunnels in the lower third of the root, and the weevil (with wider, more open tunnels) works on the shoulders. We do have the colorful striped parsley worm, but not in high enough numbers to cause economic damage. This white, yellow and black worm can be identified by (gently!) pressing down on its back. It projects a pair of orange "horns" and emits a "fake strawberry flavor" smell. Some areas struggle with wireworms, which can be caught by burying carrot slices, and daily removing the captives. If necessary, rowcover can be used to exclude flying pests. *Garden Insects of North America* by Whitney Cranshaw is a good resource. The Ontario Ministry of Agriculture, Food and Rural Affairs has good fact sheets on carrot insects. Some research has been done at Washington State University, using pathogenic nematodes and other possible parasitoids and predators.

Deer have been a major pest for us recently, eating our carrot tops down to nubbins, and I hear rowcover doesn't stop them. Fences, guns and dogs might.

Beet seedlings are susceptible to damping-off in cool wet conditions (*Pythium* fungus). To minimize the likelihood of this, help the beets germinate as quickly as possible, and cultivate between

This bunch of beets is Detroit Dark Red. Credit: Southern Exposure Seed Exchange.

the rows to keep the soil aerated. Beets can also suffer from a summer post-emergence fungal disease which can reduce survival. It may be *Rhizoctonia* root rot, a damping-off fungus which affects seedlings in warm soil (as contrasted to *Pythium*, the cold-weather damping-off fungus, and *Phytophthora* and black root rot, which affect established plants). Beets also suffer from *Cercospora beticola*, a fungus that can render the greens unsaleable, as well as possibly reducing yields of the roots. They can also get scab (the disease that potatoes can get): raised rough brown patches. Keeping beets well watered can reduce the chance of scab. The spinach leaf miner is the only insect pest of beets that I know of, and they leave the roots unaffected. Lambs Quarters is an alternate host.

Parsnips can suffer from two kinds of canker, *Itersonilia* canker and *Phoma* canker. Crop rotation can help avoid these diseases. Carrot pest insects can also damage parsnips, so take the same precautions when thinning and harvesting if the carrot rust fly is present.

Harvest

Carrots develop flavor and color at the same time, and harvest can begin as soon as they look and taste right. Carrots left in the ground too long may crack, and start to develop off-flavors. If the soil is dry, gently water just before harvest. This will ease the harvest mechanically and also improve the flavor of the carrots. How you harvest will depend on the scale of your carrot farming, the needs of your market and the equipment you have. To harvest by hand, dig or pull up the carrots, collecting them in a cart or wheelbarrows. This is our method. We then take the carrots into the shade, cut the tops off (cut at the transition from green to orange if for immediate use, or leaving a short length of greens if for storage). As we cut, we put the carrots into buckets of water. When a bucket is full we give the carrots a quick rub over and put them into clean rinse water. From there we remove handfuls and drain them in buckets with holes in the bottom, before transferring to perforated plastic bags for the walk-in cooler. Other growers might mow the tops off first, then undercut the roots, making them easier to pick from the loosened soil. See the Resources section for information on root washers.

Carrots are quite cold hardy when mature, down to 12°F (–10°C).

The harvest of beets is similar, but should be done before a hard frost. We harvest our fall beets right after the celeriac, and before the turnips, kohlrabi and winter radish, which are hardy to 20°F (–7°C). When trimming the tops, leave a short tuft of leaf stems, to avoid injuring the root, and to preserve the color when the beets are cooked.

Harvest of parsnips is a simple manual job for us, as we only grow a small quantity. They are the last root crop we dig up, as they are tolerant of temperatures down to 0°F (−18°C). There may be a market for large quantities of organic parsnips for baby food manufacturers, if you fall in love with growing this crop. Then you'll need the undercutters and washers, as for carrots. Modified potato diggers are sometimes used. Some people have skin that is sensitive to parsnip foliage and roots, and will need to wear gloves and long sleeves to avoid problems.

Storage

With traditional storage methods, where root crops are buried in sand or ashes in a root cellar, unwashed roots store better than washed ones. They also keep well in perforated plastic bags under refrigeration, and washed roots store as well this way as unwashed ones, with the advantage of avoiding the discoloration that can happen to unwashed carrots in storage. Don't store roots with fruits (such as apples or squash) as ethylene emitted by the ripe fruits can turn the roots bitter.

Store carrots in humid conditions at near-freezing temperatures (not below). Young bunched beets can be stored for 10 days at 32°F (0°C) and 95% humidity. Mature beets can be stored for winter for six months or more at 32°F (0°C) and 95% humidity. Parsnips can be stored for six months, if close to freezing point, with high humidity (90%–95%).

If your winters are mild enough, and you don't have voles, you can store carrots, beets and parsnips in the ground until spring, covering the bed with loose organic mulch (straw, tree leaves, spoiled hay). Parsnips can survive soil temperatures of 0°F (−18°C). They should be dug before growth resumes in the spring, as these biennial plants will consume the root in producing flowering stems.

Rotations

To prevent blights, practice three-year crop rotation. Carrots and parsnips are umbelliferae, along with celery, celeriac and fennel. Our fall carrots are planted where our garlic has been grown, often with a quick cover crop of buckwheat in between. This way we get two food crops in one calendar year. Our early carrots need beds with winter-killed cover crops (such as oats), or empty (but not weedy) beds. Early summer carrots often follow overwintered leeks. Late summer carrots often follow onions, which are harvested in June and July here.

Beets are chenopodia, like spinach and chard. Our summer sowings often follow onions or brassicas.

Seed saving

As biennials, carrots, parsnips and beets are more complicated crops to grow for seed than annuals. It is usually recommended to harvest and store the roots over the winter, then select the best to replant in spring, setting out at least 200 to avoid inbreeding depression. Minimum isolation distance is ¼ mile (400 m) for home use, ½–1 mile (800–1600 m) for pure seed for sale. Also, carrots and parsnips can and do cross with Queen Anne's Lace, so these are only worth trying for areas that don't have that wild plant. Likewise, avoid wild parsnip. Beets, Swiss chard, sugar beets and fodder beets (mangel-wurzels) all cross with each other.

Credit: Hildegard Ott

Celery and Celeriac

Celery and its rooty cousin celeriac grow slowly but are easy to care for once established. Celery can provide harvests all summer and fall, while celeriac is an eye-catching winter storage root vegetable. Celeriac is sometimes called "turnip-rooted celery." Its flavor is starchier and sweeter than celery with hints of parsley and nuts.

Commercial celery is one of the most-sprayed crops, so the market for sustainably grown celery is increasing, as people become more aware of the perils of pesticide use and more appreciative of the flavor of really fresh produce.

Less well-known members of this family include leaf celery (cutting celery). This plant is grown for the leaves only, to add celery flavor to salads and soups. Another is root parsley. There is also a French celery, Dinant, which produces many small, flavorful stems for soups and stews, rather than raw eating. Lovage, a strongly flavored perennial, is a distant relation.

Varieties

Relatively little commercial attention has been paid to varieties of these two vegetables for small-scale organic growers, which at least makes the choices easier. We like Ventura celery (80 days to maturity, from transplanting), a very widely adapted variety. We tried Conquistador, but it had less hardiness in both hot and cold conditions. Ventura has some tolerance to *Fusarium*, the disease which causes hollow pithy stems. In England, I used to grow Golden Self Blanching (85 days from transplant), as I couldn't be bothered with the wrapping and earthing up of traditional celery that was then considered essential. These days I find that people are very accepting of green celery, and there's no real need to trench, wrap or hill, which is how to produce white celery.

There are also red varieties of celery, although the trade-off for the attractive color can be stringier stalks. Frank Morton at Wild Garden Seeds is working to improve red varieties. Fans of red celery describe the flavor as spicier than green celery. Redventure is a cross between Giant Red and Ventura.

As for celeriac, we compared two commonly available varieties, Diamante (100 days from transplanting) and Large Smooth Prague (110 days), and are strongly in favor of Diamante as it seems to be more tolerant of warm weather, less prone to rot and easier to clean. I've heard several growers say that Diamante and Brilliant are virtually indistinguishable.

Crop requirements

These crops like rich soil with lots of organic matter, some shade from midday and afternoon sun, and ample water without waterlogged soil. We make sure to choose beds on the shadier side of the garden. The Virginia climate is actually on the warm side for these crops—they prefer cooler areas—but we have good success if we pay attention. Both crops can benefit from side dressing with compost during the growing season or a foliar spray of seaweed. A pH of 5.8–6.7 is ideal.

If you live in a warm climate, consider planting celery behind other tall plants, or between rows of something else. Challenges with this approach include the need for plentiful irrigation, and that the celery is likely to be around longer than most other crops. We roll mulch hay over the celery bed before planting, to help keep the soil cool and damp.

On the other hand, we don't mulch celeriac as it can rot if too damp. We prevent rot by keeping it weeded and removing some of the lower leaves to improve vital airflow. Celeriac requires long steady growth, so the task of the grower is to prevent checks to growth. Celeriac can tolerate frost quite well, so there is no hurry to harvest in the fall.

Sowing

The seed count for these crops is 70,000–75,000 seeds/oz or 2,500seed/g. They are quite tiny.

Rodale suggests soaking the seed overnight, then freezing it for a week before sowing. Some people suggest pouring 120°F (48.5°C) water on the seeds and soaking overnight. This can cut the germination time from three weeks to one. Because celery seed is very small, the easiest way to do this is to pour the seeds and the water together on the surface of the potting soil in the flat. The seed should be then barely covered with fine soil. Simply soaking overnight helps.

Some say that older seed has a better germination rate than new seed, because the germination inhibitor that induces dormancy has lost strength. My experience is that freshly self-sown volunteers come up very freely in our hoophouse, right next to their mother plant, so I'd suggest experimenting with, but not relying on, the old seed theory. If it works, it could have the advantage that seedborne diseases die before the seed does.

Germination is slow, typically 14 to 21 days, and these two crops take 10–12 weeks to grow to transplant size, so start in plenty of time. We sow in open flats, then spot out (prick out) into deeper flats. We sow on February 10, about10 weeks before our last frost date. Celeriac can be sown from 8–10 weeks before the last frost date to 26–29 weeks before the first fall frost date, according to Tanya Denckla in *The Gardener's A–Z Guide to Growing Organic Food*.

Sow seeds ⅛" (3 mm) deep, and keep the soil surface moist.

The minimum germination temperature is

40°F (5°C) and the optimal range is 59°F–70°F (15°C–21°C). The ideal temperature is 68°F (20°C) during the day and 59°F (15°C) at night. Fluctuating temperatures, with nights cooler than days by 9F° (5C°), help speed germination. Emergence takes at least 12 days at 59°F (15°C) and 7 days at 68°F (20°C). At 59°F (15°C), only 40% of the seeds produce seedlings, compared to 97% at 68°F (20°C). This information comes from Nancy Bubel's *Seed Starter's Handbook* and Knott's *Handbook for Vegetable Growers*.

We've tried a couple of methods to achieve this temperature fluctuation (given that we don't have thermostatically controlled germination chambers). One way is to have one germination chamber just for celery and celeriac, and to turn off (or reduce) the heating (lightbulbs in our case) at night. A max/min thermometer would be a good tool to use when trying this—the insulation of the germination chamber may prevent the temperature dropping enough! Another method is to pull the seed flats out of the chamber at night.

Once the seeds have germinated, you've succeeded with the most difficult part of celery growing! Bring the seedlings into full sunlight in the greenhouse (or windowsill) and grow them at 70°F–75°F (21°C–25°C). When they have two true leaves, the seedlings can be spotted out (pricked out) to individual cells, or a 2" (5 cm) spacing in 3" (7.5 cm) deep open flats. These crops do not develop deep roots, so deep containers are not needed.

If you have a long growing season, you could direct sow celeriac in the summer, for a late fall harvest. Perhaps put a board over the seed row to keep the soil damp and cool until the seedlings emerge.

Celeriac seedlings are slow growing, but easy to transplant. Credit: Kathryn Simmons.

Transplanting

Celery and celeriac should not be hardened off by reducing temperatures, as that can cause them to bolt. More than about nine nights below 55°F (12°C) will cause bolting. Plants can have their watering reduced to help them get ready for the big outdoors. Use rowcover if a cold spell arrives after you have planted them out, or if you know cold weather is likely to return. Falling apple blossoms are said to be a phenological sign that conditions are suitable.

Transplant when plants are 2.5"–3" (6–7.5 cm) tall, i.e., after your last frost when the weather seems settled and warm. If the weather is cold, just wait. We transplant celery around May 1 and celeriac around May 7 (our average last frost is April 30).

Celery plants need to be 6"–12" (15–30 cm) apart. We plant two rows in a bed, and position the plants 12" (30 cm) apart in all directions, along the crown of the bed. Celery does well when "crowded" like this.

Celeriac also gets 12" (30 cm) here, with four rows to a 4' (1.2 m) bed—that's about 10" (25 cm) between rows. We have tried 8" (20 cm), intending to remove some of the roots early at a smaller size than full maturity. We found close spacing doesn't work in our humid climate, as poor airflow encourages rot.

Rotations

Celery and celeriac are umbelliferous vegetables, like carrots and parsnips. Ours follow the previous year's brassicas, peas or onions. Celery and celeriac occupy the space all season and we often don't clear them in time for a winter cover crop. Hence the space is easy to prepare for early spring crops the next year, often cabbage or kale.

Pests and diseases

Celery and celeriac are prone to the same troubles as carrots and parsnips. Additionally celery is a "canary in the coalmine" indicator crop for boron deficiency. The Connecticut Agricultural Experiment Station has good information on plant health problems of celery and celeriac.

The most common pests are aphids, whiteflies, cutworms, tarnished plant bug and spider mites. Ours have sometimes been visited by the parsley worm, but not in high numbers. Carrot weevils and carrot rust flies are other possibilities.

The most common diseases include the fungal diseases *Fusarium* yellows, *Rhizoctonia* root and crown rot and leaf blights (*Cercospora* and *Septoria* types). Additionally there is the virus disease celery mosaic.

We find that Ventura celery has better cold and heat tolerance than other varieties we've tried. Credit: Kathryn Simmons.

Black heart or heart rot is a calcium deficiency caused by fluctuations in soil moisture. It first shows up as brown leaf tips, and quickly spreads to the heart of the plant, where soft bacterial rots enter. Maintain adequate calcium levels and irrigate regularly to avoid this problem.

Caring for the crop

The only care that healthy plants need, other than a good regular supply of water, is weeding. If you have mulched the celery, just hand-pull weeds on sunny days and lay them on the mulch to die. Because the roots are shallow, any hoeing of celeriac or (unmulched) celery should be done carefully. Some people hill celeriac slightly to improve the quality of the roots. We don't hill ours at all, so I can't vouch for that.

Harvest

Celery can be harvested one stalk at a time as soon as the stems look big enough, which can be six weeks after transplanting. This can be useful for making up "stir-fry packs" of mixed vegetables. To harvest individual stalks, bend the stalk out and down and twist it off, or cut with a sharp knife, being careful not to damage adjacent stalks. Once the plants get big, they will start to grow side shoots, which can become full-size bunches in their own right. We harvest by cutting out the central bunch with a knife, just above ground level, leaving the side shoots undisturbed for later harvesting. We find it best to aim to cut the central bunch from all of the celery plants by mid-July. This helps the side shoots to grow big and produce tender bunches for August, September and October. I picture that the large root mass helps the small side bunches grow quickly once the main bunch is removed. We trim off the leaves, although if they are in good condition they make a tasty addition to soups, or even hearty salads. The leaves can be dried, crumbled and used as a seasoning.

Celeriac can be harvested once it reaches two to four inches (5–10 cm) in diameter, or larger (grapefruit size), but we tend to wait and do a single harvest in the fall. Celeriac is hardy to 20°F (−6.5°C) and gets harvested here before the storage kohlrabi and turnips. A light frost can improve the flavor.

To harvest celeriac, loosen the roots with a digging fork, pull the plants out of the ground and trim off the small roots with a sharp knife. Then cut off the leaves and collect the roots in buckets or tubs for washing.

Storage

After harvesting, we stand celery bunches in five-gallon (20-liter) buckets and add an inch of water before putting the buckets in a walk-in cooler. It's important to keep celery stalks well hydrated.

We store celeriac, after washing and draining, in perforated plastic bags, tied at the neck. In a walk-in cooler, they will keep well for several months, even until spring. The keys are temperatures of 35°F–40°F (2°C–5°C) and high humidity.

Cooking

Everyone knows celery can be eaten raw, or used in soups and stews; but it wasn't until I went to college in East Anglia, in the fenland part of England, where celery grows really well, that I was introduced to cooked celery as a vegetable side dish. Celery can be chopped, braised and served in a cornstarch glaze. It also goes well with cheese sauces. The flavor becomes more mellow once it is cooked. And of course, celery makes a good quick-cooking addition to stir-fries.

Celeriac has an intriguing sweet nutty taste when cooked. It can be mixed with other vegetables, or for those who like something different, cut into ½" (1 cm) thick slices, parboiled, then fried in butter or olive oil until lightly browned. The key is to make sure it is thoroughly cooked and tender. Some people serve them cooked and mashed with potatoes, or in potato salad. They can be julienned, steamed and served, perhaps mixed with carrot. Grated celeriac and beets make a hearty winter salad. Peeling is more manageable if the roots are first cut into quarters.

Season extension and overwintering

To keep our outdoor celery going as long as possible, we cover it with large wire hoops and thick rowcover at the beginning of winter. Although celery is cold sensitive in the spring, it is fairly tolerant in the fall. It is hardy down to 25°F (–4.5°C); maybe even 20°F (–6.5°C). Thick rowcover gives about 5F (3C) of protection and piling up straw or tree leaves around the plants also insulates. We dig up twelve mature celery plants in November and replant them in the hoophouse. They provide us with stalks and leaves right through the winter. They do not seem to suffer at all from being moved as large plants. They can be planted very close together, so that twelve plants only need a 3' (90 cm) length of a 4' (1.2 m) wide bed.

Seed saving

Celery and celeriac are biennials, so anyone growing seed needs to wait for the second year after sowing to get seed. This gives the opportunity to select good plants to save seed from. My only experience with seed saving has been from the celery plants we move into the hoophouse. In the spring, after their winter indoors, they begin to flower and quickly set seed. The flowers attract beneficial insects. Once the seeds have set, tie a big paper bag over the heads until they seem ripe. We got a visitation from some small black beetles, but they were easily shaken out of the leaves after we cut the bagged stems. The bags can be hung from a beam or rafter until you get round to cleaning them. Celery seed is very easy to clean. A common kitchen sieve has the right size holes to hold the big chaff and let the seed and small dust pass through. Then a smaller sieve (tea strainer) is just right for screening out the dust and retaining the seed. Celery plants are very productive, so you'll probably find yourself with enough seed for everyone you know, or for keeping your kitchen well stocked with the seed for condiment.

The isolation distance for celery is 600' (183 m) for home use and up to 0.5 mile (0.8 km) (depending on barriers) for commercial growing. Remember that celery and celeriac will cross with each other.

For both crops, the minimum population size needed to ensure genetic diversity is 60 plants.

Chapter 40

Turnips and Rutabagas

Turnips and rutabagas are reliable root vegetables in the brassica family. Traditionally grown as winter storage vegetables, they are now also available in gourmet varieties, to be eaten small, young and tender.

A cool-weather crop that has no trouble germinating at high temperatures, turnips do best in much of the US as a fall crop, or a quick early spring crop. In colder areas sow turnips in spring and harvest in fall. In hot areas, sow in late summer and harvest in fall, winter or early spring. Because turnips are quick growing when conditions are right, they are a worthwhile winter hoophouse crop to provide an alternative to leafy greens.

These vegetables have long been cultivated, and their fortunes have risen and fallen. They spread from China before potatoes spread from the Americas, and were valued in northern Europe as a staple. They fed the Roman army, and improved the crop rotations of British farming. But at other times they were thought fit only for cattle feed.

Rutabagas are among the hardiest of vegetables, and can be left growing (or at least not dying) until all other crops have been harvested. Turnips are among the fastest crops other than leafy greens. Both have a place in providing food for the entire eating season.

The rising interest in local, sustainable, healthy food means that more people are seeking out these crops. They may be eaten raw (grated or sliced), lightly steamed, roasted or baked.

Varieties

First, let's clarify the distinction between these two crops: turnips are *Brassica rapa*, like Chinese cabbage and mustards. They come in a range of colors, white or yellow flesh, with white, purple, red or golden-yellow skins. The leaves are bright grass green, usually hairy, and not waxy. Turnips do not have a neck or secondary roots.

Rutabagas (known as Swedes in the UK) are *Brassica napus*, closely related to the main brassica crops. Botanically, rutabagas are part swollen taproots, part swollen stem—the upper portion of the vegetable which forms the neck—the distinguishing feature of rutabagas. The taproot has secondary roots. Rutabagas are mostly yellow-fleshed with a tan and reddish or purplish skin, although there are white-fleshed varieties. They all have blue-green waxy leaves.

Generally rutabagas are grown large for storage, while turnips are often preferred small. Even when cultivated for large storage roots, turnips are usually smaller than rutabagas. While all are good sources of fiber and vitamins A and C, rutabagas have twice the nutrients of white turnips. The leaves of both crops, at least while young, are also tasty and nutritious.

Maincrop turnip varieties include Purple Top White Globe (50 days, OP). The biggest turnip, can reach 6" (15 cm) in diameter, white flesh, high yields, good storer. Another popular maincrop variety is Gold Ball (45 days, OP). Smooth skin, soft golden flesh, best flavor and sweetness at 3" (7.5 cm) in diameter, good keeper.

The delicious F-1 hybrid Hakurei (38 days)—a smooth white flat-round shape, with crisp sweet flesh, and hairless leaves—is the most famous of the gourmet varieties to be eaten young and tender (35–50 days after sowing, up to two inches/ 5 cm in diameter). These turnips are so mild they can be enjoyed raw. Others include White Egg (46 days, OP), which has smooth skin and very white sweet flesh. Roots are egg-shaped and grow half out of the ground. They are ideal for bunching

before they attain full size. Surprisingly for such a tender vegetable, these are a good keeper, and the flavor intensifies in storage. Oasis (50 days, F-1 hybrid) is most similar to Hakurei, with smooth round roots, a pure white color, sweet flavor and crisp tender texture. They are best harvested small, but do retain quality for a while. Scarlet Queen (43 days, OP) has beautiful red-skinned roots with a slightly flattened shape and sweet crisp white flesh with red splashes of color in the flesh. Best when harvested at 2"–3" (5–7.5 cm) in diameter. Tokyo Cross (35 days, hybrid) produces very uniform, 1"–3" (2.5–7.5 cm) diameter, round, crunchy salad turnips. The high yields of bright white, smooth roots have good flavor raw or cooked when harvested from baby to mature size. Red Round (55 days, OP) is similar to Scarlet Ohno turnip (no longer commercially available), with the same tennis ball size, bright red skin and white flesh with some blushing.

There are also turnip varieties specifically grown for greens, such as the 45-day Seven Top and the serrated leaf Namenia.

Rutabagas come in fewer varieties. Laurentian (95 days OP) has a deep purple crown and cream yellow bottom. The uniform 5"–6" (13–15 cm) roots have pale yellow flesh. Marian (90 days, OP), looks similar to Laurentian, with the added advantage that it is somewhat tolerant to clubroot. Its shorter, wider leaves can grow to 8" (20 cm) across, so allow more space. Magres is used for very early production from modular transplants (cell plugs) in Ireland and the UK, a method that produces an attractive round root shape (see below). It has fine-grained, tasty yellow flesh, and is tolerant to bolting.

Gilfeather (85 days, OP) is sold as a turnip, but is botanically a white rutabaga. Sweeter and later to mature than turnips, it doesn't become woody even at softball size. The flavor improves after frost.

Crop requirements, seed counts and yield

The keys to growing mild, sweet-tasting roots include cool temperatures, sufficient irrigation, and no competition from weeds or overcrowding. Excess nitrogen will cause too much top growth at the expense of good roots, and may cause cracking of the roots in hot weather. Boron deficiency causes the middles of the roots to turn brown. Fall crops sown too early in the summer can develop woody roots. Root crops do best in loose fertile soil with a pH of 5.5–7.2.

Time your plantings so that the roots develop in cool weather, 68°F (20°C) max. Germination at the minimum temperature for turnips of 41°F (5°C) takes 9 days. The optimal range is 59°F–95°F (15°C–35°C), when germination takes only one to three days. At the optimal temperature of 77°F (25°C), 100% of the seedlings which emerge are normal. At 50°F (10°C), 79% seedlings are normal, at 59°F (15°C) 98% are normal. Rutabagas are similar in temperature requirements, but are a little slower to germinate.

The seed count for turnips is 320 seeds/g, 9,000 seeds/oz, or 150,000–185,000 seeds/lb.

It's ideal sowing density is 3,000 seeds/100' (30 m), 10g/100 m, ⅓ oz/100'.

For drilled turnip, plan for 702,000–1,230,000 seeds/acre, or 5.7 lbs/acre at 35 seeds/ft in rows 18" apart (6.4 kg/ha at 1 cm in-row spacing with rows 46 cm apart). For broadcast sowing plan for 1–2.5 lbs/acre or 1.1–3 kg/ha.

The seed count for rutabaga is 400 seeds/g or 12,000 seeds/oz (200,000 seeds/lb).

Drilled rutabaga will be at 1 oz/1,400'. Account for 1 lb/acre at 6 seeds/ft in rows 24" apart, 1.5–2 lbs/acre in rows 30"–36" apart, 15m/g (1.12 kg/ha) at 5 cm in-row spacing with rows 60 cm apart, or 1.7–2.2 kg/ha in rows 80–90 cm apart.

For small plantings, plan on 10' (3 m) per person. Yields can be 100 lbs of turnip greens and 50–120 lbs of roots/100', (45 kg greens and 23–54 kg roots/30 m). Yields of rutabagas can be 50 percent higher. On a large scale, reckon on turnips yielding 30,000 lbs/acre, (34,000 kg/ha) and rutabagas up to 44,500 lbs/acre (50,000 kg/ha).

Sowing

In Zone 7, we sow a small crop of turnips outdoors under rowcover March 15, or earlier if spring is mild; then a bigger fall crop for winter storage, Aug 6–Sept 15 at the latest, allowing 70 days before the first hard freeze date. In the fall and winter, we sow in the hoophouse three times, for eating fresh, young and tender turnips during the winter. These make a pleasant change from our stored maincrop outdoor turnips. Rutabagas are sown here only in late summer for winter storage. They take longer to grow to full size than turnips, so it is necessary to start earlier: July 15–Aug 4, allowing 90–100 growing days before a hard freeze. For both crops we sow four rows across in 4' (1.2 m) beds. When flea beetles or grasshoppers are a problem, we sow under rowcover.

I used to believe that these root crops had to be direct seeded, as transplanting damages the all-important root. Since then I have read of Canadian work on transplanting, using soil blocks or modules/plugs from cell flats. Never say never, about transplanting! It is important to keep transplants above 41°F (5°C) until they are 10 weeks old, or else they will bolt. As few as three nights with temperatures around 37°F (3°C) will ruin them. The rewards of this extra effort are beautiful early crops.

Turnip greens are sown in successions in the Southeast, making a new spring or fall sowing when the previous crop is 50 percent emerged. Rows should be 12"–24" (30–60 cm) apart and the plants 1"–2" (2.5–5 cm) apart in the row. Sowing dates are dependent on the climate, from late winter to May when it gets too hot. For the fall, they are sown in late July/early August, until late September/early October.

Seeds need to be 0.5" (1.2 cm) deep.

Cultivation

Early thinning is important for well-developed roots, especially for rutabagas. Small thinnings may be used for salad mix. Turnips can be thinned initially to 1" (2.5 cm), then to 3" (7.5 cm), for better greens from the thinnings. Then next time either harvest all at once, or pull the largest, leaving others to fill the space.

Rutabagas should be thinned to 4" (10 cm) when 1" (2.5 cm) tall, then to 10" (25 cm) when 2"–3" (5–7.5 cm) tall. If not well-thinned, they will grow in odd shapes and be small.

Many common weeds are in the *Brassica* genus, and could harbor pests and diseases that could attack the crop, so use crop rotations, stale seedbeds and clean cultivation to remove the weeds.

Pests and diseases

Aphids, flea beetles, cabbage worms and grasshoppers can all be a problem. Rutabagas have worse trouble with aphids than turnips do. Brassica flea beetles are not the same species as the nightshade flea beetles often found on eggplant. Rowcovers and the planting of insectaries (flowers to attract beneficial insects such as ladybugs) can help avoid the problems. Bt (*Bacillus thuringiensis*) can be used for the caterpillars, soaps for the aphids and Nolo bait for the grasshoppers (except where banned in order to preserve rare species of grasshopper).

Much good information (particularly descriptions of pests and diseases) is available in the *Prince Edward Island Department of Agriculture, Fisheries and Aquaculture Turnips and Rutabagas Production Guide*. (If you stop at page 11 you can avoid the pesticide list, which is not organic.)

The main diseases of turnips and rutabagas are clubroot, downy mildew, powdery mildew, rhizoctonia rot, bacterial scab and blackleg. All except scab are fungal diseases. Organic methods of prevention are crop rotations and field sanitation (plowing in residues promptly, removing weeds). Clubroot fungus is able to live in the soil for up to ten years, so is hard to eliminate. Avoid all brassica crops in an affected field for ten years and be vigilant about eliminating brassica family weeds. (Develop a fondness for spinach, chard and beet greens!)

Harvest

Small spring turnips can be pulled by hand—ours are ready May 20, and we clear the last of them in early June, refrigerating them till mid-July if we have enough. Fall turnips are ready from late September (greens and roots). We harvest ours as needed several times a week, then dig the last in mid-November for winter storage. Our rutabagas are ready from mid-October.

For manual harvest, loosen the roots with a digging fork as needed, then pull. Trim tops and tails in the field (or in the shade if hot). All foliage should be removed for successful long-term storage. For turnips, cut cleanly between the leaves and the root. For rutabagas, cut through the neck. Then wash, drain and store. Young turnips can be pulled, banded, washed and sold with tops intact. Prompt cooling is important to keep the leaves from wilting. For large plantings of storage roots, top cutters and undercutters can be used. Some potato diggers will also tackle turnips and rutabagas.

Cut and damaged roots do not store well. If you haven't enough humans to feed them to, but you have milking animals, you could chop them (to prevent choking) and feed them to your livestock. Even moderate quantities will not flavor the milk.

The flavor of rutabagas is pleasantly sweetened by a few frosts.

Post-harvest and storage

Prompt washing before the soil dries on the roots will make them easier to clean. Storage in perforated plastic bags under refrigeration works well for us. Turnips will keep for about four months at temperatures close to freezing and humidity of 90%–95%. Higher humidity will make them rot. Rutabagas can store for as much as six months, and do best stored above 95% humidity. They can survive lower temperatures than turnips.

In my twenties, I lived in a commune in England where we grew turnips one year for winter feed for our three Jersey cows. We lacked machinery, so we dug the turnips by hand and stored them the old-fashioned way in a "clamp" in the field: we dug a circular drainage ditch, piled the turnips in the middle, covered them with straw and slapped a final coat of soil on the sides of the mound. We left a straw "chimney" poking out, so the roots could breathe. We spent several days at this task, and came to notice that our sweat smelled of turnips, simply from handling so many. You have been warned! Oh yes, the storage system worked fine. We just uncovered one side every time we needed to extract some roots, then covered it up again.

In the UK, rutabagas are not waxed as they are in North America. In fact, they store well without waxing, and I encourage you to try skipping the petroleum product.

Season extension/overwintering

Rutabagas can be stored in the ground (unlike turnips, except in warm climates). Mulch over them with loose straw once the temperatures descend near 20°F (–7°C). If you don't manage to eat all the roots before spring, they will re-sprout and you can have an "early spring bite" of greens (a term more usually used for cattle fodder crops).

Turnips do very well in the winter hoophouse. We sow our first ones Oct 15 (around our first frost date) for harvest from Dec 4. We like Red Round and Hakurei and have tried out Oasis and White Egg to find a cheaper replacement for Hakurei (Oasis is the closest). We sow in rows 7.5" (19 cm) apart, six rows to a four-foot (1.2-m) bed. We do a second sowing Nov 9 and a small third sowing Dec 10, but these are only worthwhile if thinned promptly (or else they bolt rather than size up).

Seed saving

Isolate by a minimum of 600' (180 m) for home use, and a quarter to one mile (400–1600 m) for pure seed. Also isolate turnips from Chinese cabbage and mustards. Isolate rutabagas from all brassica crops, such as cabbage, broccoli, kale and collards. The International Seed Saving Institute recommends planting "at least six different plants to ensure a reasonable amount of genetic diversity."

Chapter 41

Summer Squash and Zucchini

Summer squash and zucchini are productive summer crops available in many colors and shapes to please the eye. The "summer squash" include yellow squash, patty pan squash and round squash as well as zucchini, a specialized kind of summer squash. They are easy to grow—the main challenges are pests, diseases and moderating the quantities available. When a succession of healthy plantings can be maintained, these crops can crank out high yields of a vegetable almost everyone likes.

Varieties

For early crops, look for fast-maturing varieties. We like the green-and-yellow Zephyr squash, yellow crookneck Gentry squash and Jackpot and Raven zucchini (all hybrid bush types). Squash is one crop where we think it pays to use hybrids; open-pollinated varieties tend to be less predict-able, less productive and less reliable (although we really like the pale green Grey OP zucchini). We avoid OP vining zucchini varieties like Cocozelle and Costata Romanesco, as they take a lot of space producing leaves and few fruit (at least for us).

Most summer squash are in the species *Cucurbita pepo*. Growers with bad pest problems can try Tromboncino squash (a *C. moschata* squash); or luffa (vegetable sponge) or cucuzza (snake or serpent gourd), both edible gourds that have more resistance to squash bugs. Just be sure to cut them relatively small (while skins can be pierced by a fingernail) for good flavor. These are very vigorous, sprawling plants, which, happily, are also prolific. One option is to plant a row of regular zucchini alternating with these bigger plants—when the zucchini succumb to disease, the sprawlers will take over.

Our market, even though captive in our dining hall, doesn't go for patty pans or yellow zucchini, based entirely on looks!

We plan to try Partenon, a parthenocarpic squash (one that doesn't require pollination), in our spring hoophouse. We have been having trouble with baby squash dropping off before developing, probably because they are not getting pollinated indoors, where there are few pollinators at that time of year. Other varieties said to produce a relatively high number of fruit parthenocarpically include Cavili, Elite, Chefini (also known as Eight Ball), Black Beauty and Black Magic. The fruit production rate without pollination varies from one variety to another—dark green zucchini seem to have a better success rate than paler squash.

Some varieties are especially attractive to cucumber beetles and may be grown as a trap crop: Cocozelle summer squash, Seneca and Dark Green zucchini. When beetles accumulate in the trap crop, flame it or till it in.

There are quite a few GMO varieties on the market, so if you want to avoid those, watch out for Liberator, Patriot II, Prelude II, Conqueror III, Justice III, Independence II and Judgement III. They all sound like B movies to me!

Crop requirements, seed specs, yield

Summer squash thrives on high levels of organic matter. Although it is only a light to moderate feeder for nitrogen and phosphorus, it requires quite high levels of potassium. Foliar seaweed spray can be used to make up for nutrient deficiencies. This crop does best with plentiful (but not excessive) moisture. That being said, be careful not to overwater the seeds before they germinate, or they will rot.

This frost-tender warm-weather crop germinates at 60°F (15.5°C) minimum, and does best at 85°F (30°C). Germination takes 5–10 days. During growth the ideal temperatures are 75°F–85°F (24°C–30°C) in the day and 65°F (18°C) at night. From sowing to harvest takes 40–50 days (60 days for Tromboncino).

There are 160–240 seeds/oz (6–9/g); 2,500–5,000/lb (6,000–9,000/kg). Zucchini seeds tend to be the smallest, yellow squash the biggest and patty pans in the middle. At a direct seeding rate of 3 seeds/ft (1 seed/10 cm), you will get 62'–100'/oz, or 1,000'–1,600'/lb (60–90 m/kg). At this rate in rows 6' (1.8 m) apart, you need 5–7.25 lb/acre (5.6–8.1 kg/ha).

Seed can remain viable for up to five years, if stored in a cool, dark, dry place. Yield can be 200–800 lbs/100' (300–1,200 kg/100 m), 4–6 tons/ac (9–13 tonnes/ha).

Rotations

We grow relatively small plantings of summer squash, so we pop a row in here and there depending on where the spaces are. If you specialize in summer squash and grow large areas you will need to pay more attention to rotations. Pigweed can be a problem in squash plantings. If you know you have pigweed seeds in the soil, flush them out first. Till in your winter cover crop, ideally including brassicas (or a preceding brassica food crop), then spread compost and pause to allow the pigweed to germinate in response to the compost application. Cultivate once or more to kill off the pigweed and follow with planting. When the squash is finished, plant summer cover crops such as sorghum-sudan or buckwheat. Ideally, plant cool-weather crops the next season, to further reduce warm-weather weeds.

To reduce cucurbit diseases, try not to plant cucurbits anywhere they have been growing in the past two years. Cucumber beetles are very mobile so on a small farm there is not much hope of moving a new planting far enough from an old one to reduce their numbers for long. For best disease and pest control, always till in cucurbit crop residues as soon as possible.

Sowing

Seeds should be sown ½"–1" (1–2.5 cm) deep, 6"–12" (15–30 cm) apart in rows which are 3.5'–5' (1–1.5 m) apart. Sow more shallowly in cool soils, and deeper in hot weather. In hot weather it helps to turn on the drip irrigation, then sow the seeds right by an emitter. Water at sowing, then not again until seedlings emerge.

Covering the row with rowcover will keep the pests away and also provide a warmer, more sheltered environment for the plants to grow in. If possible, use hoops or a rope above the row, pulled taut between stakes, to hold the rowcover above the leaves, as cucurbits are susceptible to abrasion damage. Remove the rowcover as soon as you see female flowers to allow for pollination, unless you are growing a parthenocarpic variety.

For a slightly earlier harvest, you can pre-sprout the seed. See the Cucumber chapter for details.

Transplanting

It is possible to transplant squash, taking care to disturb the roots as little as possible. We like soil blocks for our early cucurbits, as this method causes minimal root damage. Large plugs (such as 606) are another possibility. Winstrip 50-cell trays, which have vertical slits in the sides to cause air-pruning of roots, are another good choice for this crop, provided you can plant out while the squash are young. Otherwise use something larger, such as soil blocks. If you sow two seeds in each cell, be sure to pinch off the smaller one as soon as possible, to remove the competition.

As soon as the plants have one or two true leaves, at about three to four weeks of age, transplant in a mild spell, burying the entire stem up to the seed leaves (carefully!) below soil level to protect it from damage. Cucurbits grown indoors respond to lower light levels by growing long spindly stems, which are quite fragile. Make the hole deep enough, slip the plant out of the container, perhaps on the blade of a butter knife, pop it into the hole and close up firmly. Plant two to three feet (60–90 cm) apart in the row. Use rowcover to reduce the transplant shock from the more variable outdoor conditions. Keep leftover seedlings for a week to fill gaps in the row. In my experience, summer squash plants have a little more cold tolerance than cucumber and muskmelon transplants.

Transplanting allows a crop to be established earlier than would be possible with direct sowing in spring. Or a crop can be gained from a summer window in the planting schedule for a bed, after an early spring crop. Transplants do not grow as sturdy a root system as direct sown squash do, so I would not recommend transplanting if direct sowing is possible.

Caring for the crop

When transplanting, you can set the plants out into a clean bed—you probably will not need to control weeds until the plants start to flower and it is time to remove the rowcover. If direct sowing, we usually hoe once during the time the rowcover is still in place, then again after uncovering. We

An attractive and fast-maturing variety, Zephyr offers a feast for the eyes as well as the table. Credit: Kathryn Simmons.

thin to two feet (60 cm) before the plants touch each other. From then on, hand pulling a few weeds while harvesting is often enough, as this crop has a big leaf canopy and grows fast, reducing the chances of weeds taking hold. It is important to keep squash weed-free for the first four to six weeks of growth; later weeds do not impact the yield much (but could be trouble for the next crop of course!).

Succession planting

Squash plants mature quickly and are then set upon by various bugs and fungi, so it makes sense to plant successions every two to four weeks until late summer. To make the pest evasion successful, make the sowings in widely separated parts of your farm and destroy the old sickly plants as soon as the new plants are producing. We plant six times during the year, aiming to have a new planting come into production every 25 days, starting May 15 and ending with the first hard frost. See the Succession Planting chapter for more on planning successions.

The first squash is a transplanted sowing for our hoophouse, sown March 1, transplanted April 1, with rowcover. Our second sowing (March 25) is transplanted outdoors with rowcover April 20, about 10 days before the frost date, depending on the weather. After that, we direct sow May 24, June 23, July 15 and Aug 5 (70 days before our average first frost). We have a note to ourselves that if the spring is exceptionally cold, we could sow the third squash in the greenhouse May 17 (rather than direct sow May 24) and transplant June 7, but we haven't done this for many years. Lilac in full bloom is a phenological sign that the spring is warm enough to sow squash. Other sources say to sow the second planting a month after frost and the third three weeks after the second.

Pests

Squash bugs are probably the most common pest. The adults are long gray insects which emerge from overwintering in wooded areas once temperatures reach about 70°F (21°C). They also drift in on storm fronts. They are more attracted to summer squash and other Pepo squash than they are to Moschata winter squash or to watermelon and cantaloupes. They feed on young seedlings for preference and lay groups of spaced bronze eggs on the undersides of leaves. The eggs hatch in ten days and the nymphs suck sap from the plants. There are many generations each year and numbers quickly build. Drop handpicked adults into soapy water rather than crushing them, as the smell of the crushed bugs attracts more. It is also possible, but not easy, to crush the hard-shelled eggs. An alternative is to tear off egg-covered leaf pieces and drown the eggs in soapy water. Squash bugs can be trapped overnight under boards near the plants—in the morning, lift the boards and collect the bugs.

Striped cucumber beetles are our worst pests some years, even compared to squash bugs. In the hoophouse we handpick the insects in the mornings while they are still sluggish. They are found inside squash flowers. Our hope is that by battling this first generation, we will reduce numbers for the rest of the season. Alison and Paul Wiediger have had success with pheromone lures and sticky traps sold by Johnny's (more on this in the Cucumber chapter). To attract beneficial insects which are predators or parasites of the cucumber beetle, grow buckwheat, cilantro, dill or winter peas near the squash plants, timed so that they flower when the rowcover is removed from the squash plants.

For more on cucumber beetles and stink bugs, including the brown marmorated stink bug, see the Cucumber chapter. If all else fails and action is imperative, Spinosad will kill them. Neem doesn't kill them, but does deter them.

The adult squash vine borer is a fairly large orange and black fuzzy moth with glassy black wings. It is active in May and early June, and there may be a second generation in late July/early August. The small flat brown eggs are laid singly at the base of the stem. The larvae hatch after one week, and are white with a brown head. They spend four to six

A heavy infestation of leaf-footed bugs can cause significant crop loss. Adult leaf-footed bugs have a flattened appendage on their hind legs. The nymphs, which are harder to recognize, feed on leaves and fruits, causing wilting, pitting and discoloration. We had misidentified these bugs as beneficial assassin bug nymphs. Leaf-footed bugs have slender, straight, plant-sucking beaks while assassin bugs have beaks that are sturdier, shorter and curved (better for assassination). In general, predators are more likely to go at it alone rather than cluster as plant pests do. To help with identification, see the photo website bugwood.org. Credit: Kathryn Simmons.

weeks in the stem, eating upwards and growing to one inch (2.5 cm) in size. Then they dig into the soil and pupate for the winter. Plant symptoms include wilting of the whole plant and stem holes surrounded by frass (caterpillar poop). It is possible to save the plant if you act at the first sign of trouble and pick out the grub, then pile soil over the injury to help the plant recover. Injecting Bt (*Bacillus thuringiensis*) into the stem is another solution. For prevention, rotate crops and use rowcover over plants until flowering. Squash vine borer adults are reputed to be repelled by radishes sown in a 12" (30 cm) diameter circle around each squash plant, or in bands down either side of a row. It's fine to harvest some of the radishes, but make sure to leave enough that their leaves touch, to provide a continuous barrier. Watering beneficial nematodes on the soil at the base of squash plants may help reduce numbers of squash vine borer. Some growers recommend mixing wood ash or charcoal in the soil with the seeds at planting. Another technique to try is pinching out the growing tips of the vine when it is one to two feet (30–60 cm) long, which causes the plant to branch and increases the odds some part of it will survive.

One key to reducing pest numbers is to till in old plantings promptly. Flaming is another option (if the soil is too wet to till, for instance).

To repel aphids that transmit mosaic diseases, reflective foil mulch can be used. Borage and nasturtium interplanted in cucurbit rows are said to repel insects. They look pretty either way.

Diseases

- If the first flush of fruit shrivels and drops off, this is not a disease, but a lack of pollination.
- Powdery mildew (a white coating starting on older leaves in hot dry weather) can be treated with baking soda (see ATTRA, *Use of Baking Soda as a Fungicide*). Drip irrigation helps reduce PM, compared to overhead irrigation.
- Downy mildew (yellow angular spots with leaf curling and a purplish cast to the undersides of leaves) happens in cool wet weather late in the season, and plants can recover if the weather heats up.

- Anthracnose (dark brown spots on the leaves and round sunken spots on the fruit) is a fungal disease worse in warm wet weather from mid-season onwards.
- Angular leaf spot (browning of the leaf between the veins, and small round spots on the fruit) is a fungal disease more common in cool damp weather.
- Black rot/gummy stem blight (*Didymella bryoniae*) is a fungus occurring with wet soils, in cool or warm wet weather.
- Mosaic viruses can be spread by aphids. Succession cropping helps evade these. Some yellow squash varieties have the "precocious yellow" gene which masks the green mottling that virus diseases can cause.
- Bacterial wilt (*Erwinia*) causes sudden dramatic wilt and death of the whole plant. It is spread by cucumber beetles.

Harvest

First harvest will be 40–50 days after sowing (longer for slower-growing spring crops). Usually squash plants will produce only male flowers for about a week first, before changing to the productive mix we need. Flowers are only open for one day each. It takes a pollinated female flower only a few days to grow to harvestable size so frequent harvesting is needed. We pick daily, but those in cooler regions may be able to harvest only on alternate days. To preserve the appearance of the squash avoid scratching the skin on the leaves. One way to do this is to enclose the fruit in your palm as you cut through the stem. Another is to move the spiny leaves out of the way.

We harvest our zucchini at 6"–8" (15–20 cm) but smaller on Saturdays, as we don't harvest on Sundays. Zephyr squash is sometimes picked

Long-sleeved shirts can be an advantage for harvesting squash because of the scratchy leaves. Credit: Brittany Lewis.

smaller, as little as 4" (10 cm), for best flavor and gourmet appearance.

Post-harvest and storage

Summer squash will keep 10–21 days (if you are lucky) at 80% humidity and 45°F (7°C). Do not store at lower temperatures or chilling injury can occur. This shows up as a pitted surface that later grows black mold.

Squash blossoms

Squash blossoms may be used in salads or for stuffing or frying. Squash have both male and female blossoms: males have thin stems; females thick stems and a swollen base (baby fruit). If a squash crop is also wanted, harvest only male flowers, being sure to leave some to pollinate the female flowers. For the largest supply of male blossoms, remove female blossoms.

There are some varieties of squash grown specifically for the edible blossoms, not for the squash. Harvest the blossoms when fully open. Snip the flowers from the vine 1"–2" (2.5–5 cm) below the flower base with scissors or pruners. Blossoms can be stored in water in a refrigerator for up to two days.

Butter Blossom is an OP variety selected for its large, firm male blossoms (fruit may be harvested also). Tromboncino male blossoms also sell well.

Season extension

At both ends of the warm season it is possible to grow squash using rowcover. Black plastic mulch will also help to warm the soil in spring, and trap heat in the fall. Hoophouses, caterpillar tunnels and low covers can push the season a bit more in each direction too. For direct sowing with plastic mulch, poke holes in the plastic at half the final spacing and sow two seeds at each spot. Because squash are widely spaced plants, this method is not too arduous.

Seed saving

Most summer squash and zucchini are *Cucurbita pepo* and will cross with each other. They do not usually cross with other *Cucurbita* species, *moschata*, *maxima* or *argyrosperma* (*mixta*). Isolate from other *pepo* varieties by at least an eighth of a mile (200 m) for home use, and as much as one mile (1.6 km) for sale purposes. For thorough details, see the Saving Our Seeds *Cucurbit Production Guide*.

Chapter 42

Winter Squash and Pumpkins

Winter squash is a very rewarding crop to grow, providing high yields without a lot of work. It offers big results for fall or winter CSAs or farmers' markets, and those growing for food banks, schools or other institutions. Stored winter squash also come in handy to bulk out early spring offerings.

ATTRA has a very good publication on the subject, *Organic Pumpkin and Winter Squash Production*. Call them to ask for its enclosures.

Varieties

Pumpkins are squashes, even though we sometimes classify them as if they were a different vegetable. There are four main species of squash, and some hybrid crosses. Suzanne Ashworth's book *Seed to Seed* has excellent descriptions and long lists of varieties for each type.

The species *Cucurbita pepo* includes summer squash, zucchini, pattypans, acorn squash, delicata, dumplings, spaghetti squash and some pumpkins (New England Pie, Winter Luxury Pie and Connecticut Field). *Pepo* squash mature fast, do not store well (beyond a few months) and have a mild flavor. They have prickly leaves and hard, angular (non-flaring) five-sided stems. The fruits are often ribbed. *Pepos* are susceptible to vine borers.

The *C. maxima* species includes many large squash, some of which store quite well, from a few months up to a year. The flesh is fine-textured and good-flavored. *Maxima* plants, however, are very susceptible to wilts, borers and squash bug damage. These plants have soft, round stems; huge, hairy leaves; and fruits with thick, round stems. This group includes buttercups/kabochas, hubbards, bananas, the Big Max extra-large pumpkin, Jarrahdale, Candy Roaster, Galeux d'Eysines and Rouge Vif d'Etampes. We have found Jarrahdale and a hybrid kabocha, Cha-Cha, to have relatively high resistance to squash bugs compared to others in this group. They are the only *maximas* we grow.

The *C. moschata* species provides the best keepers. They usually have bright orange flesh that is sweet and tasty. The plants have large hairy leaves and fruits with flared angular stems. This species is the one to focus on if you want trouble-free squash, with no damage from borers or cucumber beetles. The tougher stems are more able to repel invaders. They need warm growing temperatures above 60°F (16°C). Butternuts and similar tan-colored squash, such as Seminole, Cheese and the large Tahitian Butternut and Lunga di Napoli, are in this species.

The fourth species, *Cucurbita argyrosperma* or *C. mixta*, includes many old-time Southern varieties. The flesh is often yellow rather than orange, and low in sugars, so these squash are often cooked with sweeteners. Plants are rampant; leaves are large and hairy. Fruit stems are slightly flared, slightly angular and hairy. This group has the best drought-resistance and also good resistance to borers and beetles. Storage is medium-term. Cushaws are *mixta* species.

Seeds needed and potential yields

Squash seeds are large, some much larger than others. On average there are 5–20 seeds per gram. Johnny's catalog divides squash seed into four size groups:

- Hubbard and Kuri maxima squash have 3–5 seeds/g, 95–125/oz, 1,500–2,000/lb, and can sow 50'/oz, 850'/lb (54 m/100 g).

- Buttercup and kabocha *maximas* and spaghetti *pepo* squash have 5–7 seeds/g, 140–200/oz, 2,200–3,200/lb, and can sow 75'/oz, 1,250'/lb (82 m/100 g).
- Acorn *pepos* and butternut *moschatas* have 10–13 seeds/g, 280–375/oz, 4,500–6,000/lb and can sow 155'/oz, 2,500'/lb (170 m/100 g).
- Delicata and Sweet Dumpling *pepo* squash have 15–21 seeds/g, 425–594/oz, 6,800–9,500/lb and can sow 225'/oz, 3,750'/lb (245 m/100 g).
- For maximum yield for a given area of pumpkins, medium-sized ones (18–22 lbs, 8–10 kg) do better than either small or large fruits. Good yields of smaller varieties are 2,000–4,000 fruit or 5–7 tons/ac (5,000–10,000 fruit, 11,200–1,570 kg/ha). The large types may yield up to 1,000–2,000 fruit or 10–30 tons/ac (2500–5,000 fruit, 22,400–67,200 kg/ha).

Crop requirements

Young squash plants are not at all frost-hardy, while mature vines can take one or two light frosts. Squash needs warm days and nights with full sun and well-drained, generously fertilized soil to grow well. Having a healthy canopy of foliage maximizes photosynthesis, which in turn increases the sugars in the squash. The plants take a lot of space, which can be problematic if the vines overrun a smaller crop next door. Some can take 90–120 days to reach maturity, so plan carefully to be sure of getting a harvest.

Squash needs at least one inch (2.5 cm) of water per week in the main season. They do have deep roots, so you can get away with irregular watering. For dryland farming, without irrigation, it is important not to move the vines to new positions: the dew and rain drips from the leaf edges encourage root growth directly below the vines (where they get the most water and shade). Long vining types have more stress tolerance than bush types, because they grow extra roots at the stem nodes (if not on plastic mulch). The critical period for water is the first 30–35 days of fruit development. Lack of watering during this period causes the stems to shrivel early in storage, leading to decay.

Bees are important for pollination: honeybees, bumblebees or the native squash bee will do the job. Insufficient pollination leads to fewer and smaller fruit, with more irregular shapes.

Sowing

Soil should be at least 60°F (15.5°C), and all danger of frost should be past. The optimum germination temperature is 86°F (30°C) but don't wait for that unless you have extremely long growing seasons. We sow (in Zone 7) on May 25 and get plenty of squash before the end of October. Fast-maturing Halloween pumpkins can be sown in late June in our area, following an earlier crop such as strawberries. Sow 0.5"–1" (1–2 cm) deep. Either "station sow" two or three seeds at the desired final spacing, or make a drill and sow seeds 6" (15 cm) apart. There are various stick planters and jab planters that can be used for this kind of station sowing. Thin later. Rows will need to be 6' (1.8 m) apart or more. Some growers plant in a square pattern so that spaces between rows can be mechanically cultivated in both directions.

Bush varieties take less space than the vining types, and rows can be 4' (1.2 m) apart. The vining ones can need as much as 9' (2.7 m) between rows, or even more.

One version of the traditional Three Sisters planting has a vining winter squash plant in the center of a circle of eight field corn plants that are 6' (2 m) from the squash and 12" (30 cm) apart from each other. Outside this circle, each corn plant has two climbing beans planted nearby.

No-till planting is an option with squash if you have grown a suitable winter cover crop that can be killed by mowing or rolling. Information is available from ATTRA.

Rowcover can be used to protect young plants from unsuitable weather and pests. If possible, use hoops, as cucurbit leaves are easily damaged by abrasion. We don't usually cover ours, as the rows are so long, and many varieties are resistant to cucumber beetles. (We don't suffer badly from squash vine borers or squash bugs.)

Transplanting

Usually, winter squash are direct seeded, which is part of what makes them so easy. If you get a horribly wet season that prevents timely soil cultivation, or your growing season is short, or your rotation is tight, it's good to know you can get very satisfactory results from transplants. Just start the seeds in cell-packs or soil blocks a week or so earlier than you would direct sow (to allow time for recovery from transplant shock). Snip off extra seedlings to leave one strong one in each cell. Cucurbits don't transplant well from open flats, as their recovery from root damage is poor. If your decision to transplant is a last minute change of plan, take a look at the days to maturity of the varieties you planned, and switch out the slow maturing ones, which might not make it in time in your shortened season.

Caring for the crop

Squash plants grow slowly at first, so you will have time to cultivate between the rows, or even plant a fast cover crop to be mowed later, such as buckwheat or mustard. Then one day the vining types will decide to start running, and you'll need to be ready with clear space for them to cover. An alternative, practiced in New York State, is to undersow with crimson clover before vining. If necessary, fold the vines over to one side to cultivate and sow, then fold them back over the other side to deal with the next aisle. Cucurbits don't like to have their vines turned upside down, so be sure to restore order before leaving the field. We tried this undersowing scheme in central Virginia and it was a hopeless failure, as the vines grew so much faster than the clover and smothered it.

After vining, the only weeding possible is the "wade-in" type of hand-weeding. Tall rubber boots (or at least long pants) are best worn for this job, as the leaves are scratchy. Pull the weeds if your soil allows, or else clip them with pruners to prevent seeding.

Hoe and thin the squash to 18" (45 cm) apart for the smaller bush varieties and 24" (60 cm) for the vining types. Some people like to use organic mulch, but this should not be applied until the soil is good and warm, otherwise growth will be slowed. If you have very wet soil when the squash are ripening, you could put pieces of wood or tile under the fruits to stop them rotting.

Rotations

Our winter squash follows the previous year's white potatoes and a winter cover crop of rye with crimson clover and/or winter peas. The clover can supply all the nitrogen the squash needs. Winter peas don't provide as much nitrogen as clover, so if we have used peas, we might supplement with some compost.

The squash gets a bit of extra growing space late in the season by overrunning a finished bed of strawberries which we till in after harvest. We finish up our winter squash by Halloween, then sow rye and winter peas. The following year, that patch is used for sweet potatoes and our last planting of sweet corn. These are late sowings that give the peas plenty of time to flower and produce nitrogen.

Pests and diseases

Squash bugs can do a lot of direct damage to leaves, and they also vector yellow vine disease. Plants attacked by squash bugs look baked to a crisp, with curled, bronzed leaves. This pest can be killed with pyrethrum or sabadilla, if your certification doesn't ban these botanical pesticides. Destroy the eggs (gold eggs spaced out on underside of leaves) and trap the bugs overnight by setting out boards (they congregate under the boards and you can squash them in the morning). *Moschata* varieties are the best ones for areas with high levels of squash bugs.

Squash vine borers are large caterpillars that chew into the main stems. They appear when the vines start to run. One approach is to pinch the tips when vines are one to two feet (30–60 cm) long, causing multiple vines to grow, increasing the chance of some escaping damage. Some growers mix wood ash or charcoal with the seed when sowing, but I haven't tried this, as we are lucky enough not to have borers. Another approach is to make a collar of aluminum foil for each stem. There are pheromone lures for this pest too. If your plants do get attacked, cut the damaged stem and remove each larva, or inject Bt (*Bacillus thuringiensis*) using a syringe. Pile soil over the damaged stem after treatment, so that new roots can grow.

Waltham Butternut is our favorite winter squash. It's relatively resistant to pests, stores for at least six months and tastes delicious. Credit: Kathryn Simmons.

Striped cucumber beetles cause direct feeding damage to the leaves and fruits. They also vector bacterial wilt, caused by *Erwinia*, which affects some varieties more than others (high beetle numbers do not always lead to disease). Initially, leaves wilt, then the vines collapse and the plant dies. Jack-o'-lantern pumpkins are unaffected, Waltham Butternut is relatively resistant, but Golden Delicious and Blue Hubbard are susceptible. Hubbard squash is suggested as a trap crop for cucumber beetles, in a row beside a plot of summer squash. Here the Hubbard seedlings barely make it out of the ground before they are devastated by the beetles and there is nothing left to trap anything! Keeping beetles from feeding on your plants is key. Rowcover can be used until flowering, to reduce damage, but then the plants must be uncovered for pollination. Currently we

are trying the sticky traps with pheromone lures. Fall cultivation, if done while the beetles are still active, can kill as many as 40 percent of them.

Powdery mildew (PM) is a fungal disease that occurs during hot dry periods—never below 50°F (10°C). PM-tolerant varieties get as much of the white mildew on their leaves as non-tolerant ones, but survive the attack. Milk sprays can be used against PM. ATTRA publishes *Use of Baking Soda as a Fungicide*, as well as *Downy Mildew Control in Cucurbits*.

Downy mildew occurs in cool, damp weather. The leaves become black or purplish and die to a crisp. If the weather dries up and gets warmer, your plants can recover.

Gummy stem blight (black rot) is another fungal disease, appearing as sunken black spots on the fruit once they are in storage.

Harvest

We start harvesting the quicker squash varieties in early September. It would be possible to grow the earlier acorn and Delicata types here and start harvesting in August, but we've decided that August is too busy already, and some crops are better appreciated later!

Not all the squash will be ripe at the same time, but they come to no harm sitting in the field (provided there is neither frost nor groundhogs). We harvest once a week, using pruners to cut the tough stems. I tell the crew about three times never to set the pruners on the ground and describe the signs of ripeness of the different types. Then we line up like a search party and cross the patch with a six- to eight-foot (2–2.5-m) "swimming lane" each. We use five-gallon (20-l) buckets for the smaller squash and separate the "Use First" squash from the storers as we go. We carry the larger squash individually to the edge of the patch. We usually designate a particular garden cart for the Use First squash and bring the truck alongside to load. Harvesters need to be reminded to handle squash as if they were eggs, not footballs. Bruising leads to rot in storage.

The first rule is to leave the squash on the vine as long as possible. Once the vine is dead, the squash will not get bigger, but it can ripen, change color and convert starches to sugars. This curing improves the flavor. Mature fruit can also cure indoors. (Acorn squash do not need curing.) With jack-o'-lantern pumpkins, the stems harden 20–35 days after fruit set, and full color is reached 45–50 days after fruit set. If the vine dies while the pumpkins are immature, their fruit will collapse. In general, squash are fully mature and storable when fingernails cannot pierce the skin.

Pepo squash usually have an orange or tan spot where they touch the ground. When they are ripe, this ground-spot is either the color of pumpkin pie filling after the cinnamon is stirred in, or else bright orange. The stem may still be bright green. Wait until at least 45 days after pollination. Harvesting too early will disappoint: the squash will be watery, fibrous and bland.

Maxima squash are ripe when at least 75 percent of their distinctive thick, round stems look dry and corky (tan, pocked, wrinkled).

Moschata squash are ready when the skin is an even tan peanut butter color, with no pale streaks or blotches. Some varieties have green lines (radiating down the squash from the stem edge) that disappear when the squash is ripe. If in doubt, cook one and see, or try a slice raw.

We harvest squash with two-inch (5-cm) stems, which helps them store best. Those who need to pack them in crates might need to remove the stems so that they do not injure their neighboring squash. If you do this you need to cure the open ends (see below). Some growers wipe each squash with a piece of burlap as they harvest, to remove potential rot organisms.

Our harvest period runs for two months, September and October. We used to harvest as late as possible in the fall, but now we prioritize getting a good cover crop established, to replenish and protect the soil, so we have a Grand Finale Harvest just before Halloween, when we bring in all the large, interesting, almost-ripe squash and give some away for lantern carving. Some go to the chickens. We harvest before the fruits get frosted, which causes the skin to appear water-soaked.

Curing and storing

If you have removed the stems, cure the squash by exposing the ends to wind and sun (or fans and warmth indoors) for five to ten days. Ideal conditions are 75%–80% humidity and 80°F–85°F (27°C–29°C), although many farms use ambient temperatures far from these ideals. Cover with rowcover or tarps if frost threatens.

We store our squash in a rodent-proof cage in a big basement, in large plastic crates or trays. Once a week, we sort through these to remove anything nasty. Spatulas are handy tools for sliding jelly-like squash into a compost bucket! Growers needing larger scale storage use pallet bins.

Store between 55°F–60°F (13°C–16°C). Temperatures below 50°F (10°C) can cause chilling injury, which will reduce storage life. Squash need good air circulation and 50%–70% humidity. Never store them together with potatoes or onions; like most ripe fruit crops, squash "exhales" ethylene, which increases sprouting in roots and bulbs.

Acorn squash. Credit: Southern Exposure Seed Exchange.

An eye-catching display of winter squash. Credit: Southern Exposure Seed Exchange.

Acorn and buttercup squash will keep for a few months—they are best used before the end of the year. *Moschata* squash such as butternuts will keep round until the next May. Seminole is a *moschata* that will keep a whole year after harvest.

Season extension

If you have a short growing season, you might want to use rowcover to get plants started, then again at the end of the season to get beyond the first couple of fall frosts and buy some more ripening time. And, as mentioned, you can get a jump-start by using transplants. Avoid the slow-maturing types, unless you want the challenge. Black plastic mulch helps speed growth considerably.

Seed saving

Crossing occurs easily between varieties of the same species, and rarely (but occasionally) across species. Check your seed catalog to be sure you are not planting more than one kind in the same group. Suzanne Ashworth's Seed to Seed is an excellent resource. Isolate varieties of the same species by 600' (200 m) for home use, and at least 1,600' (500 m) for commercial crops. It is possible to grow a *pepo*, a *moschata*, a *maxima* and an *argyrosperma* and have pure seed from each. Hand-pollination is safest.

Chapter 43

Cucumbers and Muskmelons

Cucumbers and their cucurbit relatives, muskmelons (often called cantaloupes), are warmweather crops, usually direct sown. Both are grown the same way; however muskmelons need slightly warmer conditions to do well, take a bit more space, and need a longer growing season (80–100 days, compared to 55–60 days for cucumbers). Cucumbers and muskmelons are very easy to grow, apart from dealing with pests and diseases! My information files on cucumbers are 90 percent about these challenges.

Look for resistance to diseases you know to be a problem in your area. Poinsett 76 (67 days) is an open-pollinated (OP) variety resistant to downy mildew (DM), powdery mildew (PM), *Anthracnose* (AN), scab, Fusarium, angular leaf spot (ALS), spider mites (SPM) and cucumber mosaic virus (CMV). For a maincrop slicer we like General Lee (F1 58d CMV, DM, PM, scab). This variety is fondly known to us as Generally. Previously we grew Marketmore 76 (OP, 58d, CMV, PM, DM, ALS and AN), a good dependable eight-inch (20-cm) sweet cucumber that grows well in the mid-Atlantic as well as the North. We have also had success with Olympian (F1 52d AN, ALS, CMV, DM, PM and scab). We choose Little Leaf pickler (56d, resistant to ALS, AN, BW, CMV, DM, PM bacterial wilt and scab) not only for ease of recognition but also because of its disease resistance. Another pickler with disease resistance is Edmonson (70d, resistant to ALS, AN, CMV and scab).

Cucumber varieties

Cucumbers are basically divided into two types: slicers and picklers.

Among the slicers, there are standard American varieties (open-pollinated Marketmore types and hybrids such as General Lee), European varieties such as Telegraph (60d OP originally developed for greenhouse production but suitable for growing outdoors) and Asian ones such as the 60-day varieties Poona Kheera, Suyo Long and Yamato. The Asian varieties are useful if your market likes them, but they do not have the disease-resistance of varieties bred for US climates:

- Pickling cucumbers are usually smaller than slicing cucumbers. Some kinds are harvested relatively large, and sliced or cubed to make pickles. Others are picked small and pickled whole. We like the small-leaved Little Leaf, also known as H-19 and Arkansas Little Leaf. We have lots of inexperienced helpers, and it's good to be able to make a clear distinction: "If the plant has small leaves, it's a pickler, so pick small."
- Both picklers and slicers are also available as bush and trailing (vining) varieties. For an early hoophouse crop, we grow Spacemaster (60 days, cmv and scab resistant), a bush-type, full-sized variety.
- Some varieties are distinguished by having a recessive "bitterfree" gene. Research by Todd Wehner, published in *HortTechnology* in 2000, showed that "burpless" is not genetically connected with bitterness. Genetically bitterfree varieties, such as Marketmore 80, Marketmore 97, Diva (bred by Johnny's, 58d, dm, pm, scab) and European and Dutch greenhouse varieties (which stay mild even under drought conditions), are just as likely to cause burps in susceptible individuals. Other varieties, such as oriental trellising types like Tasty Green and Tasty Bright, although genetically "normal-bitter," taste milder than the standard "normal-bitter" American varieties and are less burpy.

"Bitterfree" varieties do attract fewer cucumber beetles.

- Both slicing and pickling cucumbers are available in gynoecious (all female) varieties. To make sure the fruits will be pollinated, these seeds come packaged with 10%–15% seeds of a pollinator variety (sometimes dyed so that growers can ensure they get sown evenly in the planting). General Lee, Olympian and Diva are gynoecious. Non-gynoecious cucumber varieties initially produce only male flowers and then as many as twenty male flowers to every female flower after that. The flowers with the miniature fruit behind the petals are female.
- Parthenocarpic varieties set fruit without pollination, so the cucumbers are seedless unless cross-pollinated by another variety. As they can be kept under rowcover for the entire life of the plant, these are valuable for extending the season, or if you have a very bad pest problem, or if pollinators are an issue in your hoophouse. Little Leaf is parthenocarpic, as are Telegraph and Diva.

Muskmelon varieties

Jeff McCormack of Saving Our Seeds distinguishes eight types of melon:

- *Cucumis melo reticulatus*, a group that includes muskmelons (which we commonly call cantaloupes). They have orange or green flesh and usually have netted skin. They slip from the vine when ripe (perhaps with a nudge);
- True cantaloupes, *Cucumis melo cantalupensis*, are rare in the US. They are rough and warty rather than netted. Fedco sells Prescott Fond Blanc and Petit Gris de Rennes;

Little Leaf pickling cucumber has small leaves as well as small cucumbers, so is easy to distinguish from plants with slicing cucumbers. Credit: Kathryn Simmons.

- *Cucumis melo inodorus* is the group of winter melons: casabas, crenshaws, honeydews, and canary melons. They have a smooth rind, no musky odor, and must be cut from the vine—they will not slip;
- *Cucumis melo dudaim* includes Plum Granny and Queen Anne's Pocket Melon, grown for aroma, not flavor;
- The four groups less-common in the US are *C. m. flexuosus* (snake melons, including Armenian cucumbers), *C. m. conomon* (Asian and Oriental pickling melons), *C. m. chito* (mango melon and others named after other fruits) and *C. m. momordica* (snap melons).

In this chapter I am talking about *C. m. reticulatus*, and will call them muskmelons. Externally, they turn beige and slip from the vine when ripe. They have soft sweet orange flesh. They are sometimes divided into two types: Eastern varieties are sutured (scalloped in shape) and can have a very short shelf life, while Western ones are typically not sutured but are netted (covered with a corky mesh of lines) and will usually hold for two weeks after harvest. I see many netted and scalloped melons, so I don't use this classification. We used to grow netted Ambrosia, before Monsanto bought Seminis Seeds. Now we grow Kansas or Pike from Southern Exposure Seed Exchange. Kansas (90d) is an heirloom muskmelon with excellent flavor, 4 lb (1.8 kg), hardy, productive, with good resistance to sap beetles. Pike (85d) was bred for growing in unirrigated clay soil. It is vigorous, high-yielding, disease-resistant and (depending on irrigation) produces 3–7 lb (1.5–3 kg) fruits with great flavor. We have also had success with Edisto 47 (88d OP), about 6"–7" (15–17 cm) in diameter. With resistance to ALS, PM and DM, it exceeds the disease resistance of many hybrids.

Other types of melons such as the fast-maturing crisp-fleshed Asian melons can do well. We like Sun Jewel from Johnny's. The University of Kentucky has a publication on specialty melons.

Crop requirements, seed specs, yield

Cucurbits require a fertile, well-drained soil with a pH of 6–7 and plenty of sunshine. They have no frost tolerance. Adequate water is especially important in the seedling stage and during fruiting. Vines can sprawl and cover a four-foot (1.2-m) bed or fill even a seven-foot (2-m) row when grown on the flat. Relatively close spacings will increase yields and provide a more uniform harvest. Just what you want if you have a large contract with a pickle company. Otherwise, for a longer harvest from each planting, do not crowd them.

Cucumber seed specs: 1,000 seeds/oz, 36 seeds/g. 0.5 oz /100', 6 oz. /1,000' at 6 seeds/ft (100 seeds, or 11g/m at 2.5 cm spacing).

Cucumber yield can be around 260 lb/100' (388 kg/100 m), and the amount to grow could be 10–15 lb (5–7 kg) per person for the season, so for 100 people grow 577' (176 m). Unless you are growing parthenocarpic varieties, you will need to ensure there are adequate pollinators around. It takes ten to twenty bee visits per flower during the one day the flower is open, for a good-shaped and -sized fruit to grow.

Melon seed specs are the same as cucumbers for size and weight. Melon yields will be affected by irrigation during fruit development, but not by watering levels during vegetative or flowering stages. Marketable yields of muskmelons can be 7,000–10,000 fruits per acre (17,000–25,000 per hectare) when grown on plastic mulch, and down to half that on bare ground. Most melon plants will yield three or four good melons.

Sowing

For cucumbers, soil temperatures should be at least 60°F (15.5°C), preferably 70°F (21°C), so you might do transplants early in the year (see below). We transplant our first planting (sown March 25, planted out April 20) and direct sow the rest. Cucumber seeds will not germinate at a soil temperature below 50°F (10°C). They take 13 days to emerge at 59°F (15°C), 6.2 days at 68°F (20°C), 4 days at 77°F (25° C), and only 3 days at 86°F (30°C).

For melons, slightly warmer temperatures will work better. The seeds take 8.4 days to emerge at 68°F (20°C), 4 days at 77°F (25°C), and 3.1 days at 86°F (30°C). We sow our first ones April 15 to transplant with rowcover May 6, which is a week after our last frost date. Temperatures below 45°F (7°C) can stunt growth. If the spring is cold, just wait it out. Melons will do OK with fluctuating temperatures, provided they are not too cold. For a main crop, we direct sow May 25 and June 25.

For sowing in open ground, we make a furrow 0.5"–0.75" (1.3–1.8cm) deep, water the furrow if the soil is dry, put one seed every 6" (15 cm), pull the soil back over the seeds and tamp down. Growers commonly space seeds at two inches (5 cm), but using the wider spacing gives us no problems, and uses less seed. We cover all our cucurbit sowings with rowcover until the plants start to flower (about a month) as we have many pests and diseases. When the plants start to flower, we remove the rowcover, hoe and thin to 12"–18" (30–45 cm) for cucumbers, 18"–24" (45–60 cm) for muskmelons. Some growers thin cucumbers to only to 4"–12" (10–30 cm). Cucumber rows need to be 3'–6' (1–2 m) apart. Melons can use 7.5–15 ft² (0.7–1.4 m²) each on plastic mulch, and double that space on bare ground. Melon rows are typically up to 6'–10' (2–3 m) apart.

If you want a faster harvest than you'd get from direct sowing, you can chit (pre-sprout) the seed. Put the seed on damp paper towels, roll them up and put the bundles in plastic bags loosely closed, or plastic sandwich boxes, not sealed. Keep at 70°F–85°F (21°C–29°C). Check twice a day (this also introduces fresh air to the seeds), and sow before the root reaches the length of the seed. Seeds that are already sprouting will not need more watering after sowing until the seedlings emerge, unless the soil is dry as dust.

It is possible to sow cucurbits through plastic mulch by jabbing holes in the plastic and popping the seeds in. This method leads to earlier harvests, as the mulch warms the soil, and there will be no weeds.

Transplanting

Cucurbits are not very easy to transplant, so choose a method that minimizes root damage,

We transplant the bush variety Spacemaster, in our hoophouse for an early crop. Credit: Kathryn Simmons.

such as soil blocks, Winstrip trays or 2" (5-cm) deep cell flats that are easy to eject plants from. Sow 2–3 seeds per cell 0.5" (1 cm) deep. Single (thin to one plant per cell) by cutting off weak seedlings at soil level. Keep temperature above 70°F (21°C) during the day and 60°F (16°C) at night. Cucurbit seedlings are sometimes damaged by foliar sprays, especially ones including soaps, so avoid killing by mistaken kindness.

Sow four weeks before you intend to plant out, and harden the plants by reducing water before transplanting. "Days to maturity" in catalogs are usually from direct seeding; subtract about ten days to calculate from transplanting to harvest date.

Warm overcast conditions late in the day are best for transplanting, and rowcover (preferably on hoops to reduce abrasion) can be used to provide warmer and less breezy conditions. Cucurbit transplants are often leggy and should be planted so that the entire stem up to the base of the leaves is below soil level; otherwise the fragile stem is liable to get broken.

Caring for the crop

In bare soil, hoe soon after the seedlings emerge, and thin the plants if you sowed thickly. Larger spacing can be used later in the year when vines grow faster. Some growers trellis and prune their cucumbers, especially in hoophouses, or if they are growing cucumbers with long slim fruits, such as Telegraph, and want to ensure straight better-quality fruits.

Drip irrigation and plastic mulch can do a lot to improve the quality, yield and earliness of melons. Plastic mulches work well with rowcover, as there will not be weeds growing out of sight concealed by the rowcover during the critical weed-free period as the vines grow. Plastic mulches can also reduce cucumber beetle numbers, as they deter egg laying and larval migration. Reflective mulches especially reduce beetle populations. Avoid working the crop (including harvesting) when the foliage is wet, as fungal diseases spread this way.

Rotations

Because of the many cucurbit pests and diseases, good crop rotation is important. We have only two big cucurbit plantings, winter squash and watermelon. These are three years and seven years apart in our ten-year rotation. Cucumbers are planted in small quantities each time, and we use a fairly ad hoc planning process, simply trying not to plant cucurbits anywhere they have been growing in the past two years. Longer would be better. Because old plants are more likely to yield bitter cucumbers, succession planting is very worthwhile even if you have no pests.

Cucumber beetles are quite mobile, so rotation to a field next to last year's crop will not reduce their numbers. To minimize their chances, always till in cucurbit crops residues at the first opportunity.

Succession planting

We grow five plantings of outdoor cucumbers and one early one in the hoophouse. Our second and fourth outdoor sowings include picklers as well as slicers. Our sowing dates are Feb 21 (to plant in the hoophouse April 1 or so), March 25 (to plant outdoors April 20), May 24, June 23, July 15 and Aug 5. Aug 5 is about as late as it is worth sowing here, where the first frost can be Oct 14. We use rowcover on cold nights for this late crop. In the Green Beans and Summer Squash chapters I described recording sowing and first harvest dates for succession crops. The same technique can be used for cucumbers to determine sowing dates to get an even supply throughout the season. As a rule of thumb, in spring, make another sowing when the first true leaf appears in the previous sowing. In summer, make the next sowing when you have 80 percent emergence of the previous planting. (You'll have to keep peeking under the rowcover.) This information comes from Mary Peet in *Sustainable Practices for Vegetable Production in the South.*

We sometimes grow three plantings of muskmelons. The recommended last date for sowing melons is 100 days before the average first frost date (July 6 for us).

Insect pests

See the Resources section for links to several good publications. As always, encouraging beneficial insects and predators will reduce pest numbers. Soldier beetles (Pennsylvania leatherwings) and wolf spiders are good predators. *Heterohabditis* nematodes, available commercially, can control cucumber beetles, and may carry over into the next year.

Spotted and striped cucumber beetles cause feeding damage and transmit bacterial wilt and squash mosaic virus. Rowcover will keep beetles from vines, but will need to be removed (except for parthenocarpic varieties) when the female flowers open. For picklers it is possible to grow a densely planted crop, harvest for a very brief period, then till in the vines. Some people report good control using the yellow plastic sticky traps along with the cucumber beetle lure sachets sold by Johnny's Seeds. These can last a whole season and be moved from one crop to the next, suspended on wire hoops. To make your own sticky traps, use yellow plastic cups stapled to sticks so the cups are just above foliage height. Coat the cups with one part petroleum jelly to one part household detergent (if you are not prevented by USDA certification) or the insect glues available commercially. Cucumber beetles are attracted by clove and cinnamon oils, which can be used to lure them. Use one trap per 1,000 ft^2 (92 m^2). Another approach is to grow a trap crop of a variety particularly attractive to the beetles, such as Cocozelle summer squash, Seneca or Dark Green zucchini, along the edge of the field. The trap crop is then flamed or tilled in when pest numbers build up. If all else fails, and action is imperative, Spinosad will kill them. Neem doesn't kill them, but does deter them.

Brown marmorated stink bugs

The brown marmorated stink bug (BMSB) is emerging as a very serious problem. It is believed to have arrived on the East Coast from Asia and has reached at least thirty states. It has over three hundred host species, causes tremendous damage and, so far, has no reliable methods of control, organic or otherwise. It has several generations per year in the South. Ted Rogers at the Agriculture Research Service of USDA helped gather and disseminate information. Dr Tracy Leskey at ARS USDA at Kearneysville, WV, has collected information on anr.ext.wvu.edu/r/download/74527. After the pictures of ruined apples are two pages on the "Apparent Biology and Phenology" of BMSB. Rutgers also has a two-page factsheet on their website.

Matt Grieshop and Anne Nielsen have set up an interactive website. Go to the Grower Forum at bmsb.opm.msu.edu, where farmers are posting thoughts on dealing with this problem. Researchers are asking growers for specimens, locations and information on crops damaged and crops unaffected. Sample monitoring plans are being designed and posted there, for growers wanting to help.

BMSB feeds on almost any fruit or podded crop: tree fruit, berries, tomatoes, peppers, okra, cucurbits and all legumes including soy. This true bug releases an aggregation pheromone (especially in the fall) which causes adults to gather together in buildings and other protected places to overwinter. The aggregation pheromone of this particular stink bug is being isolated in a USDA laboratory now and will become commercially available. Until that is ready, some success can be achieved using other stink bug aggregation pheromones, especially in the fall. Pheromones are relatively expensive.

Low-cost tactics to try include growing parthenocarpic varieties that can be covered throughout growth and harvest, or other varieties with row-cover or small mesh netting on hoops, caterpillar tunnels or hoophouses with screen doors and enclosed bee colonies. Kaolin clay (Surround) sprayed early and often has shown some success for pome fruits. The company has offered samples for on-farm trials.

Russ Mizell has published a paper on trap cropping for native stink bugs in the South: see the Resources section. He recommends buckwheat, triticale, sunflower, millet, field pea and sorghum for native stink bugs. A succession of trap crops including these and others such as pumpkins, cowpeas and other small grains (which are most attractive in the milk or soft dough stage) could help. Flaming the trap crops is likely to work well. Trap crops only work if they are more attractive than the crop.

Predatory stink bugs, assassin bugs, spined soldier bugs and two native egg parasitoids will reduce the BMSB numbers, but do not give adequate control. Several egg parasitoids from China may be released from quarantine in the US, in 2013 at the earliest, to tackle the pest. BMSB are attracted to yellow, and to corrugated cardboard. Chickens and praying mantids seem to lose interest after a few bites. Hogs will eat them. There are concerns that the flavor of the stink bugs carries through to milk, meat, eggs, wine and soy products, either from ingested bugs or insect parts mixed in with the crop.

Soapy water will kill the nymphs, but be cautious: cucurbit seedlings are sometimes damaged by soap sprays. One spraying killed about two-thirds of the adult bugs in one study. Other studies found soaps ineffective.

Diseases

To minimize diseases choose disease-resistant varieties, provide favorable growing conditions, plow in or remove and compost plant refuse, and control insect pests. Diseased foliage reduces the ability of the fruit to develop sugars: melons from plants with PM will never taste good.

The Cornell University *Organic Resource Guide: Organic Insect and Disease Control for Cucurbit Crops* has information on dealing organically with most common diseases:

- Angular leaf spot (bacterial) occurs during cool, wet weather. Symptoms include interveinal browning of the leaf and small round spots on the fruit. The leaf damage is tan, rather than dark;
- Anthracnose, a fungus disease, is most common during warm rainy weather. It causes angular haloed dark-brown spots on the leaves and dark, round, sunken spots on the fruit;
- Bacterial wilt (*Erwinia*) causes sudden dramatic wilting and death of the vines;
- Black rot/gummy stem blight (*Didymella bryoniae*) is a fungus occurring with wet soils in cool or warm wet weather. It is less of a problem in cucumbers than in other cucurbits;
- Downy mildew occurs during wet, cool spells and plants may recover if the weather heats up;
- Mosaic virus causes a yellow and green mottling of the leaves and reduces plant vigor;
- *Phytophthora* blight occurs in some regions but not others;
- Powdery mildew occurs during hot, dry spells;
- Scab is not usually a problem to those growing resistant varieties. The fungus, *Cladosponum cucumerinum*, is worse in cold wet weather. Try compost teas or baking soda spray.

There are good photos in *Identifying Diseases of Vegetables* from Penn State. The University of Tennessee has a concise list of melon diseases and pests in its publication *Producing Cantaloupes in Tennessee*.

Harvest

Growers with large pickle plantings use machines to gather the harvest from concentrated plantings of 60,000 plants/acre (150,000/ha). The rest of us pick by hand, and know it to be true that harvest is the most labor-intensive stage of cucurbit growing. Once the season is up and running, we need to pick every day to prevent blimp cucumbers and split melons. Some people find the scratchy leaves very irritating and like to wear shirts with long sleeves and gloves.

To harvest cucumbers, put the hand around the fruit and use the thumb to push the stem away from the top of the fruit. Our instructions to the crew are to harvest the picklers at the size of their little finger, or bigger. For slicers we tell people to picture the cross-section of the fruit and let it round out into a circle with slight corners (not a balloon!). We try to avoid harvesting underdeveloped triangular cucumbers. Pick oversize cucumbers (to stimulate continued production) and drop them on the ground. No need to carry those useless things out of the field.

Harvesting muskmelons essentially uses the same method, although with a lighter touch. "Full slip" is the term for melons that separate cleanly from the stem with only the very lightest pressure. Waiting too long leads to rotten fruit, especially in hot weather. Look carefully at the point where the stem joins the melon—and as the melon ripens you will see a circular crack start to open around the stem. This small disk of melon stays with the stem when it slips off the vine. More usually, growers give the stems a "little nudge" to see if the melon will fall off the vine. Depending on the delay between harvest and sale, you may need to pick at half-slip or three-quarter slip (when half or a quarter of the stem disk sticks and breaks rather than slipping free). If you are growing melons on a large scale, it will be worth buying a refractometer to test sugar levels. Melons at full slip should register 12%–14%, while those at half-slip should show at least 10%. Half of the final sugars accumulate in the last week of ripening. Full flavor develops a day or two after picking, but the sugar content does not increase. I like nothing better than eating fruit fresh from the field, still warm from the sun.

Cucumbers and melons do not need to be rushed to the cooler as greens and sweet corn do, so they can be picked and set in the shade until a full load of produce is ready to be moved. Melons are subject to sunscald when left unprotected in the sun after harvest.

Also note that cucurbits should not be harvested while the leaves are wet (from dew or irrigation), as this spreads diseases.

Storage

Cucumbers can be stored at 45°F–50°F (7°C–10°C) and 90% humidity for up to two weeks. They will be damaged by temperatures that are too cold, becoming soft and slimy. Storage near other ripening fruits or vegetables can cause the cucumbers to become bitter.

Muskmelons can be stored at 36°F–41°F (2°C–5°C) and 95% relative humidity, also for up to two weeks.

Muskmelon can be bruised or split if they are dropped more than eight inches (20 cm) onto hard surfaces. When they are stacked more than six layers deep or are transported over rough roads, pressure bruising can result, leading to discolored flesh.

Season extension

Late crops can be covered with rowcover to fend off a few light frosts. Pollinators won't be able to get at the flowers, but that doesn't matter if you already have enough fruits on the plants. You can even pinch off immature fruits to concentrate the plants' energy into ripening the bigger fruits. It takes cucumbers about 45 days or more from pollination to harvest, so if you are having a few early frosts, using rowcover and pruning to get a last flush of fruit can be very worthwhile. It isn't worth it to coddle every last little nubbin of a fruit, as the smallest ones won't ripen in cool temperatures and will get killed by heavier frosts.

Early crops can be grown in a hoophouse, using transplants, with rowcover added on cold nights.

Research at Virginia Tech has shown success with Asian melons such as Jade Star, Sun Jewel and Emerald Sweet transplanted into hoophouses at the end of March at their Petersburg, Virginia location.

Seed saving

Cucumber varieties need to be isolated by an eighth of a mile or 660' (200 m) for home use and a minimum of a quarter to one mile (400–1600 m) for seed for sale. Melon varieties need to be isolated by 660' (200 m) for home use and at least a half to one mile (800–1600 m) for seed for sale. Note that all eight melon groups cross with each other.

Credit: Diana Hernandez

Chapter 44

Watermelon

Nothing beats a big slice of watermelon during a break from working in the fields. It's a good cure for dehydration, especially if lightly salted to balance the electrolytes, and helps improve heat tolerance. Watermelon is easily digested and adds fiber to the diet. Second only to tomatoes as a source of lycopene (said to prevent some cancers), watermelons are also an excellent source of vitamin C, beta carotene, folic acid, biotin, potassium, magnesium and citrulline (an amino acid important for healing wounds and removing toxins from the body). The seeds, if well chewed to break up the indigestible seed coat, can provide amino acids, fatty acids, vitamin E, potassium and phosphorus. It's good to know watermelons are nutritious, but frankly, their main claim to fame is that they are delicious, and just about everyone wants one when the weather is hot.

Varieties

After trying several varieties, we settled on Crimson Sweet, a 20–25 lb (9–11 kg), striped, 10×12" (25×30 cm) oval OP melon which takes 86 days from transplant to harvest. It has tolerance to some strains of *Anthracnose* and *Fusarium*. We are saving seed, selecting for size, earliness, disease resistance and flavor (see the Seed Growing chapter for more information). Charleston Gray is another popular large variety, with a redder color but less sweetness than Crimson Sweet.

Black Tail Mountain (OP, 73 days) was made famous by Glenn Drowns, who developed it as a fast-growing, highly productive, rich-flavored melon he could grow in a cold climate in northern Idaho. It stores for up to two months after harvest.

OrangeGlo (85 days) is an outstanding orange-fleshed variety, with large fruits and great tropical flavor. William Woys Weaver (a contributing editor to *Mother Earth News*, winner of three cookbook awards and owner of the Roughwood Seed Collection) recommends it for frozen desserts.

"Icebox" varieties are six-inch (15-cm) round, 8–12.5 lbs (3.6–5.7 kg) melons, perfect for small refrigerators. Some varieties are ready as soon as 64 days after transplanting. Seedless triploid varieties Dark Belle (F1 75d) and Fun Belle are said to be better tasting than traditional Sugar Baby (77d) and orange New Orchid (80d). See the 2003 study of icebox watermelons by Washington State University, in the Resources list. Selling portions of larger melons is a way of providing for this need if you don't want to grow icebox types.

Triploid seedless watermelon seed is expensive and harder to germinate, and the transplants are very fragile and tricky to establish. Triploid varieties are hybridized from a cross between two plants with incompatible sets of chromosomes. This results in sterility (lack of seeds).

Even smaller than the icebox size are "mini" watermelons, 3–6 lb (1.4–2.7 kg). Golden Midget is an OP seeded mini from Baker Creek 70d, 3 lbs (1.4 kg). There are also triploid seedless varieties including Petite Perfection (the smallest in a 2003 University of California trial of mini watermelons, with the highest number of melons for the area, and the lowest total yield). Extazy, Demi Sweet and Liliput are some other mini varieties. Solitaire (triploid, 88d) is a single serving watermelon from Johnny's. It has six-inch (15-cm) diameter fruit. Harvest can begin 78 days after transplanting.

Crop requirements and yield

Watermelons do best in free-draining light soils that warm quickly in spring. Ensure high organic matter content, sufficient boron and a pH of 6.5. Black plastic mulch, either the removable or the biodegradable kind, will speed growth and ripening. If you want to use organic mulches, put them around the plants after the soil has warmed up, or you will have later harvests. Drip irrigation is definitely better than overhead, as it reduces the chance of foliar diseases. Water well during fruit development, then cut back during the harvest period, for best flavor and to prevent fruit bursting. We often run our irrigation at the same time as harvesting, so we can easily check for leaks.

If drainage is an issue, make ridges or raised beds before planting. You can use straw or spoiled hay in the aisles to absorb some of the water. Watermelons easily die in waterlogged soil.

There are on average 24 seeds/g (670/oz, 11,000/lb; 24,200/kg). Large seeded varieties require 1–2 lbs/ac (1.1–2.2 kg/ha) for direct sowing. Crimson Sweet seeds are about half the size, so need only half as much.

Labor for one acre (0.4 ha) can be 20–40 hours for production (including land preparation and transplant production), 60 hours for harvest, 10 hours for grading and 10 hours for plastic removal, according to the University of Kentucky Extension. Smaller areas with less mechanization inevitably use proportionally more than that.

Yield of Crimson Sweet and other varieties can be 20,000 lbs/ac or 460 lbs/1,000 ft² (22,400 kg/ha, 227 kg/m²). For our 6,600 ft², (613 m²) we can expect an average of 3,000 lbs (1,400 kg), whether 150 melons at 20 lbs (9 kg) or 300 at 10 lbs (4.5 kg). We get higher yields than these, and have got 300 melons (unweighed) from this area, using 2' × 5.5' (0.6 × 1.7 m) spacing.

Rotations

Because watermelon is planted relatively late, there is time in the spring for leguminous winter cover crops such as crimson clover, hairy vetch or (in zones 7–9) Austrian winter peas to reach the flowering stage. For maximum nitrogen, mow and incorporate the cover crops as they start to flower. A good leguminous cover crop can provide all the nitrogen the crop will need. Another possibility is to grow watermelons following a legume food crop the previous year.

Our rotation has paste tomatoes and peppers the year before the watermelons, in no-till mulch. Those crops finish with the frost, and it takes us a little while to get all the stakes and twine out of the patch before disking. We usually sow rye and Austrian winter peas (it's too late for clovers). Following the watermelon, we like rye, Austrian winter peas and the higher N-producing crimson clover, which have time to reach flowering before being turned under for the middle sweet corn plantings the next year. Watermelon and winter squash are the only two cucurbits we grow in quantity, and so we have a gap of three years between watermelon and winter squash, then seven years between squash and melons.

Sowing

Watermelon seeds need a soil temperature of at least 68°F (20°C) to germinate, taking 12 days at that temperature, but coming up in a mere 3 days at 95°F (35°C). If direct seeding, station-sow 4–6 seeds 1–1.5" (2.5–4 cm) deep at the final spacing. Later, thin the emerging seedlings to one or two at each spot. Pests are more likely to attack plants stressed by planting in cold conditions. If in doubt, wait.

Transplanting is the way to go for early melons.

We use soil blocks for this crop, or Winstrip 50 ventilated plug flats. Soil blocks are more time consuming than plug flats, but the results are so much more dependable that I think it is well worth it. Cells should be at least 1.5" × 1.5" (4 × 4 cm). We put two seeds in each block and sow more blocks than our usual 20 percent extra, because casualties with melons are usually fatal. After emergence, we pinch off the weaker seedling.

We sow 30 percent more blocks than we hope to take to the field, which is another 30 percent more than we need to plant because of their fragility. (We expect casualties on planting day.) We sow April 26 and April 29. We split the sowing up because we don't have a big enough germinating chamber to accommodate them all at once. They come up very fast in our hot germination chamber, and a three-day sowing difference is unnoticeable later. We transplant at 15–19 days old. Four weeks old is about the maximum for watermelons—they start to get stunted if held too long.

Once the seedlings emerge, they need maximum light and warmth, but not too much watering. I have a favorite spot in our greenhouse for them: right by the south and west windows.

Transplanting

Transplanting allows the young plants to be raised in close to ideal conditions, and gives the soil time to warm up. Don't rush transplants into cold soils, it's better to wait—cold conditions can permanently stunt them. Once outdoor daily mean temperatures have reached at least 60°F (15.5°C) and the first true leaf has fully opened, you can plant them out.

We have found watermelons to be amongst the crops needing the most skill at transplanting. The stems are fragile, the roots respond poorly

Crimson sweet watermelon. Credit: Kathryn Simmons.

to disturbance and spending extra time later replacing the dead starts is frustrating and doesn't lead to early melons. It also requires the grower to produce lots of spare plants, which all take time and care. Now we aim to have only the most experienced people plant out melons, and it's so much more satisfying!

Roll out the drip tape and the plastic on top, then turn on the irrigation while planting. This helps ensure no one stabs the drip tape, and the plants can be set by the emitters. (Yes, you can still find them, even though they are under the plastic.) Watermelon transplants can easily get leggy in the greenhouse, so make holes deep enough to bury the stem as well as the roots. Some people use bulb transplanters to punch holes through the plastic into the soil. We just use pointed trowels. On a larger scale, waterwheel transplanters work well. Despite the recommendations of the mulch manufacturer, our soil surface is often uneven, so we use small rocks to hold down the edges of the holes in the plastic anywhere they could damage the plants. Provide wide margins at the edges of

the patch as the vines will invade other plantings given the chance.

We drape rowcover tent-like over ropes between stakes hammered into the soil at regular intervals along the row. This prevents the rowcover abrading the leaves and creates a volume of warm air around the plants. A week after transplanting, we fill any gaps with more transplants or with a few seeds station-sown at each spot where we want a plant. Sowing pre-sprouted seeds will help make up for lost time if something has gone wrong.

Spacing

Spacing can make a difference to size and yield, but not sweetness. There are widely varying recommendations, from 9 ft^2 (0.8 m^2) to 80 ft^2 (7.4 m^2) each! Perhaps large spacings suit backyard growers who want "specimen" plants. In commercial growing we are looking for wise use of land and other resources. It doesn't matter if 12 ft^2 is provided as 1'×12', 2'×6', or 3'×4'. The area is the important factor, so choose a row spacing that works nicely for you and adapt your in-row spacing to give the area you want for each plant.

We used to transplant our watermelon 2' (60 cm) apart in rows 10' (3 m) apart. We used spoiled hay as mulch. The melons were nice but were late coming in. Early fruit is a goal for us—not bigger plants, longer vines or more (later) melons per plant. I read a brief reference to watermelons only needing 10 ft^2 (0.9 m^2) per plant, so we switched to a spacing of 2'×5.5' (0.6×1.7 m), 11 ft^2 (1 m^2), in order to fit more plants in the space and therefore get more first and second melons. 5.5' (1.7 m) is just the right width for rolling hay bales. The new spacing seemed fine until someone complained that our melons were smaller since the change and not as sweet. I checked on total yield and that seemed to be in the right range.

Next, I researched the questions of sweetness, size and spacing. Scott NeSmith at Georgia Experiment Station in Griffin, Georgia, whose research involved Star Brite and Crimson Sweet, assured me there is no difference in sweetness between melons at different spacings.

Close spacing (down to a point) increases total yield, but decreases the weight of each melon. A Brazilian study on Crimson Sweet found that 13 ft² (1.2 m²) per plant gave the highest total yield, but 15–20 ft² (1.4–1.9 m²) gave bigger melons. NeSmith found that reducing plant spacing 50% may reduce the size of each melon 10%. Earlier research (1979, Brinen et al) had shown that reducing plant spacing 50% resulted in increases in yield of 37%–48% with only a 13% reduction in average fruit weight. Bigger varieties are more likely to have their size affected by closer spacing than small varieties are.

Here are the factors in deciding spacing:

- Total yield (by weight): reduced spacing (to a certain point) increases total yield. Reduced spacing does not increase the percentage of cull fruit;
- Number of melons/area ("fruit density"): increases with plant density;
- Size: reduced plant spacing sometimes affects melon size, but not in a linear way. Other (environmental) factors affect melon size. Small size is an advantage in some markets;
- Yield/plant: decreases at close spacing, sometimes because number of melons per plant is reduced, sometimes because the size of the fruit decreases;
- Melons/plant: the number decreases as plant spacing is reduced, but not linearly. At close spacings, the difference is negligible;
- Early yield: variety, early transplanting, good conditions and hot weather will provide more early melons. The first melon on each plant is the early harvest. More plants means more first melons. Plastic mulch produces crops a month before organic mulched crops. Spacing has no influence on the ripening rate;
- Sweetness: the flavor of watermelon is not related to the size of the ripe melon or the plant spacing. Healthy foliage and long hot sunny days are the biggest factors in building good flavor. July has longer days than August, and September's days are getting shorter, so don't expect late-season melons to be as sweet;
- Plant health: overcrowding can cause foliar diseases to take hold;
- Labor requirement: closer spacing = more transplanting. More melons = more time harvesting.

Clarify your goal and choose your variety and spacing accordingly. If your goal is the highest weight of watermelons for a given area, plant Sugar Baby at 10–11 ft² (0.9–1 m²) each. If your goal is the highest number of melons, try them even closer! But if you like Crimson Sweet and want fairly large melons, what to do? Try 15 ft² (1.4 m²) if 12 lb (5.4 kg) melons are an acceptable size (you might still get 15 lb (6.8 kg) melons!). Otherwise, use 20 ft² (1.9 m²). Go up to 30 ft² (2.8 m²) if you want big melons and can accept a lower total yield. Personally, I want a 5.5' (1.7 m) row spacing, and an easy-to-measure space between plants. We have tried some at 2½' (80 cm) in-row spacing, some at 3' (90 cm) and some at

3½' (1.1 m). These spacings correspond to areas of almost 14 ft² (1.3 m²), 16.5 ft² (1.5 m²) and 19 ft² (1.8 m²) each. We didn't keep records, and didn't notice a difference in size.

Ideally, the ground will be filled with foliage by the time the first blossoms appear so that the crops can intercept and use all the available sunlight. Given that the market is for early melons, and early ones are sweeter, having many plants (one early melon each), and having them optimally cared for, is important.

Caring for the crop

Remove rowcover from transplants after three weeks (wait longer with direct sown crops) and pull any big weeds. (Cultivate between the rows if you have bare ground.) We wait to remove the rowcover until we see a few open flowers, which indicate that pollination is now the critical step, not warming. Water regularly—drip irrigation set out at planting is the best way to go, as there will be less chance of fungal diseases than with overhead watering.

Weeding is important, and needs to be completed before the vines run. If big weeds get away from you, and pulling them endangers the crop roots, wade in with pruners and clip off the weeds at ground level. This prevents the weeds seeding, and lets the melons get more sunlight again. Flaming around the edges of the patch before the vines get there can set back perennial grass weeds.

Watermelons have separate male and female flowers on the same plant, and insect pollinators are necessary. Many species of native bees pollinate watermelons, but augmenting them with honeybees will help pollination, which means bigger, better-shaped melons as well as more of them.

Do not turn over the vines when weeding—cucurbits don't like it! Older publications refer to pruning vines, leaving only one or two fruits per plant to develop. If you are in marginal watermelon climates, you might try this. It sounds tedious, and technically difficult to find where one plant finishes and another starts. Removing damaged fruit will help the good ones grow better.

We used to keep our watermelons growing until frost, but nowadays we keep an eye on production rates, the weather and the calendar, and "pull the plug" when we think we have satisfied the need for watermelons. This lets us disk the patch and get good winter cover crops established in September or the early part of October.

Season extension

As it is possible to store ripe melons for a time, there is not usually a reason to extend the harvest later into the year. Exceptions might be for a special seed crop, or for growers in cold climates trying to ripen any melons. In these cases, protect from cold weather with rowcover, caterpillar tunnels or hoophouses.

As far as ripening melons as early as possible, use transplants, black plastic mulch and rowcover until flowering. If you are philosophically or politically opposed to dealing with removing and recycling plastics, and your USDA Organic certification doesn't prevent you, try the biodegradable black mulches such as Bio-Telo. It's amazing! Choosing fast ripening varieties is useful, too.

Watermelons take a long time to mature, and the market is for early watermelons, not melons in October, so it is only in the Deep South that more than one planting of watermelons makes much sense.

If you store watermelons outdoors, watch out for smart squirrels! Credit: Twin Oaks Community.

Pests

Compared to some crops, watermelons are not challenged by many pests.

Striped cucumber beetles are our worst pest. They eat not only the leaves (which reduces the sweetness of the melons) but also the rind of the melons, leaving an unattractive russeted surface, thinner than it was originally, and easily damaged. The action level is two beetles per plant. Cucumber beetles can also interfere with fruit set by eating the stamens and pistils of the flowers.

Aphids (usually green peach aphids) can be a problem to young plants—another reason to use rowcover. If needed, use insecticidal soap, or import ladybugs or lacewings.

Spider mites can be a problem in hot dry weather if populations are driven into the patch by mowing of bordering grassy areas. Heavy rain, vigorous spraying with water or overhead irrigation will reduce numbers.

Root knot nematodes can attack roots and produce galls. This leads to loss of vigor and wilting.

Diseases

Organic growers do not usually get many disease problems with watermelon, provided the soil fertility is well balanced and the plants are not physically damaged. There are a few diseases to watch for:

- *Alternaria* leaf spot
- *Cercospora* leaf spot
- Gummy stem blight
- Watermelon fruit blotch/bacterial fruit blotch is a serious disease that is seed-borne (another reason I like growing our own seeds—we'll never bring it in).
- Bacterial wilt—watermelon is resistant, although young seedlings could succumb.
- *Fusarium* wilt is a persistent soil-borne fungal disease that infects the roots, invading the xylem cells.
- *Anthracnose* is a fungal foliar disease that can cause loss of vigor and can cause fruit spotting.

More information about these diseases and what to do about them can be found in the Cucumbers and Muskmelons chapter.

Harvest

The skill of the harvester in discerning ripeness is a major factor affecting the taste. We switched to having just a select few people do the harvesting, and get better melons.

The first sign we look for is the shriveling and browning of the tendril on the stem directly opposite the watermelon. If this tendril is not shriveled we walk on by. Next we slap or knock on them.

According to Southern Exposure Seed Exchange, when a watermelon is ripe, it will have a hollow sound when you thump with your knuckles: it sounds like thumping your chest. If it sounds like knocking your head, it's not ripe; if it sounds like hitting your belly, it's over-ripe. There is a 10–14-day period of peak ripeness for each variety. We harvest ours from around July 25 (75 days from transplanting) to the end of August. We hope not to be still harvesting in September.

Lastly, we do the "Scrunch Test": put two hands (heels together) spread out across the melon, press down quite hard, listen and feel for a scrunch—the flesh in the melon is separating under the pressure. Rumor has it that it only works once, so pay attention!

Other growers with other varieties use different ripeness signs, such as the change in color of the "ground spot" (the area touching the ground), or the change in rind texture from glossy to dull.

I like to cut the melon stems with pruners, but some people break them off. Watermelons need gentle handling, as do the vines if you will be returning to harvest again. After harvest, we set the melons out to the side of the row for pickup. This gives time for sap to start to ooze out of the cut stem. If the sap is red or orange, the melon is ripe. If it is straw-colored, the melon was cut too soon. This is useful feedback for new crew.

Post-harvest storage

Watermelons can store for a few weeks, but then flavor deteriorates. We store ours outdoors in the shade of a building or a tree. Rotating the stored stock is a good idea (consider dating it with a grease pencil/china marker). The ideal storage temperature is 50°F–60°F (10°C–15.5°C), with 90% humidity.

Seed saving

See the Seed Growing chapter for details on how we save our watermelon seed. Isolate varieties by at least an eighth of a mile (200 m) for home use, a quarter to one mile (800–1600 m) for pure seed for sale. I mark selected watermelons in the middle of July as they size up, and collect seed during August.

Chapter 45

Garlic

There are two basic kinds of garlic, hardneck and softneck (elephant garlic is really a leek). We grow mostly hardneck because big easy-peel cloves make life easier when cooking for a hundred people. Softneck garlic is easier to braid because it does not usually produce scapes or flower stems. It stores later in the spring without sprouting, but usually has smaller cloves and tighter skins. We grow a small amount of a Polish White artichoke-type softneck, which has relatively large cloves, to use after the hardneck is finished. Some people say that softneck varieties lack the complexity and richness of flavor that hardnecks have. Softneck garlics don't do as well in colder regions as hardneck varieties.

This chapter is about bulb garlic and also the by-products of garlic growing that make good crops in their own right: the scapes from hardneck garlic, which are harvested about three weeks before the garlic bulbs mature, and garlic scallions, which can be harvested even earlier in the spring.

Ron Engeland, in *Growing Great Garlic*, identified nine phenotypically distinct types of garlic: artichoke, silverskin, porcelain, rocambole, asiatic, creole, turban, marble purple stripe and purple stripe. Of these, artichoke and silverskin are identified as *Allium sativum*, softneck garlic. Porcelain garlics, erratic scape producers, have been genetically identified as *A. sativum* also. Rocamboles are *A. ophioscorodon*, the classic hardneck type. Asiatic garlic (perceived as a softneck with some scapes) is classified as *A. pekinse*. Creole and turban types are also softnecks with some scapes. Purple stripe and marble purple stripe are *A. longicuspis*.

Dr. Gayle Volk of the USDA–ARS National Center for Genetic Resources Preservation has done DNA fingerprinting on 211 varieties of garlic. She found massive duplication (multiple names for genetically identical garlic), and created a Garlic DNA Map (see Resources). Of the above nine phenotypical types, she found seven groups significantly different from each other: artichoke, porcelain, rocambole, asiatic (genetically a hardneck), purple stripe, marble purple stripe (both genetically *A. sativum* types) and, lastly, Creole and turban types, which were too few in number to give clear results. Silverskins did not show as a distinct genetic group even though they can be distinguished by their almost vertical blue-green leaves from artichokes, which have yellow-green, more horizontal leaves.

Naturally, the quality of the garlic from suppliers can vary—some Music is better than other Music. "Phenotypic plasticity" or "biological elasticity" is the scientific name for local variation in how a given variety performs, depending on soil, climate and cultural practices, as that variety adapts to its surroundings.

Timing

Both hardneck and softneck garlic do best when planted in the fall, though softneck garlic may also be planted in the very early spring if you have to (with reduced yields). The guideline is to plant when the soil temperature at four inches (10 cm) deep is 50°F (10°C). The usual time for thermometer readings is 9 am. If the fall is unusually warm, wait a week. Instructions from Texas A&M say less than 85°F (29°C) at two inches (5 cm) deep. We plant in early November. In New Hampshire, mid-October is the time. The guideline for areas with cold winters is two to three weeks after the first frost but before the ground freezes solid for the winter. In Michigan, planting time is six weeks prior to the ground freezing, giving enough time for root growth only, to avoid freezing the leaves.

In California, garlic can be planted in January or February.

If you miss the window for fall planting, ensure that your seed garlic gets 40 days at or below 40°F (4.5°C) in storage before spring planting, or the lack of vernalization will mean the bulbs will not differentiate (divide into separate cloves).

Garlic roots will grow whenever the ground is not frozen, and the tops will grow whenever the temperature is above 40°F (4.5°C). In colder areas the goal is to get the garlic to grow roots before the big freeze-up arrives, but not to make top growth until after the worst of the weather. In warmer areas, the goal is to get enough top growth in fall to get off to a roaring start in the spring, but not so much top growth that the leaves cannot endure the winter. If garlic gets frozen back to the ground in the winter, it can regrow and be fine. If it dies back twice in the winter, the yield will be lower than it might have been if you had been luckier with the weather. When properly planted, it can withstand winter lows of –30°F (–35°C). If planted too early, too much tender top growth happens before winter. If planted too late, there will be inadequate root growth before the winter, and a lower survival rate as well as smaller bulbs.

Seed stock

Store at 50°F–60°F (10°C–15°C). Avoid temperatures of 40°F–50°F (4.5°C–10°C) during the summer, as this will cause sprouting before you are ready to plant. In other words, don't refrigerate. We keep our seed garlic on a high shelf in the shed from June to November and the conditions are perfect. If you need to store the bulbs over the winter, aim for 27°F (–3°C). If you are buying seed stock, it is usually recommended to buy from a supplier in a similar climate zone.

Having said that, I'll tell you that our hardneck garlic originally came from a bag of Chinese garlic bought at the wholesale produce market! We have been carefully selecting seed stock from this for about twenty years now, and it does great. Cloves for planting should be from large (but not giant) bulbs and be in good condition.

Quantity

A yield ratio of 1:6 or 7 seems typical, and makes complete sense when you consider you are planting one clove to get a bulb of 6–7 cloves. If you achieve a yield ratio of 1:12 you are doing very well indeed. Divide the amount you intend to produce by six to figure out how much to plant. For large areas 750–1,000 lbs/ac (842–1,122 kg/ha) are needed for plantings in double rows, 3"–4" in-row (7.5–10 cm), beds 39" (1 m) apart. For single rows, 8 lbs (3.6 kg) of hardneck or 4 lbs (1.8 kg) of softneck plants about 100' (30 m).

In the US, one person eats 3–9 lbs (1.4–4.2 kg) per year. If you love growing garlic, move to Korea, where each person reportedly eats 60 lbs (27 kg) of pickled garlic each year.

Popping the cloves

The seed garlic bulbs should be taken apart into separate cloves 0–7 days before planting. We often do this while holding our annual Crop Review, when the crew meets to make notes on the past season. This task is a good group activity. Twist off the outer skins and pull the bulb apart, trying not to break the basal plate of the cloves (the part the roots grow from), as that makes them unusable for planting. With hardneck garlic, the remainder of the stem acts as a handy lever for separating the cloves. We sort as we go, putting good size cloves in big buckets, damaged cloves in kitchen buckets, tiny cloves in tiny buckets and outer skins and reject cloves in compost buckets. The tiny cloves get planted for garlic scallions, as explained below. Don't worry if some skin comes off the cloves—they will still grow successfully.

Pre-plant treatments

Many of us do nothing special with the cloves before planting, but if you have pest and disease problems, use pre-plant soaking treatments, usually done the night before planting. Some growers find they get better yields from treated cloves even if no problem was obvious.

To eradicate stem and bulb (bloat) nematode (*Ditylenchus dipsaci*), soak the separated cloves for thirty minutes in 100°F (37.7°C) water containing 0.1% surfactant (soap). Or soak for twenty minutes in the same strength solution at 120°F (48.5°C), then cool in plain water for ten to twenty minutes. Allow to dry for two hours at 100°F (37.7°C) or plant immediately. Anytime your garlic grows poorly and you can't tell why, send a sample with the soil it's growing in to your Extension Service to be tested for nematodes.

Mites can eat the skins of the cloves, survive the winter and multiply all spring long, seriously damaging or even killing your crop. To kill mites (which hide between the wrappers) before planting, separate the bulbs into cloves and soak them overnight (up to sixteen hours) in water. Possible additions to the water include one heaping tablespoon of baking soda and one tablespoon of liquid seaweed per gallon (around 8 ml baking soda and 4 ml liquid seaweed per liter). Just before planting, drain the cloves and cover them in rubbing alcohol for three to five minutes, long enough for the alcohol to penetrate the clove covers and kill any mites inside. Then plant immediately. The long soaking will loosen the clove skins so that the alcohol can penetrate. Mite-infested garlic soaked like this does much better than unsoaked infested garlic.

The solution used to kill mites can also be used to kill various fungal infections. The cloves need only fifteen to thirty minutes soaking. In trials comparing treated and untreated cloves, treated cloves were larger and healthier than untreated ones.

See the Diseases section below for descriptions of various diseases. *Fusarium* levels can be kept down by adding wood ashes when planting and then possibly dusting the beds with more ashes over the winter (use moderation—don't add so much that you make the soil alkaline). Or you could soak the cloves in a 10% bleach solution, then roll them in wood ash (wear gloves for handling ashy cloves). The wood ash soaks up the dampness of the bleach and provides a source of potassium. This information came from the Garlic Seed Foundation. Join GSF to find out all the details!

Crop requirements

Garlic does best with a sandy or clay loam with very good drainage and a pH of 6.0–8.4 (with 6.8 being optimal). Onion maggots thrive if the soil is alkaline, so it pays to watch the acidity. A rotation of at least five years away from alliums is a good practice to reduce the likelihood of disease. Generally one to two inches (2.5–5 cm) of water per week during the growing season (not during the winter) is about right, until the leaves start to yellow and the bulbs start to dry down, when irrigation should be stopped.

Fertile soil with lots of organic matter and a full range of nutrients is needed to grow good

We plant our hardneck garlic at 5" (13 cm) spacing in the row, and 8"–10" (20–25 cm) between rows, usually with four rows in a bed. Credit: Kathryn Simmons.

garlic, and so is full sun. Most growers spread compost or soybean meal at planting time. Foliar feeding, although recommended by some sources, provides no gain in yield if the soil had adequate fertility at planting time. Also, it is tricky to get foliar fertilizers to stick on the waxy near-vertical garlic leaves—it tends to run off, so a good spreader-sticker is essential (some kind of soap). Foliar feeding (or side dressing with compost or organic fertilizers) is wasted after the fifth leaf, and certainly after the bulb starts to enlarge. If soil fertility is uncertain, northern growers may feed every two weeks in early spring until there are four leaves. In the South, spring is too late for foliar feeding, as garlic reaches a four-leaf size before winter. It is unwise to over-fertilize in the fall or the growth will be too fast and tender to survive cold conditions, and the storage life of the garlic will be reduced. So if your garlic typically reaches four leaves before winter, forget about foliar feeding and side dressing.

Spacing and depth

We plant at 5" (13 cm) spacing in the row, and 8–10" (20–25 cm) between rows, usually with four rows in a bed. The beds are 3.5'–4' (1–1.2 m) wide. That's 40 in² (258 cm²) each. The minimum is 32 in² (206 cm²), and 72 in² (465 cm²) is recommended for very large bulbs (which might win ribbons at the fair, but might not give you the highest yield for the area). Many growers plant at 6" (15 cm) in-row. Colorado State University Specialty Crop Program research on garlic in 2004 and 2005 found that 3" (7.5 cm) was too close; the shading of one plant by another reduces the yield. For best use of drip tape, you can run a length of tape and plant a double row, one row each side, with all plants 6" (15 cm) apart in all directions, and 40" (1 m) or less between drip lines.

Cloves are usually planted with 1.5"–2" (4–5 cm) of soil over the top of the cloves in the South, and 3"–4" (8–10 cm) of soil in the North. Planting depth in Michigan is six inches (15 cm). (The deeper planting helps prevent too much top growth and also moderates the soil temperature the clove is growing in.) In Arizona, some growers set the cloves on the soil surface, then cover with six inches (15 cm) of straw. This makes for a clean crop and an easy harvest. Organic mulch can be added immediately after planting, or if you live in a colder area than we do, after the tops get frosted off. Avoid planting deeper than necessary, as any mold problems you have may get worse.

If you are planting a hardneck variety, make sure the cloves are planted the right way up (pointy end up)! Hardneck cloves planted with the points down suffer a 30 percent reduction in yield. Softneck cloves can be planted any way up, so are easier for mechanical planting. Our method is to make furrows with pointed hoes, then lightly press the cloves into the furrows at the chosen spacing, using pre-cut measuring sticks. After that we pull soil over the cloves using regular hoes or rakes, and tamp the soil down with the back of the tool. Some other growers who also plant by hand make a planting jig to make four or more holes at a time in loose soil, rather than make a furrow. Plant a clove in each hole and cover with the right depth of soil.

If you can't squat to hand plant, or you are planting from the seat of a tractor, use a three-foot (1-m) length of pipe to drop the cloves into the furrows. Dropped from that height, through a tube wide enough for the garlic to tumble end-over-end, the cloves will land the way they need to be.

In *The Natural Farmer* magazine in fall 1992, Grace Reynolds of Hillside Organic Farm in New York described how she converted a Cole one-row corn planter on the toolbar of her tractor to plant garlic. She attached a long tube to the planter and an angel food cake pan to the top of the tube. She set the tractor in crawler gear and walked behind it, dropping cloves through the pan into the tube. She also added a mark on the turning plate in the corn planter, dropping a clove down the tube each time she saw the mark, which made for regular spacing.

No-till planting

There were trials at Virginia Tech to develop no-till planting methods for garlic, planting in the fall into a frost-killed cover crop. Sorghum-sudan hybrid, lablab bean and sunn hemp were planted in the first week of August in raised beds. As soon as frost had killed the cover crops (Oct 24) the beds were rolled to flatten the crop residue, and garlic cloves were planted five to six inches (12–15 cm) deep in holes made with a soil probe. (This seems surprisingly deep to me, and a slow method.) Some plots were then covered with thick straw. All were given organic fertilizers. The disappointing results were that no-till caused a 32–44 percent bulb loss, with sorghum-sudan by far the worst cover. So don't reinvent the wheel on that one. The speculation was that the cover crop residues tied up the available nitrogen. Adding straw mulch was found to be beneficial, always.

David Stern in upstate New York successfully plants into oats that have reached six inches (15 cm) tall. He cuts slots through the oats with a disc-furrower and plants the cloves in the slots. The oats continue to grow until winter-killed, and they continue to protect the garlic. Timing is obviously critical and site-dependent.

Nurse crop

Another idea is to over-sow the garlic with oats to hold the soil and reduce erosion. The oats grow in the early winter and then die at 10°F–20°F (–10°C to –7°C), and the dead plants continue to hold the soil in place. This involves sowing oats much later than you would for a good stand as a winter cover crop, because the garlic is planted first. In our climate I think the oats would not have time to make much growth at all before dying. This idea may work best in places with a longer fall moving more slowly into winter; it works for Rich Sisti in Wantage, New Jersey.

To mulch or not to mulch?

We like to roll round bales of spoiled hay over our beds immediately after planting. We come back a

Garlic scallions are a valuable early-season crop, making use of otherwise wasted small cloves. Credit: Kathryn Simmons.

couple of weeks later and free any shoots trapped by clumps of over-thick mulch. Then we leave it all alone until late February, when we start weeding (once a month for four months). In the South organic mulches help keep the soil cool once the weather starts to heat up. It is also possible to add mulch after the garlic has started to grow. This is more difficult than rolling bales across the bed, but if you have planted while it is still warm and want to allow the soil to cool before mulching, in order to prevent too much top growth before winter, this is an option. Myself, I would just plant later.

Organic mulches will protect the cloves from cold winter temperatures to some extent. It is also possible to use thick rowcover to protect garlic over the winter—even a double layer of rowcover in very cold areas—whether or not you use mulch.

Garlic scallions

Garlic scallions are small whole garlic plants, pulled and bunched like onion scallions. They are chopped and cooked in stir-fries and other dishes. They are mostly green leaves at this point, although the remains of the clove can also be eaten. Hardcore garlic lovers eat them raw like onion scallions. They provide an attractive early spring crop.

To grow garlic scallions, save the smallest cloves when you plant your main garlic crop, and plant these close together in furrows, simply dropping them in almost shoulder-to-shoulder, any way up that they fall. (If you've just finished a large planting of maincrop garlic, you'll probably be too tired to fuss with them anyway!) Close the furrow and mulch over the top with spoiled hay or straw.

You could plant these next to your main garlic patch, or in a part of the garden that's easily accessible for harvest in spring. Or you could plant your regular garlic patch with cloves at half the usual spacing and pull out every other one early. Think about quantities, though. If we double planted, we'd have over seven thousand scallions, far more than we could use. The danger with double planting is stunting the size of your main crop by not thinning out the ones intended for scallions soon enough. We plant our small cloves for scallions at one edge of the garden, and as we harvest, we use the weed-free area revealed to sow the lettuce seedlings for that week.

With a last frost date of 20–30 April, we harvest garlic scallions from early March until May, depending on how long our supply lasts out, and when we need the space for something else. Harvesting is simple, although depending on your soil, you may need to loosen the plants with a fork rather than just pulling. Trim the roots, rinse, bundle, set in a small bucket with a little water, and you're done! Some people cut the greens at 10" (25 cm) tall and bunch them, allowing cuts to be made every two or three weeks. We tried this, but prefer to simply pull the whole plant once it reaches about 7"–8" (18–20 cm). The leaves keep in better condition if still attached to the clove. Scallions can be sold in small bunches of three to six depending on size. If you do have more than you can sell in the spring, you could chop and dry them, or make pesto for sale later in the year.

Weed control

As with all alliums, removing weeds is important. Otherwise, yield can decrease by a phenomenal amount (as much as 50 percent). Because garlic is an overwintering plant in most regions, it will be necessary to kill the spring cool-weather weeds, and later kill the summer weeds.

Growers not using mulch will need to cultivate fairly frequently to deal with weeds. Hillers will deal with the between-row weeds and some of the in-row weeds, but be careful not to cover too much of the foliage as this will reduce yields. Many growers use hand hoes, and those with mulch hand-weed. Keep the leaves in good shape as best you can—take care when hoeing or cultivating. Each leaf you damage or remove will reduce the yield by about 17 percent.

Five applications of 10% acetic acid vinegar spray during the growing season has been shown to be a useful technique in controlling broadleaf weeds, but has no effect on grass weeds. Start when the garlic is 18" (46 cm) tall and spray about every ten days. Spraying from both sides of each row is the most effective. Wear a mask and gloves, as well as long sleeves and long pants, when spraying this caustic strength of vinegar. It is possible to reduce labor by 94 percent using vinegar rather than hand-weeding, so if broadleaf weeds are what you get, this is a good solution. See the 2004 SARE Grant report by Fred Forsburg in the Resources section.

Flame-weeding

Growers who prefer not to mulch need to deal with weeds sooner. Flame-weeding can achieve as good results as hand-weeding using one-third of the labor. Flame-weeding can be used for relatively mature garlic, but young plants (four or fewer leaves) are too easily damaged. The flame is directed at the base of the plants, in the morning, when the plants are turgid. Naturally, if you have used straw or hay mulch, flame-weeding is not such a smart idea!

Diseases

The major diseases are mostly fungal: white rot, *Fusarium*, *Botrytis*, rust, penicillium molds, purple blotch, powdery mildew and downy mildew. Use pre-plant clove treatments to reduce these diseases. Bacterial soft rots are also sometimes seen. Remove isolated sick plants as soon as you see them. Always remove garlic debris from the field at the end of the season, or till it in and plant a non-allium crop. In summer, soil biological life is very active, and soil organisms will quickly break down the debris.

White rot is most active below 75°F (24°C) and leads to yellowing and dying of older leaves, tipburn, destruction of the root system and rotting of the bulb. This fungus can persist in the soil for ten years, and requires assertive action to reduce the problem. A clever trick is to spray garlic extract on the soil when the temperature is 60°F–70°F (15°C–21°C) and you have no garlic growing. The fungal mycelium may grow and then die off in the absence of food. Several weeks later, garlic can be planted and will escape the rot.

Fusarium fungi usually attack plants that are under stress (in our garden it is the plants on the gravelly edge of the patch). *Fusarium* infections grow during hot weather, with symptoms similar to white rot, but slower to develop. The fungal disease produces small brown spots on the cloves, yellowed leaves and stunted browned roots. The discoloration of the leaves spreads from the tips. The main organic approaches to controlling it are good sanitation (and pre-planting treatments) as well as fostering strong plant growth.

Botrytis symptoms include "water-soaked" leaves, and can lead to bulbs rotting, sometimes during storage. This fungus grows best (worst!) in warm wet weather. Good airflow during growth, curing and storage will reduce the chances of botrytis problems.

Rust shows up initially as small white flecks on the leaves, developing into orange spots. Favorable conditions include temperatures of 45°F–55°F (7°C–13°C), high humidity but low rainfall and low light. Stressed plants are the most likely to be stricken. Infected bulbs may shrink, yellow and die. Once again, good sanitation and rotations are the organic approaches.

See the Bulb Onions chapter for more on allium diseases.

Pests

Pests include nematodes, thrips, onion maggots, cutworm, armyworm and mites. Weekly scouting is a good practice. Use pre-planting treatments against nematodes and mites. Caterpillars can be killed with Bt.

Nematode infestations show up as distorted, bloated, spongy leaves and bulbs, perhaps with brown or yellow spots. Top growth yellows and may separate from the root system.

Farmscaping (planting flowers to attract beneficial insects that feed on pests) can work for thrips, which are on the menu for ladybugs and minute pirate bugs.

Beneficial nematodes can be effective against onion maggots; ground and rove beetles, birds and braconid wasps all prey on some life stage of the onion maggot. Rowcover can exclude the fly (mother of the maggots).

Stages of growth

It is important to establish garlic in good time so that roots and vegetative growth are as big as possible before the plant turns its attention to making bulbs. The start of bulb formation (and the end of leaf growth) is triggered by days length exceeding 13 hours (April 10 here at 38°N). Air temperatures above 68°F (20°C) and soil temperatures over 60°F (15.5°C) are secondary triggers. We all have 12 hours of daylight on the spring equinox. After that, the farther north you go, the longer the day length is. Northern latitudes reach 13 hours of daylight before southern ones, but garlic does not start bulbing there at 13 hours because temperatures are not yet high enough. For example, in Michigan, bulbing begins in mid-May. In warmer areas, temperatures cause harvest dates to be earlier than in cooler areas at the same latitude. We have no control over when garlic starts to make bulbs, only over how large and healthy the leaves are when bulbing starts, and how large the final bulbs can be. Small plants here on April 11 will only make small bulbs!

Garlic can double in size in its last month of growth, and removing the scapes (the hard central stem) of hardneck garlic can increase the bulb size by 25 percent. Watering should stop two weeks before harvest to help the plants dry down.

Hot weather above 91°F (33°C) will end bulb growth and hasten maturation or drying down. It is important to get plenty of good rapid growth in before hot weather arrives.

Garlic scapes

Garlic scapes are the firm, round seed stems that grow from hardneck garlic and start to appear three weeks before harvest, as the bulbs size up. If these are removed, the garlic bulbs will be easier to braid, if you want braids from hardneck varieties. Scapes also make an early-season visually attractive crop. Contrary to ideas mentioned by some sources, leaving scapes in does not increase the storage life of the garlic. Most people who remove scapes cut them where they emerge from the leaves. We prefer to pull ours, to get the most out of them. We don't wait for the top of the scape to loop around, as the scapes will have begun to toughen and reduce the final yield of the garlic. As soon as the pointed caps of the scape have cleared the plant center, grasp the round stem just below the cap and pull slowly and steadily vertically upwards. The scape emerges with a strange popping sound and you have the full length of the scape, including the tender lower portion. Sometimes the scapes will snap rather than pull right out, but the remainder of the stem can be pulled next time, when it has grown taller. We gather into buckets, with the scapes standing upright, so we can put a little water in the bucket and the scapes are aligned, easy to cut up. They will store well in a refrigerator for months if needed. Late morning is a good time to pull scapes (or early afternoon).

Garlic harvesting is a good job for a group, although it is important to train everyone clearly to avoid damage to the bulbs. Credit: Marilyn Rayne Squier.

The wound heals over in fifteen to twenty minutes in the heat of the day, whereas otherwise it could drip for up to 24 hours, increasing the risk of disease, and losing water from the plant.

We harvest scapes two or three times a week, for about three weeks in May. The crew always enjoys this task, partly because it's a stand-up job and partly because we encourage a friendly competition to see who can get the longest scape of the day. This encourages everyone to perfect their technique too. Scapes can be chopped and used in stir-fries, pesto, garlic butter, pickles and other dishes in place of bulb garlic. They can also be frozen for out of season use. Searching the Internet will reveal lots of recipes. Scapes sell in bunches of six to ten. One acre (0.4 hectare) of hardneck garlic can produce 300–500 lbs (140–225 kg) of scapes.

Harvest date

With hardneck garlic, scapes will start appearing about three weeks before the bulbs are mature.

(Day length as well as accumulated degree days determines when scapes appear as well as when bulbs are ready to harvest.) This is a good time to pay more attention to your garlic crop, and what better way than walking through pulling scapes? We take the opportunity to remove any diseased plants from the patch at the same time. Three weeks before the expected harvest, we remove the mulch to help the bulbs dry down, and to prevent fungal diseases. In our rotation, the spring broccoli is usually next door to the garlic, and we move the old garlic mulch to the broccoli to top up the mulch there. It helps us stay on track with getting the broccoli weeded too.

Our garlic is mature in early June, or even at the end of May. It has less than 50 days in which to grow the bulb. In cooler regions, it is possible to plant garlic in spring, as the bulbing conditions are not reached until later in the year.

While harvesting scapes, monitor the plants for signs of maturity. Garlic is ready to harvest when the sixth leaf down is starting to brown on

50 percent of the crop. See Ron Engeland's excellent book for more on this. For some years I was confused about which was the "sixth leaf," and I confess that I was counting up instead of down. The point is to have five green leaves still on the plant, to provide the protection of five intact skins over the bulb. Each leaf corresponds to one wrapper on the cloves or bulb; as the leaf dies, the skin rots away. Keeping five intact skins on the garlic is a challenge in our humid climate, and because we are not shipping our garlic anywhere, it seems less crucial. So I also use a second method of deciding when to harvest: I pull three or four plants and cut the bulbs across horizontally and look at the center of the bulb. When air space becomes visible between the round stem and the cloves, it's time to harvest. Usually that's June 7–June 14 for our main crop of hardneck garlic, but it has been as early as May 30, and as late as June 18. Harvesting too early means smaller bulbs (harvesting way too early means an undifferentiated bulb and lots of wrappers that then shrivel up). Harvesting too late means that the bulbs may "shatter" or have an exploded look, and not store as well.

It is possible to take apart any small or shattered bulbs and replant them immediately after harvest to grow more scallions. They won't start growing immediately if the weather is really hot, they'll just stay dormant until cooler weather arrives. We don't do this, because we're busy enough at that time of year.

Harvesting

Growers with large amounts of garlic use a tractor-mounted undercutter to loosen the bulbs, or a root-harvester to completely dig them up. Subsoilers, European leek harvesting machines, or homemade undercutters fashioned from an old snow plow blade bent into a rectangular shape, have all been used. We don't have an undercutter so we harvest by hand, with digging forks to loosen the soil.

In drought years we've needed to use some overhead irrigation the evening before, to loosen the soil enough to harvest without damage to the bulbs. When harvesting, it's important to treat the bulbs like precious sun-sensitive eggs! Bruised bulbs won't store well, nor will sun-scalded ones. It's better not to wash the bulbs when harvesting, as drying is what's needed. We shake off the soil, without banging the bulbs, and harvest into buckets to keep the bulbs shaded. We also use buckets because they can more easily be carried to our curing area, the upstairs of an old tobacco barn. Others might use crates. Some growers sort in the field into small, medium and large bulbs (and compost material) and cure the three sizes separately. If it's hot, as it is where we are, I recommend getting the garlic out of the field quickly, hanging it up and getting it drying, leaving sorting to do later (indoors!). Don't let garlic get above 121°F (49°C) as it will start to cook.

Curing

Depending how hot it is outside, the indoor job of hanging the garlic to cure as it comes in from the field may be more or less preferred. It takes us several morning shifts to get our 3,200 row feet (975 m) of garlic harvested and hung up. In cooler climates some people cure garlic outdoors, but ours would bake! In less humid climates, people don't need to pay as much attention to airflow as we do. Some growers tie the plants in loose bundles of about eight to twelve plants and hang the bundles under cover. If you can size the bunch so it ends up around one pound (500 g) in weight,

This close-up shows how we put one garlic plant in each hole of the vertical netting to cure. Credit: Marilyn Rayne Squier.

you may save yourself a task later. Whatever method you are using, get the garlic spread out immediately. Don't leave it in plastic containers where the heat and moisture will incubate fungi!

We prefer to hang our garlic in netting around the walls of the upstairs barn. The netting is nylon, with a diamond mesh about two inches (5 cm) wide. We start at knee height and thread a bulb in each diamond, by bending the tops of the leaves and feeding them through the space. The weight of the bulb causes the netting to stretch downwards and hold the garlic in place. People take a section of netting and work upwards in rows, back and forth, covering the walls in garlic. Yes, sometimes we have overloaded the netting and had the nails pull out of the walls! We try to start out by adding nails and rope to hold the netting up before we start harvesting. A wall of garlic roots is quite a sight! We use fans to move the air, which you should consider if your climate is also humid.

Snowfencing (slats and wire, or the plastic kind) can also be used to hang garlic. Or you can make horizontal racks and lay the garlic on top.

People who intend to braid their softneck garlic will need to start on the braiding within the first week of curing, before the leaves become too brittle. They will also need to clean their garlic.

We leave the garlic to cure for two to four weeks. Meanwhile we till the old garlic area and sow buckwheat. We have about six or seven weeks before we'll use these beds to sow our fall carrots at the very beginning of August.

Snipping and sorting

We test the curing garlic by rolling the neck of a few sample bulbs between thumb and forefinger. If it feels dry, rather than moist, it's ready. We use scissors to cut off the roots close to the bulb and the tops a quarter to half an inch (5–10 mm) above the bulb. Some growers brush any mud-covered bulbs with toothbrushes, but we find enough of the dirt drops off during storage to save us this tedious task.

We measure the bulb and assess whether it's a good bulb for seed garlic, or whether to eat it. We save for seed all bulbs 2"–2.5" (5–7 cm) in diameter, with an even shape and cloves that are tight together, not opening up. We have many marks on posts, chairs and fans around the barn, for people to measure the bulbs against. Of course, after a while people only need to measure the borderline ones, as they've developed a sense of what's too small or too big. Some growers use jigs with two foam-lined battens tapering towards each other on a board to measure sizes.

The reason we don't save all the biggest bulbs is that they tend to be uneven ("rough") in shape and quality, with cloves of all sizes. We really value large cloves! We put the seed bulbs in green net bags and the eating ones in red net bags ("Green for Growing"). We also have a "Use First" category for non-storable bulbs, and compost buckets for all the tops and roots and any disasters. If we drop a bulb on the floor, we make it a "Use First" as the bruising would probably cause it to rot in storage. We weigh the filled bags for vanity reasons and to monitor the amount we are saving for seed. We have been selecting our seed garlic in this way for many years now and have no trouble getting plenty of seed garlic, so once we have enough we stop measuring and selecting and simply use all red bags.

Storage

The seed garlic goes on a high shelf in the garden shed, at quite variable ambient tempera-tures, and does fine until early November, when we plant it. The ideal storage conditions for seed garlic are 50°F–65°F (10°C–18°C) and 65%–70% relative humidity. Storing in a refrigerator is not a good option for seed garlic, as prolonged cool storage results in "witches brooming" (strange growth shapes), and early maturity (along with lower yields). Storage above 65°F (18°C) results in delayed sprouting and late maturity. The eating garlic is stored in a dry, coolish basement at 60°F–70°F (15.5°C–21°C) over the summer. In late September or sometime in October we move our eating garlic from the basement to the walk-in cooler at 35°F–38°F (1.5°C–3°C) to make space available for the winter squash harvest. By this time most of the apples have gone from the walk-in cooler, and space is available there. Also there is no longer the problem of ethylene emitted by the apples, which causes garlic to sprout. Ideally they would never be in the same storage space. Do not store peeled garlic in oil, as garlic is low in acidity and the botulin toxin could grow.

I hear that garlic can be stored for up to nine months at 27°F (–2.7°C), but I have not tried that myself. It does not freeze until 21°F (–6°C). At 32°F (0°C) it will store for six to seven months. There is a middle temperature range of 40°F–56°F (4.4°C–13°C) at which garlic should not be stored, as this encourages sprouting. This is another reason why we move garlic out of the basement in the fall—temperatures there are dropping below 56°F (13°C).

Chapter 46

Bulb Onions

Onions are cool-season plants. They have three distinct phases of growth—vegetative, bulbing and blooming—and the switch from one to the next is triggered by environmental factors. It does not work to plant onions at a random date in the year without taking account of these environmental factors. Success depends on understanding what this crop needs during each of the three phases, so I'll start by describing that.

Vegetative phase

The first stage is vegetative growth (production of roots and leaves). To grow large onions it is important to produce large healthy plants before the vegetative stage gives way to the bulbing stage. Each onion leaf represents one ring of the future onion bulb: more leaves means more rings. The larger the leaf, the fatter the ring becomes. If plants are small when bulbing starts, only small bulbs are possible. Cool, but not cold, weather and adequate irrigation encourage heavy leaf growth.

It's important to grow varieties that are suitable for the latitude of your farm. The further north you are, the more hours of daylight you have in summer. Onion varieties are often described as "northern/long day" or "southern/short day." There are also "intermediate day" types and a few genuinely "day neutral" varieties. The names refer to the relative day length needed before the plant will start to make a bulb. The dividing lines between short day (south of 35°N) and long day (north of 38°N) varieties leave a gap where neither type is ideal. Living at 38°N, we have had the opportunity to learn about onions by many trials and many errors. Your latitude can be found from maps or on weather forecasting websites. I found convenient sunrise and sunset tables in the hunting regulations from my Extension Service.

Bulbing phase

Bulbing is initiated when the daylight length reaches the number of hours critical for that va-

A bed of young onion plants. We are trialing several varieties to determine the best ones for our farm. Credit: Kathryn Simmons.

riety. (Actually, according to Dixondale Onion Farms and other sources, it is the length of darkness that is important, although varieties are normally classified by their day length.) Temperature and light intensity also determine when vegetative growth stops and bulbing starts. It takes a daily average temperature of 60°F (15.5°C), or even 70°F (21°C), to trigger bulbing (depending on the variety).

The rate of bulbing is more rapid with high light intensity and increased temperature. The optimum temperature for rapid bulb development is 75°F–85°F (24°C–29°C).

Long day onions are bred to start bulbing at 14–16 hours of daylight, depending on the particular variety. The pungent storing onions tend to be long day onions, so if you live north of the 38° N, you can happily grow these kinds. In cooler northern regions, the day length trigger may be reached before the temperature trigger. In these places, bulbing is delayed until warmer weather. Onions then bulb during the summer and are harvested in the fall. Further south, the temperature trigger is reached before the day length trigger, so bulbing starts as soon as the days are long enough, and finishes early in the summer.

Here at 38° N, our longest day (summer solstice) has 14 hours and 46 minutes of daylight. We have 14 hours of daylight six weeks earlier, on May 6. A few varieties of long day onions can be grown here, but those requiring 15 or 16 hours of daylight will never form bulbs at this latitude. Catalogs are starting to indicate how many daylight hours each variety needs, which is helpful.

South of their ideal growing region, long day onions don't start bulbing until day length-triggered near summer solstice, and the bulbs are exposed to hot conditions as they mature. Soils dry out fast under hot conditions, and if water supply is insufficient, growth will be stunted. The leaves may die, and the bulbs can get sunscald or even start to bake if temperatures are 90°F (32°C). Our summers here are humid and we contend with a lot of fungal diseases. Drip irrigation (so leaves stay dry and don't get diseased from overhead watering) help the 14-hour long day onions survive here at the southern end of their range, through our non-ideal conditions in June and July. We want tasty, pungent, storing onions, so we try to extend the range of long day onions as much as we can.

Short day onions start to bulb at 10–12 hours of daylight, provided temperatures are warm enough. In the South, below 35° N, they are sown in September or October, grown through the winter, and are harvested in May. If short day onions are grown too far north (where they cannot be overwintered and must be started in spring) they will bulb before much leaf growth has occurred. Only small bulbs can result from this. We reach 10 hours of daylight on January 21, 11 hours on February 20 and 12 hours, naturally enough, at the spring equinox, March 21. Here, bulbing initiation for short day onions gets delayed beyond the day length trigger, until temperatures are higher than 60°F–70°F (15.5°C–21°C), which is early April. It's a waste of time to sow short day onions here in spring, as they have an impossibly brief time (January to early April) to grow a decent-sized plant before bulbing starts. One way to deal with this is to start the seedlings in the late fall/early winter, let them make some vegetative growth, and keep them alive indoors over the winter, to continue growth in the spring. Further north or inland, it gets harder to keep onion plants alive through the winter. Onions can survive at 20°F (–7°C), but not under colder conditions. We have tried overwintering onions outdoors, but even with rowcover, our losses were too high.

Short day onions are mild flavored and do not store. What makes an onion sweet is the same as what makes it not store well—high water content. Where they can be grown, short day varieties can provide early season onions for immediate use. Cipollini, also known as pearl or boiling onions, are varieties of short day onions sown in spring, planted at high density, which form small bulbs and mature in a couple of months.

Intermediate day onions start to bulb at 12–14 hours of daylight. (We reach 13 hours on April 10 and 14 hours on May 6.) As yet there are not many intermediate day varieties. They are relatively sweet (not as pungent as long day onions) and mostly do not store well. This is not a concern for growers selling all their crops soon after harvest, but my goal is to provide onions for as much of the year as possible.

Day neutral onions bulb independently of day length. Candy is the only one I've found that stores at all. It's a medium-sized onion. Superstar is another day neutral variety, but we don't grow white varieties as they are prone to sunscald.

Blooming phase

Onions are a biennial plant, which means that they normally grow leaves and produce a bulb in the first year; in the second year they send up more leaves, followed by a thick central flower stem. When onion plants experience an extended period of cooling temperatures, such as winter, they go dormant. When temperatures rise, they start growing again. After being exposed to cold temperatures, smaller seedlings with a diameter less than pencil thickness (⅜" or 1 cm) and fewer than six leaves will resume growth and not usually bolt (bloom). The trigger for the transition from bulbing to flowering is temperatures below 50°F (10°C) for 3–4 weeks, after the plants have six leaves or more. This can happen, if you are unlucky, after an unexpected cold period in spring. Bolting is to be avoided because the flower stems are tough and inedible, and the bulbs start to disappear to feed the growing flower stems. Bolted onions will not dry down to have tight necks and so will not store.

It is possible to sow onions in the fall and plant the seedlings out in the early spring, for bigger vegetative growth and therefore the chance of bigger bulbs. The temperature-and-size trigger limits how early in the fall seeds can be sown; if the seedlings have made lower stems larger than a pencil in diameter when winter closes in, the plants are likely to bloom in the spring rather than forming bulbs. A few onion plants will likely always bolt, especially if the spring is long with alternating warm and cool spells.

Starting seedlings in a hoophouse in early November works well for us. Previously we sowed outdoors in late September and protected the plants with rowcover and coldframes, a system that would work fine somewhere warmer than Zone 7. The hoophouse works much better for us because the plants get much better airflow, are protected better from very cold temperatures and can be easily seen and cared for. The plants grow faster in the hoophouse than outdoors, so we can start them later. Outdoor sowings tend to suffer some winter killing and varying degrees of mold. The colder the temperatures the plants experience, the more likely it is for the larger ones to bolt before growing large bulbs. Hence a more moderate microclimate, such as a hoophouse, reduces the rate of bolting. In colder zones, a slightly heated greenhouse might work better for overwintering.

Onions are outbreeding plants, that is, they cross-pollinate each other. Open-pollinated varieties can suffer from "inbreeding depression": saving seed from too small a population can cause problems with vigor and yield. Inbreeding depression is caused by the reduction of diversity in the genetic material. The qualities that make onions susceptible to inbreeding depression also make them benefit from the hybrid vigor resulting from crosses.

Sets

Onion sets are small dormant bulbs in arrested development that are planted in spring in hopes of growing bulbs. Often when sets are sold, the variety is not stated. Onions grown from sets are more likely than transplanted seedlings to go to seed before forming a bulb, because they are in their second year and are big enough to have flowering triggered by low temperatures. Storing sets before planting needs to be at 32°F–35°F (0°C–2°C) or above 65°F (18°C). Store-bought sets may well not have had suitable storage before you buy. We have given up on sets for our farm. In areas where spring temperatures have a steadier progression, sets can make renewed vegetative growth before temperature and day length trigger bulbing. The likelihood of cold temperatures after the sixth leaf stage is smaller. Further south, it may be possible to plant sets in the fall.

Seeds

Use fresh seed of a variety suitable for the latitude and the time of year (fall or spring sowing). Store seed in an airtight container in a cool place. Seed from the previous year is as old as you should go. Yellow onions are easier to grow than whites or reds, and have tougher skins. Whites are more susceptible to sunburn, and to green streaks if rained on near harvest. Reds are slower growing and therefore tend to make smaller onions. One gram (225 seeds) equals 75' (23 m) of transplants. The approximate yield is 100–150 lbs/100' (15–23 kg/10 m).

Timing and choice of variety

North of 38° N: Use only long day or intermediate day varieties, as short day varieties will bulb before reaching a good size. Sow indoors in late winter (12 weeks before likely suitable weather for transplanting). If you can keep seedlings above 20°F (−7°C), sow indoors in late fall. In mild coastal locations, it may be possible to sow outdoors in the late summer or fall. Transplant in spring, but avoid having transplants thicker than a pencil.

Between 35 and 38°N: Sow suitable long day varieties (ones which bulb at 14 rather than 16 hours of daylight) indoors from November to early January; or outdoors in a mild spell in January or February, 12–16 weeks before the frost-free date. Use rowcover. Sow intermediate day varieties in mid-October–late November (or January if necessary). If temperatures don't go below 0°F (−18°C), some intermediate day varieties will survive outdoors (e.g., Walla Walla or Juno, which have good cold tolerance). Sow short day varieties in the fall, but keep seedlings from getting too cold (e.g., sow in greenhouse and transplant in February). Choose only the most cold-tolerant short day varieties. Short day onions will harvest May–June; long day varieties June–July.

South of 35° N: Sow in the fall (Sept–Oct), using only short or intermediate day varieties, as long day varieties will likely get scorched by hot weather before finishing bulbing, and may never make decent bulbs if the longest day is still too short for that variety. If you have drip irrigation and some afternoon shade, you could experiment with long day varieties if you are between 32° N and 35° N. South of 32° N, stick with short day varieties.

Sowing

Sow seeds a quarter to half inch (6–13 mm) deep, a quarter to a third of an inch (6–8 mm) apart if you plan to transplant later. Sow every two inches (5 cm) if direct sowing. Keep the soil damp, as the first roots will be very near the surface. Use a temperature of 57°F–86°F (14°C–30°C) to germinate. The ideal is 75°F (24°C). Use rowcover if there's a very cold snap. If you sow too late you will get small bulbs; if too early your onions will bolt.

Onion seedlings take 3.6 days to emerge at 77°F (25°C) and 7.1 days at 59°F (15°C). Onions will still germinate at 95°F (35°C), but take 12.5 days and only produce 73 percent normal seedlings. Our November sowings in the hoophouse, at lower temperatures, take 10 days to come up.

Soil

Rotate your onion beds with at least three years between alliums. A pH of 6.0–6.5 is ideal. Onions need fairly rich soil, moist but well drained, with plenty of organic matter and a loose crumb structure. Waterlogged soil promotes disease, so make raised beds if your soil does not drain well. Onions need 145 lbs/ac (168 kg/ha) of nitrogen during the growing season. The pungency of onions relies on soil sulfur. Ensure adequate potassium or the onion necks will not dry down well, and the onions won't store. Onion plants compete poorly with weeds, so choose an area with low weed pressure (from the weed seed bank left from the previous year).

Transplanting

Transplant fall-sown onions as early in spring as possible, and those sown after New Year once they have at least three leaves (four or five is better). The final bulb size is affected by the size of the transplant as well as the maturity date of that variety. The ideal transplant is slightly slimmer than a pencil, but bigger than a pencil lead. Onion seedlings are slow-growing: even in spring they can take ten weeks to reach a size suitable for transplanting. Overly large transplants are more likely to bolt. If seedlings are becoming thicker than a pencil before you can set them out, undercut two inches (5 cm) below the surface to reduce the growth rate. Onion roots are tough and thick, not thread-like.

Some books recommend trimming the tops at transplanting time, but I believe it reduces the yield. Transplant 4" (10 cm) apart for single seedlings or 12" (30 cm) for clumps of three or four (not more than four). Set plants with the base (stem plate) ½"–1" (1.3–2.5 cm) below the soil surface. Some books recommend as deep as two inches (5 cm). Don't plant too shallowly. Give plenty of water to the young transplants: keep the top 3"–4" (8–10 cm) of the soil damp for the first few weeks to prevent the stem plate from drying out.

Caring for the crop

Keep weeds under control—onions do not compete well with weeds as their leaves have a relatively small surface area and easily get shaded out. Yield is reduced 4% per day by weeds, or 50% in two weeks. Some growers use plastic mulch. Those growing on bare ground cultivate frequently (shallowly, as the roots are near the surface). In colder areas, some growers add straw or hay mulch after the soil has warmed a bit in spring to combat weeds; in warmer areas this would encourage too many fungal diseases, and is not helpful. Unmulched crops can be flame-weeded; see ATTRA for details.

The ideal is to grow at least 13 leaves before bulbing starts. Give one inch (2.5 cm) of water per week until the leaves start to die back. If needed, side-dress to provide extra nitrogen in spring while the plants are still growing.

When the leaves begin to brown, consider drawing some soil away from the bulbs to help them dry out. Do this gradually, once a week, two or three times. If the weather is very hot, drawing soil away from the bulbs might not be a good idea as it could lead to sunscald. Start to restrict watering, as far as you have control over it. Do not break the tops over as some sources suggest—this can harm the storage quality and encourage fungal diseases.

Pests

Rowcover can be used to exclude pests. The most likely problems are root maggots, aphids and thrips. To control thrips, attract ladybugs and lacewings by planting flowers they are attracted to. Mow the field edges to reduce thrips habitat. Onion maggots (larvae of the onion fly) are attracted by the smell of rotting alliums, therefore do good crop clean-up. The small grey-brown onion fly lays eggs at the base of the onions. When the larvae hatch, they feed on the onion roots. There can be three generations each year in Virginia. To disrupt the onion maggot lifecycle, consider an early spring allium-free period (if you don't grow garlic) or a July/August gap (if you don't grow leeks).

Harvest

Be patient and watchful. If harvested too early, the neck area will not have softened enough to allow shrinkage and tightening and bacteria will enter, causing the center of the bulb to rot.

In the South, when 30–40 percent of the tops of that planting have died, lift the onions gently and put in partial shade to cure. Leaving bulbs in the ground too long risks sunscald as well as fungal and bacterial diseases. Bruising can seriously

Onion Diseases Chart

Disease	Type	Symptoms	Likely Conditions	Longer Term Effects	Prevention and Control
Downy Mildew	Fungus *Peronospora destructor*	Yellowish spots on upper half of leaves, soon developing to a furry, bluish gray mildew. After two days, tops weaken and yellow.	Midseason, poor drainage, poor air circulation. Damp weather.	Winters between scales (layers) of onions. Breeds in soil. Wind-borne. Bulb becomes spongy and of poor keeping quality.	Practice good crop sanitation, especially fall cleanup. 4–5 year rotation.
White Rot	Fungus *Sclerotium cepivorum*	Yellowing and dying back of leaves, from tips down. Roots rot away. Semi-watery decay of scales. Base of bulb has fluffy white mycelia, later black spores.		Breeds in soil. Not seed-borne.	Destroy infected plants. 4–5 year rotation.
Purple Blotch	Fungus *Alternaria porri*	Greyish, deep lesions with dark centers, on leaves. Dark centers spread into purplish blotches which can spread and kill the plants. Enters neck area usually, not always. Semi-watery rot of scales, which darken and dry up.	Warm humid weather may contribute. Starts when onions are close to maturity. Also occurs in storage, from bruising during harvesting.		Waxy foliaged varieties (*Danvers, Red Creole, Abundance*) are more resistant than "glossy" ones (*Sweet Spanish, Grano, Bermuda*)
Pink Root	Fungus *Phoma terrestris*	Roots turn pink, shrivel and die. Plants' food supply is cut off, so bulbs are very small.	Southern and western US. Throughout growing season, all stages of growth. Attacks other crops too.		Resistant varieties (*Creole, Grano, Granex*). *Sweet Spanish* has some resistance. Never replant in infected areas.
Neck Rot	Fungus *Botrytis alii*	Starts as softening of scales at neck and works down. Can start at a wound or bruise. Mushy "cooked" areas later become sunken and covered with greyish feltlike mat or powdery mass. Sometimes there are black spores on and between scales.	Infestation occurs during growth, but manifests and worsens after harvest. Moist, cool conditions. Air-borne. Most susceptible time is just before harvest. Can be caused by premature harvesting.	Very serious. Spores live on dead onion foliage.	Remove onions at first sign of trouble. Burn infected plants—the fungus can survive freezing. Prevention: 1) Don't bend onion tops over; 2) Avoid bruising; 3) Protect bulbs from rain and dew while curing; 4) Provide good ventilation in storage. 32°F (0°C) or just above. 65% humidity.
Basal Rot	Fungus *Fusarium oxysporum f. sp. cepae*	Semi-watery decay starts at base of bulb and works up, totally destroying bulbs. Damage is already underway when symptoms are seen—yellow leaf tips—all foliage is dead in two weeks.	Appears as soil warms in spring. Widespread infection enters damaged tissues. Worse damage in warmer weather.	Soil-borne.	4–5 year rotation. Harvest before soil gets too hot. Handle bulbs gently. Burn infected plants.

Onion Diseases Chart (cont'd.)

Disease	Type	Symptoms	Likely Conditions	Longer Term Effects	Prevention and Control
Smut	Fungus *Urocystis cepulae*	Dark swollen area on first leaf, spreading to later leaves as they appear. Bulbs, if any, have black blisters between outer scales. Plants die before producing bulbs, within 3–5 weeks of infection.	Above 36°N. Cool weather. Direct-seeded crops only. Soil-borne. Appears soon after germination.		Use transplants (plants beyond seedling stage have complete immunity).
Black Mold	Fungus *Aspergillus niger*	Starts at sites of injury as patches of black mold in the outer skins, especially along the veins. Progresses inwards, leaving sooty black mold between the scales (layers). Later the skins shrivel. May also be followed by secondary bacterial soft rot.	Warm or hot climates, both humid and dry. Grows in live or dead plants. Soil-borne. Seed-borne.	Infection can also grow in storage, in onions that appear fine from the outside. Black spores appear. Can cause respiratory infections in people.	White varieties are resistant. 4–5 year rotation. Avoid injuring onions during cultivation. Avoid bruising onions at harvest.
Smudge	Fungus *Colletotrichum circinans*	Infection confined to unpigmented tissues of neck area. Begins with small green and black dots, often with concentric rings on outer scales. Later penetrates fleshy scales and develops into deep sunken areas on bulb.	Central and NE states.	Considerable damage in stored onions—bulb shrinkage and premature sprouting.	White varieties susceptible. Others resistant. 4–5 year rotation. Practice good crop sanitation.
Damping Off	Fungi *Fusarium, Pythium, Rhizoctonia solani*	Sudden development of dark rotted area on root and lower stem along soil line. Seedling topples and dies. Sometimes seed rots before germination.	Spreads more rapidly in cool cloudy weather, in thickly planted onions and in still air.	Can wipe out seedbed overnight.	Don't over-water. Water in morning. Keep good air circulation by weed control and appropriate spacing.
Soft Rot	Bacteria *Erwinia carotovora* subsp. *carotovora*	Not always evident before harvest—develops rapidly if storage temperature too warm. Begins at neck, works down, mostly in center. Affected area has water-soaked glossy appearance, becomes mushy with a foul odor. Often only 1 or 2 scales affected—the rest may be salvaged for processing.	Warm humid temperatures. Biting pests or other injuries introduce bacteria. Very destructive and common in most soils.	Destroys onions in storage. Long lived in soil. Although it develops rapidly, it does not spread rapidly to other plants.	Cull out damaged bulbs. Cure and store in dry, well-ventilated place. 4–5 year rotation.
Sour Skin	Bacteria *Pseudomonas Burkholderia cepacia*	First symptom is glazed or water-soaked appearance on outer scales. These then disintegrate into slimy yellow mass with sour smell. Skin slips off, leaving center of bulb firm and salvageable for processing.			Remove affected bulbs from storage to prevent spread of disease.

Onion Diseases Chart (cont'd.)

Disease	Type	Symptoms	Likely Conditions	Longer Term Effects	Prevention and Control
Aster Yellows	Virus	Stunting, yellowing foliage, elongated pedicels, distorted flower heads.	Spread by 6-spotted leafhopper.	Also affects other vegetables—lettuce, carrots/umbelliferae, tomatoes, spinach. Infected bulbs shrink and sprout in storage.	Choose resistant varieties (*Sweet Spanish*). Pull up and burn infected plants. Kill aphids (use farmscaping).
Yellow Dwarf	Virus	Appears as first leaf emerges. Short yellow streaks at base. All later leaves affected. Later whole plant is yellow, wrinkled, twisted, droopy.	Spread by aphids, but not other pests. Also spread by tools and hands. Most severe in potato onions.		As for Aster Yellows.
Leaf and Tip Blight	Environmental	Tips of older leaves become bronzed as if heat seared. Affected leaves then brown and die.	Prolonged rainy periods and cool, cloudy weather. Thickly planted onions.		Prevention: use wider crop spacing.
Blast/ Leaf Scorch/ Sun Scorch	Environmental	Succulent foliage undergoes rapid dehydration and dies.	Follows abrupt change from warm, moist period to a hot, dry one.		Irrigate more when hot dry weather follows abruptly after a damp spell.
Sunscald	Environmental	Affected tissues appear bleached and slippery. Dries to become leathery.	Hot sunny south. Immature bulbs most susceptible. Also, most common after harvest, while curing outdoors.	Damaged plants are later susceptible to soft rot, neck rot, and other storage diseases.	Don't harvest during high temperatures and bright sunlight. Keep bulbs of growing plants covered with loose soil until harvest. Cover bulbs with tops if curing outdoors.

damage the bulbs, as can stressing the necks, so do not pull the onions. If they don't lift easily from the ground, use a spading fork (digging fork, not hay fork) to raise them. One key to organic production is to handle the bulbs in ways that minimize physical damage. Handle and place onions gently. Think eggs, not tennis balls!

In the North, where intense sunlight and humidity when onions mature is not such an issue, it is fine to wait until most or all of the tops have fallen, then lift the onions and dry in the field for two to three days, as long as heavy rain is not expected. Compost all onion crop residues. Don't leave any residue in the onion bed.

You can start eating the onions fresh right away, and cure the rest.

Curing

Curing takes two to three weeks. Some books recommend curing in the outdoor sun. These books are written in the North. This doesn't work in the mid-Atlantic, where we need to provide partial shade, moderate temperatures and good airflow (and no rain). Spread the untrimmed harvested bulbs in a single layer in a warm dry place, and check every few days. The ideal conditions are 85°F–90°F (27°C–32°C) with constant air movement and no strong direct sunlight. We use racks in a barn, with fans to keep the air moving. Further south than us, in Georgia, onions mature much earlier than they do here (before the sun is too intense and the humidity too high) and they usually can be field cured, if the weather cooperates.

Storing

Rub the neck between finger and thumb to detect any remaining slidey slipperiness, which would indicate that the onion is not yet fully cured. When the necks are dry, clip the tops to one to two inches (2–5 cm) and gently brush off dirt and loose scales. Minimize the removal of skin, though, as the skin serves to protect the onion in storage. Remove any onions that are not curing properly, for immediate use or processing.

Bag the cured, trimmed onions in nets and label with the variety name (to ensure less storable varieties are used first and to compare storage quality).

Store in a dry place at either 60°F–90°F (15°C–32°C) or 32°F–40°F (0°C–4.5°C). Avoid 45°F–55°F (7°C–13°C), and do not move onions

Curing onions at Twin Oaks on a netting rack. Credit: Wren Vile.

from cold storage into warm storage or you will promote sprouting. You could use a barn, shed or basement at first, until the temperature there drops too close to 55°F (13°C), and then move the onions to a refrigerated cooler. If at all possible, do not store onions with fruits, including squash, as these exude ethylene, which also promotes sprouting.

Chapter 47

Potato Onions

Credit: Southern Exposure Seed Exchange

Potato onions are a type of multiplier onion, a traditional heirloom crop that also has curiosity value for market booths, CSAs and growers themselves. Gourmet chefs can provide a local demand. The National Gardening Bureau claims that multiplier onions can produce a larger yield per area than any other vegetable except staked tomatoes. Other types of multiplier onion include the Egyptian walking onion (topset, tree onion) and shallots. Potato onions are also known as hill onions, mother onions and pregnant onions. Topset onions are more cold-tolerant than hardneck garlic, which in turn are more cold-hardy than potato onions. Shallots are less hardy again (I have lost those mid-winter in Virginia).

Planting a mix of sizes of potato onions ensures that the grower gets both an edible crop and seed bulbs for the next year. Barring disaster, you need never buy new seed stock. Potato onions do not start from seeds, so the whole job of starting seeds and transplanting is avoided. This crop has big appeal to those interested in local food, sus-

tainability, permaculture and perennial crops. It means independence from Monsanto and other large seed conglomerates!

The larger seed onions are planted in the fall. They grow and divide to produce a cluster of small bulbs, which can be replanted the next fall or winter. Small onion bulbs are planted in late fall or winter and most simply grow larger and produce a single bulb. Some small bulbs may grow and divide into a cluster of medium-sized bulbs. The largest bulbs are the food crop and the smaller ones are mainly a seed crop.

Another advantage of including potato onions in your "crop portfolio" is a spreading out the workload—planting happens in fall, winter or early spring, and given good mulch, little work is needed until harvest. We harvest our maincrop hardneck garlic at the very beginning of June, then our potato onions, followed by our bulb onions, which start about two weeks later. Potato onions provide some early onions for sale and, as I will explain, it's good to get the largest of your potato onions eaten right away, as they don't store well. The rest of the crop is spread out on racks to slowly mature, and can be picked through for more eaters/sellers once a month or more. The curing and sorting stage is a low-stress job, and the work can be done on rainy days (we don't seem to get those any more), or during the heat of the day to help you stay out of the sun.

Yields can be three to eight times the weight of the seed stock, depending on growing conditions. Individual bulbs can be grown indoors in a pot to produce a steady supply of green onions during the winter.

Varieties

There are various regional varieties, and I have grown a red one in the past. Now I grow the most common one, the Yellow Potato Onion from Southern Exposure Seed Exchange, and I have grown some as a seed crop for them. Their Multiplier Onions Starter Package includes a four-page "Garlic and Perennial Onion Growing Guide." The guide is also available on their website.

Crop requirements

Required soil conditions for this crop are the same as for other onions. The main distinguishing factor is that the ground needs to be available in fall, so avoid sowing winter cover crops where you intend to plant. In our area, we can grow one or more food crops that mature by the frosts in mid-October, then prepare the soil and roll out hay mulch to prevent weed growth. At planting time we roll back the mulch, plant the onions, put the mulch back on top, then go back to hibernation-with-seed-catalogs. After the potato onions are harvested in early June, so long as we keep the soil fertility up by spreading compost, we can plant

more food crops (lettuce, fall greens, late carrots, beans, squash or cucumbers), and so get high productivity for those two consecutive years.

Planting

The smallest potato onions—below 1.5" (4 cm) diameter—store longest, and are best planted after the winter solstice and the hardest weather, as the small plants are more prone to winter-killing if planted too early. We plant ours in a mild spell in late January or early February, 4"–5" (10–13 cm) apart. This way not everything is lost if we have an unusually cold winter. Larger onions, conversely, are best planted earlier, or they will sprout and rot before you get them in the ground. We now plant our largest ones—over 2" (5 cm) in September, 8" (20 cm) apart and our medium-sized ones—1.5–2" (4–5 cm) diameter—in mid-November, at 6" (15 cm). We have tried other dates and other spacings, and this is what seems to work best here. For comparison, we plant our hardneck garlic November 3–6. Plant in October in northern areas; December in southern areas. Florida and south Texas are too hot.

If you are the pioneer for this crop in your area, and have no old-timers to consult, you'll probably have to experiment a bit at first. Always save some for spring planting, even though yields from fall planting are significantly higher. Don't risk having all your onions killed by unusually cold weather. Initially, you might want to maximize the yield per onion from your bought-in starter pack by using wider spacings, and getting larger onions. The following year you can plant these large ones and get lots of smaller ones.

We draw furrows deep enough to cover the onions with one inch (2.5 cm) of soil, set them by hand, then hoe to cover and roll out hay mulch.

As they grow, larger potato onions divide and form a cluster of small and medium onions. Credit: Kathryn Simmons.

The onion shoots will make their way up through the mulch, and we find it worthwhile to "rescue" any shoots that are trapped under over-thick clumps of hay, once it looks like most of them should be emerging. Potato onions can withstand subfreezing temperatures in every area of the continental US when suitably planted and mulched. For spring planting, leave the top third of the bulb exposed, to reduce the chance of rot.

Pests and diseases

Potato onions have been trouble-free in our garden. Apparently they are less prone to attacks by the onion fly, compared to onions from seed. Certainly they are no more prone to pests or diseases than other onions while growing, and they dry down and cure in Virginia humidity better than most regular bulb onions.

Harvest

When the tops start to die, lift the onions gently and take them to cure. If planted at the correct depth, the bulb clusters will be at the soil surface, and you won't need a fork to dig them. They will not all be ready at once. Harvest the mature clusters once every few days. Leaving mature bulbs exposed to hot sunshine will lead to rot. The fall planted ones will tend to be ready a week or two before those planted in late winter. Do not break up the clusters at this stage, as it encourages sprouting. We do not cure onions outdoors here, as they can get sunscald. We cure on slatted racks that stack on each other, leaving space for air, in a barn with fans, for at least a couple of months. Onions bigger than 2.5" (6.5 cm) diameter are best moved along to dinner plates within a couple of weeks of harvest. If you want to plant these largest ones, refrigerate them immediately and plant in September.

Curing and storage

Curing for a couple of months is important to develop flavor and hardiness. The weight after curing for one week is about twice the final dry weight after trimming. I aim to sort through our potato onions once a month, starting a few weeks after harvest. I remove any rotting or sprouting onions each time I sort. At the late June sorting, I take out any onions larger than two inches (5 cm) diameter for eating. If you are on a fast track to increase your crop, you can refrigerate these along with any you stored immediately after harvest, for September planting. Eat them or plant them, the large ones won't keep long.

At the late July sorting, I merely remove any rotting onions, either individual onions from a cluster, or whole clusters.

In late August I separate the clusters, trim the tops to one inch (2.5 cm) and sort the bulbs by size. Sorting by size is not essential, but I do it to help me figure out what to save for planting and

Seed Stock Needs

	Size (ins)	Bulbs/lb	Lbs/360' bed	Size (cm)	Bulbs/kg	Kg/100 m
Very small	<1	30–60	30 @ 4"	<2.5	70–140	12–24 @10 cm
Small	<1.5	20–33	60 @ 4"	<4	44–73	25 @ 10 cm
Medium	1.5–2	8	90 @ 6"	4–5	18	37 @ 15 cm
Large	2–2.5	6	120 @ 8"	5–6.5	13	50 @ 20 cm

what to eat or sell as seed. We sort smalls (<1.5" or 4 cm), mediums (1.5–2.0" or 4–5 cm) and compost material. The rack space required after this stage is only a third of what it was before. I make initial plans about what to do with surplus planting stock.

In mid-September I decide how much to keep for planting. I used to store the onions in net bags, but I found I get better results if I just leave them in a single layer on the racks. The small ones stay there till late January, through freezing conditions (or more accurately, alternating freezing and thawing conditions). They can appear to be frozen solid, but are in fine condition. Ideal conditions are 32°F–40°F (–1°C to 4.5°C), 60%–70% humidity, with good ventilation. Layers should not be more than 4" (10 cm) deep.

Beds of yellow potato onions in early May. Credit: Kathryn Simmons.

Seed saving

This is a natural part of growing potato onions. We started with half a pound (200 g) of seed stock, planted one spring. We only harvested a pound and a half (700 g) that first year, partly because of the late start. We kept all for seed stock and planted more each year in a 1:2 or 1:3 larger:smaller ratio by rowlength, to get enough small and medium onions to plant the same area

the next year, and to get lots to eat as well. When our focus changed to growing for seed stock we started to sell all the small ones and plant a bed each of large and medium ones, in order to get the most seed stock.

The table above gives rough figures to help with calculating seed stock needs for planting. Add a margin for decay during curing. Depending on your conditions this could be 20–30 percent.

Chapter 48

Leeks

This wonderful, tasty, hardy, attractive vegetable is easy to grow, won't bolt (because it's a biennial) and provides an eye-catching change from winter roots at your market booth or CSA. Leeks have a mellow onion-like flavor with no pungency, and come to maturity in late fall when bulb onions may be long gone. Unlike onions, leeks will grow independently of day length and will stand in the field at temperatures below what many other vegetables can handle, increasing in size until you choose to harvest them. A flexible harvest date during fall and winter is a boon to growers wanting a steady supply of produce. Planting dates can be chosen to suit your climate.

Leeks are not just for leek and potato soup! Include some recipes and cooking suggestions for your customers. In general, leeks go very well with white sauces, cheese, mushrooms, piecrust and doughs. British recipe books often offer a lot of ideas. Both the white and the green parts of the leek are delicious. Only the tougher parts of the outer leaves need to be composted. Chop the leeks into short pieces and soak them in water to release the soil from them. If soil has got between the layers, cut the sliced leeks lengthwise too, before soaking, so that the dirt can be removed more easily.

Varieties

Leeks come in two main types: the less hardy, faster-growing varieties, often with lighter green leaves, which are not winter-hardy north of Zone 8, and the blue-green hardier winter leeks. In the first category, we like Lincoln (50 days to slender bunching leeks, 75 days to mature leeks), King Richard (75 days, fast-growing) and Giant Bulgarian. American Flag has not worked well for us. Giant Musselburgh (105 days) is bolt-resistant, for overwintering in milder climates. For winter leeks we like Tadorna (100 days), Jaune du Poiteau, King Sieg (84 days, a cross between King Richard and the winter-hardy Siegfried, from Fedco) and Bleu de Solaize (105 days, very hardy).

Crop requirements

Leeks do best in well-draining soil rich in nutrients, with a pH of 6.5, and good sunlight. Ideal growing temperatures are 55°F–75°F (13°C–24°C). Growth is slow above 77°F (25°C), but the plants do not deteriorate and will resume growth when cooler weather arrives. Some varieties are hardy to 10°F (–12°C), or even to below 0°F (–18°C) if protected by a one-foot (30 cm) mulch of straw or hay or by thick rowcover.

Sowing

Leek seed keeps for only one more year after it's sold, so don't make false economies there. There are 10,000 seeds/oz (360/g). Leeks are slow-growing but easy to care for and frost tolerant. Seed will germinate between 52°F–73°F (11°C–23°C). Count back from your expected first harvest date, using the time to maturity (7–17 weeks) in your calculation. Add 12 weeks for the time from sowing to transplanting to these times, plus 1–2 weeks for the seed to germinate. Usually this means sowing indoors in January, February or early March. Seeds may be sown ¼"–½" (6–13 mm) deep in open flats, or channel trays at 3 or 4 seeds per inch (2/cm), or in plug flats in clumps of 4 or 5. Seedlings in open flats may later be spotted (pricked out) to 2" (5 cm) apart—this is probably more worthwhile for those with a short growing season. Harden off the seedlings before the transplant date.

If you have a long enough growing season (zones 6 and 7) and don't want leeks in summer, you can delay sowing till March. We forego the indoor planting, for simplicity, and sow in an outdoor nursery bed on March 21. We sow ½"–¾" (13–20 mm) deep in rows 3" (8 cm) apart, 10'/100' (10 m/100 m) of the final row at 6" (15 cm) spacing. As needed, the seedlings are thinned to ½"–1" (13–25 mm) apart, and weeded (leeks do not compete well with weeds). We transplant ours in late May or early June, in beds cleared of early spring crops or finished strawberries.

People with a longer growing season (zones 8–9a) can plant two crops: the first 12–14 weeks before the last spring frost, and the second in mid-July, to transplant in late September or early October. In zones 9b–11, sow only in July, and use a bolt-resistant variety for leeks to harvest in the new year. If sowing in hot weather, chill the seed overnight before sowing and keep the seedlings cool but brightly lit. Spring sowings in hot climates won't make much growth before pausing for the summer. It really isn't worth tying up the space for the length of time they'd need to size up, and meanwhile weeding every four weeks.

Transplanting

The ideal size for transplanting is between a pencil lead and a pencil in thickness. We plant at 6" (15 cm) spacing, with four rows to a 48" (1.2 m) bed. People wanting really huge leeks use wider spacings. Leeks can also be planted in clumps of

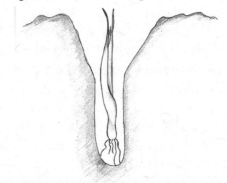

A combination of deep furrows and dibbled holes allows leek seedlings to be planted deeply for long white shafts. Fill the holes with water but not with soil. Credit: Pam Dawling.

four to six, either at 10–12" (25–30 cm) in-row spacing for easier hoeing or at 6" (15 cm) for smaller bunching leeks. We use a special planting technique for our bare root transplants, in order to develop long white shanks, which are prized more than the equally edible green parts. A similar technique can be used for seedlings from flats or plugs. We find it efficient to divide the crew up and specialize in one part of the job.

First, if the soil is dry, water it well, preferably more than an hour ahead. Then one person makes parallel V-shaped furrows, 3" (8 cm) deep, along the bed. Next, a couple of people make holes 6" (15 cm) apart in the furrows. Tools for this job include hoe handles, purpose-bought "dibbles" or dibblers, or ones homemade from broken digging fork handles, with the end sharpened to a point. The tool needs to have a diameter of 1½–2" (4–5 cm). The depth of the holes is determined by the height of the transplants, and is likely to need to be 3" (8 cm) or more. If the holes cave in, you need to water the soil more before proceeding. Meanwhile another person digs up some of the transplants from the nursery bed and transfers them to a small bucket containing an inch or so of water. We make useful little buckets from one-gallon (four-liter) plastic jugs with the top cut off. A rope handle knotted into holes at the top of the new bucket makes it easy to carry. Resist any temptation to trim either the roots or the tops of the leeks.

To transplant, take a leek, shake it free from its neighbors and decide whether to plant it. Discard the ones thinner than pencil leads. If the plant is a good size and looks healthy, twirl it as you lower it into the hole to prevent the roots folding back on the plant and pointing at the sky—they need to grow downwards. This works best if the roots are still wet and muddy from the water bucket.

Bobbing the plant up and down as you settle it in the hole will help a transplant that has slightly bunched roots. If at first you don't succeed, remove the plant from the hole, dip it back in the water and try again. Soon you will develop this quirky planting skill, and will be able to move along the row at a good clip. Ideally the tops of the leaves will poke out of the furrow, not more. Get the depth of the hole-making adjusted to suit the prevailing plant height. This creates the depth for growing a long white shank. Surprising as it may sound, it is not necessary or desirable to fill the holes with soil (you don't want to bury the seedlings). The soil fills in naturally as the plants become tall enough to survive the depth.

Next someone gently waters each hole, either from a low-pressure hose, watering can or using an overhead sprinkler, once everyone else is out of range. The goal is to water the plant roots, adding only a little soil to each hole. The shelter of the hole helps the plant get over the transplant shock, and because leeks have slender tough leaves, they do not lose a lot of water by transpiration. This means that transplanting in quite hot weather is possible, as is transplanting in the mornings.

Keep the soil damp for several days after planting, and then give one inch (2.5 cm) water per week as needed. Like other alliums, leeks do not compete well with weeds, so cultivate as needed. Hoeing will help fill the holes. Some people hill up their leeks, but if you do, be careful not to get soil above the point where the leaves fan out from the stem, or they will be very hard to clean later.

If you have grown seedlings in plug flats or channel flats, you could instead transplant leeks in bunches of up to six leeks, with bunches 10–12" (25–30 cm) apart. This makes for easier weeding, and is fine for small leeks for summer harvesting.

Pests

There are few pests to trouble leeks in the US. Some onion pests (onion maggot fly and thrips) can damage them; see the Bulb Onions chapter. The leek moth is moving south from Canada. It makes pinholes in the leaves and tunnels into the material of the leaf. See the Ontario Ministry of Agriculture, Food and Rural Affairs for more information.

Diseases

Leek diseases include many found on onions such as pink root, white rot, purple blotch, downy mildew, botrytis leaf spot, botrytis neck rot and smudge (see Bulb Onions chapter). In general, leeks are sturdier and less likely to succumb to disease than bulb onions. Leek rust (*Puccinia porri*) is a fungus that can occur in mature crops in dry weather. Rust-colored spores appear on both upper and lower leaf surfaces. Rust can seriously reduce yield and marketability of the crop.

Harvest

Leeks can be harvested whenever they seem big enough. Some people plant leeks in rows much closer together than we do, with the plan of harvesting out alternate rows in late summer, leaving every other row to grow to full size, possibly hilled up.

When harvesting leeks, remember how deep you planted them and try to avoid spearing them. Put the tines of a digging fork (spading fork) vertically down in the ground 2"–3" (5–8 cm) away from the leeks. I try to dig up two at once for efficiency. Step on the fork and lever back until the leeks move. Impatient pulling of unloosened leeks leads to broken ones. Remove one leek, chop off the roots, invert the plant and cut the leaves in a V shape, so that the tougher outer leaves are shortest

These Tadorna leeks have overwintered well. Once the damaged outer leaves are removed, large, healthy leeks are revealed for eating in February and March. Credit: Twin Oaks Community.

and the younger inner leaves are longest. Clean up any obviously inedible outer layers, then put the leek in a bucket. We put an inch (2.5 cm) of water in the bottom of the bucket (to keep the leeks hydrated) before taking the leeks to the cooler.

If the ground is frozen too deep to pierce the crust with the fork, you may be able to harvest a few leeks for immediate use by pouring boiling water along the row at the base of the plants. This does not seem to damage the leaves.

Storage

Leeks are best stored at 33°F (0.5°C) and 65% relative humidity. We use a walk-in cooler and keep the root ends in water. It is also possible to store leeks with the roots packed in soil, shoulder to shoulder in a crate or box in a root cellar, where they will keep for two to six weeks. They can be stored in plastic bags for two to three months at the right temperature, or frozen. Another possibility is to leave them in the garden, mulched with a foot (30 cm) of straw or hay as well as rowcover, if temperatures are below 10°F (–12°C). Our winter temperatures fluctuate a lot, so inground storage doesn't work well for us.

Chapter 49

Tomatoes

Almost everyone wants tomatoes! A fresh-from-the-vine, unsprayed tomato tastes infinitely better than a tired, much-traveled industrial tomato. The range of colors, shapes and flavors is enormous. Although this crop practically sells itself, it might not be easy money: tomato production can be labor-intensive—360 hours/acre (900 hours/hectare) just for harvesting—and then there is a range of pests and diseases to avoid.

Varieties

Tomatoes have a range of uses: there are slicing and cherry varieties, paste (processing) tomatoes for sauces and even special varieties for drying. Varieties can be divided into two growth types.

Determinates are varieties which can only reach a certain height. The number of flower clusters is part of the genetic makeup of that variety. The number of leaf nodes between one truss (cluster) and the next decreases by one each time a truss is produced. When the terminal bud is reached no more leaves or flowers form. Then the plant expands the existing leaves and produces fruit. Once all the flowers are pollinated, no more leaf growth happens: fruit ripens and the plant starts to die back. Harvest can be as long as three months from start to finish, (although some resources for home gardeners claim the harvest is all over in two weeks!). Because they tend to be faster to mature than indeterminates, determinate varieties are often chosen for earlier crops. Some determinates, such as Roma and Celebrity, are quite tall and produce for quite a long season, but for most you'll need to plant a succession in order to have tomatoes all summer.

Indeterminate varieties can continue to grow taller and produce more trusses of fruit as long as the weather is warm enough, and as long as they don't get struck down by a plague. The number of leaf nodes between one truss and the next remains the same all the way up the vine. If you have don't have space to plant a succession of tomatoes and you want to plant once only and get a long harvest season, choose indeterminates. This only works if your farm is not prone to many tomato diseases. Yields from most indeterminates are much higher than from most determinates.

Factors to consider when selecting varieties include the preferences of your market, suitability for your climate, resistance to or tolerance of prevailing diseases and suitability for your preferred growing system. We choose fast-maturing determinates such as Glacier and quick-maturing indeterminates like Stupice for one row in the hoophouse, and our favorite standards (Tropic, Jubilee) with unusual varieties like Green Zebra and Cherokee Purple for the second row. We pull these plants in early August, once the outdoor ones are producing well. Our maincrop outdoor varieties include lots of Tropic, a tasty disease-resistant round red type, and a mix of a few cherries such as Sun Gold and Black Cherry, and oddities like the lovely Garden Peach, orange Jubilee and the exuberant Striped German. Because diseases take down our maincrop, we plant a late crop of disease-resistant varieties that don't take too many days to mature. We also grow 250–300 Roma paste tomatoes that I have selecting from for several years (Virginia Select).

Soil fertility and nutrient requirements

Tomatoes do best on well-drained, slightly acid soils (pH 6.0–6.8). They need moderate amounts of nitrogen (N)—too much can produce excess foliage and over-soft fruit. 75–100 lbs N/ac (83–112 kg/ha) is needed to supply or replace what the crop will use. This can be provided by one shovelful of compost per plant, or five tons per acre (2242 kg/ha). Fairly high levels of calcium (Ca), phosphorus (P) and potassium (K) are also needed. Insufficient K reduces the quality and quantity of fruit. Boron (B), iron (Fe), zinc (Zn) and molybdenum (Mb) are also important.

Biologically active soils ("soils in good heart"), if given regular inputs of compost and cover crops, can provide most or all of a tomato crop's nutrient needs, and unless one of the micronutrients is insufficient, there is no need to add concentrated organic fertilizers.

Temperature and water requirements

Tomatoes like daytime temperatures between 62°F–90°F (17°C–32°C) and night temperatures between 55°F–70°F (13°C–21°C). Frost will kill them, low temperatures will reduce fruit set, and high temperatures will reduce fruit set as well as quality.

Tomatoes use a lot of water, although they can survive dry spells because of their large root systems. If soil moisture levels vary widely, fruits can split or suffer from blossom end rot. If too much water is given, flavor deteriorates. Drip irrigation is a good way to ensure regular watering without risking fungal diseases on the leaves. My first venture into drip irrigation was for our long rows of paste tomatoes as a step towards reducing losses from Septoria leaf spot. I admit to being a reluctant adopter, but after that first year I was convinced. The costs are quickly recovered by the reduced cull rate as compared to sprinkler systems.

Mulch, either plastic or organic, will help conserve soil moisture as well as reduce weeds. Organic mulches should be applied after the soil has warmed to 70°F–75°F (21°C–24°C), so as not to slow down production by keeping the soil cool.

Plastic mulches require the use of drip tape, as they exclude rainfall. Black and brown paper mulches reduce soil temperature, while oiled paper and clear polyethylene may raise the temperature too much.

Yields

Tomatoes provide high yields from a small area, and are labor-intensive, so start small rather than get overwhelmed. Labor estimates are in the order of 350 hours for each staked acre (875 hours/ha).

Yields average 180–250 lbs/100' of row (270–400 kg/100 m). With good conditions, organic growers can get yields as high as 750 lbs/100' (1,200 kg/100 m), or 15 lbs (6.8 kg) per plant.

Per capita US consumption of fresh tomatoes is 20 lbs (9 kg) per year. We grow 250 paste tomatoes and 90 maincrop slicers for 100 people. We can about 500 gallons (2,000 l) of sauce.

Cherry tomatoes are fun, and a few plants go a long way. The bad news is they take a long time to harvest, and some, though delicious (Sun Gold), split easily.

Sowing

Just how early you start your first tomatoes will depend not only on your climate, but also what facilities you have for keeping seedlings and young plants warm enough, and whether you will be planting them under cover or outdoors. Tomatoes struggle with cold winds and are usually grown as transplants, to take advantage of warmer protected conditions for seedlings, and also because many growers like to grow many different varieties rather than long rows of the same kind. If you do want to direct sow, the usual method for small plantings is to "station sow" up to five seeds at each point where you want a plant, and thin to one plant when the seedlings are a few weeks old. Direct-sown plants can catch up with transplants started a month earlier. Another lower-labor method is to sow in the soil in a coldframe (once temperatures are suitable) and do bare-root transplanting into the field. Yield is just as high.

Sow seeds about half an inch (1.3 cm) deep, 2–3 seeds/cell, or in open flats, aiming for 5 seeds/inch (2/cm). Plan on six weeks to transplanting (eight weeks in late winter) and work back to figure out your seeding date. For early sowings, use a seedling heat mat or other heat source to keep the plants at 75°F–85°F (24°C–29°C). When the seed leaves spread open and true leaves start to appear, either single the seedlings (in cells), pot them up individually in 3" (7.5 cm) pots or spot them out into bigger flats, about two to three inches (5–7 cm) apart.

The plants are ready to plant out when about 6"–8" (15–20 cm) tall, sturdy and dark green. If the plants need to be held in the greenhouse or coldframe longer than expected, you could pot them up into 4.5" (11.5 cm) pots. Or space the small pots out so that each plant gets plenty of light and doesn't grow leggy. If you do need to delay planting, consider a foliar feed if your potting compost could be running out of nutrients. After a transplanting shift, we often pot up leftover plants from 3" (7.5 cm) pots into 4.5" (11.5 cm) pots and keep them for a couple of weeks more until we see if we need to replace any casualties.

While starts are in the greenhouse, they may be troubled by aphids. Be sure to deal with these in a timely way—a foliar spray of insecticidal soap every five days works well. Sometimes two sprayings will do the job, sometimes three are needed.

Tomato plants are usually hardened off for transplanting by reducing the watering and ex-

Potted tomato seedlings growing in the greenhouse.
Credit: Kathryn Simmons.

posing the plants to more direct sunlight and breezier air, rather than by exposing them to lower temperatures.

Succession planting

We start our first tomatoes (to be grown in the hoophouse) on January 17, for planting out March 15, a month or more before our last frost date. These plants need a fair amount of cosseting during their eight weeks in the greenhouse. We use a heating mat and a small clear plastic mini-tunnel inside the greenhouse, which has no additional heating. It may well be sensible to have a small amount of heating in the greenhouse on very cold nights. These plants will provide crops from the end of May until the end of July (longer, if we need them) and disease control is not too bad.

For early outdoor crops, grown under row-cover, sow in late February (Zone 7), transplant at eight weeks, and harvest eight weeks after that.

We sow our maincrop tomatoes March 15, transplant at six or seven weeks old and harvest from mid-July until the frosts. Yields do go down after two to three months of harvesting, even if frosts don't bring the plants to an abrupt end.

We sow May 14 for a late outdoor crop, to produce fruit from mid-August until frost. In our area, people who want more successions pull diseased plants and sow on Feb 25, March 15, April 3, April 17 and May 1.

In the past we have also sown June 18 and transplanted in the hoophouse at a relatively young age (tomatoes grow quickly by that point of the year) for a late crop to take us beyond the first frosts. Determinate varieties usually bear lightly the first month of harvest, heavily the second month, then lightly for the third month, so it is not very productive to plant crops so late that they don't reach their second (main) month of production before frosts.

Transplanting and plant spacing

We space our plants 2' (60 cm) apart in the row, whether in the hoophouse or outside. Our rows are 5' (1.5 m) apart in the hoophouse and 5.5' (1.7 m) outside (because we often roll bales of hay out in between outdoor rows as mulch, and this is the best spacing: tight but doable). Small determinate varieties can be planted closer together. Plants more widely spaced can give higher per-plant yields. Some growers advocate 3' (1 m) in-row spacing, if trellising or using cages.

Tomatoes have a great capacity to grow new roots from the stem, and this trait can be used to advantage if you are planting out over-large or leggy plants. Dig a short diagonal trench and lay a plant with the root ball at the deeper end and the top tuft or more of leaves above ground. (The diagonal planting method is better than simply digging a deep vertical hole, except perhaps late in the season in hot climates, because the soil would be too cold lower down.) I recommend orienting the diagonal holes all the same way, so that when

Rows of juvenile Roma paste tomatoes. On the left are plants with biodegradable plastic mulch; on the right is a no-till killed mulch of rye, hairy vetch and Austrian winter peas. Credit: Bridget Aleshire.

you install stakes, you know which side of the plants the root-balls are, and don't damage them. Cover the roots and stem back to normal ground level. In a few days, the plants will be growing upright again, and will be stronger for the extra roots they grow along the buried stem. This does not leave fragile spindly stems out in the air at the mercy of stiff breezes, bouncing puppies, gamboling groundhogs, etc. In addition, if a late frost wipes out the above-ground growth, new shoots will grow from the below-ground stem.

Repairs can be made to broken tomato stems with electrician's tape or duct tape, using a splint if necessary.

Cultivation and management

Care of tomatoes involves setting out drip tape or other irrigation; staking, trellising or caging; mulching, sooner or later; weeding; and likely some training or tucking of wayward branches. Monitoring and managing pests and diseases also takes time.

Tomato plants usually start to flower and to grow side branches after they have grown 10–13 leaves, when the plant is manufacturing more sugar than a single stem can use. It is often recommended to pinch out all but one of the suckers (sideshoots) below the first fruit truss (cluster), especially if using the stake-and-weave system. This helps improve airflow around the plants, reducing the chance of fungal diseases at the price of increasing the risk of sunscald. Overzealous pruning can result in earlier and larger fruit, but lower total yields. Frankly, we never prune ours. Sometimes when they get too tall we chop them off with hedge clippers. But we live in a climate where tomatoes grow like weeds all summer. When I lived in the north of England, we grew our tomatoes up vertical strings in a glass greenhouse. I pinched out all the side shoots, making a regular weekly job of it. The amount of pruning you do will reflect your climate and your goals. In general, more stems means more but smaller fruit and a longer fruiting season.

Many growers use black plastic mulch, as it warms the soil, stops fungal diseases caused by "splash-back" from the soil and prevents weeds from growing. We use biodegradable plastic for some of ours. Later in the season, we top the disintegrating plastic with hay mulch. We also use hay or straw mulch for our outdoor maincrop, adding it in the middle of June when the soil has warmed. For our large planting of paste tomatoes, we use a no-till method, sowing winter rye, Austrian winter peas and hairy vetch in the fall, mowing it at flowering in early May and immediately transplanting into the stubble. Austrian winter peas are said to reduce the incidence of *Septoria* leaf spot. Winter rye is allelopathic, releasing compounds that inhibit seed germination for several weeks after the cover crop is killed. Yields are good with this system.

No-till cover crops do prevent the soil from warming as fast in spring, so they are not good choices for crops you want to harvest early. But for our processing tomatoes, earliness is not important, so we can benefit from the advantages of no-till: reduced tractor passes on the field, reduced soil inversion, grown-in-place nutrients (including enough nitrogen to see the crop through till frost) and increased soil organic matter for the next year.

Another approach to mulch is to use landscape fabric with holes burned at the appropriate spacing. This can be rolled up at the end of the year and reused for many years, although good sanitation is needed to prevent spread of disease. It also has the advantage of being permeable to rainfall. White on black plastic is sometimes used late in the season, and red plastic mulch is reputed to ripen tomatoes faster.

During the first month after transplanting it is important to control weeds to ensure the crop gets enough light and water. Tomato plants grow vigorously and later in life small weeds do not directly threaten their yield. It is, of course, still important to hoe, pull or cut weeds to prevent seeding, and to maximize the airflow around the tomato plants.

Plant support

Most growers use trellises, at least for indeterminates and large determinates. Although it might not seem like it on planting day, it's usually easier to put stakes in soon after planting, while the soil is still soft and everyone remembers where the drip tape is (and which side the roots are, if you planted in diagonal trenches). We use six-foot (1.8-m) steel T-posts, with rows up to 150 feet (45 m) long without any special bracing at the ends. Some people strengthen their trellis by tying an extra stake at an angle to the end stakes as a brace. Most of our tomatoes are large determinates, and although our indeterminates grow tall, we don't much like the extra effort it takes to use seven-foot (2-m) T-posts, so we live with "rolled over" tomato plants later in the season.

Some growers like to use staked cages made from five-foot (1.5-m) tall galvanized wire fencing with 4" (10 cm) square holes. Others use 16-foot (5-m) wire stock fence panels, either vertically or curved into tunnel-shapes. One way to use vertical panels is to have two panels as "walls," held apart by wood stakes about three feet (one meter) long. Three or more plants are inside each rectangular cage. This system is more suited to a scale of a hundred plants than a thousand. The initial cost for fence panels and the extra spacing may be justified by the high yields produced.

For small determinate plants, it is possible to make a fairly quick support system by curving stock panels lengthwise into low tunnels and putting these over the rows of tomato plants. The plants then grow up through the panels and need no tying in. To protect early plantings from spring frosts while the plants are small, the tunnel can be covered with rowcover.

We like the "Florida stake-and-weave" or "string-weaving" system for indeterminate varieties, large determinates (Roma) and even small determinates, and peas too. According to ATTRA, stake-and-weave is not the best support method on any one criterion, but is equal best or second best on most: earliness, fruit size, yield, quality, protection from sunburn and pest control. It is worst as far as labor cost. If earliness, large fruit size and good pest control are important factors for you, choose a high wire with one string per plant attached, and prune out suckers. For high yield and protection from sunscald and cracking, use cages.

Stake-and-weave support system

For the stake-and-weave method, set one T-post after every two plants along the row. We use a stringing tool made from a two-foot (30-cm) length of broken canoe paddle with a hole drilled through close to each end. Any comfortable length of wood will do, or a piece of pipe (pipe doesn't need holes drilled as the twine can be threaded down through the pipe). With a wood handle thread the twine through one hole and back out the other; the twine is not tied to the tool but moves through it the whole time. The tool functions as an extension of the worker's arm, to get the twine over tall stakes, and it can be given a quarter turn to pull the string tight. (Pulling twine tight against your hand for several hours can cut through the skin.) Once you have made your stringing tool, proceed with the stake-and-weave method as follows (photo credits: Kathryn Simmons).

1. Start when the plants are 12"–15" (30–38 cm) tall. Tie the twine onto one end stake, about 8"–10" (20–25 cm) above the ground.

2. Sweep the twine past the two plants in front of you, then cross in front of the next stake and loop the twine round the back of the stake and pull it tight, perhaps twisting the tool to help the tightening.

3. Then use the thumb or forefinger of the other (non-tool-holding) hand to keep the tension you have created and loop the twine around the stake again, going a tiny bit lower so that the second loop crosses over the first and locks it in position. At this point you should be able to let go of the tool without losing the tension on the twine.

4. Proceed along the row, then flip the twine over the end post.

5. Weave back down the other side of the same row, putting another row of twine at the same level as the row on the first side. You are creating two "walls" of twine with the plants in the middle.

6. Once a week, as the plants grow, add a round of twine every 8" (20 cm) up the stakes until you reach the top of the stakes a couple of months later. You can measure the growth of the plants with your hand, as depicted in the photo below.

7. I find that when I think I'm doing the job for the last time that year, it pays to put the row of twine an inch (2.5 cm) down from the top of the stakes, leaving room for a real last round the following week. By that point the previous week's twine will have stretched (if, like us, you use sisal) and the plants will have grown some more, and will benefit from another round of string-weaving.

The first year we did this we used heavy spools of baler twine, and the instruction to carry the twine with you along the row was rather daunting. We got a little trolley, but then discovered a better method. Park the twine in a bucket at one end of the row, with a pair of scissors. String-weave the first side, letting a length of unused twine play out behind you (get yourself inside the loop when you start, to avoid tangles). At the end of the row, flip the twine over to the far side of the row and work back, using up the free length of twine from the ground. This technique does require flipping the twine over each stake (and the plants) before wrapping it, but this is a great trade-off for not needing to haul a big spool of string around. Laughing Stalk Farm in Missouri has posted a video of this method on YouTube.

Disease reduction strategies

Because nightshades have a lot of fungal, bacterial and viral diseases, it pays to take action to minimize the chance of diseases attacking your plants. Here are some strategies:

1. Choose disease-resistant varieties. Consult seed catalogs for details. Save your own seed and select for disease resistance—you can get noticeable results in just a few years;

2. Practice good crop rotation. It's best to rotate away from nightshade crops for at least three years. We don't manage this ideal one year in four. In our ten-year rotation, three of our ten years are nightshades (one paste tomatoes and peppers, two plantings of potatoes);

3. Add compost and cover crops to help increase the diversity of soil microorganisms and build naturally disease-suppressing soil, as well as building fertile soil to support strong plant growth;

4. Use foliar sprays of seaweed extract, compost tea or other microbial inoculants to boost general disease resistance;

5. Practice good sanitation. Avoid smoking, especially near tomatoes, and have smokers wash hands with soap or milk before working with tomatoes. Tobacco can spread tobacco mosaic virus (TMV) to nightshade plants. Avoid handling tomato plants while the leaves are wet. Avoid touching the diseased parts of plants except to remove them. Remove and destroy diseased plants and rotten fruit, especially for late blight, TMV and tomato spotted wilt virus (TSWV). Clean tools in between use in one field and another. When the tomato harvest is finished, till the plants into the soil to speed decomposition, or remove and compost or burn them. Scrub posts if they will be used for tomatoes next year;

6. Remove nightshade weeds (e.g., horsenettle, jimsonweed and black nightshade), which can be alternate hosts for pests and diseases;

7. Improve soil tilth, drainage and aeration. Chisel plow to break hardpan, or grow deep-rooting cover crops ahead of your tomatoes. Grow tomatoes in raised beds and hill soil around stems so that excess water drains away from plants;

8. Maximize air circulation around plants. Choose a bright, breezy location (avoid frost pockets as they also collect dew), orient the rows parallel to prevailing winds and give the plants plenty of space. Remove any leaves touching the ground. Prune the suckers below the first flower cluster;

9. Provide vertical support for plants rather than sprawling them on the ground;

10. Prevent soil splashback onto leaves, to reduce outbreaks of soil-borne diseases. Use mulch. Use drip irrigation rather than overhead sprinklers;

11. Plant a succession to reduce the impact of disease by providing fresh, healthy plants, allowing you to remove diseased earlier plantings;

12. Monitor the weather to minimize sprayings of copper compounds to times when fungal diseases are most likely. Copper fungicides cannot cure an existing disease but can slow down most fungal and bacterial diseases. They must be applied before symptoms are observed and must cover all plant surfaces, including the undersides of leaves. This strategy will involve spraying every 7–10 days (8–12 times per season), so new foliage is protected as plants grow. Because of the toxicity, if this can be avoided by use of other strategies, so much the better. Chitosan is a spray additive that can be used with copper products to stimulate plant defenses and reduce the amount of copper needed. Copper compounds are toxic to earthworms, blue-green algae and some other soil microbes. Even though some organic certifiers still permit some copper compounds, we don't use them;

13. Try biofungicides for use against some tomato diseases. F-Stop, T-22G Biological Plant Protectant Granules or other forms of *Trichoderma* can control damping off (*Pythium*), *Rhizoctonia*, *Fusarium* and *Sclerotonia*. Soil-Gard (*Gliocladium virens*) can work against *Pythium* and *Rhizoctonia*. *Bacillus subtilis* works against *Pythium*, *Rhizoctonia*, *Fusarium* and *Sclerotonia*. Mycostop (*Streptomyces griseoviridis*) can be used against *Phytophthora*, *Alternaria*, *Pythium* and *Fusarium*. 35% hydrogen peroxide diluted to a 0.5–1% foliar spray solution may help control early blight. 1% solution = 3.7 oz in 124.3 oz water to make one gallon (1 ml:33 ml). There are commercial products such as Oxidate that are based on hydrogen peroxide, which is corrosive and challenging to handle;

14. Use biorational controls if needed, including

Full-size Roma plants with hay mulch. Credit: Twin Oaks Community.

AQ10, baking soda or milk against powdery mildew. 4 teaspoons of baking soda and 4 teaspoons of soap per gallon (4 liters) of water. Or 1 milk : 9 water, sprayed twice a week for a cure; 1 milk : 19 water for mild cases;

15. Disinfect seeds against diseases that are seedborne. Dip them (in a cloth bag) into hot water at 122°F (50°C) for 25 minutes, or into a 1:10 bleach:water solution for 90 seconds. Follow by four rinses. These seed treatments are best done immediately before planting, else you need to dry the seeds well enough to prevent germination;

16. Consider getting a high tunnel for tomato production. It can keep leaves dry and greatly reduce the introduction and spread of diseases;

17. Consider saving your own seeds for an important variety, selecting for resistance. Use a fermentation process when saving seed as this kills some diseases. Tolerance to some diseases (*Fusarium*, *Verticillium*, *Septoria*) can be bred into seed in five to six years;

18. Consider grafting disease-prone tomatoes onto disease-resistant rootstock.

Diseases: wilts, leaf spots and blights, fruit spots and rots, viruses

- *Verticillium* wilt: a soil-borne and seed-borne fungus that clogs the plant's vascular system, causing whole branches or plants to suddenly wilt. Older leaves yellow, dry and drop. The necrotic lesions on the leaves usually extend from the leaf-tip in a V shape, surrounded by yellow tissue. Internal stem tissue is darkened. Worse in zones 3–5 and higher elevations further south.

- *Fusarium* wilt (several races): also a soil-borne and seed-borne fungus, usually not appearing until after fruit set. The plant wilts during the heat of the day, and one side of the plant only, or one side of a leaf, becomes bright yellow. The cut stems reveal a brown coloration in the xylem. Worse in the South.

- Bacterial wilt: a soil pathogen which causes a dramatic wilt of the whole plant in a short time, after fruit set, without any yellowing. In severe infections, cut stems ooze with milky bacteria when suspended in water. Unlike with *Verticillium* and *Fusarium* wilts, the pith at the center of the stem becomes discolored.

- Bacterial Canker: causes lower leaflets to wilt; leaves then develop brown edges and the wilt proceeds inward from the edges of the leaves. After the leaves drop, the petioles remain attached to the stem. Small white spots appear on the fruit, quickly developing raised dark centers with white haloes turning brown. Open cankers appear on the stem. Canker is seed-borne and can survive in debris.

- Septoria leaf spot (*Septoria lycopersici*): a common fungal disease of cool to moderate temperatures and moist weather. It is seed-borne and survives for three years on plant debris. The disease starts to appear after fruit set. Numerous uniform, small, round tan spots with a brown margin start on the lower leaves and spread up the plant, causing leaves to wither, die and drop off, thus reducing yield. Small black fruiting bodies of the fungus may be seen in the center of the spots, using a hand-lens. *Trichoderma* does not help, but anti-transpirants may. Potato, eggplant and nightshade weeds are also susceptible.

- Early blight (*Alternaria solani*): another common fungal disease, which first appears on lower leaves as irregular-shaped medium or large brown spots, with distinct concentric rings and a yellow halo around each one. During warm humid conditions, the fungus steadily defoliates the plants, reducing yields, exposing fruit to sunscald and sometimes causing black, sunken lesions on the fruit near the stem end. On stems, small lesions enlarge to form elongated tan or brown areas with concentric markings. The disease is seed-borne, soil-borne and airborne, surviving on plant debris and nightshade weeds. It can appear late in the season, not just early. The beneficial fungus *Trichoderma harzianum* can give good results.

- Late blight (*Phytophthora infestans*): a fast-moving and devastating disease caused by a species of water mold (previously considered fungi, now reclassified as protozoa) that blows in on the wind. It is worse in warm wet weather with cool nights. Late blight starts as "water-soaked" spots on the leaves. These expand into gray-black "scorched" areas, sometimes with a dotted white mold growth, especially on the underside of the leaves.

Fruit develop greenish-brown, firm, shiny or greasy-looking lesions. Cut stems reveal a dark circle of infected tissue. The disease spreads rapidly, turning plants black, as if badly frosted, and can kill an entire planting in ten days unless stopped by hot dry weather. The disease spreads via cull piles and nightshade plants and petunias—it needs live plant material to survive. Preventive action may be taken with sprays every five days of copper products, hydrogen peroxide, *Bacillus pumilus* or *Bacillus subtilis* products.

- Gray leaf spot (*Stemphylium solani*): dark brown spots, enlarging to gray glazed areas that dry and fall into holes. This fungus affects only the leaves, which can yellow, wither and drop. Soil-borne and survives in crop debris. Worse in the humid Southeast, or elsewhere with sprinkler irrigation.
- Bacterial spot (*Xanthomonas vesicatoria*): small dark brown greasy spots on leaves and stems, becoming brown and scabby. Affects tomatoes and peppers.
- Sclerotinia stem rot: a white fluffy mold that attacks seedlings early in the season, when soils are cool and wet. The stem collapses near the soil. During extended damp conditions, the mold produces sclerotia—black fruiting bodies—which survive in the soil for many years. It can also become airborne.
- Anthracnose fungus: mainly attacks fruit (especially those lying on the ground), causing flattened or sunken, water-soaked circular spots with dark centers. Seed-borne, soil-borne and overwinters in residues.
- Bacterial speck (*Pseudomonas syringae*): tiny rough brown/black flecks on fruit (sometimes surrounded by skin that remains green) and small spots (often with yellow haloes) on leaves. Worst in cool conditions. Plants can outgrow the problem as hotter weather arrives. Seed-borne and survives in residues. Affects only tomatoes.
- Buckeye rot (*Phytophthora parasitica*): water-soaked fungal spots form on the fruit where it touches the soil. These enlarge and develop alternate dark and light concentric rings. Unlike late blight lesions, these have a smooth surface and lack a sharply defined margin. Soil-borne.
- Tobacco mosaic virus (TMV): a serious disease causing small distorted mottled leaves and a stunted plant. It is spread by people who handle tobacco. It affects all nightshades and is long-lived.
- Cucumber mosaic virus (CMV): affects many plant species, including tomatoes. Plants are stunted and distorted, with "shoestring" leaves. It is spread by aphids and does not persist in crop residues.
- Tomato spotted wilt virus (TSWV): shows up as small dark spots on leaves and stems. The leaves turn bronze in color, and dark streaks appear on the stems. The plants then yellow and wilt. Some plants survive, although stunted, and the fruit has raised red and yellow target rings on the skin. Yield is greatly reduced and fruit is deformed as well as spotted. Heirlooms suffer most, cherries and small tomatoes survive best. Among maincrop varieties, Rutgers, Viva Italia and First Lady show some tolerance. Amelia and Johnny's BHN-444 are resistant. TSWV is vectored by thrips, often found in many ornamental flowers. For this reason, it is best not to grow bedding plants and tomatoes in

the same greenhouse. In Zone 7, peak flights of thrips occur in early July.

Physiological disorders

These are caused by stress, not by pathogens, and can make tomato fruit unmarketable:

- Blossom end rot (dry black flat spot at blossom end): caused by a calcium deficiency in the fruit. This can happen in cool weather, or during a shortage of water, even when the soil calcium level is sufficient. It usually clears up as the season progresses, provided irrigation is adequate. Waterlogging, root damage or excess soil nitrogen or salinity may contribute;
- Catfacing (pinched-in scars at the blossom end): aggravated by temperatures lower than 60°F–65°F (15°C–18°C) by day and 50°F–60°F (10°C–15°C) by night for a week or more, or by very hot dry weather;
- Puffiness and zippering (contracted linear scars): can be caused by incomplete pollination due to low or very high temperatures;
- Cracking: increased by heavy rain or excess irrigation after dry spell;
- Sunscald (bleaching and decay of fruit): caused by exposure to hot sun, usually after defoliation;
- Green shoulders: more common in some heirloom varieties than in hybrids, and worse in hot weather;
- Nutrient deficiency: symptoms can sometimes look like a microbial disease. Yellowing of older leaves (nitrogen deficiency), purple-red leaves (phosphorus deficiency, worse in cold weather), bronze spots between leaf veins (potassium deficiency) and small, twisted leaves (boron deficiency) are all examples.

Pest management

Tomatoes have many insect pests, but few are really serious as the plants grow fast and can tolerate 20 percent defoliation without their yields suffering. (What is pruning, after all, but human defoliation?) Farmscaping can help keep pests down: attract beneficial insects with alyssum, buckwheat, peas, beans and sunflowers.

- Aphids carry viral diseases such as CMV. Rowcover can protect young plants, and insecticidal soap can be used every fifth day three times on older plants.
- Thrips vector TSWV. Solutions include rowcover and soap, as for aphids.
- Flea beetles are unlikely to reduce yield unless populations are very high or they arrive when the plants are still small. Note that nightshade flea beetles are not the same as brassica flea beetles. Try Spinosad.
- Plant-pathogenic nematodes (minute worms), such as the root-knot nematodes, can attack tomatoes, reducing vigor and yield. The roots have elongated and round swellings.
- Cutworms can be prevented from doing damage by pushing cardboard collars (or split lengths of plastic drinking straws) into the soil around each plant, or by planting out older, thicker-stemmed transplants.
- Hornworms can be huge, up to four inches (10 cm) long, and can strip entire branches. Handpicking works, and Bt is reassuringly effective. Often these worms are parasitized by a braconid wasp and you need do nothing. If you see caterpillars with little white cocoons sticking up out of their backs, just step back and marvel.
- Other caterpillars that eat the leaves or fruit include tomato fruitworm (aka corn earworm), beet armyworm, southern army-

worm and fall armyworm. Bt works if you can get it on them soon after hatching, before the caterpillars burrow into the fruit. Corn earworms much prefer corn in the silking stage, so if that is available, they will not trouble your tomatoes.

- Colorado potato beetles (CPB) sometimes chew tomato leaves, but unless populations are very high or the CPB arrives when the plants are smaller than eight inches (20 cm) tall, they don't have much impact. Monitor weekly for CPB, counting 50 plants' worth of egg masses, larvae and adults, and assess defoliation. Action to control CPB is only needed if the number of adults or larvae is higher than 1.5 per plant or egg masses exceed one per ten plants, or if defoliation is worse than 20 percent. Spinosad works well.
- Blister beetles can be trapped in crops of beets or chard next to the tomatoes. The beetles are easier to see and catch in the trap crops than in tomato foliage. Blister beetles contain cantharadin, which can cause blisters on the skin. If there aren't too many it may be worth putting up with them, as their larvae are carnivorous and eat grasshopper eggs.
- Stink bugs can damage fruit and make it unmarketable, but tomatoes are not their preferred food. Clearing weeds around the tomato patch will make it less attractive to these pests.
- Brown marmorated stink bugs (BMSB) arrived in the US in 2009. They are serious pests, and as I write this, have no reliable methods of control, organic or otherwise. BMSB feeds on almost any fruit or podded crop: tree fruit, berries, tomatoes, peppers, okra, cucurbits and all legumes including soy. See the Cucumber chapter and bmsb .opm.msu.edu for more information. Low-cost tactics include rowcover for small plants, hoophouses with screen doors and soap sprays to kill nymphs (and perhaps a third of the adults). Ants eat BMSB eggs. A succession of trap crops could also be grown and then flamed.
- Mammal pests of tomato plantings include deer, groundhogs (woodchucks) and box turtles. Tomato leaves are fairly poisonous to humans but not to deer or groundhogs, who love the ripe fruit and can eat young tomato plants down to nubbins. Good fences or good dogs can keep deer away. Groundhogs are hard to fence out because they dig, so I have resorted to trapping them. Box turtles will eat tomatoes if they can reach them, although I can't quite imagine them as a serious threat.

Harvest and post-harvest handling and storage

For fresh use, pick tomatoes fully ripe (they can be picked half-ripe and ripened off the vine, but the flavor won't be as good). Processing varieties can remain on the plant for up to a week once ripe, without deterioration. We harvest two or three times a week.

Tomatoes can be cooled to 55°F–70°F (12.8°C–21°C) to prolong shelf life to a week or so. They should not be cooled below this temperature, as the flavor deteriorates.

At the end of the season, harvest partly-ripe fruit the day a frost is expected and finish the ripening process indoors. Contrary to rural myth, tomatoes do not need light to ripen. In fact they need warmth—70°F (21°C)—so putting them on a drafty windowsill is not so good. We stack our Romas in pulpboard eggtrays which hold thirty eggs or tomatoes each. (We put bigger fruits in pulpboard apple trays.) Once a week, check through and remove the ripe and composting fruits. One or two ripe or nearly ripe tomatoes in each tray will help the others ripen (because of the ethylene they off-gas). Garden Peach is an extraordinarily good storage tomato, much better than those called Longkeepers.

Season extension

Because tomatoes are not frost tolerant, possibilities for season extension into the colder parts of the year are limited. In spring, rowcover can be used over early plantings, or a bucket can be put over each plant on cold nights. In the fall it is possible to keep tomatoes alive through a frosty night by running overhead sprinklers all night until the sun comes up and temperatures rise above freezing. We often do this once or twice, as we can get an extended warm spell after the first couple of frosts. Obviously it's not a good system if you have four frosty nights in a row, as your soil will be waterlogged and the days are likely to be chilly in between. An alternative is to give the frozen plants an overhead watering from before sunrise until after the sun reaches them. Setting a timer would save you from having to get out of bed to turn on the water.

Hoophouses offer the best way of extending the season. Under conditions of fast growth and high yields, such as in a hoophouse, plants can run out of potassium, which they need to sustain full flavor and good quality. A study by Steve Bogash at Penn State showed that this can happen with the Luminance THB diffusion plastic used on Haygrove tunnels—because it diffuses light more evenly than Tufflite IV or Solarig 172 plastics, the lower leaves get more light and plants grow bigger, requiring more nutrients to support the potential high yields. Get a soil test before the season starts and add potassium fertilizers as needed. Consider adding more halfway through the season.

Single-skinned temporary "field houses" are the next best way to extend the season. Hooped structures can also be made to cover several beds, and a layer of hoophouse plastic or rowcover will protect the plants in spring or fall.

When faced with long spells of daytime temperatures over 85°F (29°C) and — even more importantly — nights above 72°F (22°C), tomatoes may fail to set fruit. Providing afternoon shade can reduce this problem. There are both heirloom and hybrid varieties selected for their ability to produce in hot conditions. They are referred to as "hot set" or "heat set" varieties. Southern Exposure Seed Exchange sells many open-pollinated ones: Hazelfield Farm, Homestead 24, Illinois Beauty, Neptune, Tropic, Arkansas Traveler, Ozark Pink and Eva Purple Ball. Some early (cold-set) varieties are also able to function in hot weather: Stupice is one. Some cherries also do well: Lollipop and Yellow Pear are two. Among commercial varieties there are BHN 216, Solar Fire from Harris, Sun Leaper from Rogers, Sunpride, Sunchaser, Sunmaster, Solar Chaser, Heat Wave, Floraset and Florida 91. These varieties are also less likely to crack. I have not tried any of the hybrid varieties myself. Usually we don't have long spells with hot nights, thank goodness, as we don't use AC! See the Resources section for a couple of useful links.

Seed saving

See the Seed Growing chapter for full details. As well as the usual *Solanum lycopersicum* species, there is a second species: *S. pimpinellifolium*, the currant tomatoes. The two species can cross.

Chapter 50

Peppers

Peppers are warm-weather vegetables in the nightshade family (*Solanaceae*). They are not at all frost-hardy and require a fairly long growing season to ripen fully. They occupy their space for the whole growing season.

Varieties

There are many kinds of peppers within the basic divisions of sweet peppers and hot peppers. There are many green-to-red peppers and also several fancy colors, which eventually ripen to yellow, orange, red or chocolate brown.

Among sweet peppers there are:

- Bell peppers, the most familiar type;
- Frying peppers: long, tapered, thin-walled, sweet fruity-flavored peppers (Bull's Horn, also called Corno di Toro);
- Cubanelles: shorter fryers with mild, non-pungent flavors. Also used for pickling;
- Banana peppers: long, pointed, mild-flavored peppers that tend to be ripen earlier than bells. Hungarian Wax is a hot banana type;

- Cherry peppers: 1"–1.5" (2.5–4 cm) round, sweet or hot peppers used for pickling and gourmet snacks;
- Paprikas: can be sweet or hot, often 4"–5" (10–13 cm) long, 1.5" (4 cm) wide. Can be dried and ground;
- Pimientos (cheese peppers): ribbed, flattened small globes. Juicy and full-flavored. Good for eating raw (Apple, Cheese Pimiento).

Our favorite varieties of sweet peppers include the red bell peppers Fat N Sassy, Turino, World Beater, Napoleon, Bullnose and Super Shepherd, and orange bells Valencia, Giant Szegedi, Corona and Gourmet. Others I've heard recommended are Chinese Giant (red bell), Honeybelle (gold bell), Jupiter (red bell), Flamingo (cubanelle) and Giant Marconi (frying type).

Factors to consider when choosing varieties are productivity, earliness, size, shape, flavor, leaf cover (to avoid sunscald) and fruit wall thickness.

Small-fruited varieties tolerate hot humid conditions better than large-fruited ones.

Seed savers should note that most peppers are *Capsicum annuum* and will cross with each other. Habanero, other Scotch Bonnets and the milder Aji Dulce and Trinidad are *C. chinense* and will not cross with *C. annuum*. Unusual peppers include *C. frutescens* (Tabasco) and *C. baccatum*.

We grow relatively few hot peppers, so this chapter focuses mostly on sweet peppers. Our basic hot peppers are jalapeño and habanero, with a few Thai Dragon, Fish, Serrano, Anaheim and Cayenne peppers for drying. See the interesting chart of over fifty hot peppers and their heat ratings at usHOTstuff.com. See also the Chile Pepper Institute and the Chile Man in Virginia.

Crop requirements

A pH of 5.5–6.5 is ideal. Peppers require soils with fairly abundant nitrogen, phosphorus and potassium, but not too much nitrogen, or the fruit set may be reduced and the foliage too lush. During fruit growth, foliar feeds of fish emulsion may be used.

Peppers grow slowly unless the nights are warmer than 55°F (13°C). Below 52°F (11°C) they get stunted for a day or two, and 50°F (10°C) shocks them more seriously. Hot peppers get hotter with hotter drier weather, and also with less fertile soil.

Plant four or five peppers for every person you are providing for. We generally plant 45 in our hoophouse and about 350 outdoors (only 40 are hot peppers) for 100 people.

Crop rotations

We often plant our maincrop sweet peppers into a no-till mulch of mowed overwintered rye, hairy vetch and Austrian winter peas (as we do for our paste tomatoes). See the Tomato chapter for more on this. We plant an early crop in our hoophouse so the delayed ripening of the no-till outdoor crop is not a problem. We have also used biodegradable black plastic mulch covered with straw or hay mulch when it starts to disintegrate.

After the frosts end the pepper harvests, we sow rye with crimson clover or Austrian winter peas, which have time to flower before we plant watermelons the next year. Although tomatoes and peppers appear only once in our ten-year rotation, potatoes appear twice, so the time between nightshade crops is usually three years.

Sowing

We sow peppers as early as Jan 24 for planting out April 1 in the hoophouse, giving us a nine-week-old transplant. We used to sow Jan 17, but it's hard to keep peppers going in very cold weather so we've decided to start a week later. Even so, we have lost peppers to unseasonably cold weather early in March. For outdoor planting we sow indoors on Feb 8 and transplant May 8 (12 weeks). We could probably wait another week or two to sow and use a younger transplant. Pepper seedlings do take longer to grow than tomatoes.

There are a few seed-borne diseases of peppers, and you can heat-treat your seeds to kill these diseases: immerse them in 122°F (50°C) water for 25 minutes, cool in cold water, drain and sow. Some people soak seeds in a 1:4 bleach solution (1.05% sodium hypochlorite) for 40 minutes with constant agitation. Then rinse with vinegar, then water.

Pepper seeds are fussier than tomato seeds and have a short shelf life, so be sure to use fresh seed. Peppers are slow to germinate: they take 25

Pepper transplants almost ready to go outside. Credit: Kathryn Simmons.

days at 60°F (15°C), 12.5 days at 68°F (20°C), 8.4 days at 77°F (25°C) and 7.6 at 86°F (30°C), so the ideal range is 68°F–86°F (20°C–30°C), with 60°F (15°C) as a minimum and 95°F (35°C) as a maximum.

Start in cells 2"–2.25" (5–6 cm) diameter, 2.25"–3" (6–7.5 cm) deep. We use tray inserts with 38 round cells in a 1020 sheet, or else deep nine-packs. Pot up to 3"–3.5" (7.5–9 cm) pots as needed. If we start ours in nine-packs, we pot up, but if we start them in the round 38-cell size, we can plant out straight from those.

To do well, peppers need to grow without any checks. Keep the soil moisture low and water with warm water. Grow the new seedlings at 70°F (21°C) in the daytime and 62°F (17°C) at night. Promote the uninterrupted growth of the seedlings until it's time to give them a "cold conditioning" (which is said to increase the number of fruit). When the third true leaf appears, reduce the night temperature to 53°F–55°F (12°C–13°C)

for four weeks, then raise the temperature to 70°F (21°C) day and night. We make a hooped tent of slitted plastic sheeting inside our greenhouse to house pepper seedlings for the first few weeks, then move them out to the greenhouse proper. (Ours get cooler than the ideal, and I almost manage to convince myself that we are doing something close to this conditioning!)

Prepare for planting out by hardening off the transplants over seven to ten days. Do this by reducing moisture slightly but never let the plants wilt. Don't harden off nightshades by reducing temperatures below 53°F (12°C). Gradual exposure to sun and wind is helpful.

Transplanting

Eight weeks is the minimum transplant age for peppers. Older plants can be used, as long as they are still actively growing, not root-bound, and not yet fruiting. Peppers love sun and heat, so if necessary, move your transplants up to larger containers and wait for the right weather. Bigger plants can take conditions one to two degrees cooler than smaller ones. Average soil temperatures of 65°F (18°C) are best. The full or falling blossoms of the American dogwood (*Cornus florida*, also known as flowering dogwood) are a good phenological indicator for suitable outdoor planting temperatures. This is usually a couple of weeks after the average last frost date. Spacing of 18"–24" (45–60 cm) in the row and 30"–36" (75–90 cm) between rows is usual. Another method is to plant double rows 12" (30 cm) apart in beds on 60" (1.5 m) centers. We plant at 18" (45 cm) in the row and 66" (1.7 m) between rows, so that we can unroll big round hay bales as mulch between the rows. In wet areas, we have planted peppers on ridges. Our no-till mulch becomes thin pretty quickly, so we top it up with spoiled hay. This also reduces the soil temperature, but the saving in dealing with weeds is worth it. We set out irrigation drip tape before we roll the mulch. In the hoophouse, where the plants grow huge, we plant a single row at 24" (60 cm) spacing in the 4' (1.2-m) beds.

Growing

In the hoophouse we use 5' (1.5-m) stakes every two or three plants and do Florida string-weaving as we do for tomatoes, to provide the sturdy support that the huge plants there will need. Outdoors we use short 2' (60-cm) stakes every three plants and do two rounds of string-weaving to help prevent branches from breaking off.

Peppers are drought-tolerant to some extent, but drought-stressed plants can produce thin-walled fruit that may even be slightly bitter. An inch (2.5 cm) of water per week throughout the season is about right. Drip irrigation, or overhead watering which finishes before noon, will reduce fungal diseases caused by wet foliage in cooler temperatures.

Foliar feeding with fish or seaweed emulsion can be used once fruits have started to set. Foliar calcium sprays may strengthen the fruit walls and help against pests. Boron side dressing may also help. Maintain high levels of phosphorus for sustained yields. (Perhaps like us and many other organic growers, you are dealing with an excess of phosphorus in your soils and don't need to worry about maintaining high levels!)

By the way, it is untrue that if hot peppers are planted next to sweet peppers they'll cause the sweet peppers to become hot. But water-stress can heat up sweet peppers. Obviously, you want to keep hot peppers separate from sweet peppers when it comes to harvesting and marketing!

Pests and diseases

We do not have many bug issues on peppers here. Cutworms, flea beetles and aphids can present problems, as can fruit worms, corn earworm, fall armyworm and European corn borer (ECB). ECB is worse where fruits are crowded and can introduce soft bacterial rots, so thinning overloaded plants is wise.

Blossom end rot can occur due to calcium uptake shortages, especially in dry weather. Sunscald can happen if leaf coverage is inadequate.

Diseases of peppers include bacterial speck, *Verticillium* wilt, *Anthracnose*, *Septoria* leaf spot, tobacco mosaic virus (TMV), tomato spotted wilt virus and other viruses. Monitoring and controlling aphids and thrips can help prevent diseases they vector. We try to prevent smokers from handling our pepper plants, as TMV can spread from dried tobacco to nightshade plants. Soft bacterial rots can follow injury caused by insects or mechanical damage. Use mulch and drip irrigation to reduce fungal diseases caused by soil splash.

Season extension

At the beginning of the season, rowcover can protect a first outdoor planting from a surprise frost. Because peppers are not at all cold tolerant, it's better to wait rather than to risk too early a planting. A hoophouse or a greenhouse offers great opportunities for growing early crops, or a cold-frame can be used for slightly early transplants after the spring seedlings free up space there. We have transplanted peppers in our coldframe, then set flats diagonally between the small peppers until the flats were ready for transplanting outside and the peppers could grow to full size there.

At the other end of the season, fall frost threatens to end pepper harvests. Often there is a period of mild weather after the very first frost, and if you can protect your plants from major damage, they can continue to ripen fine peppers for several more weeks. Throwing rowcover over your plants for a night or two may be worthwhile. When the first frosts threaten, we start a special harvesting style, picking all the peppers exposed to the sky regardless of ripeness. The first frosts often just nip the tops of the plants, and peppers protected by leaves will be fine. Running overnight or pre-dawn sprinkler irrigation can protect from frost—just be sure to keep the water on until the sun is up and warming the plants. After that, each time frost threatens, we harvest all the exposed peppers as well as the ripe and damaged ones. Be sure not to pick the dull-skinned very immature ones, which rot quickly. Some people pull up whole plants and hang them upside down in a frost-proof building to continue ripening indoors. Fruits will keep for four to six weeks in a cooler. Refrigeration definitely compromises the flavor of peppers, but if your only other option is no peppers, it can be worth it. Another alternative for a small number of plants is to pull the plants, stand them in buckets with the roots in water and store them for up to four weeks.

During very hot weather flowering plants need to be cooled below 90°F (32°C) to reduce stress, or pollination may be poor and yields reduced. Shadecloth, or overhead irrigation for half an hour at noon, may help.

Harvest

All peppers will eventually ripen to some shade of red, orange, yellow or chocolate brown. In addition, some varieties have long, partially ripe stages of different colors (yellow, purple or white). There is a market for green peppers, even though

We prefer to harvest most of our peppers at the fully ripe color and flavor, rather than green. This is Lady Bell. Credit: Kathryn Simmons.

fully ripe ones are more nutritious. Of course, if you harvest all your peppers at an earlier color stage, you'll never get ripe ones before the end of the season. Some growers harvest a few green peppers from each plant if the plant looks overloaded, then wait for the rest to ripen. Others reserve some rows for harvesting red. Red peppers bring a better price than green ones, but total yield/plant is lower. Everyone has to find their own balance between quantity, quality, demand, supply and price.

Most peppers require 65–80 days from transplanting to full-size immature fruits. Expect to wait another two to four weeks for ripe peppers. Yields will likely be 5–18 lb/10' (7.5–26 kg/10 m) of row. If you harvest partially ripe (multicolored) peppers they can finish ripening off the plant. Harvest at least twice a week, to reduce losses to pests and diseases. We pick ours three times a week. We harvest all the fully ripe ones, all the damaged ones for processing and a few green ones if the plant is heavily laden. We don't pick the partially ripe ones. This is a straightforward instruction we can give our crew, so that most of the peppers get a chance to ripen fully.

Harvest by grasping the pepper with a thumb at the top and the fingers at the bottom of the fruit. Then snap the pepper upwards and break the stem. Some growers recommend cutting the stems, to minimize damage to the plants. Certainly everyone harvesting needs to know that the branches are very brittle and care needs to be taken.

Rotting fruit should be removed from the plant to reduce the spread of disease. As with many crops which are prone to foliar diseases, it is better to avoid handling the plants while the foliage is wet.

If you need to wash hot pepper off your skin, you can use poison ivy remedies such as Tecnu skin cleanser. If you rashly bite a raw habañero to see how hot it is (as a crew member did here one year), the best remedy I know is milk, yogurt or ice cream.

Post-harvest and storage

Don't chill! Peppers are best kept at temperatures above 46°F (8°C), with high humidity. This is not always possible, and most people do refrigerate them. For processing, peppers can be stored by freezing (frozen pre-cut rounds for pizza?), drying or by canning in salsa, relish, pickles or hot sauce. Keep hot peppers separate from sweet peppers, and clearly labeled. Paprika peppers can be smoked, dried and then ground. For powdered dried peppers, hang the fruit until the skin is leathery and the ribs dry, then grind in a blender or coffee grinder. Exclude the seeds, or include half the seeds of hot peppers. One pepper makes about half a tablespoon (7–8 ml) of powder. Store in a glass jar or a ziplock bag in a refrigerator. Ristras are strings of dried red peppers such as paprikas. Use a needle and thread to string through the calyx (cap end) of the fruit, then hang the strings to dry in the shade, protected from rain. Peppers can also be dried in electric food dehydrators.

Nutrition

A red sweet pepper contains about ten times as much vitamin A and one and a half to two times the amount of vitamin C as an immature green pepper. One cup of chopped green pepper has twice the RDA of vitamin C and 100 g of sweet raw green pepper (less than one large pepper) contains 128 mg of vitamin C. A whole orange contains only about 50 mg of vitamin C, so even a green pepper has a lot more. A great selling point to compete with citrus fruit!

A cup of green sweet pepper has 213 mg of potassium raw and 149 mg if boiled before being eaten. For comparison, an average banana contains 370 mg.

Seed saving

Those interested in seed growing should visit SavingOurSeeds.org to read the *Organic Pepper Seed Production Manual*. Peppers are self-pollinating, and an isolation distance of 40'–75' (12–23 m) for home use, or 75'–150' (23–46 m) with barrriers for commercial sale, is enough. Use 300'–600' (90–180 m) without barriers, depending on the variety. A population size of thirty to a hundred plants will be needed, depending on whether the variety has been well managed in the immediate past.

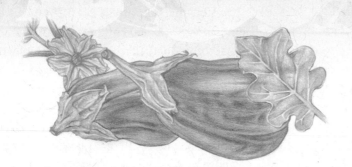

Chapter 51

Eggplant

Eggplant is a warm-weather vegetable, so if your climate is marginal for this crop, choose a sunny sheltered microclimate and a fast-maturing variety and use season extension techniques (see below). It's culture is similar to that of peppers.

Types and varieties

There are two basic types of eggplant. The first is the classic large-fruited, pear-shaped dark purple kind, often with a green calyx (cap end). The second type is often called "specialty eggplant." This includes a huge array of shapes and colors and sizes. "Oriental" eggplant generally has a purple calyx. We have found very little demand for green, orange, lavender, white, red, stripy or odd-shaped varieties. Although we can grow nice long eggplants, we can't get our cooks to use it, so we have given up. We had high hopes for Ping Tung Long (60 days), but it had more entertainment than culinary value here: one crew member misheard the name as Pink Tongue Long! The smaller long varieties can produce more fruits per plant, but the same total yield per plant as the classics. Specialty eggplant is no more difficult to grow than the classic kind, but may be more difficult to market!

Most people seem to prefer the standard dark purple glossy pear-shaped kind. After trying many varieties and deciding to stick with traditional types, we moved from growing Dusky (63 days) to Nadia (67 days), which we like a lot. Now that we have a reliable productive variety, we have been able to reduce our planting to a third of the size it was.

Crop requirements

Eggplant benefits from fertile, well-drained soils high in organic matter, with a pH of 6.0–7.0, (ideally 6.0–6.5). Average moisture with plenty of warmth and sunshine are also needed. Ideal daytime growing temperatures are 70°F–85°F (21°C–29°C).

Yield

Expect about 200 lb/100' of row (30 kg/10 m), and allow about 12 lb (5.5 kg) per person over the season. On a larger scale, be ready for 7–11 tons/ac (8–12 kg/ha).

Sowing

Sow in the greenhouse eight to ten weeks before your last frost date, at temperatures of 70°F–90°F (21°C–32°C). Starting too early is counterproductive—wait until you can guarantee a greenhouse minimum of 60°F (15.5°C). Emergence takes 13 days at 68°F (20°C), 8 days at 77°F (25°C) and only 5 days at 86°F (30°C). At the low end of this range, only 21 percent of normal seedlings will emerge, compared to 60 percent at the top end. Eleven grams of seed will produce about a thousand starts.

You can sow in plug flats or open flats and separate the plants later into 3"–4" (7.5–10 cm) pots. Protect the seedlings from flea beetles, either in the greenhouse or on benches outside. Flea beetles cruise at low altitudes, so setting your flats three feet (one meter) above the ground may be all you need to do to keep them away.

Transplanting

To harden seedlings off for planting out, reduce moisture levels rather than drop the temperature, as this crop is easily stunted by cold temperatures. Transplant the 8–12 week-old plants 1–2 weeks after the last frost date. The transplants should be 6"–10" (15–25 cm) tall, without any buds, flowers or fruit. We allow our eggplant to be one of the later crops to be set out, after the tomatoes and peppers. To help warm the soil, you could spread black plastic mulch two weeks before transplanting. Avoid organic mulches at planting time, as they cool the soil. Placing rowcover on hoops is a good idea for new transplants, both to keep the plants warm and to keep flea beetles away. See below for our technique to dislodge flea beetles while transplanting.

Plant spacings of 18"–24" (45–60 cm) in-row and 30"–36" (75–90 cm) between rows—or more to accommodate machinery—are usually recommended. We used to grow two staggered rows in our 4' (1.2-m) wide beds, aiming to have the plants 30" (76 cm) apart. This created crowded aisles, so we now plant a single row in each bed, with closer in-row spacing of 20"–24" (50–60 cm) to create a "hedge" and make the paths more accessible. This fits with the approach that considers the area each plant has, rather than the intensive planting approach that favors equal space in all directions.

Some growers pinch off the growing tips to encourage branching, although ours branch just fine without, and I hate removing bits of healthy crop plants.

Caring for the crop

Once the soil is fully warm, you can cultivate and apply a layer of organic mulch. Because the plants will be in the ground for a long season, and organic mulches break down, we use our eggplant beds as convenient places to drop off finished crop residues or weeds—only healthy, non-seeding material of course. In hot dry weather, weeds can be pulled from the mulch and laid on top to die.

Any stress from cold weather, disease or low fertility will cause the skins to thicken and become bitter. If your soil fertility is low, feed monthly with fish emulsion or side-dress with compost. Don't overdo the nitrogen or you will get lots of leaves but few fruit.

As with most crops, the critical time for irriga-

Established eggplant needs very little care other than irrigation. Credit: Kathryn Simmons.

tion is during flowering and fruit formation. Insufficient water during this stage can lead to blossom end rot, misshapen fruit and reduced yield.

We don't usually stake our Nadia, but if your area is windy, you could stake tall varieties, with three- to five-foot (1–1.5 m) stakes every third plant around the perimeter and twine every twelve inches (30 cm) up the stakes to corral the bushes. This may produce better-shaped fruit as well as more upright bushes. When the branches threaten to take over the aisles, snip them off as you harvest.

Growers in marginal climates for eggplant sometimes prune low branches and leave just two main stems for the ripe fruit. (Fruits on lower branches can rot where they touch the ground.)

In the fall, the big plants can continue on maturing existing fruits, if rowcover is used to keep the first few frosts off. No new fruit will set once the temperature drops below 70°F (21°C).

Rotations

Three years between nightshades is a wise crop rotation. We grow our eggplant in our permanent raised bed area, as it benefits from close attention, and we grow a relatively small amount of it. If you are growing eggplant on a larger scale, you could group it with other nightshades to make crop rotation easier in future years.

In the South it may be possible to get double use from plastic mulch by first growing an early crop of spring cabbage or collards. When this crop is finished, the plastic can be left in place and reused — provided it is still in good shape.

For years I have kept a note that planting crimson clover in the fall before transplanting eggplant (into the clover) will reduce flea beetle outbreaks, but I have yet to try it.

Ratooning

Succession planting is not usually a possibility: most growers start plants early and keep them growing all season. In the South, some growers ratoon their eggplant—i.e., mow it down to a height of 6"–8" (15–20 cm), being sure to leave two or three leaf axils. This is usually done in July or early August, after the first flush of fruits, when the plants start to decline. The plants are then side-dressed with compost, which encourages vigorous new growth and stimulates flowering. Fruit will be ready again four to six weeks later and should carry on until frost.

Pests

The main scourge of eggplants is flea beetles. These are small shiny black insects that jump when poked. Early Black Egg and Orient Charm (both 65 days) are said to be more tolerant of them than other varieties. Note that nightshade flea beetles are not the same species as brassica flea beetles, and if you have one of these pests you don't always have the other. Covering the soil with plastic mulch does reduce the problem. Beneficial nematodes can be bought to help prevent future attacks. A trapping method I have not yet needed to try is to fill clear or white plastic bottles with water, cap them, coat their outside with sticky Tanglefoot compound around the middle two-thirds, and set them every fifteen feet (5 m) along the row under hooped rowcover. The water warms during the day and attracts the flea beetles at dusk. In four days you can expect to have caught almost all of them.

Our technique for minimizing flea beetles while transplanting is to set out hoops for rowcover and sticks to hold it down on either side of the bed. The rolled rowcover is at the ready. One or two people transplant while a third person with a hose and spray head gives the plants a strong spray, directing the flea beetles out of the bed. A fourth person follows close behind, unrolling the rowcover and battening it down quickly.

Colorado potato beetle can be a big pest, although in my experience they much prefer potatoes. Handpicking will deal with small numbers; Spinosad with large numbers.

Aphids, lace bugs and red spider mites can be a problem, especially to young plants. Check twice a week. The most effective natural enemies of aphids are parasitic wasps, predators such as ladybugs and fungal diseases. A strong water spray can dislodge most aphids and get numbers down to a manageable level. Spraying with insecticidal soap or three tablespoons (45 ml) of Murphy's Oil Soap per gallon (four liters) of water several times five days apart will generally deal with aphids. Although aphids can transmit many diseases, eggplant is not as susceptible as other vegetable crops.

Mites are tiny, and feed on the leaf undersides, so they are hard to see. The damage they cause appears as a white stippling of the upper surface of the leaves, which later become bronzed in appearance. Soap sprays (as for aphids) will also deal with mites.

Many other insects may be seen feeding on eggplant but few of them are a threat. Armyworms can be killed with Bt (*Bacillus thuringiensis*) if numbers become high.

Diseases

Eggplant is susceptible to several fungal diseases, each a reason to practice good rotation. Remove and destroy all diseased plants and promptly till in crop debris.

Verticillium wilt is a soil fungus that attacks many crops. Leaves wilt and yellow and can even die.

The less common *Phomopsis* blight is a problem in hot humid areas. It causes leaf spots, canker-like stem lesions and dark sunken areas on the fruits. Stem lesions can encircle the stem and kill the plant. Leaf spots are one-inch (2.5-cm) circles with brown margins. Light-colored centers develop as they age, and black fungal reproductive structures may appear in the centers. Diseased leaves may yellow and drop. Fruit spots are large, sunken, pale or brown lesions that make the fruit unmarketable.

A *Phytophthora* blight (not the potato and tomato late blight one, but the tomato buckeye rot one) can infect eggplant, causing damping off of seedlings, spotting of leaves, collar rot of the main stem, fruit rot and death of the plant. It is only active above 86°F–90°F (30°C–32°C) and in wet weather.

Southern blight is caused by the soil-borne fungus *Sclerotium rolfsii*, which develops in moist conditions above 86°F (30°C). Leaf yellowing and wilting and sunken stems are followed by the appearance of a mat of white fungal growth at the soil surface. This disease can spread down the row, killing one plant after another.

Alternaria leaf spot is another fungal disease, also known as early blight. It is more common in tomato and potato, but it can also do damage to eggplant. The leaf spots, usually clearly zonate (target-like) on tomato leaves, are less clearly marked on eggplant leaves.

Harvest

Eggplant is mature 100–150 days from seeding, or 65–85 days from transplanting. Here we harvest from July 10 until frost. I notice that this is only 57 days from transplanting and I conclude that Nadia's usual 67 days is probably counted in a cooler climate than ours. Harvest once or twice a week.

Deciding when eggplant is ripe is quite a skill. This is a task we prefer to delegate to a few trained people. It helps to know the usual mature size for the variety you are growing. We tell people to look for large shiny eggplant and then squeeze it gently. Ripe eggplant has some give or springiness to it. You may see a slight indentation remaining after you press on the side of the fruit. Unfortunately, immature eggplant goes through a springy stage when quite small, before a firmer adolescence and then full ripeness. In the past, we too often had someone cut immature eggplant with extreme thoroughness, all along the row. It then took quite a long time before replacement fruit reach any level of maturity!

We like to use pruners (secateurs) to cut eggplant rather than trying to pull it from the plant, as this causes less damage to fruit and plant. Be sure to also get harvesters to pick any damaged fruit, any calyx parts remaining on the branches, and any overly mature fruit. This will ensure that the plants keep cranking out more new fruit and don't senesce. Overly mature eggplant loses its shine and develops bronzy-green streaks. Later the streaks become golden patches. Only seed growers want golden eggplant!

Storage

Eggplant should ideally not be refrigerated. Below 45°F (7°C) it suffers from chilling injury: surface pitting and decay. Large quantities will require forced cooling with fans. Do not use ice or chilled water to cool this crop. Eggplants harvested in midsummer can be stored at 54°F (12°C) for up to a week with 90%–95% humidity. When harvested in the fall under cooler growing conditions, it can be stored for up to ten days at 46°F (8°C) and the same humidity.

Seed saving

Isolate varieties by a minimum of 150' (46 m) for home use, or by 600' (185 m) for pure commercial seed. When the seed is ripe, the fruits will be bright yellow. It may be shredded in a food processor (magically, the seeds don't get cut), before starting a wet fermentation process (just 2–3 days; less time than tomato and pepper ferments).

Chapter 52

Potatoes

Potatoes are a rewarding crop to grow, with a lot more flexibility about planting dates than the traditional instruction to plant on St Patrick's Day might have you believe. As Carol Deppe points out in *The Resilient Gardener*, potatoes provide more carbohydrates per area than any other temperate crop, and more protein per area than all other crops except legumes. Many people are surprised to learn this. (The Atkins Diet has a lot to answer for!) A 2,000-calorie all-potato diet contains considerably more protein than a 2,000-calorie all-rice diet. Potatoes contain 10.4 grams of protein per 100 grams dry weight, and are a good source of vitamin C and carbohydrates.

Potatoes have four development stages:

1. First the plant produces roots, shoots and leaves. This vegetative state lasts 30–70 days. Bigger plants have more yield potential, so the goal for this stage is to produce robust large plants;

2. Tuber formation (a two-week process) and branching of the stems comes next. During this stage, leaf growth continues. Flowering can happen too, but it's not essential;

3. Third, the tubers grow larger, but don't increase in number. Adequate water and nutrients are important during this critical stage which lasts up to 90 days, until the plant reaches maturity for that variety;

4. Finally, the tops naturally yellow and die. The skins of the tubers thicken. No more growth is possible.

Varieties

The many varieties of potatoes are generally divided into four categories. Early potatoes take 55–65 days from planting to harvest—the more famous ones include Yukon Gold, Irish Cobbler, Red Pontiac and Caribe. Mid-season potatoes mature in 70–80 days, and include Kennebec, Katahdin, Desirée and Yellow Finn. Late-maturing varieties take a full 85–120 days to mature and include Russet Burbank, Butte and Green Mountain. The fourth category is fingerling potatoes, which are small, attractive and have a high market value. They are prolific and no harder to grow than other potatoes.

Some suppliers of organic seed potatoes are listed in the Resource section. Farms not certified organic have the option of buying non-organic seed potatoes locally, which saves money on shipping. Be sure, though, to buy seed potatoes that are certified disease-free. Late blight is a disease not worth risking. Some growers buy "B" potatoes that are small enough to plant without cutting. For most growers, "B" potatoes are not available, and we settle for larger seed potatoes, which have fewer eyes for the weight than small ones do.

Crop rotation

This is very important for potatoes, which are nightshades like tomatoes, peppers and eggplant. Colorado potato beetles emerge from the soil in spring and walk (they don't fly at this stage) towards the nearest nightshades they can detect. Give them a long hike! A distance of 750' (230 m) or more from last year's nightshade plots should keep them away. A three- or four-year rotation out of nightshades in each plot is ideal.

Suitable cover crops before potatoes include brassicas (which can help reduce root knot nematodes and *Verticillium*), Japanese millet (which can reduce *Rhizoctonia*) and cereals in general. Beware beets, buckwheat and legumes such as red and crimson clovers, and some peas and beans, as these can host *Rhizoctonia* and scab.

Planting dates

Potatoes are a cool-weather crop, but the tops are not frost tolerant. A good guideline for suitable spring planting conditions is three consecutive days with a temperature at a depth of four inches (10 cm) exceeding 43°F (6°C). A traditional phenology sign is that the daffodils should be blooming. The spring planting is usually timed with the goal of having the shoots emerge after the frosts. A light frost will only nip the tops of the leaves and do no real damage, so a small risk is worth taking. It takes a temperature of 29°F (–2°C) to kill the shoots, and even then regrowth is possible. The practice of hilling soil over most of the leaves once the plants are six inches (15 cm) tall will protect against frost. So if you have plants growing and a frost is predicted, a hilling that day may save them. Frost will nip back above-ground growth, but the plants will regrow. In the fall, frosts will kill the foliage and growth will stop, so late plantings should be timed to get the tubers to maturity before the expected frost date. Some late varieties do not bulk up until the last moment, so if you are pushing the late end of your planting season, plant early varieties or fingerlings.

We plant our first crop in mid-March, about four weeks before our last spring frost, and plant a second crop in mid-late June, which allows three and a half to four months before our average first frost date. We could plant any time mid-March to mid-June and harvest mature potatoes. If we wanted to deal with a later harvest, we could plant a fast-maturing variety in July. A grower specializing in many kinds of fingerlings might want to plant once a month during their season, for a continuous supply of fresh new potatoes. Growing in a hoophouse offers another option for growing for a late market, for example new potatoes for winter holiday dinners.

If you want to plant at "unusual" times of year, you may need to plan a long way ahead, buy your seed when it's available and store it in a cool dark place below 50°F (10°C), such as a refrigerator, until you need it. Many suppliers only ship in March and April. Growers in zones 8–10 may need to buy their spring seed potatoes in the previous fall. We buy our seed potatoes for the June planting in April, before local suppliers sell out of spring stocks. An advantage of summer planting is that the harvested crop need only be stored from October or November, not over the hotter months.

Dormancy

Potatoes have a dormant period of four to eight weeks after harvest before they will sprout, so if you plan to dig up an early crop and immediately replant some of the potatoes for a later crop, it won't work. Get around this problem by refrigerating them for sixteen days, then pre-sprouting them in the light for two weeks. Apples, bananas or onions will help them sprout by emitting ethylene.

Physiological age of seed potatoes

Seed potatoes can act differently depending on their "physiological age." The warmer the conditions are after dormancy ends, the quicker the sprouts grow and the faster the tubers "age." When we buy seed potatoes the storage conditions they have already received are beyond our control. As a guide, the length of the longest sprout is a good measure of physiological age (if the sprouting has taken place in the light). See *Know and Grow Vegetables* by Salter, Bleasdale, et al. for more on this complex topic. Evidence suggests varieties do not all show these effects to the same degree. Deliberately adjusting storage temperatures is a way of manipulating the physiological age.

Physiologically, "old" tubers mature quickly, and are therefore good for an early harvest, or a fall crop close to the frost date. They emerge faster, start tuber formation sooner and die sooner. The final plant size will be smaller and the plants will be more susceptible to drought. The total yield will be lower but earlier than from "younger" seed. To age seed potatoes, buy the seed in late fall or early winter before they break dormancy and store them rose (eye) end up in daylight at 50°F (10°C) until just before the planting date. In spring, reduce the temperature just before planting, to minimize the thermal shock.

Physiologically, "young" tubers will give a later harvest (low yields if dug early) and sub-optimal final yields. If this is your goal (I'm not sure why it would be) keep the tubers below 40°F (4°C) until planting. "Young" seed yields higher than "old" seed, though lower than "middle-aged."

"Middle-aged" tubers give the best yields for maincrop plantings (27 percent higher than young or old tubers). The pre-sprouting instructions which follow aim to produce "middle-aged" seed.

Pre-sprouting

Pre-sprouting, also called chitting or greensprouting, is a technique to encourage seed potatoes to start growing sprouts before you put them in the ground. Advantages include:

- getting an earlier start on growth in the spring;
- being less dependent on outdoor weather conditions;
- giving the potatoes more ideal growing conditions early on and so increasing final emergence rate;
- bringing harvest forward 10–14 days;

Potato plants emerging in spring. Credit: Kathryn Simmons.

- increasing yields by optimizing the number of sprouts per plant;
- making the cutting of seed potato pieces easier for new crew or CSA volunteers (the sprouts are more obvious than eyes);
- enabling cover crops or food crops to grow longer before the land is needed for the potatoes;
- giving you the chance to irrigate the soil if needed before planting.

To start the sprouting process, bring seed potatoes into a warm well-lit room, around 65°F–70°F (18°C–21°C), and set them upright in shallow crates or boxes, rose end up, stem (belly button) end down, for 2–4 weeks in spring, or 1–2 weeks in summer. If you have no space or time for chitting, warming the potatoes for a couple of weeks (maybe even just a couple of days) will be beneficial. Some people like to warm the potatoes in the dark for two weeks, then spread them out in the light for the last two weeks before planting. I don't know if the two-part process offers advantages, because I've never tried it. In the light, the growing shoots will green up and not become leggy and fragile. Make sure the potatoes have a moist atmosphere so they don't shrivel while they are sprouting. At this point don't worry if a few sprouts break off; more will grow later.

For extra-early spring planting, aim to sprout relatively few eyes per potato, so that the seed pieces can be relatively big with enough nutrients to get the sprout above ground. Do this by priming the seed potatoes at 65°F (18°C) until the eyes at the rose end just start to sprout. Store at 45°F (7°C) until two weeks before planting time, then finish the sprouting in warmth and light. The early sprouting of the rose-end eyes suppresses the sprouting of the other eyes.

In spring, the sprouts will grow considerably faster with indoor warmth than they would if planted unsprouted in cold ground, where they could take as long as four weeks to appear. Once planted, chitted potatoes will emerge sooner, which is always reassuring, and the weed competition will not be as fierce. Fewer seed pieces will die before emerging. And if weather prevents soil preparation when you had planned, just wait and know that your plants are growing anyway.

For summer planting, encourage sprouting success by storing seed pieces in a cool place 45°F–50°F (7°C–10°C) until two weeks before planting time, then sprouting them. This encourages the lower eyes as well as those at the rose end to sprout. For warm-weather planting, one sprout per seed piece is usually sufficient. Tubers with many sprouts can be cut into many seed pieces, which can save money. Planting seed pieces with many sprouts will cause only small potatoes to grow, as each stem is effectively a single plant and will be competing with the others for light and nutrients. Also, overcrowding can force tubers up through the soil, and they will turn green if they reach the surface.

Cutting seed pieces

Before planting, cut the seed potatoes into chunks about the volume of a ping-pong ball and weighing 1–2 oz (30–60 g) each, with the smaller fingerlings at ¾–1 oz (20–30 g). For cold-weather planting early in the year, have two sprouts per piece, which allows one for insurance if the first one gets frosted off after emergence. For warm-weather plantings, one sprout per piece is enough. The cuts should not be too close to the eyes. Avoid cutting thin slices or slivers, as these may dry out and die rather than grow. Reject any potatoes with no sprouts. The actual size of the seed piece has little effect on the final yield, so long as it doesn't shrivel before growing, and has enough food reserves to get the stem up into the sunshine.

Some people cut their potatoes a few days ahead of planting and put the pieces back into the crates to allow the cut surfaces to heal over. Some coat the cut surfaces with sulfur or bark dust to help suberization (toughening of the cell walls). We usually cut ours the day before planting, and I have not found healing to be a real problem. If using 10" (25 cm) spacing, we buy enough to plant 16–17 lbs/100' of row (around 1.2 kg/10 m). For 12" (30 cm) spacing, the recommendation is to allow 10–12 lbs/100' (7–9 kg/10 m). In practice, we need a higher seed rate, maybe 15 lbs/100' (11 kg/10 m). 12" (30 cm) spacing is more common, providing bigger potatoes than at 10" (25 cm), although yields may be lower.

Crop requirements

Potatoes benefit from generous amounts of compost or other organic matter (they use 10 tons/ac, 22,400 kg/ha) and will grow in soils with a pH of 5.0–6.5. They use high amounts of phosphorus (P) and potassium (K), and need adequate soil levels of iron and manganese. They are less affected by low levels of copper and boron. Hay mulch can be a good source of K. See the ATTRA publication *Potatoes: Organic Production and Marketing*. As Carol Deppe points out, potatoes will still produce an OK crop in poor soil, where you might not be able to grow much else.

Vegetative growth of potatoes is favored by warm, 80°F (27°C) moist weather, but tuber growth is favored by cooler soil conditions of 60°F–70°F (15.5°C–21°C). This combination can be achieved either by planting in spring, when the increase in soil temperatures lags behind increasing air temperatures, or by adding organic mulches to keep soil cool when planting in early summer.

The number of tubers produced per plant depends on hours of daylight, temperature and available water in the 1–2 week period of tuber initiation. Watering stimulates the production of more tubers. 5 gal/yd² (22.8 l/m²) is a good amount to supply when tuber formation begins. Short day length is optimal, with a night temperature of 54°F (12°C). If temperatures at night are 68°F (20°C), initiation will be reduced; and at 84°F (29°C), will be inhibited. High nitrogen also inhibits initiation.

The size of the tubers depends on subsequent growing conditions. Two or three weeks after flowers appear, the baby potatoes will be 1–1½" (2.5–4 cm) across. The best temperature is around 65°F (18°C), and I've read that the size decreases by 4 percent for every Fahrenheit degree (7 percent per Celsius degree) above the optimal. Spacing is another factor—we got large potatoes one

Planting and hilling

Potatoes need to have a good final depth of soil and/or organic mulch above the seed piece. All the new potatoes will grow out of the stem that grows up from the seed piece. None will grow below the seed piece, so be sure to plant deep enough and hill up and/or lay on thick organic mulch to provide plenty of space for your crop. A soil temperature of 55°F–60°F (13°C–16°C) is best for planting; 43°F (6°C) is a minimum. Row spacing of 32"–45" (80–115 cm) is common, with in-row spacing of 10"–15" (25–38 cm).

In early spring, when the soil is cold — if you want fast emergence and can hill up two or three times — you could plant shallow: as little as one inch (2.5 cm) deep in the North and four inches (10 cm) deep in the South. This technique helps avoid *Sclerotinia* problems. When the chilliness of deeper soil is not an issue, plant deeper, especially if your chances to hill might be restricted.

Start hilling (pulling soil up over the plants in a ridge) when the plants are six inches (15 cm) high. Hill again after two or three weeks and two weeks after that, if the plant canopy has not already closed over, making machinery access impossible. You can also increase the effective depth of planting by covering with straw or hay mulch. This is easiest to do immediately after planting, before the plants emerge (as we do in June). We don't mulch our spring-planted potatoes because we want the soil to warm up. During warm weather, deeper planting, hilling and thick organic mulches will help keep the plants cooler, as will irrigation.

We currently use a semi-manual planting method, making single furrows with our BCS walk-behind tiller, planting by hand in the furrows, then using the tiller again to cover the seed

For our June-planted potatoes, we hill immediately after planting, then unroll hay mulch bales over the rows the next day. The mulch prevents Colorado potato beetle, reduces weeds, keeps the soil cooler, and provides vertical space for the crop to grow. Credit: Kathryn Simmons.

summer because we had poor emergence and therefore wide spacing! The summer was certainly hot. Try to ensure they get at least one inch (2.5 cm) per week, up until two weeks before harvest. Avoid very uneven watering, or overwatering, as hollow heart could result.

For the earliest possible crop in a dry climate (but not the highest yield), plant "old" seed in early spring, hold off on watering until the tubers are marble size, then give a single good watering at 5 gal/yd² (22.8 l/m²).

pieces and hill. Tractor-mounted furrowers are less work, for sure, as they can do two or more furrows in each pass.

Weed control

Potatoes are sometimes said to be a "cleaning" crop, as if they did the weeding themselves. Not so! Any cleaning that takes place is a result of cultivation. Hilling can deal with lots of weeds in a timely way, especially if the machine work is followed up by the crew passing through the field hoeing. Organic mulches also reduce weeds. Potatoes produce a closed canopy that discourages more weeds from growing until the tops start to die.

·As with many plants, the initial growth stage is the most critical for weed control. Mary Peet reports that potato yields were decreased 19 percent by a single red root pigweed per meter of row left in place for the entire season.

In wet weather it can get impossible to hill when you'd like to, and this is where flaming can save the day. Potatoes may be flamed at 6"–12" (15–30 cm) tall. Beyond that, flaming is not recommended. See ATTRA for more on flame weeding.

Pests

Farmscaping with sunflowers, peas, vetch, buckwheat or small grains, to encourage ladybugs and lacewings, can make insect control unnecessary in a good year. Ground beetles and bats can consume surface and air attackers before you even need to look.

Colorado potato beetle is the most common pest potato growers get to deal with. The pink blob-like larvae of this beetle can eat enormous amounts of potato leaves while growing into bigger pink blobs. Left alone they can kill a planting. Acceptable amounts of defoliation are 50%–75% of the top leaves on a 6"–8" (15–20 cm) plant, 25% on a 12"–16" (30–40 cm) plant, a mere 10% at the critical full bloom stage, and up to 25% once full grown. As with many pests, having a few of them is not important—it's all about the numbers. Potato plants can tolerate up to 30% defoliation (depending on the variety) without loss of yield.

Mulching with hay or straw can prevent CPB from finding your potato plants—we never have to kill them on our summer planting. Our unmulched spring planting is a different matter. I scout the field once a week, counting adults and larvae on a hundred randomly selected plants. As soon as I see more than 50 adults or 150 large larvae or 400 small larvae per 100 plants, I unpack the sprayer. I do a spraying with Spinosad, a fermentation product of a soil bacterium. It kills insects by overstimulating the beetles' nervous systems. Usually one spraying is enough, although I continue weekly checks. Prior to using Spinosad, we used Bt, but the version for CPB that was generally available in small quantities was genetically modified, so we stopped using it, not wishing to be part of any support for GMOs. Neem and *Beauvaria bassiana* can also kill larvae. In the South, there can be three generations of CPB each year, so stay vigilant.

Flaming when the potatoes are less than eight inches (20 cm) tall is another effective control measure. Choose a warm sunny day when the pests are at the top of the plants. Flaming can kill 90 percent of the adults and 30 percent of the egg masses, according to ATTRA.

Leaf hoppers can cause leaves to shrivel and die. *Trichogramma* wasps parasitize the eggs. Garlic with insecticidal soap, sprayed early in the morning, especially on the undersides of the leaves, can control hoppers.

Wireworms (click beetle larvae) can tunnel through the tubers. Avoid planting potatoes the first year after turning under pasture. If you expect to have wireworms, use small whole seed potatoes rather than cut pieces. Cut slices of potato can be used to trap wireworms (dig up the trap pieces each day and kill the wireworms).

Cutworms can be a problem, eating the leaves from the bottom of the plant up (the opposite direction from CPB larvae). Once the plants are full grown, up to 75 percent loss of lower leaves is unimportant. At earlier stages, if you see any cutworm damage, dig around the stem and kill the worms.

Nematodes can be deterred by choosing appropriate preceding cover crops, or by applying 1–2 tons/ac (2.2–4.5 t/ha) of crushed mustard seed meal to the soil before planting. This will also reduce early weeds and act as a fertilizer.

Blister beetles can cause trouble later in the season, skeletonizing leaves and causing blisters on the skin of unwary and unlucky harvesters. We try to endure them, as the larvae eat grasshopper eggs.

Diseases

Late blight (*Phytophthora infestans*) is by far the worst disease to afflict potatoes. This is the disease that contributed to the famine in Ireland (caused by the profiteering of the English landowners, who sold the barley and left the tenant farmers to subsist almost entirely on potatoes). The disease, which starts as dark grey-black areas on the leaves, spreads fast and causes the foliage to die. The best defense is to always remove volunteer nightshades from your fields and compost or bury all crop debris. (See the Tomato chapter for more on late blight.) If it occurs late enough in

the season, you can save your crop by mowing off the foliage, raking it off and disposing of it, then leaving the field untouched for two weeks before harvesting whatever tubers you have. This prevents the spores getting into the soil and infecting the tubers. Disposing of large amounts of blighted foliage is no easy task. When I had to deal with late blight, back in the '70s, we made a fire and gradually added more tops as the previous ones burned. The fire was smoky and polluting — no doubt a contributor to global warming. Digging a big hole and burying the foliage would probably have been a better option.

Early blight (*Alternaria solani*) affects stressed or older plants, starting with the older leaves. The disease progresses fastest after the potatoes flower. It starts as small brown spots, which conglomerate into blotches that are restricted by the leaf veins, and so can be angular in shape. The lesions have a bullseye appearance (concentric circles). The manifestation of blight symptoms can be minimized by growing strong healthy plants, supplying sufficient water and spraying with compost teas. Brew one part compost to five parts water by volume for fourteen days before spraying. Consult ATTRA for compost tea information.

Rhizoctonia solani is commonly known as black scurf or stem canker. It is worst in cold wet soils. Early in the season it can cause sprout death. On older plants, red-brown stem lesions develop into cankers and the infection can spread to the tubers, which then become cracked and misshapen, and may have dead tissue at the stem end. There may be firm black sclerotia (small dried reproductive bodies) on the tuber. If you have trouble with *Sclerotinia*, you could dust the seed pieces with the commercially available fungal antagonists *Trichoderma viride* and *Trichoderma virens*.

The first time we tried our new Checchi and Magli SP100 potato digger we didn't realize we needed to remove the mulch first. Credit: Twin Oaks Community.

Harvest

Harvest for immediate use anytime you'd like after the tubers reach a big enough size. When the leaves start to turn pale, the plant has finished its leaf-growing stage and will be putting energy into sizing up the tubers under the ground. To grow for maximum yield, and to harvest for storage, wait until the tops are completely dead—if possible wait two more weeks to allow the skins to toughen up. If necessary, you can bring about an early vine death by mowing or flaming. This will also remove weed growth that could interfere with your digging equipment. To test for skin maturity, dig up a few potatoes and try rubbing the skin off with your thumb. If the skin abrades,

wait a little longer. Avoid irrigating at the end of the growing period or the potatoes may develop hollow heart, make knobbly secondary growths or even crack. Yields are likely to be 150 lbs/ac (168 kg/ha); 200 lbs/ac (224 kg/ha) is a good yield.

See the Resources section for more on various potato diggers. We use a Checchi and Magli single-row side delivery SP100 harvester. It does a good job in clean soil and an excellent job in clean fairly dry soil but gets stuck if we have a lot of organic material on the soil (weeds or mulch). The 1-row Potato Digger from US Small Farm Equipment, which a neighboring farm bought, has the same challenge. For our unmulched March-planted July harvest, we mow two weeks before our planned harvest, because we have a tight turnaround, with fall brassicas to follow— we need the potatoes to finish up on schedule! In hot weather the cut tops and weeds dry up. If it rains we mow again the day before harvest. For our fall-harvested crop, planted and mulched in June, we need to remove the mulch after the mowing and before the harvest. This is a slow job, but necessary, and the mulch makes a good addition to our compost pile.

We do as much preparation as possible the day before harvesting: we get the crates beside the patch and clean and air the root cellar. During harvest, someone walks alongside the tractor with a long-handled hook/claw tool, to clear blockages and hook any potatoes from the path of the tractor wheels. Then the rest of the crew follows, picking up and sorting the potatoes. If they are wet, we leave them to dry for a short time. We sort the damaged ones into "Use First" buckets and crate up the good ones. We try not to leave any potato parts in the field, to reduce the chance of spreading diseases.

After we removed the mulch, the machine ran well. Here are our Kennebecs ready to pick up. Credit: Twin Oaks Community.

Summer planted potatoes with hay mulch (after a windy day). Credit: Wren Vile.

In hot weather we aim to work until done and not leave any potatoes in the field baking for long. In cold weather we aim to get done before nightfall and not have any freeze overnight! During summer we don't put the crated potatoes straight into the cellar but leave them overnight covered with a tarp, under a tree. This lets them cool down, rather than heat the cellar.

Potato harvesting raises rocks to the surface, so we try to find time to collect them for use in road repairs and construction.

Post-harvest and curing

After harvest, potatoes need to cure for two weeks at a surprisingly warm temperature, 60°F–75°F (15.5°C–24°C), and 95% humidity. This will help heal over any cut surfaces. After two weeks, we sort through the crop for rot, then reduce the temperature to 50°F (10°C) for long-term storage, and to 40°F (4.5°C) in the winter. We aim to keep the potatoes below 50°F (10°C) once they are more than a month from harvest, so they don't sprout. Fluctuating temperatures can cause stress and physiological aging, which also leads to sprouting. See the chapters on Winter Vegetable Storage and Root Cellars for more about potato curing and storage conditions.

Mary Jo Frazier has researched the use of essential oils of mint and cloves to inhibit sprouting in storage. Frequent and repeated applications are needed, and then there is the issue of flavors carrying over into the tubers.

Chapter 53

Okra

Okra is a tropical annual in the mallow family, and is widely adapted where the frost-free season is long enough. Those in cold climates should choose fast-maturing varieties and transplant into black plastic. In areas with cold nights, okra can only be grown in a hoophouse. Those in hot climates will need to deal with its exuberant growth in midsummer. Okra is heat and drought tolerant and has few serious pests or diseases.

Varieties

We like Cow Horn okra from Southern Exposure Seed Exchange. It is one of a few varieties that can grow relatively large pods without their becoming tough. We are sometimes not good at finding all the pods when harvesting, so it is an advantage to us if they are still good to eat when bigger than normal. We have educated our cooks and diners to know not to make assumptions about toughness without testing with a thumbnail. If you are selling at market, to a changing group of custom-ers, this might not work. We get good yields and sturdy plants in our Zone 7 climate. The Kerr Center did an heirloom trial in 2008, described in *Growing Heirloom Okra*.

High-yielding varieties include Cow Horn (55 days), with 8"–14" (20–36 cm) pods on 7'–8' (2.1–2.4-m) tall plants. The large pods remain spineless and tender up to 10" (25 cm), but are best quality at 5"–6" (12–15 cm) long. Jade (55 days) is a high-yielding, early-maturing heirloom with few side branches and good ability to mature in late plantings. Plants average 4.5' (1.4 m) tall and the straight, dark-green pods remain tender to 6" (15 cm). Cajun Jewel (50 days) is another high-yielding, early-maturing, good-flavored heirloom. Short plants, 3'–4' (90–120 cm) tall, produce an early crop of spineless, tender 1" (2.5-cm) diameter pods up to 7" (18 cm) long.

Spineless (easy to harvest) varieties include Clemson Spineless (56 days) with dark green pods. Plants with few side-branches reach 4'–5'

(120–150 cm). Medium-sized ribbed pods grow to 8" (20 cm) long, but are best harvested when 3" (8 cm) or smaller. Evertender (50 days) is an unbranched, spineless variety growing 5.5' (1.7 m) tall and producing 7" (18-cm) pods of good eating quality, similar to Clemson Spineless.

Red-podded varieties include Burmese (58 days), a high-yielding dwarf heirloom that starts to bear when plants are 18" (45 cm) tall. It has huge leaves and slightly curved, almost spineless pods 9"–12" (23–30 cm) long that are tender and sweet enough to be eaten raw in salads. Pods are less glutinous than other okra. Red Burgundy (49 days) has magenta stems, emerald green leaves and 5"–8" (13–20-cm) slim spineless maroon pods. The short bushy plants are 3'–4' (90–120 cm) tall. Red Burgundy okra is reported to do well in "cooler" areas, although it will not do much until daytime temperatures reach 80°F (27°C).

An unusual heirloom okra is Star of David (61 days), which has a distinctive flavor that might be appreciated by okra lovers seeking something different. It can grow 8'–10' (2.4–3 m) tall, with few side branches. It bears a moderate yield of fat pods that can mature to 6" (15 cm) long, but are best harvested small. Star of David is tolerant of root-knot nematode.

Seed and crop requirements, and yield

Okra does best in well-drained, fertile, loamy soils with high organic matter. Wet clay soils can drown the plants. It grows best with a pH between 6.5 and 7.0, although as high as 7.6 is still OK.

Optimum soil temperatures for okra are 70°F–95°F (21°C–35°C). Germination takes 27 days at 59°F (15°C), 17 days at 68°F (20°C), 12.5 days at 77°F (25°C), 7 days at 86°F (30°C), 6.4 days at 95°F (35°C) and back up to almost 7 days at 104°F (40°C). The highest percentage emergence of normal seedlings is at 77°F (25°C).

There are approximately 20 seeds per gram; 5 g sows 50' (15 m) at 6" (15 cm) spacing and 6–7 lb of seed are required to plant one acre (7–8 kg/ha).

Average yields are about 3–4 tons/acre (6.8–9 t/ha) on bare ground in the South. Yields on black plastic are considerably higher. For smaller plantings, expect 50–100 lb/100' (7.6–15 kg/10 m). We grow 90' (27 m) for 100 people, which provides enough for some pickling too.

Sowing

According to a Rodale publication, *600 Answers*, germination speed can be improved by freezing the seed overnight, then soaking in hot water for 30–60 minutes before sowing. Otherwise soak the seed at room temperature for 24 hours. Another approach is to rub the seed in sandpaper, or nick the seed coat with a knife. For larger quantities, roll a piece of coarse sandpaper with the grit to the outside and use it to line a jar. Put in the seed, between the sandpaper and the glass, close the lid and swirl the seed around in the jar for a while.

When we are sowing, we "station-sow"—i.e., we put three seeds half to one inch (1–2.5 cm) deep at each spot where we want a plant to grow. We do this on May 1, with rowcover, as this is around our last frost date, and we want to avoid disasters! When seedlings have three to four leaves, we thin to the strongest seedling.

Usually we transplant — especially if we are intercropping with cabbages that have hay mulch, which cools the soil below what okra likes best. See below about intercropping. For transplants, we sow on April 15, using soil blocks or Winstrip 50-cell flats.

We harvest our Cow Horn okra every day except Sunday. There can be big surprises on Monday! Credit: Kathryn Simmons.

Transplanting

Okra has fragile taproots, so care is needed when transplanting. We transplant 3–4-week-old starts—a plant with three or four leaves is ideal—at 18" (45 cm) spacing in a single row down the middle of a bed. We transplant May 11, ten days later than the direct-sowing date. In the past we used wider in-row spacing, but found we could get a higher yield with the "hedge-like" closer spacing.

Some growers plant as close as 6" (15 cm) in the row, with 5' (1.5 m) between rows, or plant double rows with 12" (30 cm) between plants, and wider spacing between the beds. Overly thick planting requires very fertile soil, and risks diseases from poor air circulation. Overly wide spacing can lead to better-branched plants but perhaps a later start to the harvest, as flowering is delayed while the plant grows bigger.

Immediately after the initial transplanting, we bump up (repot) any leftover seedlings into bigger pots to fill any gaps a couple of weeks later.

Caring for the crop

Okra is a long-season crop, so cultivation or black plastic mulch will be needed to keep down the weeds. We hoe until hot weather arrives and then mulch with spoiled hay (using organic mulches earlier cools the soil and delays the plants). During the rest of the growing season, the okra bed becomes a useful repository for any mulch-like materials from other beds.

Okra has a great ability to withstand drought compared to other vegetables, but for good growth and production you'll need to water at least one inch (2.5 cm) a week. If there is an extended dry period and you can't water everything, okra will be the last to suffer.

Rotations

Okra is about the easiest vegetable to fit into a rotation, as you are unlikely to have any other mallow family crops to worry about, and are not so likely to be growing a huge amount. We find it easy to locate a spot that hasn't grown okra for the past several years. Cotton, hibiscus and the fiber plant kenaf are also in the mallow family, but do not cross-pollinate with okra.

Intercropping

Okra grows slowly until hot weather arrives. We sometimes take advantage of this and its upright growth habit to transplant it into a bed of early cabbage. We transplant cabbage in two rows along a 4' (1.2 m) bed on March 10 and the okra in a single row down the middle on May 11. At first the cabbages are relatively small and the okra uses the open space in the middle of the bed. As the plants grow, we remove any outer leaves of the cabbage that might overshadow the okra. Finally we harvest the cabbage and leave the okra to grow to full size. This method saves space and efficiently uses our time to help two crops with one weeding.

I have read of intercropping cucumbers and okra, giving each plant 3 ft² (0.3 m²). Again, this uses the very different growth habits of sprawled cucumbers and tall okra to get more crops from the same piece of land. Good soil fertility is needed if the two crops are not to stunt each other.

Succession planting

Generally only one planting is made each year, although it is possible to sow in spring and fall in hot climates, using a fast-maturing variety for the later sowing. Or to choose to sow only in the summer for a fall crop, and not fight cold spring soil temperatures.

Pests and diseases

Cool weather will stress okra, and the plants may then suffer from *Verticillium* and *Fusarium* wilts, soil-borne diseases that cause plants to wilt and die. Fight soil-borne diseases as usual—avoid soil splashing onto the plant by using drip irrigation and mulching. Foliage blights may occur, but generally they do not reach serious levels. Blossom blight can be a problem in long rainy periods.

Old varieties of okra tend to have deeper root systems and are more tolerant of root-knot nematode, to which okra is very susceptible. If you have nematodes, choose heirloom varieties. Okra will do fine after grains, such as sweet corn, or following a winter rye cover crop.

Insect pests that you may run into include Japanese beetles, stink bugs, aphids, corn earworms,

A basket of Burmese pods. Credit: Southern Exposure Seed Exchange.

flea beetles, blister beetles, and cucumber beetles. You may see ants climbing the plants to drink nectar but they don't seem to cause damage. Grasshoppers may strip the leaves in late summer in bad grasshopper years.

Harvest

Once the pods appear, we harvest six days a week, using pruners or a small, serrated knife. The stems are quite tough. Some people find their skin is irritated by the spiny leaves and like to wear long sleeves when harvesting. We tell crew to cut Cow Horn at five inches (13 cm) or bigger, and try to have the same few people do the picking on a regular basis for consistent results. Some years we have attached a piece of card to special pruners for the job, showing the size to cut. New crew sometimes lack a sense of size! We've also been surprised to find it necessary to show new people the difference between a rounded "empty" pointed flower bud and an angular firm pod.

For maximum yields it is important to harvest at least every second day. If pods are missed, they will mature and limit the future flowering and therefore the yield. It takes only a few days for a bloom to open, close, and produce a pod ready for harvesting.

Our harvest starts in mid-July and runs until frost, a period of twelve weeks or so. At the end of the season, we find we need to dig out the massive trunks and consign them to a spot on the edge of the woods that we call "The End of the World." This is where we put all our woody plant waste, as well as evil weeds such as wiregrass that we don't trust to the compost pile.

Post-harvest

Okra does not like to be chilled! Here we need to deal with the American tendency to store almost everything in the fridge. Chilling injury of okra causes dark damp spots on the pods, which lead to pitting and slimy breakdown. Okra can be stored at 45°F–50°F (14°C–15°C) in unperforated plastic bags for up to fifteen days. This compares favorably with only two days without a bag or seven in a perforated bag. The plastic bags keep the humidity high.

Okra is very tasty pickled, if you have high yields, time to do the pickling, and a demand for off-season value-added products.

Season extension

Black plastic mulch, rowcovers and transplanting can all help get the crop going earlier.

In hot climates, such as Texas and Alabama, okra plants benefit from being rejuvenated in the middle of the summer, by cutting them back to a six-inch (15-cm) stub after the first heavy harvest (perhaps in the fifth week of harvesting) and after market prices start to decline. For large areas, use a mower. The plants will produce again in the fall, when prices have risen again. Alternatively, to avoid losing all your crop at once, do this progressively along the row, perhaps cutting 5 percent of the plants each day, starting after they reach shoulder height. As well as revitalizing the productivity, this process keeps the plants at a manageable height. Side dressing at this stage can also boost production. This process is called "ratooning" in the Southeast. Fall yields of ratooned plants can be higher than spring crops.

Other growers tell of cutting back plants to a height of four feet (1.2 m), which causes the plants to branch more.

Seed saving

Saving seed from okra for your own use is a simple matter if you are only growing one variety. We have saved seed by decorating chosen plants with colored plastic surveying tape, like tinsel on a Christmas tree. Then we don't pick from those plants for eating. We wait until the pods are big and dry, then harvest them before they split and shed their seeds and put them in bags or cardboard boxes indoors to dry further and until we have time to deal with them. Mature okra seeds are greenish black. For small quantities, just twist the pods and break them open over a container of some sort. Seeds can be screened out of the mixture and/or winnowed. Empty pods make good weed-free mulch.

Okra is outcrossing but can self-pollinate. Varieties need to be isolated from each other by an eighth of a mile (200 m) for home use or a quarter to half a mile (400–800 m) or more for seed for sale. A plant population of 10–20 plants is needed for genetic diversity, and more than 20 plants is better for seed for commercial sale. It takes 30–50 pods to provide one pound of seed (66–110/kg).

Chapter 54

Sweet Corn

Sweet corn presents some challenges for organic farmers. Some of these can be easily overcome; others warrant careful consideration. There's no denying that sweet corn takes up a lot of space, so if you are really short of land, you may decide to forego it. On the other hand, corn doesn't take a lot of work, so if you have the space but are short of help, it's a good choice. The rewards are probably obvious: devoted, satisfied customers! USDA Certified Organic hybrid varieties (such as Luscious, Sugar Pearl and Brocade) are available in bulk quantities.

Genotypes and varieties

With most food crops, cross-pollination with other varieties or types is only a problem for those growing seed. With sweet corn, the seed *is* the food crop. Your neighbor's GMO sweet corn will cross with your corn, if it's close enough for the wind to bring the pollen in. The flavor of sweet corn can also be completely ruined by un-wanted cross-pollination. Isolate these three categories—super sweet/shrunken sweet corn varieties (sh2); all other types of sweet corn; and all other corn types (e.g., popcorn, field corn)—by at least 250' (75 m) from each other. Or isolate by time, i.e., sow on dates to achieve at least a 12–14 day gap between maturity of the different plantings.

Open-pollinated (OP): Most OP sweet corns are noticeably less sweet than modern sweet corn, so consider using hybrids for your customers even if you like OP corn for yourself. OP varieties also deteriorate faster after harvest than hybrids, becoming starchier. Luther Hill is said to be the sweetest OP variety.

There are several genotypes of hybrid sweet corn. Nearly all rely on one of two recessive genes, su or sh2. Cross-pollination with other corn groups will produce the dominant genetics of field corn, i.e., failure to isolate can lead to starchy unpleasant-flavored corn.

Normal sugary (su or ns): (su) types have old-fashioned corn flavor but are sweeter than OPs, although the sweetness disappears fairly rapidly after harvest. Most can germinate well in cool soil.

Sugary-enhanced (se) and sugary enhanced homozygous (se+ or se-se): (se and se+) types are more tender than (su), and slower to become starchy after harvest. Most, especially the (se+) types, are sweeter than (su) types.

Triplesweet sugary enhanced (se-se-se): was created to be sweeter than se-se.

Super sweet (sh2), also known as shrunken: (sh2) varieties are very sweet and slow to become starchy. If not isolated from all other types of corn, they will be very starchy and unpleasant. They have very poor cold soil germination. The kernels are smaller than other corns, giving this type its name. The seed needs careful handling, to prevent mashing between a seeder plate and the hopper.

Synergistic (se-se-se-sh2): these types are combinations of genetics from the three previous genotypes. Each ear has 75 percent (se) kernels and 25 percent (sh2) kernels. They are flavorful, tender and sweet, but only when they are ripe. If picked too soon, they are a watery disappointment.

Augmented shrunken: these newer types contain the sh2 gene and some of the tenderness from the se types.

"Bt corn" has been genetically modified by incorporating Bt (*Bacillus thuringiensis*) genes so the corn includes its own insecticide. There are many reasons not to grow GMOs, including the spread of random bits of genetic material by cross-pollination with previously non-GMO crops and the likely consequence of Bt-resistant insects, so I won't give them more space here. Note that Fedco Seeds tests for the presence of "drifted-in" GMOs, and will pull seed from sale if it exceeds a threshold.

Some catalogs (Johnny's, for example) indicate which varieties are suitable for certain latitudes (corn flowering is day length sensitive). Many companies run cold-germination testing and can tell you which varieties that year have good potential for early sowings.

For our location, we rely on the first three in this list, supplemented with some of the others for variety:

- Bodacious, 77-day (se) yellow, great flavor for one this early;
- Kandy Korn, 89-day (se) yellow workhorse;
- Silver Queen, 96-day (su) white longtime favorite with some drought tolerance and insect resistance;
- Luscious, 77-day (se-se-se) bicolor (organically grown, good cold soil emergence);
- Tuxedo, 80-day (se) yellow (tightly-wrapped, earworm resistant);
- Sugar Pearl, 72-day (se+) white (very early, on short plants);
- Argent, 86-day (se) white (tasty with tight earworm-defeating husks);
- Spring Treat, 66-day (se+) yellow, one of the earliest yellow sweet corns with good cold soil tolerance.

Crop requirements

A pH of 6.0–6.5 is ideal. Very fertile soil is needed, including high phosphorus. P deficiency shows up as purple leaf tips and margins; N deficiency as pale spindly stalks, yellow leaf tips and shriveled kernel tips; Mg deficiency as white-yellow striping between veins, with older leaves reddish-purple, perhaps with dead tips. Corn is sensitive to deficiencies of zinc or copper, but less so to low

We plant corn six times through the season, often with three varieties in each sowing. Silver Queen features almost each time. Credit: Kathryn Simmons.

levels of boron. When looking for deficiencies, it helps to know what is normal for that variety, and to also consider water shortage.

If you use legumes in the winter cover crop preceding the later sowings of corn, a good stand can provide all the nitrogen the corn needs (100–125 lbs/acre; 112–142 kg/ha). When the legume reaches its flowering point, the nitrogen nodules on the roots contain the maximum nitrogen.

The number of rows of kernels on the cob is set five weeks after emergence (although each variety has a number that is usual, under good conditions). Ear length and number of plants with double ears is established nine weeks after emergence. There's the feedback on your farm's fertility plan.

Sweet corn needs warm soil. Catalogs usually indicate the soil temperature (measured at 9 AM) recommended for each variety. 50°F (10°C) is the absolute minimum, and applies to treated seed and OP or (su) varieties only. 60°F (15.5°C) is better for most, and 65°F (18°C) or higher is required by some varieties. Natural II is an organic seed coating containing nutrients that help the seed get growing quickly and strongly. It performs almost as well as the old standard fungicide treatment, without the disadvantages. Common phenology signs for the season being advanced enough to sow corn are that oak leaves are the size of squirrels' ears and that ragweed is germinating. For us the first corn sowing date is usually around April 26, which is also our average last frost date.

Corn has no tolerance to frost. However, escape from a late spring frost is possible if the seedlings are less than two weeks old and not yet very tall, as the growing point may still be underground. Thus, in a spring that promises to be warm and dry, it is possible to risk an early planting as much as 2–3 weeks before the last frost date. Having some transplant plugs for a backup helps reduce the risk level.

Emergence takes 22 days at 50°F (10°C), 12 days at 59°F (15°C), 7 days at 68°F (20°C) and 4 days at 77°F (25°C).

Sowing

2 oz/50' (55 g/15m) or 1 lb/400' (370 g/100 m) are generally required. The (sh2) types have more seeds for the weight, because their shrunken seeds are lighter than other types—200/oz (7/g), 2,500–5,500/lb (5,600–12,300/kg); 1 lb/1,250' (118 g/100 m).

Corn is usually grown in rows 36" (1 m) apart. We sow fresh seed 6" (15 cm) apart, and if using seed from the year before, we sow at 4" (10 cm) apart. Depth of sowing can vary with the soil temperature, from a very shallow 0.5" (1 cm) in spring to one inch (2.5 cm) in summer, when soils are warmer lower down and seeds benefit from the extra moisture. Seedlings are thinned to 8"–12" (20–30 cm) apart.

Research by Ray Samulis at Rutgers showed that the yield-reducing effect of shading by the leaves of adjoining plants is much less at distances greater than 8" (20cm), confirming that this spacing is usually optimal. Upper leaves get seven to nine times the light of lower leaves, therefore it is important that the upper leaves are in good condition, to photosynthesize well. The lower ones get much less light, so may be broken off for easier harvesting.

Corn is wind pollinated (although you will find plenty of bees collecting pollen). For best pollination, plant in patches at least four rows wide. Inadequate pollination leads to ears with flat undeveloped patches among the kernels.

Corn seed must have moisture to germinate. If you have tractor equipment that supplies water while seeding, all well and good. If you use a push seeder, irrigate after sowing. Because we sow small areas of many different varieties, and because people love to plant corn, we sow by hand. Our method has the advantage of delivering water right where the seed needs it. We measure and flag the rows, put out ropes on stakes along the rows and make furrows (drills). The rope and its shadow make useful guides for keeping the rows straight. Next we have someone flood the drills with water from a hose, then hand sow into the mud. After covering the seed and tamping the soil, we ignore the patch until the seed germinates. The watering in the furrow reliably provides enough moisture to get the plants up out of the ground. The ropes (about 12"/30 cm above the ground)

deter the crows, making it hard for them to land near the buried seeds. This method is slower than machine planting, but for us the advantages outweigh the extra time.

We use overhead irrigation for corn. If you use drip tape, you might set out the tape, turn on the water for long enough to mark the soil with damp spots, then sow those spots with a jab planter.

Transplanting

It is quite possible to transplant sweet corn, so those in marginal climates don't need to give up hope. We usually prepare some plugs the same day we sow our first corn outdoors and use these to fill gaps at the first cultivation. We use 200-cell Styrofoam Speedling flats (1", 2.5 cm cells). We float these in a tank of water until we set them out. Some vegetable seedlings would drown if continuously in water, but corn does not. It is important to transplant the corn before the plant gets too big and the taproot takes off. Two- to three-inch (5–7.5-cm) plants seem OK. The plugs transplant easily using butter knives.

Caring for the crop

Generally, corn needs cultivating at least twice: once two weeks after sowing and once at four weeks. Even better are four cultivations: one at 7 days, a second at 14, a third around 21 days (when the plants are 6"–12" / 15–30 cm tall) and finally one around 35 days when they are 18"–20" (45–50 cm) high. We use a walk-behind tiller and follow up with hoeing and thinning. A wheel hoe can be a useful tool. After about 30 days, corn plants get too big to get machinery between the rows.

At tight spacing, adequate irrigation becomes more important. Never allow soil in corn plantings to dry out. More than one inch (2.5 cm) per week may be needed for maximum productivity, although corn is more drought tolerant than some crops. The most important times for watering are silking (when the silks first become visible outside the husks) and ear-filling.

There used to be a belief that it helped production to remove the suckers that came from the base of the plant. This practice has been tested and was found to damage plants and possibly even reduce yields (reports from Clemson in 2002 and Colorado State in 2004).

Flame-weeding can be used after planting, pre-emergence or, with care, after the crop is two inches (5 cm) tall, using a directed flame. Consult ATTRA for more details.

Undersowing

No-till planting into strips tilled in a white clover living mulch sounds good but has been found tricky, especially during the grower's learning curve. Jeanine Davis addresses this in NCSU's *Organic Sweet Corn Production*. The clover may outcompete the corn, becoming invasive and hard to get rid of. Soil temperatures will be lower (a disadvantage in spring) and slugs and rodents may abound.

A more successful approach is to sow a cover crop into the corn at the last cultivation, 28–35 days after emergence. We undersow our corn with soybeans (oats and soybeans for our last planting). Although they don't supply the highest amount of nitrogen, they are cheap, quick, somewhat shade tolerant and can withstand the foot traffic during harvesting. Other growers sow forage brassicas. Research has shown that this does not depress corn yields. The brassicas can be harvested for forage after the sweet corn harvest is

finished. Undersowing with white clover is also possible.

Successions

In order to have a continuous supply of sweet corn all summer, a bit of planning and record keeping is called for, so that each year's plan can be fine-tuned. The easy and approximate method of getting a good supply is to sow more corn when the previous sowing has three or four leaves or is one to two inches (2.5"–5 cm) tall. For a more even supply, sow several different varieties, with differing days to maturity, on the same date. We sow Bodacious, Kandy Korn and Silver Queen on the same day, and get over two weeks of harvests.

To fine-tune to get the most even supply, nothing beats real information about what happened, written at the time it happened. We have a Planting Schedule on a clipboard in the shed, and we write down actual sowing dates (next to the planned sowing date) as well as harvest start and finish dates. Having graphs of sowing and harvest dates for each crop has been very useful for planning effective planting dates. See the Succession Planting chapter for details on how to calculate best planting dates and intervals for a continuous supply.

We make six plantings to provide fresh eating every two weeks: April 26, May 19, June 6, June 24, July 7 and July 16. The planting intervals are 23, 18, 18, 13 and 9 days. Because we plant three varieties, new corn comes in three times during each two weeks. To calculate the last worthwhile sowing date, add the number of days to maturity and the length of the harvest window (7–14 days) and subtract this number from your average first frost date. For our October 14 frost date, using an 80-day corn as an example, 80 + 7 = 87 days, which brings us back to July 19 for our final sowing date.

Rotations

Three out of ten row crop plots in our garden are partially or wholly filled with corn. We make sure we have two or three years without corn between the corn years. Happily, corn is the only crop member of the grass family that we grow. Here's the crop sequence for each of our six sowings:

Plantings 1 & 2: Winter-killed cover crop of oats or sometimes a clover patch which has been growing all the previous year. Early corn, followed by garlic planted in early November in half the patch and oats and soy in the other half.

Plantings 3, 4 & 5: Watermelons the previous year, followed by a winter cover crop of rye with crimson clover (if sown before Oct 14) or Austrian winter peas (if later in October). Corn followed by more rye and crimson clover in October.

Planting 6: Winter squash the previous year, followed by a winter cover crop of rye and winter peas. This corn is undersown with oats and soy as its following cover crop. We mow high after harvest and leave the oats and soy to grow until winter-killed. This makes the patch easy to bring into use for potatoes in March.

Pests and diseases

Crows and other birds can be troublesome, removing the seed before it even grows. We leave the row-marking ropes in place (when hand sowing), or put some sticks and string in after machine sowing. Bird-scaring flash-tape may be even more effective. Rowcover would also work.

Some say interplanting corn with big vining squashes deters raccoons and other critters, but I think it deters crew too!

There are several caterpillar pests. An integrated organic approach to keeping pest numbers below economically damaging levels includes crop rotations, tillage, choosing resistant or tolerant varieties, encouraging beneficial insects and ensuring adequate fertility and water. The next step is to scout for pests regularly and take action as required.

Corn ear worm (CEW) is the most common pest. There may be six generations a year in the South. These caterpillars can bite—it's just a nip, but it can be a shock! A first line of defense is to choose corn varieties with tighter husks, which are harder for the worms to get into (Bodacious, Tuxedo, Silver Queen). Natural predators can be encouraged by planting alyssum or other small, open-flowered plants. You could buy *Trichogramma* wasps. Caterpillars can also be dealt with putting a few drops of vegetable oil in the tip of each ear. Mixing with Bt gives better results, when applied two to three days past the full-brush stage of silking. The Zea-later, available from Johnny's, is a tool for applying this treatment. If pest numbers are not too high, you could simply cut or snap the ends off the ears.

European corn borer (ECB) drills through the whorl of leaves of the young plants, leaving a pattern of large holes as the plant develops. Bt and Spinosad will kill these, as will *Trichogramma* wasps. To reduce damage in future years, be sure to mow and disk old corn stalks into the soil at the first opportunity. Organically farmed soils have less of a problem with ECB.

Fall armyworms (FAW) are also killed by Bt and Spinosad. These three pests (CEW, ECB, FAW) can be monitored in a single program, starting when the corn plants are at the whorl stage. At that point, scout for FAW and treat if

Silver Queen is a reliable white variety. Credit: Kathryn Simmons.

For a more complete description of corn insect pests, I recommend the 2004 *Organic Insect Management in Sweet Corn* by Ruth Hazzard and Pam Westgate. It includes good photos of the beasties and detailed instructions on applying vegetable oil and Bt for CEW, as well as treatment for ECB and FAW. Cornell has a good *Resource Guide for Organic Pest and Disease Management*. Search under "Crop Management Practices" for Sweet Corn. The 2001 video *Farmers and their Ecological Sweet Corn Production Practices*, produced by Ruth Hazzard and Vern Grubinger, is available from the UMass Extension Bookstore.

Corn smut fungus (*Ustilago maydis*), known in Mexico as *huitlacoche*, is edible at the stage when the galls are firm and tender. The flavor is sweetish. Silver Queen is the variety "best" at producing this fungus, should you wish to grow it. We carefully harvest the infected ears (or pieces of stem) into a special Smut Bucket, trying not to scatter the spores. Because none of us like this delicacy, we take it to the compost pile.

Harvest

Harvest corn before daybreak for best flavor, because the sugars manufactured in the plant the day before become concentrated during the night. We're not that dedicated. We harvest ours in the morning, and hurry it to the walk-in cooler.

Harvest may start 18–24 days after half the ear silks show, if the weather has been reasonably warm. Judging corn's ripeness is a skill, based on information from many of the senses. The first sign we look for is brown dead silks. If the ear has passed that test, we investigate further. All ears should look and feel plump and rounded to the tip. Each variety is a little different, so close attention is needed. Some varieties exhibit "flagging"

more than 15 percent of your plants are infested. At the pre-tassel and tassel stage scout for ECB and FAW. If infestation exceeds 15 percent, make a foliar spray with Bt or Spinosad. Check again in a week and repeat if needed. Then at the early silk stage, look for CEW and if needed, inject oil in the tips. If you also see ECB moths, apply Bt or Spinosad.

Cutworm can be a problem following sod, or if there are adjacent grassy areas. Bait them with bran, cornmeal or hardwood sawdust mixed with molasses and water—these swell inside the pests and kill them.

Corn rootworms are best controlled by rigorous rotations.

Young sweet corn (and a fiber banana plant). Credit: Bridget Aleshire

of the ear, meaning it leans away from the stalk as it matures and gets heavier. New crew can test for ripeness by opening the side of the husk with thumbnails and puncturing a kernel: the kernels should look filled-out and squarish, not round and pearly; the juice should be milky, not watery or doughy. The advantage of opening the side of the husks is that it is possible to close the gap if the ear is not ripe, without risk of collecting dew or rainfall. If the ear is ripe, we bend it downwards, give it a quarter-turn twist, then pull up away from the plant.

We harvest every other day, which balances getting the amount we need with not spending more time than needed picking. Such a schedule can work well for CSA farms. Other growers could well need to harvest every day, if daily fresh corn is what your market needs. Leaving a three-day gap risks poor quality starchy ears and a lower total yield.

Take steps to keep the crop cool while harvesting. Never leave buckets of corn out in the sun. Even at room temperature, harvested OP ears lose half their sweetness in 24 hours.

After harvest, cool the corn quickly. Hydro-cool if you have a large operation: drench or immerse the crop in near-freezing water. Otherwise, simply refrigerate and keep the corn cool until it reaches the consumer.

Season extension

Transplanting can provide an earlier harvest, as already mentioned. Clear plastic mulch is sometimes used to increase soil temperature and germination rate and to conserve moisture, producing earlier maturing corn. The plastic is spread over the seeded beds and slit when the seedlings emerge. It can be cut and removed thirty days after emergence. Weed-free seedbeds are needed for this method to work organically, and plastics disposal is an issue. Rowcover is another way to warm soils (and keep birds off).

Seed saving

Isolate corn varieties by 600' (180 m) for home use and half to one mile (0.8–1.6 km) for commercial sale. Save at least 500 seeds from at least 200 plants (or 10 percent of the plants in the field, if that's a higher number) in order to maintain vigor and genetic diversity. Don't be tempted to save only two big ears!

Credit: Lisa Dermer

Sweet Potatoes

Sweet potatoes are related to morning glory, in the genus *Ipomoea*. They are not yams, even though they are often called yams! True yams are a tropical species of tuber (genus *Dioscorea*). Sweet potatoes are roots, not tubers, and will not even cross with yams. So forget yams. Unlike white potatoes, which have the annual plant sequence of vegetative growth, flowering and dying back, sweet potato plants would continue growing forever if the weather was warm enough.

Fitting sweet potatoes into a rotation is easy because it is unlikely that you are growing anything else in that family. They are a frost-tender crop, so there is time before planting them to grow a winter cover crop mix including legumes to the flowering stage. They thrive in hot weather and are fairly drought-tolerant once established. Sweet potatoes fit easily into the timing of growing other crops. Started after the first flush of seed-starting, the slips (young plants) grow with little attention. Planting out comes later than most spring crops.

After the vines cover the ground they need little care during the summer (apart from irrigation) until harvest. They leave you free when there's lots to do elsewhere. There are modern varieties that grow to a good size in as little as 90 days, so they are not just for the South!

You can provide a long-keeping crop of delicious roots for filling those CSA bags at the end of the year, or for sale in late fall and winter. At Twin Oaks we aim to grow enough to supply us from November to early May. One baked 4 oz (115 g) sweet potato has 185% the recommended daily allowance of vitamin A, 28% the vitamin C, 100% of vitamin E, lots of antioxidants, and 160 calories, none from fat.

Sweet potatoes do not require a lot of organic matter in the soil, nor high fertility levels.

The mystique of sweet potato slips

Sweet potatoes are not grown from seed or from replanted roots, but from "slips," which are pieces of stem with a few leaves, grown from a mother root. We used to buy bare-root sweet potato slips to plant, believing growing our own would be very tricky. The collapse of our supplier and our desire to have organic plants (plus a need to reduce our expenses one year), pushed us into growing our own. We had some problems initially, so I can warn you about how not to do it. Now we have a system we really like, and we've found several advantages of homegrown slips over purchased ones.

With purchased slips, we had to specify a shipping date months ahead, then hope the weather sprites would be kind. We had to jump to when the plants arrived, and get them all in the ground pronto, to keep them alive as best we could (because their roots needed moisture). We accepted as normal a certain amount of drooping. We can have late frosts, spring droughts or El Niño wet springs, and climate change is only adding to the uncertainty. With homegrown slips we can delay planting if that seems wise; we can plant them in stages rather than all on one day. The transplants don't wilt. We can grow them big and plant them with three to five nodes underground, giving more chance of survival in heat or frost. We can keep some spares on hand to replace casualties. The sturdy plants get off to a strong start, which could be an even bigger advantage further north where the season of warm-enough weather is on the short side for a 90–120-day plant. And we are self-reliant—we never have to spend money on them.

Planning ahead

Decide how much space you want to devote to sweet potatoes, or how many you want to grow. One slip will produce a cluster of 4–10 roots, each weighing 3–17 oz (80–500 g). The yield range is 276–805 lb/1,000 ft² (14–40 kg/10 m²), or 6–17.5 tons/ac (13.5–39 t/ha). Yield depends on climate, spacing and length of growing season.

Planting space is 6"–18" (15–45 cm) in the row—wide spacing gives larger roots, mostly out of favor these days, as families are small. We do 16" (40 cm) as we have drip tape with emitters every 8" (20 cm) and we like to get some jumbos (our cooks cater for a hundred people). 12" (30 cm) would be a good space to start with your first year. The space between rows can be 32"–48" (80–120 cm). The vines are rampant once they get going. Calculate how many slips you'll need and add 5–10 percent. Each mother root can produce ten or more slips, depending how much time you allow.

Selecting mother roots

Ideally, seed roots would have been selected from high-yielding plants at harvest. If you didn't do that, choose small to medium-sized roots of typical appearance (no rat-tails!). Each root will produce about the same number of slips, regardless of size, so there is no advantage in selecting jumbos. If you are buying sweet potato roots and you are in the North, choose a variety with a short number of days to maturity. There are two tests we do before the initial two weeks conditioning, to choose good virus-free roots, so allow an extra 20 percent at this stage.

Timing

Figure out your ideal planting date and work back to find your starting date. Planting is usually done about two weeks after the last frost. The soil temperature should reach at least 65°F (18°C) at 4" (10 cm) deep on four consecutive days. For us, that's around May 12. It takes eight weeks to grow the slips, and the roots produce more slips if conditioned for two weeks (or even four), before you start to grow slips. So start ten to twelve weeks before your planting date. We now start March 1. Here's where I made my first big mistake—following directions written for much further south, I tried to start growing slips in mid-January. Dismal fight against nature!

Testing

First test the roots in a bucket of water—the ones that float are said to yield better and produce better-flavored roots. Next, test for viral streaking—also known as color breaks or chimeras, where paler spots or radial streaks appear in the flesh—and discard roots with pale spots or streaks wider than a pencil lead. Cut a thin slice from the distal end of each root—the stringy root end, opposite the end that was attached to the plant stem. All the sprouts will grow from the stem end, so don't cut there! If you can't tell the difference between the ends, you can ignore this step and plan not to propagate your own slips for more than a couple of years (so the virus load doesn't get too high). Or if you are a home gardener dealing with a small crop, you could keep the slips from each root separately and cut up the mother root before planting and then discard the slips from streaked roots. An option for commercial growers is to check some, which become your seed stock, and are planted in a different plot from the market stock (the unchecked ones, for sale).

Conditioning

Put the chosen roots in flats, boxes or trays, without soil, in a warm, moist, light place for two to four weeks. Ideal conditions are 75°F–85°F

The conditioned mother roots are planted in flats of compost and kept warm and moist. Slips are cut off each day as they become tall enough. Credit: Kathryn Simmons.

(24°C–29°C) at 95% humidity. This can double or triple the number of sprouts the root will produce in a timely manner. We use our germinating chamber, which is an old glass door refrigerator heated by a light bulb. Conditioning after testing allows the cut surfaces to heal before they are covered by compost. The environment for sprouting the roots is similar, so you can probably use the same location.

Sprouting

At this point, I'll mention another path not to take—the first time I tried this (reading information from further south) I was puzzled by talk of using coldframes. Ours were freezing cold at that time of year. I set up a soil warming cable in a cinder-block-enclosed bed on the concrete floor of our greenhouse. This is how I discovered that most soil warming cables have thermostats set to switch off the heat at 70°F (21°C). I just couldn't get the soil warm enough.

Set up a place with light, humidity and ventilation at 75°F–85°F (24°C—29°C) and with about 12" (30 cm) of headroom. Plant the selected roots almost touching each other, horizontally in free-draining light potting compost in flats or crates. Water the boxes and put them to sprout. Once again, we use our ex-fridge germinator. Using boxes is much more manageable than having the roots loose in a big coldframe. Indoor spaces are much easier to heat than the great outdoors! Boxes can be insulated and put on a bench at a decent working height, with lights or heat lamps over them. Keep the compost damp, and if your planting medium is without nutrients, give liquid feed occasionally once sprouting starts.

For small quantities of slips, it is possible to sprout the potatoes half-submerged in water, either in trays of water or by suspending a sweet potato impaled on toothpicks, resting on the top of a glass of water. For larger quantities I recommend our method.

Cutting and spotting the slips

After 5–7 days, the roots will begin to produce slips. Ideally, wait until the slips are 6"–12" (15–30 cm) tall with 4–6 leaves, then cut them from the root and stand them in water. If necessary, cut them a bit shorter. Some people pull or twist the slips from the roots, but this could transfer diseases by bringing a small piece of the root with the sprout. I cut the slips daily, bunch them in a rubber band and stand them in a small bucket of water. The slips will grow more roots while they

The cut slips are stored in water until roots develop. Credit: Kathryn Simmons.

are in water for several days, which seems to be an advantage.

Once a week I spot (plant) the oldest, most vigorous slips (with good roots) into 4" (10-cm) deep wood flats filled with compost. The spotted flats require good light in a frost-free greenhouse and sufficient water. If you are two weeks away from your planting date and are short of slips, you can take cuttings from the first flats of slips that were spotted, to make more. The slips planted in flats become very sturdy, allowing flexibility about planting dates and a longer slip-cutting season. About ten days before planting, start to harden off the flats of slips by reducing the temperature and increasing the airflow.

It's also possible to skip the spotting stage and transplant the slips outside directly from the water, but I don't think this is as good as spotting them into flats of good compost for a few weeks.

Once a week we plant the rooted slips in flats.
Credit: Kathryn Simmons.

Planting

It works well for us to do two plantings a week apart, using the older slips first. Then we make a third visit after another week to fill any gaps. It's better to wait for the spotted slips to grow four leaves or more in the greenhouse before planting, rather than rush them outside. Sweet potatoes are often hilled, to reduce flood damage. You can do this before planting or a little later, once the plants are established. It is important to plant into damp soil, and drip irrigation is a real advantage here. Set out the drip tape, turn on the water and then plant by each emitter (or every other one, depending on your situation). For big potatoes, plant the slip vertically. For average size roots but larger crops, plant horizontally 2"–3" (5–7.5 cm) deep. Have 3–5 leaf nodes underground and only the tips above the ground—this gives the plants a second chance if frost strikes. If, on the other hand, you are planting in hot dry weather, water the soil first, and keep the roots enclosed in damp or wet compost as you plant.

In colder areas, black plastic mulch can be used to warm the pre-formed ridges for about three weeks before you plan to plant. This will increase the rate of growth and the yield.

Development

The first month after transplanting is the root development stage. Roots can go 8' (2.4 m) deep in 40 days, so don't be alarmed at the lack of above-ground action. Give one inch (2.5 cm) water per week as needed and cultivate to remove weeds.

The second month or so is the vine growth stage. The roots begin to store starch and sugar close to the stem base. Cultivate until vines cover the ground, after which very little weeding will be needed. We have sometimes done one walk-through to pull or clip pigweed. One year we sowed buckwheat between the rows, intending to till it in before the vines really "ran." You can guess what happened. We ended up wading in to hand pull the buckwheat to prevent it seeding. I'm not sure whether re-seeding would have become a problem or not. Possibly the vines would have prevented germination of the self-seeded buckwheat.

During the last month of growth for that variety (third or fourth month), the potatoes develop and the vining stem stops growing. They have no pre-programmed finish date for root maturity, so you can choose when to dig them up, ahead of cold weather. The longer you wait, the bigger the potatoes, but you are gambling with the weather. Most varieties need 90–110 days from planting out to reach a good size.

Afflictions

Sweet potatoes are often relatively problem-free, but here are a few afflictions you might come across and the reasons they happen, so you know what to do differently next year:

- Round chunky roots, low yield, purple color: planted too early, when it was too cold;
- Rough irregular shaped roots: heavy clay soils or organic matter above 2%;
- Low yield of storage roots: flooded or crusted soil, particularly 6–7 weeks after planting;
- Souring: tissue breakdown caused by poor soil aeration, for instance due to flooding;
- Water blisters: small whitish raised bumps around the lenticels (breathing holes), caused by wet soil;
- Cracking: uneven water supply or too much late-season water;
- Long, slender malformed roots, reduced yield: potassium deficiency;
- Small raised bumps appearing several months into storage—"blister": boron deficiency;
- Fine hairline cracks: also a boron deficiency symptom;
- Brownish surface discoloration, worse in wet years; superficial damage, but roots shrivel in storage: scurf, a fungal disease more likely to occur if compost applications were too generous;
- Metallic black lesions on the root surface, maybe covering most of the root surface: black rot fungus, *Certocystis fimbriata*. Internal decay is not deep, but the fungus may impart a bitter flavor;
- Sunken brown lesions that sometimes encircle the root: ring rot, *Pythium* fungus;
- Sunken lesions that dry and may fall out: circular spot, *Sclerotium rolfsii*. The underlying tissue may taste bitter;
- Hard, dry, black, sunken spots developing in harvest wounds: *Fusarium*. Spots may be-come larger than two inches (5 cm) diameter, but damage is not deep;
- Pitting: caused by soil rot or soil pox fungus in the presence of water stress. Roots will be small and malformed. *Streptomyces* root rot bacterium causes a similar rot;
- Numerous fine or coarse irregular cracks and browning of the surface; dry, corky, dark-colored clumps of tissue scattered throughout the flesh, becoming worse if roots are stored at temperatures higher then 60°F (16°C): russet-crack/internal cork, feathery mottle virus (yellowing of leaves in feathery patterns). Do not use as seed stock.

See the North Carolina Sweet Potato Commission commercial growing page for more diseases and a great selection of photos.

When to harvest

Usually sweet potatoes are harvested the week the first frost typically occurs in your region. Aim to harvest on a mild day—above 50°F (10°C)—to avoid chilling the potatoes. Don't wait until soil temperatures have been below 55°F (13°C) for several hours. If frost does strike before you have harvested, waste no time—get them up within a few days. If the days are warm, a couple of light frosts will not harm your crop. Contrary to myth, there is no toxin that moves from frozen leaves down into the roots. On the other hand, cold injury can ruin the crop, and roots without leaf cover are exposed to cold air temperatures and have lost their method of pulling water up out of the soil. Cold wet soil can quickly rot sweet potatoes (I know, it's happened here). If your soil is very dry, irrigate the field before harvest, to avoid scratching the skin with hard clods of soil.

How to harvest

Before you start your harvest, if your winters are relatively short, you could consider propagating sweet potatoes for the next year by taking vine cuttings from the growing tips in the fall, rooting them in water, then potting them up as house plants for the winter and keeping them warm, to provide early slips next spring. (But perhaps, like me, you don't "do" houseplants, because you appreciate an indoor space where you don't need to think about keeping plants alive!)

To harvest, first remove the vines from the area to be harvested that day. If there is more than one day's digging, leave live vines to protect the rest of the crop. We use pruners to snip the vines where they emerge from the soil, leaving stumps to show where to dig. We roll the vines into the spaces between the rows—ours are four feet (1.2 m), so if you have close rows you may need to roll the vines further away. Pitchforks or digging forks (spading forks) can be useful tools for moving vines, but rakes just don't seem strong enough. Mowing may be a possibility, but it's important not to damage the roots, which sometimes stick up out of the ground.

Using digging forks, we carefully dig up the roots, which grow in the ground in a bunch-of-bananas shape. We want to select good potatoes for seed, and we grow several different kinds (Georgia Jet, Beauregard and a couple of heritage varieties whose names we don't know), so we make sure not to mix potatoes from different rows. As we dig, we set the potatoes out beside the spot they've grown, one clump per plant, so it's easy to identify the most productive plants. It's important not to drop, throw or in any other way bruise the roots.

Don't leave roots exposed to temperatures higher than 90°F (32°C) for more than half an hour or they will get sunscald; below 55°F (13°C), they'll get chilling injury. Avoid any abrasion of the skin, which is very fragile at this stage. Do not wash the potatoes, and never scrub them. I've even seen sources recommend wearing cotton gloves to handle the roots, but we don't go that far! Leave them to dry on the ground for one to three hours, unless the weather is unsuitable.

Selecting seed potatoes

Save at least one root per five slips wanted. Do not save for seed any roots exhibiting disease symptoms. Damage due to poor growing conditions can look like a disease, but as it isn't, it will not carry over to the next crop.

Choose plants with a high yield and no string (rat-tail) roots. From these plants, choose small-to medium-sized potatoes with typical appearance (shape and color). If you want to be sure to avoid saving roots with color breaks, you can cut a small slice from the distal end (the end distant from the plant) for examination (see the Testing section above for more on this). The cut surfaces will heal over during curing.

Sorting and crating

When grading and crating the roots in the field, first select seed potatoes, if you want to grow your own slips next year. Next, sort storable from "Use First" roots. Large open broken surfaces will cure and can be stored, but any roots with soft wet damaged areas or deep holes (whether from bugs or fork tines) will not store and should be graded out, for composting or immediate use at home. We sort into wood flats for curing and buckets for the "Use First" category. Mechanical harvesting is done with a modified disk or moldboard plow with a spiral attachment. Potato harvesting machines usually do too much damage if used for sweet potatoes.

After harvest is complete, we disk the area and sow cover crops. It's usually too late in the year for oats. Sweet potatoes inhibit the formation of nitrogen nodules, so following with a legume may not be very successful. With our average first frost on Oct 14, we can sow wheat, winter rye and Austrian winter peas up till Oct 31; crimson clover or hairy vetch till Oct 10. We prefer winter wheat after the sweet potatoes, sown around Oct 25, because we use that area next year for spring white potatoes in mid-March, and rye takes too long to break down early in the spring.

Curing

Cure the sweet potatoes immediately after harvest, field drying, sorting and crating. Curing allows the skin to thicken, cuts to heal over and some of the starches to convert to sugars. Uncured "green" sweet potatoes are not very sweet at all, and are better used in dishes where they combine with other foods. A baked uncured sweet potato is a sad disappointment. Immediately after harvest, bring the boxes of sweet potatoes into a warm damp basement or other indoor space. Ideal conditions are 85°F–90°F (29°C–32°C) and 80%–95% humidity for 4–7 days. There also needs to be some airflow and ventilation. Curing takes longer if conditions are less then perfect. The length of the curing period also varies with the dryness of the soil just prior to harvest. We usually reckon on 10–14 days.

To test if curing is complete, rub two sweet potatoes together. If the skins scratch, they need longer. Once sufficiently cured, the roots can be stored until needed.

In the past we used our greenhouse to cure sweet potatoes, but it really is too hot, sunny and dry. Nowadays we use a basement with the door to the adjoining furnace room open. We stack our 4" (10 cm) deep flats of roots on pallets, and put wooden spacer sticks between boxes in each stack, to ensure airflow. We get quite good temperatures, but keeping humidity up is difficult for us. We cover the flats with newspaper to hold in some moisture. Some people use perforated plastic. Perhaps rowcover would work? We have also used domestic humidifiers and we've tried hanging strips of wet cloth from the ceiling. The best result seems to come from splashing water on the concrete floor several times each day. We use box fans to improve the airflow, and the basement already has some natural ventilation.

Storage

Ideal storage conditions for sweet potatoes are 55°F–60°F (13°C–16°C) at 85%–90% humidity, with one air change each day. Above 60°F (16°C), shrinking and sprouting may occur, and below 55°F (13°C), permanent chilling injury with the nasty name of hard core can happen. We had a disastrous case of this one year. The potatoes remain hard no matter how long you cook them, and are useless. Never let the temperature drop below 50°F (10°C). As far as I can tell, sweet potatoes do not need to be stored in the dark.

We store our sweet potatoes in a rodent-proof "cage" in our basement. We stack the boxes directly on top of each other and this seems to keep enough moisture in. This way, sweet potatoes will store into May and early June, by which time interest in them is waning, and we have plenty of newer fresher kinds of vegetables to offer.

Chapter 56

Seed Growing

The demand for organic and heirloom seeds is growing, and seeds can bring a good financial return for the time and land invested. There are both practical and political reasons to grow seed crops, and perhaps new skills to learn along the way. Growing seed allows you to improve on a variety, or simply maintain it for bulk sale. In this chapter I'll write about getting started growing a few seed crops alongside vegetable crops.

Starting small

You could start small, by growing one or two seed crops for yourself. Growing seed for use on your own farm is a valuable project, as you can select plants that grow especially well under the conditions on your farm, as well as save on seed costs.

When you are ready to grow a commercial seed crop, contact seed companies before the start of the season to agree on a contract. Read up about seed growing and the isolation distances re-quired for your particular crop. Cucurbits need as much as a half-mile (800-m) distance from other flowering plants of the same type, whereas tomatoes only need 180 feet (55 m) at the most. Be sure to have a large enough population of plants to ensure a diverse genetic pool. With self-pollinators (inbreeders) such as beans, twenty plants may be enough, but for out-breeders (cross-pollinators) you will need at least a hundred to avoid inbreeding depression.

USDA Certified Organic seed is much in demand, but uncertified sustainably/ecologically grown seed also has a market, especially for heirloom or heritage varieties.

It's possible to sell your seeds directly by joining Local Harvest and selling on their website, where there are over a thousand entries on seed for sale. You can create a free listing with Local-Harvest.org if "You are a direct marketing family farm that does not grow GMOs, a producers' farmers' market, a business that sells products made from things grown locally by family farms, or an organization dedicated to promoting small farms and the 'buy local' movement." Even eBay now has heirloom seeds! Most seed growers continue to grow a mix of crops—a seed crop, like any other crop, could fail. But if seed growing really suits you, you could move more towards growing seeds and away from other crops. See Ira Wallace's chapter on the business side of growing seeds, which follows this one. Vegetable growers can fit seed growing into their schedules because seed crops have a longer, slower season and a less pressured pace most of the year. (After all, while the stuff is growing, you don't harvest till the seeds are ripe.) Some seed crops (okra, winter squash) can sit around for a while drying, with no particular hurry. Others will need more immediate attention.

What type of crop to grow?

Choose an open-pollinated variety of the crop you want to grow. Hybrids, which are produced by deliberately crossing two genetically different varieties, do not "run true"—that is, seed saved from hybrids produces very mixed progeny.

Know if you are saving seed from a self-pollinated crop or a cross-pollinating crop. Self-pollinated crops often have good open-pollinated varieties, so they are a good place to start. Self-pollinators use their own pollen to set seed, without any transfer of pollen from other plants. Because avoiding unwanted cross-pollination is not an issue, isolation distances are smaller. Plant populations do not need to be very high, because a self-pollinated variety is already genetically quite homogenous.

Cross-pollinating (outcrossing) plants may be wind- or insect-pollinated. It is important to prevent unwanted pollen (from compatible plants other than your seed crop) reaching your crop. Isolation distances (from other crops which could pollinate your crop) are sometimes large. Large plant populations are needed to maintain genetic diversity.

Hybrids of cross-pollinators such as corn exhibit "hybrid vigor" (meaning hybrids undeniably have an edge as far as productivity goes), but hybrids of self-pollinators don't show this trait as much. While experienced seed growers can develop stable strains from a hybrid, this requires several years of work and is not the place to start.

Choose a crop you can easily grow to maturity in your climate. Avoid unfamiliar crops, biennials or too many different seed crops in your first year. Choose something you tend to only grow one variety of, or can easily grow far enough from others.

Choose something that interests you. Maybe you'd love to see a host of orange cosmos flowers brighten your vegetable field! We have been growing Roma paste tomato seeds because we suffer from *Septoria* leaf spot. The reward for developing a strain of Roma that is resistant to *Septoria* is of great value to us. We started growing Crimson Sweet watermelon seed because we wanted larger earlier melons. I'm also crossing my fingers hoping that never introducing other watermelons into our gardens means we can avoid seed-borne watermelon fruit blotch disease.

Our experience getting started with seed growing

You may already be doing some small-scale seed saving, or perhaps saving clones of vegetatively produced bulbs. Like most garlic growers, you may save your own seed garlic, as we do. We also grow yellow potato onions and shallots and save some for replanting, and select and save peanuts to replant.

Now we grow, select and save seed from paste tomatoes, watermelon, edamame, cowpeas, okra and some flowers grown to attract beneficial insects (cosmos seed is very easy to collect). From time to time we also save seed from other crops, if a need or an opportunity presents itself.

Roma paste tomatoes

We grow Roma paste tomatoes because we make a lot of sauce, juice and salsa, but for years our yields were much reduced by *Septoria* leaf spot. There didn't seem to be any commercially available *Septoria*-resistant variety when we looked, so I decided to develop our own resistant strain. Here's how we do it: We put out our 250 transplants as usual, two feet (60 cm) apart in late April or early May. We make sure when we plan our plot layouts that we don't put any tomato plants of any other varieties within 180 feet (55 m) of any of our Romas. We use the Florida string-weave system, with a metal T-post after every two plants and a new round of twine each week. See the Tomato chapter for more on this.

As soon as the tomatoes start to ripen, we start monitoring the plants, selecting those with healthy foliage and those apparently yielding very well. We use flagging tape (plastic or crepe paper marking tape): green to indicate healthy foliage with reasonable yield and red for abundant fruit with reasonable foliage (so the best plants get both a green and a red ribbon!). We tie the tape on the T-post next to the chosen plant, with the bow or knot on the side of the post facing the plant. We do this once a week, on the day before the crew comes through to do a harvest (no good looking for high yields when they've already been picked!). We add or remove ribbons according to what we see that week.

After one or two weeks of monitoring and flagging, we start picking for seed, on those same just-before-bulk-harvest days. We assess the flagged plants and take one or two tomatoes from each plant that has both green and red flags, or has one flag and is not worse than average on the other factor. If the plant no longer looks so great, we remove its ribbon. If a plant without a ribbon starts to excel in the healthy foliage department as the season wears on, we add a ribbon. (We don't add many red ribbons after the start of the harvest, because we want to keep selecting for early fruit, and plants that yield well later are not really what we want.) Our method combines well with crew harvesting most of the fruit as food. If you are growing the variety only or mainly as a seed crop, you would save all the seed from the chosen plants, or from the whole row after pulling out any unpromising plants ("roguing").

We usually pick about five gallons (19 L) for seed each week. We store those buckets of tomatoes in a secret location, where no one will eat them, for five days, which lets the fruit get completely ripe. Because our main focus is food production, we do our seed saving in a way that fits in with getting maximum food as well as seeds. We cut the dead-ripe fruits in half, rejecting any diseased ones. Next, we scoop out the seeds with a tablespoon. This lets us use the "shells" for tomato sauce for our own use. The seeds ferment in the bucket in a shed for two to three days, nominally at 70°F (21°C). We stir two or three times a day.

When fermentation is over (no more bubbling), we take several clean buckets and a sieve and wash the seed clean. This art gets easier and quicker with practice. The good seed sinks to the bottom. Pour off the top half of the bucket (mostly no good) into another bucket. Add water to both buckets, stir, let things settle and then pour off the junk (tomato pulp and no-good floating seeds) from both. Consolidate the better stuff in one bucket, the worse stuff in another, and pour away the seedless water. When the "worse stuff" bucket no longer has any seed in it, throw that bucketful out. Once the seed looks fairly clean, strain it through the sieve and put it in clean water. When the seed is really clean, strain it and spread it to dry indoors with a fan, on a window screen or paper towels. After six to twelve hours, scrape the clumps of seed off the surface with a putty knife, turn them over and crumble the clumps by hand. After two days, once the seed is thoroughly dry, gather it into a paper bag and add some desiccant. We hold back on storing in an airtight container until we're absolutely sure the seed is dry. (See the Storage section below.)

Here's how we fit it together in an efficient, easy-to-remember way: we harvest seed on Monday and scoop on Friday, which is the day the Food Processing crew are making sauce. We start the fermentation on Friday, wash on Monday, set those seeds to dry and harvest the next batch. We usually do four or five batches of seed, all in August. It's not good to save seed from plants in decline, so get started as soon as you can, and quit while the going is good. We sell this seed to Southern Exposure Seed Exchange.

Tomatoes are self-pollinating, so planting 200 or more, pulling out any off-types, and making a selection of 80–100 of those plants gives plenty of genetic diversity. Twenty gallons (76 L) of Roma tomatoes makes 130 grams of seed.

Crimson Sweet watermelon

We had a long way to go, improving early ripening of our watermelons, because we had been using hay mulch for weed control, which cools the soil, delaying ripening! We now use the biodegradable black plastic mulch (Bio Telo Mater-Bi and Eco-One—we're not USDA certified, so the fact that they are not yet OMRI listed doesn't matter). This change alone gives us melons almost four weeks earlier than when we used hay. In the middle of July, as the melons reach ripeness, I walk through the plot with a grease pencil (china marker) and number my selected melons. I've tried other ways to mark the selected melons (flags, magic markers) but the grease pencil works nicely. Having a big number right there on the skin of the melon works to stop any crew about to harvest it. We have already done some selection for healthy plants when we transplant. Now as I walk through, I look for big melons with vigorous healthy vines. I aim to select thirty to thirty-five melons. Cucurbits, although cross-pollinators, show relatively little inbreeding depression, and a population of twenty-five plants will be enough. Crew are told not to ever pick any melons with numbers on. (Confession: one year when I suspected our cantaloupes were being "browsed," I put numbers on them too. It looked so official and scientific, it worked a charm!)

Once a week I harvest for seed, taking melons that are about a week over-ripe and discarding any that don't look healthy. I like to only deal with six to eight melons each time. I keep notes of which numbers I harvest each week, and assess them for size, ripeness and, once I open them, taste. I take a big knife, several clean buckets and a big spoon (and a damp cloth—it's messy!). I cut the melon across the middle and taste a chunk from the heart. If I don't like the taste, I don't save seed from that one. If the taste is good I scoop out the heart, which is seedless, into a very clean bucket, for eating later. Then there is a layer that is thick with seeds. I scoop this into the seed bucket. Lastly I scoop the outer flesh, also relatively seed-free, into the food bucket. One time, I was sitting on the edge of a garden cart with my horde of melons and my big knife and spoon, when a ten-year-old came by. His eyes popped out on stalks at the sight of me enjoying my private feast, with giant silverware! The scooped-out watermelon flesh makes great smoothies and sorbets.

These seeds get fermented for about four days, then washed similarly to the tomato seeds and dried. For example, if I harvest and scoop on Tuesday, I wash on Saturday and set the seeds to dry for several days. One Crimson Sweet watermelon yields 22 grams of seed; 22 melons yield one pound of seed.

It can get hard to find all the numbered melons (that's where the notebook comes in handy, so I don't waste time looking for one that I already harvested). I abandon any numbered melons that don't ripen early, and I sometimes add any huge melons that pop up after the initial numbering. Earliness is important to us, though, so I only harvest four or five times and then stop. For us that's an August task.

Hoophouses for seed growing

There is not much material available yet about growing seed crops in hoophouses, but this is a very promising frontier. Where it is humid or rainy, it is hard to grow dry seed crops such as legumes, lettuce, spinach and beets (as opposed to "wet" seed crops inside fruits, like tomatoes, melons, peppers, squash and cucumbers) outdoors, but using covered space opens new possibilities. Inside a hoophouse the hotter air can hold more water without causing damp plants. The hoophouse walls also provide a partial physical barrier to prevent cross-pollination.

Compared to outdoor crops in our climate, legumes grown in the hoophouse have very clean, unspotty beans and pods. We have found that summer legumes make a great class of summer hoophouse crops (either as produce or as seed crops). Most bean species are largely self-pollinating, so you will likely have pure seed with 100 feet (30 m) isolation distance, or barrier crops of flowers to distract pollinators. We sow in mid-July, when we pull up our early warm-weather crops such as cucumbers, early tomatoes and squash. The seed crops mature in late October or early November, just when we want to transplant our winter salads and greens. According to Nancy Bubel, in *The Seed Starter's Handbook*, bean seed is ready when your teeth can scarcely make a dent in a sample bean. Maturing happens fast in a hoophouse.

Cowpeas can be sown later than they would be outdoors, to provide seed or dry soup beans in early November. We really like the flavor of Mississippi Silver. We sow in mid-July and harvest to eat in late September through early October. Leaving them till late October/early November provides the seed crop. We tried a July 27 sowing, but it was a bit too late—we didn't harvest much.

Edamame does particularly well in our hoophouse. Growing it under cover means we get beautiful pods (an important feature for edamame!). An advantage of this crop it is that is picked all at once—take it from me that you do not want a crop that requires daily harvesting in high summer in your hoophouse, unless you live in the Far

North. We like Envy edamame, a short bush type that matures quickly. We sow July 27 and harvest Oct 4–13, or Nov 9 for seed.

Shelling beans or soup beans have given good results in the hoophouse. We grew King of the Early, sown July 13 and harvested from late September until mid-October, when we let the last pods dry out for seed. We've found that we don't like putting up big trellises, so we now choose bush type beans. Bush varieties also allow the sunlight to better reach the north side of the house.

Leaf beet seed was commercially unavailable one year, so I dug up a few outdoor leaf beet plants in the fall and replanted them in our hoophouse. Leaf beet is biennial, so in the spring tall flower stalks grow up and make seed. For better seed growing, a bigger population of plants would be needed to guarantee genetic diversity. The second year I did this, I learned a trick of beet seed growing: cut down the first tall stems that appear and you get many more flower stems, at a shorter height—two advantages: more seed, less stem. An unexpected benefit was the wonderful smell of the flower heads!

Okra

We only grow one kind of okra (Cow Horn—it gets big without getting tough), and nothing that would cross with it. So saving seed is just a matter of flagging choice pods on good sturdy plants, and letting those pods ripen. We collect them before they shatter and dry the seed in a mouse-proof place. It's a large seed, and the dry pods shatter easily, so it's simple to separate the seed from the pod pieces. And we don't need a lot of seed, so it's not a major undertaking. It's true that leaving pods or fruit on a plant to ripen will decrease the yield of food from that plant, so it is important to balance your goals and not lose your cash crop for the price of a small packet of seeds.

Flowers

We plant "islands" of flowers to attract beneficial insects to our vegetables. Some of these flowers have very easy-to-collect seeds. Cosmos, French marigold, calendula and dill can all be left growing until dry seeds appear in their heads. These can then be rubbed off into a paper bag. Clearly this way of growing seeds uses only a small population of each plant, so seed saved this way cannot be sold, but it is fine for home use for vegetable growers doing a bit of farmscaping. We like to dot sunflowers throughout our vegetables, and sometimes we save seed from those. It's best to leave the developing head on the plant as long as possible, so that it dries down well and the seed fully matures. The trick is to keep the birds off. Our answer is to tie a bandana over the head, knotted at the back. You can decorate your garden with red and blue bandanas, or find color-coordinated fabrics. One year we had a bundle of silk sari scraps, many in gold, orange and yellow. They blended in nicely! When the sunflower stalk dries out, remove the seed head and complete the drying in a place protected from mice.

Isolation distances

The isolation distance required for a particular species depends on whether the plants are cross- or self-pollinated. Bees fly a long way, so cucurbits (crossers) have long isolation distances of 1,500 feet (460 m), or even as much as half a mile (800 m) if you have no physical barriers in place. Tomatoes (selfers), especially heirloom varieties, mostly self-pollinate, and only require an isolation distance of 75 to 180 feet (23–55 m).

Some cross-pollination can still occur with selfers. Choose something your next-door neighbors don't have growing the other side of your fence line!

Once you know the required isolation distance for the crop you plan to grow, make sure your planting map gives you this space. Early each year we write a "Seed Saving Letter" to others who grow plants in the area, to tell them where we plan to grow our seed crops, and asking them not to plant anything that could cross-pollinate with that crop within the isolation distance. In return, we offer our surplus transplants, and we also ask them if they have any seed saving plans we need to know about. We grow the same dill variety in our insectaries as the Herb Garden grows for dill seed. We plant our flashy calendula far away from the medicinal ones.

Barriers such as buildings, including hoophouses, and tall crops such as corn or sunflowers, can help a borderline isolation be more certain, especially for insect-pollinated crossers. Collecting seed only from the middle of a planting block, rather than at the edges, can also help. Wind-pollinated crossers usually have the most genetic diversity and pollen that travels furthest, so they need long isolation distances and large minimum populations. There are more advanced tricks to help isolation, involving bagging, caging and hand pollinating, for those determined to find a way to get the crop they want.

The need to pay attention to isolation distances can restrict what you can grow for food, but if your growing season is long enough you may be able to have a zucchini crop for early market, then sow pumpkins for seed, and ruthlessly pull up the zucchini before the pumpkins flower. This is known as isolating by time: plant an early crop for food, and then a later crop for seed—or if your season is long enough and the crop maturity quick, plant two seed crops. The various squash varieties can sometimes cross within the same family, and if you want pie pumpkin seeds, you need to keep the zucchini away. SavingOurSeeds.org and Suzanne Ashworth's *Seed To Seed* give everything you need to know on this aspect.

How much to grow? How much space? How long will it take?

Never save seed from just one plant (unless it's the second to last on the planet). Grow a big enough population of plants to keep enough genetic diversity for future adaptability and to prevent a genetic "bottleneck." In practice this means a minimum of twenty plants for inbreeders (self-pollinators), a hundred for out-breeders (cross-pollinators). Some crops need much bigger populations: two hundred to a thousand for corn, for example. Self-pollinators (e.g., beans) are naturally already inbred and contain little diversity. For these, a small population is enough. Inbreeding depression occurs when seed is saved from too small a planting. It leads, over time, to lower quality, less vigorous plants. To avoid it, cross-pollinators need a bigger population.

There is, as yet, no published table of time from sowing to seed crop maturity. Lettuce can take up to two months beyond the eating stage to get to the mature seed stage (I found this information in Fedco's charming *Seed Growing School Curriculum*). Crops where we eat the ripe fruit take very little extra time to mature seeds—just make sure the fruit is really ripe. Cucumbers are eaten as under-ripe fruits, and the seed is mature when the cucumber reaches the yellow blimp stage. Biennials (onions, carrots, most other root crops) need a second growing season to mature seed.

Selecting "mother plants"

If you are selling seed, you will need more plants than if you are just keeping seed to resupply yourself. You have a responsibility to maintain that variety and all the genetic diversity it contains. Grow enough to allow for roguing if you are maintaining a variety for a seed company. Roguing involves removing off-type plants as well as existing fruits from the immediate neighbor plants. Also rogue out diseased plants and any early-bolting plants of crops you don't want to bolt. If you are improving a variety, selecting for certain desirable traits, you will be even more selective about choosing good mother plants.

Biennial plants

Biennials (such as onions and many root crops) need to be replanted in spring (or left over winter in the field). In the second year the flower heads and seeds will form. Leaving the roots in the ground over the winter is easier, but if your climate gets very cold, or fluctuates a lot (ours does), or if you have lots of voles (we do), then digging the roots and storing in a cool, damp root cellar is wiser. It also gives you the chance to select well-shaped roots as your seed stock.

Seed cleaning

There are two types of seed processing: wet and dry. Wet seeds are embedded in the fruit—tomatoes, cucumbers, melons, eggplant. Dry seeds are found in pods, husks or ears, and dry down on the plant—beans, okra, corn, radish. Wet processing has several steps: scooping out the seed or mashing the fruit, fermenting the seed pulp for a few days, washing the seed and removing the pulp and then drying the washed seed. Wet-processed seed is naturally cleaned during the fermentation and washing. The challenge is to ferment the seed long enough to release the clean seed, without waiting so long that the seed starts to sprout.

Dry seed processing involves harvesting the pods or the entire plants, completing the drying indoors if needed, then cracking or breaking the pods to release the seeds. Surprisingly, hoophouses with shade and good ventilation can be good places to quickly dry seeds. After drying, the seeds and chaff are then sieved through at least two different gauge mesh screens: the larger one keeps back the big pieces of chaff and lets the seed pass through, and the smaller one keeps back the seed while letting the small chaff pass through. After screening, the seed is winnowed, perhaps using a box fan and a sheet of cloth or plastic to catch the seed.

There are good details on seed cleaning methods in the *Seed Processing and Storage Guide* from Saving Our Seed (see the Resources section). While small quantities of seed can be cleaned with little equipment, if you move into growing larger quantities, you will want to buy some of the specialized equipment available, or make your own.

Storage

Make sure your storage places are mouse-proof. Initial storage can begin when seeds are down to 8 percent moisture. At this level, seeds break or shatter when you try to fold them or hit them with a hammer. They don't bend or mash. Put them in a jar (optionally with an equal weight of a desiccant such as silica gel) for seven days. Then remove the desiccant and put seed in a labeled bag inside a labeled glass or metal container with an airtight lid. Seeds need to be stored dry and cool and airtight once dry. For long-term storage, put this airtight jar or can in the freezer. When removing seeds from the freezer, allow the container to warm to room temperature for a day before opening. This prevents moisture from condensing on the seeds. For USDA Certified Organic, check the OMRI list before using desiccant, to ensure you only use allowed materials.

Germination testing

To find out how well your seeds will do, test their germination. Take a thick paper towel, fold it lengthwise, unfold it and spread fifty or a hundred seeds along the inside of the fold. Close the fold, dampen the towel with water and roll it up loosely. Put it inside a loosely closed plastic bag and set the bag somewhere at a suitable temperature; often the top of the fridge is suitable. Beware the top of water heaters that use natural gas: this inhibits tomato seeds and other nightshades. A light bulb can be a suitable heater. 75°F (24°C) is good for most vegetables, 80°F (27°C) is better for tomatoes and peppers, 85°F (29°C) for melons. See Nancy Bubel, *Seed Starter's Handbook*, for ideal temperatures for different crops. Check twice a day (the air change will help the seeds even if you know it's too early to see sprouts). Count the number of sprouted seeds after seven days and remove the sprouted ones. Repeat after another seven days and add this count to the first one to calculate your percent germination.

Double benefits

In some cases, your crop could produce both food and seed, as we do with our tomatoes and watermelon. But getting two crops from one plant does take more time, compared to simply mashing the whole tomatoes, for instance.

There are other ways to have your crop and eat it too. Harvesting a few leaves from greens grown for seed will not detract from seed production. Eat the produce from the edges of a block planting and save seed from plants in the center—this helps preserve the purity of the seed without "wasting" the edge plants. Eat the earliest fruit and save seed later, or save seed first and eat the later fruit. Seed should not be saved from plants past their prime, however, and you would not want to risk your seed crop by reducing the time it has to mature by too much. Rat-tail radish can be grown for seed and some of the pods can be eaten first.

You may be able to grow seed for flowers that attract beneficial insects or pest-eating birds to your crops. Ira Wallace suggests that biennials such as carrots or onions for seed can be interplanted in a hoophouse with overwintered greens. As the biennial plants start to grow bigger, remove the early spring greens and let the seed plants grow. Keep records of your dates, as the timing might get critical and some crops will work better than others. See the next chapter for more ideas.

Chapter 57

The Business of Seed Crops

by Ira Wallace

IRA WALLACE lives and gardens at Acorn Community Farm, home of Southern Exposure Seed Exchange, where she coordinates variety selection and seed growers. Ira is also a co-organizer of the Heritage Harvest Festival at Monticello, VA. She serves on the board of the Organic Seed Alliance and is a frequent presenter at the Mother Earth News Fairs and many other events throughout the Southeast. Her first book, *The Timber Press Guide to Vegetable Gardening in the Southeast*, will be available in early 2013.

Growing seed crops has quickly become a great opportunity for ecological farmers to diversify their incomes, add to their bottom lines, and help preserve the genetic diversity of the food supply. Southern Exposure Seed Exchange and other companies like Fedco, Baker Creek, High Mowing and Sow True Seeds have steadily increased the number of seed varieties they buy from small family farms over the last decade. Fedco Seeds went from contracting four varieties from small farms in 2002 to 125 varieties only five years later. The trend is the same for all the small and medium-sized companies I contacted. Southern Exposure listed 277 varieties on our list for potential growers in 2011. Farmers can easily combine seed growing with other farm enterprises, while taking a stand against the increasing control of our seed

supply by petrochemical giants like Monsanto, Dow and Bayer Life Sciences.

This chapter draws on my experience working with contracted seed growers at Southern Exposure Seed Exchange, conversations with growers and other seed companies, and replies to my surveys of other seed companies over the last ten years. There is a clear need for more contract growers, especially growers who are currently or can easily become certified organic. There are also many opportunities for uncertified "eco" growers, starting with small contracts of local and heirloom varieties.

The previous chapter outlined a number of ways that seed saving contributes to and is integrated into the gardens at Twin Oaks. It emphasized benefits to the farm and the intangibles like maintaining genetic diversity, keeping access to desirable OP varieties, and maintaining control of our seed supply. The majority of new seed growers are motivated (at least in part) by these causes and benefits, but farmers who continue and expand their contracts also figure out how to make seed growing profitable.

Contract seed growing

See the Resources section for advice and equipment you may need to save seeds. Growers can sell seeds directly to the consumer through the Web, farmers markets or seed racks in retail stores. This chapter is specifically about contract growing, i.e., selling bulk seeds to seed companies by contracts agreed before the growing season. This is an easier option for most beginners, as it guarantees that you will sell your seed crops. The grower has a separate contract for each seed crop to be grown. Contracts be should be clear regarding the following points:

- Who will provide stock seed for growing the crop? The seed company usually provides the seed or pays the grower for it (eg, family heirloom).
- The price per pound being paid for seed and reasons for adjusting the rate (varieties with low seed yields may merit a higher rate).
- Timing of payments (often 30–90 days after delivery of seed).
- Quality standards for full payment (germination rate, who is responsible for germination testing and its cost, assurance of GMO-free seed).
- What happens in the case of crop failure?
- Packing and shipping requirements. Who pays for shipping?
- Which crops will cross with the seed crop and require isolation distances? Are there any common weeds or GMO crops that might cross with the seed crop?
- Are there any special treatments like hand pollination, extra roguing?
- The level of cleaning of the seed, and what happens if additional cleaning is required?
- In addition, the seed grower should be familiar with growing the crop and be sure that they have enough labor for timely weeding, irrigation, harvest, drying and seed cleaning throughout the entire time the crop will be in the ground.

Drying and cleaning your seeds in a thorough and timely manner is very important and may affect the price paid for them. Most companies need to know the status of your crop in September and have representative samples for germination testing in October to determine if it will be listed in the new catalog. Most contracts have penalties for late delivery up to and including not buying the seed after a certain date. Be aware of these deadlines and keep in good contact with your seed company rep about any problems or delays.

Getting started with contract growing

Inexperienced seed growers will have an easier time acquiring small contracts when they first begin growing commercially. Larger companies like High Mowing and Johnny's are generally only interested in working with experienced seed growers unless there are special circumstances, such as an experienced grower nearby willing to mentor the new grower and able to act as a de facto field agent for the seed company.

Seed production is in many ways a cooperative venture between the seed company and the grower. Seed companies endeavor to provide growers with enough information and support to produce a good crop of high-quality seed. Growers agree to grow the crop following agreed-upon standards and are expected to communicate with the seed company about any problems in a timely manner. It is the quality of this communication, as much as the quality of the seeds produced, that builds trust and leads to larger, more profitable contracts for the grower. Seed growers are not held responsible for a crop failure, but do take the risk that if the seed does not reach the specified germination standard, the seed company will probably not buy it.

How does a farmer get started with contract seed growing? I've provided a chart with points of contact for several companies that contract with small growers. Send the grower contact person a brief introduction, telling about your farm and any experience you have with seed saving. Mention any experienced growers you know who are willing to mentor you or seed saving workshops you have attended. Ideally you would send an introductory email sometime between February and August. This leaves time for the company to request additional information and get to know you before contracts are written (usually late November to January, although small contracts may be available for summer crops as late as April or May). In September through mid-November seed companies are busy producing their catalogs and receiving seed grown the previous summer.

You may be asked to fill out a grower questionnaire, provide a copy of your organic certificate and give a phone interview. It is important to answer correspondence in a timely and professional way. At Southern Exposure we screen for communication skills as much as for experience with seed saving. A 2008 survey conducted by the Organic Seed Alliance found communication was of primary importance to more than half the seed companies surveyed.

It is good to find out if the company rep prefers to be contacted by phone or email. In early November follow up by email or phone to show your continued interest in contract growing. It is good to mention any particulars from your earlier correspondence.

Every year in late November or early December we send a list of varieties we need for the coming year to new potential growers, asking what they are interested in producing. We indicate on this list the varieties we've already contracted for the coming year. We include these to give new growers a feel for what is available to a more experienced grower. At this time we are happy to talk with the potential grower and help them make choices about which crops to consider. A number of other companies work strictly from one-on-one conversations with the seed company contact. In either case, a contract will be sent for each crop agreed upon.

The economics of seed growing

There is not a lot of detailed information available on the financial side of small-scale seed growing, as this is a new and growing income source for small farms. Fedco found that one year's contracts, 137 in all, ranged from 100 percent success meeting the contract through various shortfalls to total failure, with individual growers meeting 66 percent of their contracts on average. Southern Exposure generally has a higher rate of fulfillment, although success may be this low for first-time seed growers. This is why we encourage new growers to start with small contracts and increase the size and number of contracts as they gain experience.

At Fedco, which is based in Maine, very successful crops included summer squash, arugula, mustard, tatsoi, rapa and tomatillo, while crop failures included melons, burdock and kale. Some 38 percent of growers had at least one shortfall on the contracted amount and 32 percent had at least one complete crop failure. Reasons given for crop failures included deer, weather, birds, pollen contamination and hail.

When you will receive the income from seed crops is an important factor to consider with contract seed growing. The down side is that generally, payment won't be sent until the crop has been grown and the seed has been cleaned, shipped and tested. The benefit is that this income is arranged in advance. Don Tipping of Siskiyou explains how this works for his farm:

> We can prearrange contracts early in the winter and go into the growing season with income figures to budget with, similar to the way a CSA ensures the grower up-front cash. However, with seed crops, the farmer doesn't get paid until the seed is harvested, cleaned and germination tested, which means that often we don't receive payment until January or February. So, for us, having some fresh market income is crucial to keeping cash flow happening year round. We have learned to work around the late payment for seed crops by budgeting it as our start up money for the year.

CR Lawn of Fedco shared some information he presented at the Carolina Farm Stewardship Association Conference in 2005 illustrating the economics of seed growing. Several Fedco growers have achieved good results, one making $9.74 per hour after overhead on tomatoes (a fairly typical result), one making $9.28 an hour on three pepper varieties, one realizing $12.63 per hour on a 38-lb (17-kg) crop of Long Pie pumpkin. Two years later the same pumpkin grower averaged $8.82 per hour for a market basket of six crops; but for a time-consuming failure with a difficult onion crop, she would have achieved $14.16. Many of these crops check in with high gross per acre, with Stearn's mizuna at $26,800 and several tomato varieties exceeding $30,000. The trouble is, where are you going to sell an acre of tomato seeds?

Integrating seed growing with other farm enterprises

Learning to grow seed crops efficiently is fundamental to financial success, but there are also opportunities for auxiliary income as a result of seed growing. Winter squash, pumpkins, cantaloupe melons and peppers easily lend themselves to these enterprises: the seed grower can remove the seed and then either sell the pre-cleaned vegetables to local chefs or process them to sell as salsas, jellies, pumpkin butters, etc. Processing these crops in commercial kitchens rather than out in the field may require extra time. Tomatoes can also yield sauce as well as seed, although keeping the tomato flesh for sauces requires more labor and slightly reduces seed yield.

During the winter seed grower Scott Paquin contacts local chefs and gives them samples of his Candy Roaster melon squash. Most like it and arrange for weekly deliveries to be used in featured "local squash" specials the following fall, when Scott processes his seed crop. This arrangement doubles the value of the seed crop. Another

2011 Seed Price Volume Comparison Table

Crop Type	Southern Exposure Seed Exchange Ira Wallace 540-894-1470 Ira@southernexposure.com Virginia (certified or not) Lbs needed	$/lb	Fedco Seeds Nikos Kavanya 207-426-9005 nikos@fedcoseeds.org Maine (certified or not; pays 20% extra for certified) Lbs needed	$/lb	High Mowing Seeds Dr Jodi Lew-Smith 802-472-6171 jodi@highmowingseeds.com Vermont (certified only) Lbs needed	$/lb	Baker Creek Heirloom Seeds Randal Agella 417-924-8917 seeds@rareseeds.com Missouri (certified or not) Lbs needed	$/lb	Sow True Seed Monica Williams 828-254-0708 Monica@sowtrue.com North Carolina (certified or not) Lbs needed	$/lb
Beans/ cowpeas	10–400	$4–$12	25–100	$1.80–$3.60	500–3,000	$2.50–$5	50+	$8	10–150	$2–$8
Collards	1–10	$60–$160			25–100	$30–$50			1–10	$35–$75
Corn	10–1000	$6–$8	50–500	$3–$7	500–2,000	$4–$7			25–200	$4–$8
Rare Corn	2–30	$10–$20					50+	$7.50		
Cucumbers	3–30	$50–$60	5–20	$30–$60	25–300	$12–$35	4+	$55	1–10	$30–$60
Eggplant	¼–10	$320–$560	1–5	$395	2–10	$150–$350	0.4+	$320	¼–1	$250–$350
Lettuce	¼–10	$60–$160	1–10	$60–$90	10–100	$50–$150	2+	$240	¼–3	$50–$90
Melon	1–10	$60–$80	3–12	$35–$50	50–200	$25–$45	4+	$60–$65	1–5	$40–$80
Peas	10–500	$4–$8			500–5,000	$2.50–$4.50			25–150	$3–$10
Peppers	¼–10	$320–$560	½–2	$240–$480	3–10	$200–$300	0.4+	$440	¼–1	$200–$420
Okra	5–60	$30–$32					10+	$30	1–15	$20–$30
Squashes	3–50	$50–$60	5–25	$20–$60	50–1000	$15–$35	5+	$55	1–10	$35–$60
Tomatoes	¼–10	$320–$560	½–2	$300–$500	2–10	$150–$300	0.4+	$320	¼–1	$300–$375
Flowers/ sunflowers	0.1–5	$30–$320					5+	$35		$20–$200

grower on the eastern shore of Virginia has a similar arrangement with sweet peppers and a local salsa maker. Other growers donate food from seed production to local programs to feed the needy, often in exchange for volunteers who help remove the seeds. What these arrangements have in common is thinking ahead, networking and planning to maximize the value of the crop.

Some crops, like popcorn, grinding corns, soup beans and black-eyed peas, lend themselves to being grown for seed with any overproduction being sold as food. These crops have great shelf life and can diversify your offerings at farmers' markets. One of our seed growers alternates growing a white and a yellow variety of gourmet popcorn each year, for seed and direct sale.

What and how much to grow

It is important to consider which seeds crops will produce well on your farm. Many factors determine this, but Dr. John Navazio of the Organic Seed Alliance points to what he calls hot season, wet-seeded crops as a great place to start for new growers in the Southeast. These crops include tomatoes, peppers, watermelons, squash, cucumbers and eggplants. Hot season, dry-seeded crops like okra, sorghum, crowder peas and limas are also good choices. Talking with experienced seed

Notes for two years of growing sweet pepper varieties from seeds

Variety	Year grown	Area (sq ft)	Seed yield (lbs)	Yield (lbs/acre)	Notes
Napoleon Sweet	2008	396	1.2	132	Drought stressed.
Bullnose	2008	396	0.78	86	Hard roguing.
Sweet Chocolate	2008	297	2.04	299	
Royal Black	2009	297	1.2	176	
Keystone	2009	495	3.3	290	
Corona	2009	495	0.48	42	Late start.
Orange Bell	2009	660	3.04	201	

growers in your region is the best way to make informed decisions about which seed crops to grow on your farm. Your seed company contact can also help.

Another question that comes up frequently is how to calculate how many plants to grow to meet your seed contracts. The short answer is that it depends. Timing, soil fertility, weather conditions and the particular variety you're growing all influence seed yields. That's why it is good to start with varieties that you are already familiar with and to get as much information as you can from the seed company. Talk to seed growers who have worked with the same varieties (ask the seed company for contacts). Patrick Steiner's booklet *Small Scale Organic Seed Production* and Bryan Connolly's *The Wisdom of Plant Heritage* both have tables with estimated seed yields per acre grown at recommended spacing. Convert for smaller plantings using the formula: 27.2 lbs/acre = 1 oz/100 square feet. Multiply lbs/acre by 1.12 to give kg/hectare.

The above table shows that one grower had very different seed yields for peppers grown in two years of seed production. This may have been due to differences across varieties, but could also result from different soil conditions across fields,

differences in how the crops were irrigated, when they were planted or the spacing that was used, among other factors. It is very important to take notes on every crop!

We find that sixty lettuce plants generally yield one pound (half a kilo) of seed (in a good year it is double), a hundred feet (30 m) of pole beans produce 20–30 pounds (9–14 kg) and tomatoes can vary from 0.1–0.5 oz (3–14 g) per plant (cherry tomatoes yield more and paste types less). We always emphasize that yield estimates are truly only estimates. We write contracts for a wide range of deliverables, from minimum quantity that needs to be met up to a large maximum quantity we guarantee we'll buy, so growers can fill the contract without worrying about overproducing.

I hope this information about the needs of seed companies, the best times to make initial contacts and who to contact, as well as some general information about preferred climate zones and average yields for crops, will help farmers enjoy the financial rewards and job satisfaction of growing seed crops. You'll be doing your part to increase biodiversity and add financial resilience to your farm in uncertain economic times.

Resources

Books

The Four Season Harvest, Eliot Coleman, 1999, Chelsea Green

The New Organic Grower, Eliot Coleman, 1995, Chelsea Green

The Winter Harvest Handbook, Eliot Coleman, 2009, Chelsea Green

The New Seed Starter's Handbook, Nancy Bubel, 1988, Rodale Books

The Resilient Gardener, Carol Deppe, 2010, Chelsea Green

The Hoophouse Handbook and *The Hoophouse Update*, 2010, both published by *Growing for Market*: growingformarket.com/store/1

Extending the Season: Six Strategies for Improving Cash Flow Year-Round on the Market Farm, a free e-book download for online subscribers to *Growing for Market* magazine

Market Farming Success, Lynn Byczynski, published by *Growing for Market*: growingformarket.com/store/1

The Organic Farmer's Business Handbook, Richard Wiswall, 2009, Chelsea Green

Sharing the Harvest: A Citizen's Guide to Community Supported Agriculture, Elizabeth Henderson and Robyn Van En, revised edition 2007, Chelsea Green

Sustainable Vegetable Production from Start-up to Market, Vern Grubinger, 1999, Plant and Life Sciences Publishing, (formerly NRAES), palspublishing.cals.cornell.edu/nra_order.taf?_function=detail&pr_booknum=nraes-104

Gardening When it Counts, Steve Solomon, 2006, New Society

Growing Vegetables West of the Cascades, Steve Solomon, 2007, Sasquatch Books

The Flower Farmer, Lynn Byczynski, 1997, Chelsea Green/*Growing for Market*: growingformarket.com/store/1

The Complete Know and Grow Vegetables, John K. Bleasdale, Peter J. Salter, et al., 1991, Oxford University Press

Knott's Handbook for Vegetable Growers, Donald Maynard and George Hochmuth, 5th edition 2006, Wiley

How to Grow More Vegetables, John Jeavons, 8th edition 2012, Ten Speed Press

The Vegetable Growers Handbook, Frank Tozer, 2008, Green Man Publishing, distributed by Chelsea Green

Sustainable Practices for Vegetable Production in the South, Mary Peet, 1996, Focus Publishing/R. Pullins

Websites

ATTRA—the National Sustainable Agriculture Information Service, a program of the National Center for Appropriate Technology—has excellent publications, mostly free for digital copies: attra.ncat.org 1-800-346-9140

eOrganic: eorganic.info and eXtension.org. Click Farm, Organic Agriculture

Debbie Roos has a wealth of searchable information on her site: growingsmallfarms.org

United States Department of Agriculture (USDA) plant hardiness map: usna.usda.gov/Hardzone/ushzmap.html

The American Horticulture Society heat zone map: ahs.org/publications/heat_zone_map.htm

SARE (Sustainable Agriculture Research and Education): sare.org

Listservs

Market Farming Listserv: lists.ibiblio.org/mailman/listinfo/market-farming. A great list—growers and farmers discuss many topics, and help each other out.

Suppliers of tools, rowcover, shadecloth, greenhouse supplies, cover crop seed, biopesticides

Seven Springs Farm, VA: 7springsfarm.com

Nolts Produce Supplies: 152 North Hershey Ave, Leola, PA 17540. Phone 717-656-9764. (No website).

Peaceful Valley, CA: peacefulvalleyfarmsupply.com

Wood Creek Farm, VA: woodcreekfarm.com

Gemplers, WI: gemplers.com

Farmtek, IA: farmtek.com

Johnny's, ME: johnnyseeds.com

Klerk's greenhouse plastics, SC: klerksusa.com /products/gfilm.htm

Robert Marvel Plastics, PA: robertmarvel.com. Solaroof and other hoophouse plastics

Purple Mountain Organics, MD: purplemountain organics.com

Seeds

Abundant Life Seed Foundation, OR: abundantlife seeds.com, 541-767-9606

Baker Creek, MO: rareseeds.com, 417-924-8917

Bountiful Gardens, CA: bountifulgardens.org, 707-459-6410

Evergreen Seeds, CA: evergreenseeds.com. More than 120 varieties of Asian greens

Fedco Seeds, ME: fedcoseeds.com, 207-873-7333

Gourmet Seed International, NM: gourmetseed.com, 505-398-6111

High Mowing Seeds, VT: highmowingseeds.com, 802-472-6174

Johnny's Selected Seeds, ME: johnnyseeds.com, 877-564-6697

Kitazawa Seed Company, CA: kitazawaseed.com, 510-595-1188. Asian vegetable seeds

Roughwood Seed Collection, PA: williamwoysweaver .com. Regional foods

Sand Hill Preservation Center, IA: sandhillpreserva tion.com, 563-246-2299. Heirlooms

Seed Savers Exchange, IA: seedsavers.org, 563-382-5990. Non-profit membership organization that saves and shares heirloom seeds

Seeds from Italy, KS: growitalian.com, 785-748-0959

Southern Exposure Seed Exchange, VA: southernexposure.com, 540-894-9480

Sow True Seeds, NC: sowtrueseed.com, 828-254-0708

Territorial Seeds, OR: territorialseed.com, 800-626-0866

Terroir Seeds, AZ: underwoodgardens.com, 888-878-5247. Heirlooms.

Thompson and Morgan: tmseeds.com

Tsang and Ma International/Good Earth Seed Company, P.O. Box 5644, Redwood City, CA 94063, 415-595-2270. (No website)

Turtle Tree Seeds, NY: 800 620 7388. Biodynamic.

Wild Garden Seeds, OR: wildgardenseed.com, 541-929-4068. (No phone orders)

Tools and equipment

ATTRA, *Market Gardening: A Start-up Guide*, includes recommendations on equipment, attra.ncat.org /attra-pub/summaries/summary.php?pub=18

Roxbury Farm, in their Information for Farmers section, has a good article about purchasing equipment, roxburyfarm.com or directly at sfc.smallfarmcentral .com/dynamic_content/uploadfiles/942/Equipment %20article.pdf

Potato Diggers

Ferrari Tractor Cie, Gridley, California: ferrari -tractors.com

Checchi and Magli single-row side-delivery SP100 harvester; rear delivery SP50.

Market Farm Implement, marketfarm.com, has two small machines: Spedo 3-point hitch mounted two-row digger for tractors that straddle two rows of hilled potatoes; Zaga three-point hitch PTO powered one-row digger, designed for small tractors that straddle one row.

Willsie, willsie.com, 800-561-3025, used and new small and medium-sized equipment.

US Small Farm, 1-888-522-1554, ussmallfarm.com, D-10T drawbar hydraulic one row harvester, conveyor chain link bed; D-10M 3-point offset single row harvester, conveyor chain link bed.

Root Washers

The Organic Farmer's Business Handbook, Richard Wiswall, 2009, Chelsea Green

Iowa State Extension, extension.iastate.edu/valueadded ag/info/bulkrootsandtubers.htm

Willsie, willsie.com

Grindstone Farm, grindstonefarm.com/Go.aspx/Root
-Crop-Washers

Wyma, wymasolutions.com/products/washing-and
-brushing/barrel-washer Large ones

Tew Manufacturing 800-380-5839, tewmfg.com/TEW
Machines.html

Rototillers and Hand Tools
Earth Tools BCS, earthtoolsbcs.com
Purple Mountain Organics, purplemountainorganics
.com, Children's tools too

Manual Tools
Valley Oak Wheelhoes, valleyoaktools.com/wheelhoe
.html
DIY wheelhoes, planetwhizbang.blogspot.com
Johnny's Seeds, johnnyseeds.com
Cobrahead, Cobrahead.com hoes and Brook and
Hunter forks
Hooke 'n Crooke hoe, holdredgeenterprises.com
Circle hoe, circlehoe.com
Wetsel, wetsel.com
Clarington Forge, claringtonforge.com/forks Tools for
adults and children
Spear and Jackson forks, gardentalk.com and
gardenhardware.com/spj1550ac.html
WW Manufacturing, wwmfg.com, call 800-452-5547
for prices
Corona pruners, coronatools.com
Gemplers tools and gloves, gemplers.com
All-Pro trowels, wilcoxallpro.com/wilcox3_tools.html
Green Heron work clothing and tools for women,
greenherontools.com
Ergonomic tools and tools for people with disabilities,
gardeningwithease.com
Red Pig (Denmans), redpigtools.com/servlet
/StoreFront Handmade tools
Rogue Hoe, roguehoe.com, handmade tools
Gulland Forge, gullandforge.com, broadforks, Red Pig
tools, RedPigTools.com
Replacement handles, househandle.com
Reviews of tools and seeds, Dave's Garden,
gardenwatchdog.com

One-legged milking stools
Lehmans, Lehmans.com

Nasco, enasco.com
Homesteaders Supply, homesteadersupply.com
Mother Earth News instructions to build your own,
motherearthnews.com/do-it-yourself/strawberry
-picking-stool-zmaz84zloeck.aspx

Chapter 1: Year-Round Production
ATTRA: *Scheduling Vegetable Plantings for Continuous Harvest*, attra.ncat.org/attra-pub/summaries
/summary.php?pub=20
Season Extension Techniques for Market Gardeners,
attra.ncat.org/attra-pub/summaries/summary.php
?pub=366
Market Gardening: A Start-up Guide, attra.ncat.org
/attra-pub/summaries/summary.php?pub=18
Organic Greenhouse Vegetable Production, attra.ncat
.org/attra-pub/summaries/summary.php?pub=45
Intercropping Principles and Production Practices
(mostly field crops, but the same principles apply
to vegetable crops), attra.ncat.org/attra-pub
/summaries/summary.php?pub=105

Chapter 2: Create Your Own Field Manual
SARE: *Local Harvest: A Multifarm CSA Handbook*,
sare.org/Learning-Center/Books/Local-Harvest
Iowa State University Extension: *Determining Prices
for CSA Share Boxes*, extension.iastate.edu/agdm
/wholefarm/pdf/c5-19.pdf
ATTRA *Organic Crops Workbook: A Guide to Sustainable and Allowed Practices*, attra.ncat.org/attra-pub
/summaries/summary.php?pub=67
*Sustainable Farming Internships and Apprenticeships
Directory*, attra.ncat.org/attra-pub/internships
AgSquared is developing free online planning software:
agsquared.com
Free open-source database crop planning software in
a partially finished stage, code.google.com/p/crop
planning
Small Farm Central: smallfarmcentral.com. Provides
help developing farm websites and runs Member As-
sembler, a web-based CSA support system
Brookfield Farm offers planning spreadsheets:
brookfieldfarm.org/CropPlanning.cfm
Market Farm Forms: back40books.com/store
Roxbury Farm information: roxburyfarm.com
Slow Hand Farm: slowhandfarm.com. Josh Volk offers

consultancy and spreadsheet-based crop-planning workshops; he wrote about spreadsheets in *Growing for Market*, Nov/Dec 2010, see growingformarket .com/articles/crop-planning-spreadsheets

Growing Small Farms: growingsmallfarms.org. Click Grower Resources, Farm Planning and Record-keeping for Joel Gruver's free spreadsheets on crop scheduling, field production and harvest planning.

Web-based CSA Management Systems allow farmers to reduce their workload and members to sign up for a CSA: csaware.com is from Local Harvest; farmigo .com helps farmers collaborate regionally in marketing and helps customers buy local, sustainably produced food

Chapter 4: Crop Rotation for Vegetables and Cover Crops
The New Organic Grower, Eliot Coleman, 1995, Chelsea Green

Chapter 7: How Much to Grow
Crop Planning for Organic Vegetable Growers, Daniel Brisebois and Frédéric Thériault, 2010, Acres USA, Canadian Organic Growers: cog.ca

ATTRA: *Market Gardening: A Start-up Guide*, attra .ncat.org/attra-pub/summaries/summary.php?pub =18

Holistic Management: A Whole-Farm Decision Making Framework, attra.ncat.org/attra-pub/summaries /summary.php?pub=296

ATTRA *Resources for Beginning Farmers*, attra.ncat.org /attra-pub/local_food/startup.html

Roxbury Farm *100 Member CSA Plan*, roxburyfarm .com/content/7211

Planning for Your CSA, Mark Cain: Slideshare.net (search for Crop Planning)

SPIN-Farming (Small Plot Intensive Farming): spinfarming.com

Teaching Direct Marketing and Small Farm Viability, Center for Agroecology and Sustainable Food Systems at the University of Santa Cruz. *Unit 4.5 CSA Crop Planning*, 63.249.122.224/wp-content/uploads /2010/05/4.5_CSA_crop_plan.pdf

Johnny's Seeds Harvest Date Calculator: johnnyseeds .com/t-InteractiveTools.aspx

Kansas State, *Farming a Few Acres of Vegetables*, ksre.ksu.edu/library/hort2/mf1115.pdf

Chapter 9: Transplanting Tips
Southwest Florida Research and Education Center, swfrec.ifas.ufl.edu/vegetable_hort/transplants. Detailed information covering age of transplants, container size, biological control for pests, diseases, hardening off, plant size, planting depth and temperature.

ATTRA *Plugs and Transplant Production for Organic Systems*, attra.ncat.org/attra-pub/summaries /summary.php?pub=55

Healthy Farmers, Healthy Profits: bse.wisc.edu/hfhp /tipsheets_html/dibble.htm. Instructions for a rolling dibble.

Chapter 10: Direct Sowing
Using Manually-Operated Seeders for Precision Cover Crop Plantings on the Small Farm, Mark Schonbeck and Ron Morse, Virginia Association for Biological Farming Infosheet, vabf.files.wordpress.com/2012 /03/seeders_sm.pdf

Earthway Seeder Fix, Chris Jagger, Oregon, wannafarm .com/earthway-seeder-fix, bluefoxorganics.com

Chapter 11: Summer Germination of Seeds
Seed Germination versus Soil Temperature: tomclothier .hort.net/page11.html

Chapter 12: Succession Planting for Continuous Harvesting
ATTRA: *Market Farming: A Start-up Guide*, attra.ncat .org/attra-pub/summaries/summary.php?pub=18

Scheduling Vegetable Plantings for a Continuous Harvest, attra.ncat.org/attra-pub/summaries/summary .php?pub=20

Chapter 13: Season Extension
Solar Gardening: Growing Vegetables Year-Round the American Intensive Way, Leandre Poisson, Gretchen Poisson and Robin Wimbiscus, 1994, Chelsea Green

Extending the Season, available as a free download for online subscribers of *Growing for Market*: growing formarket.com

Fall Vegetable Gardening, Virginia Cooperative Extension Service, pubs.ext.vt.edu/426/426-334/426-334 .html

ATTRA *Season Extension Techniques for Market Farmers*, attra.ncat.org/attra-pub/summaries /summary.php?pub=366

Scheduling Vegetable Plantings for a Continuous Harvest, attra.ncat.org/attra-pub/summaries/summary .php?pub=20

Fall and Winter Gardening Quick Reference, Southern Exposure Seed Exchange, southernexposure.com /growing-guides/fall-winter-quick-guide.pdf

Chapter 14: Cold-Hardy Winter Vegetables

Oregon State University Extension Service: *Introduction to Winter Gardening*, LC 322, extension.oregon state.edu/lane/sites/default/files/documents/lc322 wintergardeningrev.pdf

Fall/Winter Gardening Tips, Brett Grohsgal, southern exposure.com/even-star-organic-farms-fallwinter -gardening-tips-ezp-37.html

Fall/Winter Vegetable Gardening, Ken Bezilla, southernexposure.com/southern-exposures-fall winter-gardening-guide-ezp-38.html

Wild Garden Seeds catalog: wildgardenseed.com

How to Grow Winter Vegetables, Charles Dowding, 2011, Green Books (UK)

Chapter 15: The Hoophouse in Winter and Spring

Walking into Spring, Paul and Alison Wiediger: aunaturelfarm.homestead.com

The Hoophouse Handbook and *The Hoophouse Update*, Growing for Market, growingformarket.com

University of Vermont: *High Tunnels*, Ted Blomgren, Tracy Frisch and Steve Moore, uvm.edu/sustainable agriculture/hightunnels.html

HighTunnels website and listserv: hightunnels.org/for growers.htm

ATTRA *Season Extension Techniques for Market Farmers*, attra.ncat.org/attra-pub/summaries /summary.php?pub=366

Specialty Lettuce and Greens: Organic Production, attra.ncat.org/attra-pub/summaries/summary.php ?pub=375

Noble Foundation research and information about hoophouse growing, noble.org/ag/Horticulture /HoopHouse/Index.htm

Kerr Center for Sustainable Agriculture report and slideshow, *How to Build a Low-Cost Hoop-House*, kerrcenter.com/publications/hoophouse/index.htm

West Virginia University: anr.ext.wvu.edu/commercial _horticulture/high_tunnels

Washington State University: mtvernon.wsu.edu /hightunnels

Penn State University: plasticulture.psu.edu

Chapter 16: The Hoophouse in Summer

Soil Solarization Homepage, agri3.huji.ac.il/~katan

Soil Solarization for Gardens & Landscapes Management, UC IPM step-by-step instructions, ipm.uc davis.edu/PMG/PESTNOTES/pn74145.html

University of California, *Soil Solarization Informational Website*, ucanr.org/repository/cao/landing page.cfm?article=ca.v059n02p84&fulltext=yes

Soil Solarization for Control of Soilborne Diseases, Craig Canaday, West Tennessee Research and Education Center, organics.tennessee.edu/pdf/Soil%20solariza tion%20for%20control%20of%20soilborne%20dis eases%20(and%20weeds%20too).pdf

University of Missouri, *High Tunnel Melon and Watermelon Production*, extension.missouri.edu/p/M173

SARE Reports on *Ginger*, mysare.sare.org/mySARE /ProjectReport.aspx?do=viewProj&pn=FNE06-564 and mysare.sare.org/mySARE/ProjectReport.aspx ?do=viewProj&pn=FNE07-596

East Branch Ginger cultural information and sales of roots: eastbranchginger.com

Alison and Paul Wiediger: aunaturelfarm.com/High -Tunnel-Ginger.html

Chapter 17: Maintaining Soil Fertility

Elaine Ingham, Soil Food Web: soilfoodweb.com

Nature and Properties of Soils, Nyle Brady and Ray Weil, 14th edition 2007, Prentice Hall

Soil: The Erosion of Civilization, David Montgomery, 2007, University of California Press

SARE *Building Soils for Better Crops*: sare.org/Learning -Center/Books

USDA NRCS *Soil Biology Primer*: ctenvirothon.org /studyguides/soil_docs/Soil_Biology_Primer.pdf

USDA NRCS Soils Website: soils.usda.gov

Mark Schonbeck VABF newsletters, 2009, 1st Quarter, 2nd Quarter and 3rd Quarter: vabf.org/information-sheets/newsletters

Center for Environmental Farming Systems, NC State, *Soil Fertility on Organic Farms*, cefs.ncsu.edu/resources/organicproductionguide/soilfertilityfinaljan09.pdf

Composting on Organic Farms, cefs.ncsu.edu/resources/orginicproductionguide/compostingfinaljan2009.pdf

Soil Quality Considerations for Organic Farmers, cefs.ncsu.edu/resources/organicproductionguide/soilqualityfinaljan09.pdf

Cornell Waste Management Institute, *Large Scale Composting Calculation Tools*, cwmi.css.cornell.edu/composting.htm

Cornell University, *On-Farm Composting Handbook*, compost.css.cornell.edu/OnFarmHandbook/onfarm_TOC.html

NCSU *SoilFacts Nutrient Content of Fertilizer and Organic Materials*, soil.ncsu.edu/publications/Soilfacts/AG-439-18

Clean Washington Center, *Basic On-Farm Composting Manual*, cwc.org/wood/wd973rpt.pdf

Virginia Cooperative Extension Service, *On-Farm Composting*, pubs.ext.vt.edu/452/452-232/452-232.html

Extension, *Soil and Fertility Management in Organic Farming Systems*, extension.org/pages/59460/soil-and-fertility-management-in-organic-farming-systems

ATTRA *Sustainable Soil Management*, attra.ncat.org/attra-pub/summaries/summary.php?pub=183

Drought Resistant Soil, attra.ncat.org/attra-pub/summaries/summary.php?pub=118

Notes on Compost Teas, attra.ncat.org/attra-pub/summaries/summary.php?pub=125

USDA ARS National Soil Dynamics Laboratory, *Managing soil for sustainable and profitable agricultural production*, ars.usda.gov/msa/auburn/nsdl

Chapter 18: Cover Crops

SARE, *Managing Cover Crops Profitably*, sare.org/Learning-Center/Books. Choose pdf or html.

ATTRA *Overview of Cover Crops and Green Manures*, attra.ncat.org/attra-pub/summaries/summary.php?pub=288

CEFS *Organic Production: Cover Crops for Organic Farms*, under Resources, Guides, Organic Production Guide: cefs.ncsu.edu/resources/organicproductionguide/covercropsfinaljan2009.pdf

NCSU Department of Horticultural Sciences Horticulture Information Leaflet 37, *Summer Cover Crops*, ces.ncsu.edu/depts/hort/hil/pdf/hil-37.pdf

VABF, *Cover Crops for all Seasons; Cover Cropping: On-farm Solar-powered Soil Building; Manual Seeders for Cover Crops; Reduced Tillage and Cover Cropping Systems for Organic Vegetable Production*, all at vabf.org/information-sheets

eOrganic Agriculture Resource Area of the eXtension website, *Cover Cropping*, extension.org/pages/59454/cover-cropping-in-organic-farming-systems

USDA/ARS *Cover Crop Chart*, ars.usda.gov/Main/docs.htm?docid=20323. The crop "tiles" can be clicked to access more information about 46 cover crops.

University of California Davis, *Cover Crop Database*: sarep.ucdavis.edu/cgi-bin/ccrop.EXE

Cornell University, *Cover Crops for Vegetable Growers*, nysaes.cornell.edu/hort/faculty/bjorkman/covercrops

Virginia Cooperative Extension Service, *Cover Crops*, pubs.ext.vt.edu/426/426-334/426-344table.html

Kansas State University, *Cover Crops for Vegetable Growers*: ksre.ksu.edu/library/hort2/mf2343.pdf

University of Georgia College of Agricultural and Environmental Sciences, *Pearl Millet for Grain*: caes.uga.edu/publications/pubDetail.cfm?pk_id=7172

Information Resource Portal for Lupins (Australian): lupins.org

Non-GMO Sourcebook, a searchable database of non-GMO suppliers: nongmosourcebook.com

Chapter 20: Sustainable Disease Management

Why Things Bite Back, Edward Tenner, 1997, Vintage Books. See the chapter on vegetable pests, accidentally introduced weeds and deliberately introduced exotics to better understand the "revenge of unintended consequences."

Identifying Diseases of Vegetables, Pennsylvania State University. Good photos and symptom lists.

ATTRA, *Use of Baking Soda as a Fungicide*, attra.ncat
.org/attra-pub/summaries/summary.php?pub=126

Sustainable Management of Soil-borne Plant Diseases,
attra.ncat.org/attra-pub/summaries/summary.php
?pub=283

Elaine Ingham: soilfoodweb.com

eOrganic, *Disease Management in Organic Systems*,
extension.org/pages/59458/disease-management-in
-organic-systems

*Biopesticides for Plant Disease Management in Organic
Systems*, extension.org/pages/29380/biopesticides
-for-plant-disease-management-in-organic-farming

Ohio State *Commercial Biocontrol Products Available in
the USA for Use Against Plant Pathogens*, oardc.ohio
-state.edu/apsbcc. Click Product List.

Hot Water and Chlorine Treatment of Vegetable Seeds,
ohioline.osu.edu/hyg-fact/3000/3085.html

University of Massachusetts, *IPM Guidelines*,
extension.umass.edu/vegetable/publications/ipm
-guidlines

Purdue University, *Disease Management Strategies
for Horticultural Crops Using Organic Fungicides*,
ces.purdue.edu/extmedia/bp/bp-69-w.pdf

Debbie Roos has a wealth of searchable information on
her site, growingsmallfarms.org

Disease management links, ces.ncsu.edu/chatham/ag
/SustAg/diseaselinks.html

Environmental Protection Agency List of Biopesticides, epa.gov/pesticides/biopesticides

Damping Off, tomclothier.hort.net/page13.html Tom
Clothier (not organic)

Cornell University, *Treatments for Managing Bacterial Pathogens in Vegetable Seed*, vegetablemdonline
.ppath.cornell.edu/NewsArticles/All_BactSeed.htm

University of Illinois Extension, *Vegetable Seed Treatment*, ipm.illinois.edu/diseases/rpds/915.pdf

Mushroom Mountain, information and mycorrhizal
fungi granules for plant roots, mushroommountain
.com

Chapter 21: Sustainable Weed Management

*Manage Weeds on Your Farm: A Guide to Ecological
Strategies*, Charles Mohler and A. Ditommaso, to be
published by Sustainable Agriculture Research and
Education (SARE). This book will likely become the
classic go-to book on the subject. I went to a work-
shop by Chuck Mohler and found it very inspiring
and informative. He includes profiles of many weeds.
2009 Project ENE06-099, mysare.sare.org

eOrganic *Weed Management*, extension.org/pages
/61887/weed-management-topics

ATTRA *Principles of Sustainable Weed Management
for Croplands*, attra.ncat.org/attra-pub/summaries
/summary.php?pub=109

The University of Maine, *Weed Management* (Quackgrass), umaine.edu/weedecology/weed-management

Center for Environmental Farming Systems (CEFS),
*Weed Management on Organic Farms: Cultivation
Practices for Organic Crops*, cefs.ncsu.edu/resources
/organicproductionguide/weedmgmtjan808acces
sible.pdf

Midwest Organic and Sustainable Education Service
(MOSES), *Managing Weed Seedbanks in Organic
Farming Systems*, mosesorganic.org/attachments
/research/10forum_weeds.pdf. A wonderful document based on pictures and charts (see the earthworm snag the weed seed on page 52!).

Ohio State University, *Commercial Biocontrol Products
for Use Against Weeds*, oardc.ohio-state.edu/apsbcc
/BioHerbicidesList2005USA.htm

SARE, *Vinegar for Weed Control in Garlic*, Fred Forsburg, 2004 project FNE03-461, mysare.sare.org/my
SARE/ProjectReport.aspx?do=viewProj&pn=FNE
03-461

Chapter 22: Sustainable Pest Management

Garden Insects of North America, Whitney Cranshaw,
2004, Princeton University Press

Farmscaping Techniques for Managing Insect Pests,
Brinkley Benson, Richard McDonald and Ronald
Morse, a six-page pdf available online

eXtension, *Integrated Pest Management in Organic
Farming Systems*, extension.org/pages/19916/intro
duction-to-integrated-pest-management-in-organic
-farming-systems

Insect Management in Organic Systems has many useful
short articles on specific facets, extension.org/pages
/59455/insect-management-in-organic-farming
-systems

*Farmscaping: Making Use of Nature's Pest Management
Services*, extension.org/pages/18573/farmscaping
:-making-use-of-natures-pest-management-services

Farmscaping, Richard McDonald: drmcbug.com

ATTRA, *Farmscaping to Enhance Biological Control*, attra.ncat.org/attra-pub/summaries/summary.php?pub=145

Organic IPM Field Guide, an attractive nine-page poster format introductory document, attra.ncat.org/attra-pub/summaries/summary.php?pub=148

Biointensive Integrated Pest Management, attra.ncat.org/attra-pub/summaries/summary.php?pub=146

Integrated Pest Management for Greenhouse Crops, attra.ncat.org/attra-pub/summaries/summary.php?pub=48

Nematodes: Alternative Controls: attra.ncat.org/attra-pub/summaries/summary.php?pub=149

SARE Handbook, *Manage Insects on Your Farm: A Guide to Ecological Strategies*, sare.org/content/download/29731/413976/file/insect.pdf

Building Soils for Better Crops, sare.org/content/download/841/6675/file/BSBC%203.pdf

Cornell University, *Resource Guide for Organic Insect and Disease Management*, web.pppmb.cals.cornell.edu/resourceguide

Biological Control: A Guide to Natural Enemies in North America, biocontrol.entomology.cornell.edu

IPM Fact Sheets for Vegetables, nysipm.cornell.edu/factsheets/vegetables

Growing Small Farms, growingsmallfarms.org, click Production, Organic Pest Management, chatham.ces.ncsu.edu/growingsmallfarms/pestmanagement.html

Web Resources, Insect Management, chatham.ces.ncsu.edu/growingsmallfarms/insectlinks.html

NRCS Conservation Practice Standard 595 – 1, *Integrated Pest Management*, nrcs.usda.gov/Internet/FSE_DOCUMENTS/nrcs143_025930.pdf

Beneficial insects, BioControl Network, TN: biconet.com

Listing of suppliers of beneficial insects in the United States, cdpr.ca.gov/docs/pestmgt/ipminov/bensuppl.htm

Brown Marmorated Stink Bug Control, Grower Forum, interactive website: bmsb.opm.msu.edu

Rutgers University, *Monitoring for the Brown Marmorated Stink Bug*: njaes.rutgers.edu/stinkbug

Biology and Current Spread of the Brown Marmorated Stink Bug in North America, northeastipm.org/neipm/assets/File/BMSB%20Resources/ESA%20Eastern%20Branch%202011/03-Biology-and-Current-Spread-of-the-Brown-Marmorated-Stink-Bug-in-North-America.pdf

Russ Mizell, *Trap Cropping System to Suppress Stink Bugs in the Southern Coastal Plain*, fshs.org/Proceedings/Password%20Protected/2008%20vol.%20121/FSHS%20vol.%20121/377-382.pdf

Chapter 23: Manual Harvesting Techniques

"Efficient Harvesting Techniques" video (out of print), CSA Works brookfieldfarm.org. Check sustainablemarketfarming.com for updates on making this work available again.

Roxbury Farm, *Harvest Manual*, sfc.smallfarmcentral.com/dynamic_content/uploadfiles/942/Harvest%20Manual%202010.pdf

ATTRA, *Equipment and Tools for Small-Scale Intensive Crop Production*, attra.ncat.org/attra-pub/summaries/summary.php?pub=373

Market Gardening: A Start-up Guide, attra.ncat.org/attra-pub/summaries/summary.php?pub=18

Healthy Farmers, Healthy Profits, *Post Harvest Handling for Best Crop Quality*, bse.wisc.edu/hfhp/tipsheets_html/postharvest.htm; bse.wisc.edu/hfhp/multimedia/Website%20power%20point.pdf Includes packing shed ideas.

Growing for Market articles on packing shed layout and construction: growingformarket.com/articles/Packing-Shed; growingformarket.com/articles/vegetable-packing-shed

North Carolina State University, *Postharvest Handling*, bae.ncsu.edu/programs/extension/publicat/postharv

Plant and Life Sciences Publishing (formerly NRAES), *Produce Handling for Direct Marketing*, palspublishing.cals.cornell.edu/nra_order.taf?_function=detail&pr_booknum=nraes-51

Society of St. Andrew, endhunger.org, a volunteer group who glean to feed the hungry.

Chapter 24: Winter Vegetable Storage

Washington State University Extension, *Storing Vegetables and Fruits at Home*, cru.cahe.wsu.edu/CE

Publications/eb1326/eb1326.pdf Includes adapted material from the out-of-print NRAES Bulletin #7: *Home Storage of Fruits and Vegetables*, 1984, Susan MacKay

Growing for Market, August 2009, *How to Store Fruits and Vegetables*: growingformarket.com

USDA Agriculture Handbook 66, 1986, *The Commercial Storage of Fruits, Vegetables and Florist and Nursery Stocks*, Draft 2004 revision: ba.ars.usda.gov/hb66

Postharvest Publications, Postharvest Technology Research and Information Center, Dept Pomology, UC Davis: postharvest.ucdavis.edu/libraries/publications. Covers around seventy crops and several aspects of storage.

Chapter 25: Root Cellars

Root Cellaring, Mike and Nancy Bubel, 1991, Storey Publishing

Two Country Wisdom bulletins, Garden Way/Storey Publishing, from the late '70s: *A 87 Cold Storage for Fruit & Vegetables* and *A76 Build Your Own Underground Root Cellar*

USDA Home and Garden Bulletin #119, *Storing Vegetables and Fruits in Basements, Cellars, Outbuildings and Pits*, lightly revised in 1978. It's online thanks to the National Agricultural Library, Agricultural Network Information Center, agnic.msu.edu/hgpubs/modus/morefile/hg119_78.pdf

Chapter 26: Green Beans

Cornell University, *2011 Production Guide for Organic Snap Beans for Processing*, nysipm.cornell.edu/organic_guide/bean.pdf Soil, plant nutrition, organic IPM strategies, pesticides, links to pest ID photos and more information. Not to be missed!

Oregon State University publication for commercial bean growers (not organic), nwrec.hort.oregonstate.edu/snapbean.html

Chapter 27: Southern Peas, Asparagus Beans and Limas

Roto-Fingers sheller, rotofingers.com/id21.htm

Lehman's Texas Pea Sheller, lehmans.com

North Carolina State University Extension, *Southern Peas*, ces.ncsu.edu/depts/hort/hil/hil-20.html

Chapter 28: Fava Beans

Nancy Berkoff, "Cooking with Fava Beans," in the *Vegetarian Journal*, 2011, vol 30, issue 3

Nutritional information, nutritiondata.self.com/facts/legumes-and-legume-products/4321/2

Chapter 29: Edamame

ATTRA, *Edamame: Vegetable Soybean*, attra.ncat.org/attra-pub/edamame.html

Organic Control of White Mold on Soybeans, attra.ncat.org/attra-pub/viewhtml.php?id=124

The Pacific Northwest Extension Service, *Edamame*, cru.cahe.wsu.edu/CEPublications/pnw0525/pnw0525.pdf

Washington State University, cru.cahe.wsu.edu/CEPublications/pnw0525/pnw0525.pdf and edamame.wsu.edu

SARE, sare.org/Project-Reports/Search-the-Database, enter "edamame."

The University of Kentucky New Crop Opportunity Center, *Soybeans*, uky.edu/Ag/NewCrops/soybeans.html

Mechanical Harvesting of Edamame, uky.edu/Ag/NewCrops/edamameharvest.pdf

Strategic Marketing Institute, *The Edamame Market*, expeng.anr.msu.edu/uploads/files/39/2-1203.pdf

Chapter 31: Peanuts

North Carolina State University, ipm.ncsu.edu/production_guides/peanuts/main.pdf An excellent site.

Virginia Cooperative Extension Service, *Virginia Peanut Production Guide 2011* (not organic), pubs.ext.vt.edu/2810/2810-1017-11/2810-1017-11_pdf.pdf

Clemson Extension Service peanut factsheet, clemson.edu/extension/hgic/plants/vegetables/crops/hgic1315.html

University of Florida, "Producing Quality Peanut Seed," edis.ifas.ufl.edu/AG190

The New Farm, "Are They Nuts? Southern Researchers and Farmers Tackle Organic Peanuts," newfarm.org/features/2005/1105/peanuts/culbreath.shtml

Home Power solar food dehydrator design, homepower.com/article/?file=HP69_pg24_Scanlin

DIY peanut sheller (hand operated, 100 lbs/hr). Contact the organization for information, peanutsheller.org

Chapter 32: Broccoli, Cabbage, Kale and Collards in Spring

Oklahoma State University publication for commercial growers, pods.dasnr.okstate.edu/docushare/dsweb /Get/Document-1388/F-6027web.pdf

ATTRA, *Cole Crops and Other Brassicas: Organic Production*, attra.ncat.org/attra-pub/summaries /summary.php?pub=27

Chapter 33: Broccoli, Cabbage, Kale and Collards in Fall

Virginia Cooperative Extension Service, *Fall Vegetable Gardening*, ext.vt.edu/pubs/envirohort/426-334/426 -334.html

Symbiont, drmcbug.com, organic broccoli production and farmscaping

Chapter 34: Asian Greens

Grow Your Own Chinese Vegetables, Geri Harrington, 1984, Garden Way Publishing. Includes alternative names in different cultures.

Growing Unusual Vegetables, Simon Hickmott, 2006, Eco-Logic books, UK.

Oriental Vegetables: The Complete Guide for the Garden and Kitchen, Joy Larkham, revised edition 2008, Kodansha USA

Chapter 35: Spinach

Cornell *IPM Organic Guides, Spinach*, nysipm.cornell .edu/organic_guide/spinach.pdf

Seed Alliance *Spinach Diseases Field Identification*, seedalliance.org/uploads/pdf/SpinachDiseases.pdf

Chapter 37: Lettuce All Year Round

ATTRA, *Specialty Lettuce and Greens: Organic Produc-tion*, attra.ncat.org/attra-pub/summaries/summary .php?pub=375

North Carolina University Extension, ces.ncsu.edu /depts/hort/hil/veg-index.html; *Lettuce Production*: ces.ncsu.edu/depts/hort/hil/hil-11.html Contains much useful information.

Colorado State Specialty Crops Program, *Lettuce Bolt-ing Resistance Project*, specialtycrops.colostate.edu /scp_exp_demo/lettuce_bolting.htm

How to Have Fresh Lettuce Year-round, Without a Greenhouse or a Cold Frame (Zone 5a), Tom Clothier, tomclothier.hort.net/page23.html

Chapter 38: Carrots, Beets and Parsnips

Virginia Cooperative Extension Service, *Root Crops*, pubs.ext.vt.edu/426/426-422/426-422.html

The Ontario Ministry of Agriculture, Food and Rural Affairs, *Carrot Insects*, omafra.gov.on.ca/english /crops/facts/93-077.htm

Commercial Parsnip Production in Ontario, omafra.gov .on.ca/english/crops/facts/parsnip.htm

Washington State University, *Carrot Rust Fly Study*, agsyst.wsu.edu/CRFreport03.pdf

Excelsior wood wool batting, American Excelsior, Wisconsin 715-234-6861 americanexcelsior.com /erosioncontrol/products/clblankets.php

Chapter 39: Celery and Celeriac

Connecticut Agricultural Experiment Station, *Celery, Celeriac*, ct.gov/caes/cwp/view.asp?a=2823&q =377614.

Chapter 40: Turnips and Rutabagas

Prince Edward Island Department of Agriculture, Fisheries and Aquaculture, *Turnips and Rutabagas Production Guide*, gov.pe.ca/agric/index.php3?num ber=69770&lang=E

Innvista nutritional information, innvista.com/health /foods/vegetables/turnips.htm

Chapter 41: Summer Squash and Zucchini

Alabama Cooperative Extension Service, *Guide to Commercial Summer Squash Production*, aces.edu /pubs/docs/A/ANR-1014/ANR-1014.pdf

Saving Our Seeds, *Cucurbit Production Guide*, saving ourseeds.org/pubs/cucurbit_seed_production_ver _1pt8.pdf

Chapter 42: Winter Squash and Pumpkins

ATTRA, *Organic Pumpkin and Winter Squash Produc-tion*, attra.ncat.org/attra-pub/summaries/summary. php?pub=30; *Downy Mildew Control in Cucurbits*, attra.ncat.org/attra-pub/summaries/summary.php ?pub=122

Chapter 43: Cucumbers and Muskmelons

Melons for the Passionate Grower, Amy Goldman
Identifying Diseases of Vegetables, Penn State. 300 color pictures of disease symptoms, 814-865-6713, AgPubsDist@psu.edu or pubs.cas.psu.edu/

Saving Our Seeds, Jeff McCormack, information on melon types and seed saving, savingourseeds.org /pubs/cucurbit_seed_production_ver_1pt8.pdf

Oregon State University, nwrec.hort.oregonstate.edu /cucumber.html

Midwest Vegetable Production Guide for Commercial Growers 2011, btny.purdue.edu/Pubs/ID/ID-56 /cucurbit.pdf

The University of Tennessee, *Producing Cantaloupes in Tennessee*, utextension.tennessee.edu/publications /Documents/PB962.pdf

What are Burpless Cucumbers? Todd Wehner, cuke.hort.ncsu.edu/cucurbit/wehner/articles/art 090.pdf

ATTRA, *Cucumber Beetles: Organic and Biorational Integrated Pest Management*, attra.ncat.org/attra-pub /summaries/summary.php?pub=133

Oklahoma Cooperative Extension Service publication E-853, *Cucurbit Integrated Crop Management*, lane -ag.org/wm-world/cucurbit_manual/e-853.pdf

Cornell University, *Organic Resource Guide: Organic Insect and Disease Control for Cucurbit Crops*, web.pppmb.cals.cornell.edu/resourceguide/pdf /cucurbit.pdf

Chapter 44: Watermelon
Washington State University, photos of 126 varieties, agsyst.wsu.edu/WatermelonPhotos.html

Washington State University 2003 study of icebox watermelons, agsyst.wsu.edu/Watermelon2003.html

University of California 2003 and 2006 trials of mini watermelons, sfp.ucdavis.edu/pubs/brochures/RM _WTRMLN_RSRCH.pdf; ucanr.org/sites/Small _Farms_and_Specialty_Crop/files/90277.pdf

University of Kentucky, uky.edu/Ag/CDBREC/intro sheets/watermelon.pdf

Virginia Cooperative Extension Service, *Organic Production of Watermelon*, pubs.ext.vt.edu/2906/2906 -1342/2906-1342.html

Purdue University Extension, *Diseases and Pests of Muskmelons and Watermelons*, mdc.itap.purdue.edu /item.asp?item_number=BP-44

Chapter 45: Garlic
Growing Great Garlic, Ron Engeland, 1991, Filaree

ATTRA, *Organic Garlic Production*, attra.ncat.org/attra -pub/summaries/summary.php?pub=29

The Garlic Seed Foundation, garlicseedfoundation.info An organization of growers and eaters. Their website lists suppliers and resources, including the ARS Germplasm Resource, which supplies small amounts of plant material to growers. They have an extensive library and information on building your own harvesting equipment.

Dr Gayle Volk's Garlic DNA Analysis, garlicseedfound ation.info/allium_sativum_DNA.htm

Nematodes, garlicseedfoundation.info/bloat-nematode -new-york.htm

garlicseedfoundation.info/images/nematode-cce.jpg

Gourmet Garlic Garden growing instructions, pests and diseases, growing in the South, and more, gourmetgarlicgardens.com/index.htm

Colorado State University Specialty Crop Garlic Project, specialtycrops.colostate.edu/scp_exp_demo /garlic_2004_spce_flme_scpe.htm

Sources for Garlic Seed
Gourmet Garlic Gardens, gourmetgarlicgardens.com /index.htm 325-348-3049, 73 varieties

Filaree Farms, WA, filareefarm.com 509-422-6940, over 100 varieties

Territorial Seeds, OR, territorialseed.com 800-626-0866, 27 varieties

Southern Exposure Seed Exchange, VA, southern exposure.com 540-894-9480, 17 varieties

Irish Eyes Garden Seeds, WA, irisheyesgardenseeds .com 509-964-7000 or 509-925-6025, 19 varieties

Chapter 46: Bulb Onions
Onions, Leeks and Garlic: A Handbook for Gardeners, Marian Coonse, 1995, Texas A&M University Press

Golden Gate Gardening, Pam Peirce, 3rd edition 2010, Sasquatch Books. Excellent descriptions of onion growth phases.

Garlic, Onion & Other Alliums, Ellen Spector Platt, 2003, Stackpole Books

Dixondale Onion Farms, dixondalefarms.com. Helpful information.

ATTRA, *Organic Allium Production* attra.ncat.org /attra-pub/summaries/summary.php?pub=25; *Farmscaping* attra.ncat.org/attra-pub/summaries /summary.php?pub=145

Flame Weeding for Vegetable Crops attra.ncat.org/attra -pub/summaries/summary.php?pub=110

North Carolina University Extension, *Bulb Onion Production in Eastern North Carolina*, ces.ncsu.edu/depts/hort/hil/hil-18-a.html

University of Georgia, *Onion Production Guide*, caes.uga.edu/Publications/pubDetail.cfm?pk_id=7749&pg=np&ct=Onion%20Production%20Guide&kt=&kid=&pid=

Organic Vidalia Onion Production, caes.uga.edu/Publications/pubDetail.cfm?pk_id=7704&pg=np&ct=Onion%20Production%20Guide&kt=&kid=&pid=

Chapter 47: Potato Onions

Southern Exposure Seed Exchange, VA, southernexposure.com 540-894-9480.

Garlic and Perennial Onion Growing Guide, Jeff McCormack, southernexposure.com/garlic-and-perennial-onion-growing-guide-ezp-29.html

Chapter 48: Leeks

Commercial Leek Production Guide for Oregon, hort-devel-nwrec.hort.oregonstate.edu/leek.html

Ontario Ministry of Agriculture, Food and Rural Affairs, *Leek Production Factsheet*, omafra.gov.on.ca/english/crops/facts/91-004.htm

Leek Moth, omafra.gov.on.ca/english/crops/hort/news/vegnews/2010/vg0810a1.htm

Chapter 49: Tomatoes

The Heirloom Tomato: From Garden to Table, Amy Goldman. Detailed information on flavor and disease resistance, plus mouthwatering photos.

ATTRA, *Organic Tomato Production*, attra.ncat.org/attra-pub/summaries/summary.php?pub=33; *Notes on Compost Teas*, attra.ncat.org/attra-pub/summaries/summary.php?pub=125

"What is that green stuff on tomato plants?" Lynn Byczynski, *Growing for Market*, growingformarket.com/articles/green-powder-on-tomato-plants

USDA ARS, *Sustainable Production of Fresh-Market Tomatoes and Other Vegetables with Cover Crop Mulches*, ars.usda.gov/is/np/SustainableTomatoes2007/SustainableTomatoes2007Intro.htm

Virginia Cooperative Extension Service, "Tomato Spotted Wilt Virus," pubs.ext.vt.edu/2906/2906-1326/2906-1326.html

"Septoria Leaf Spot of Tomato," ext.vt.edu/pubs/plantdiseasesefs/450-711/450-711.html

"Early Blight of Tomatoes," ext.vt.edu/450/450-708/450-708.html

Cornell University, *Organic Insect and Disease Control for Solanaceous Crops* in their *Organic Resource Guide*, web.pppmb.cals.cornell.edu/resourceguide/cmp/solanaceous.php

Sun-drying, homecooking.about.com/od/howtocookvegetables/a/sundriedrecipe.htm

Luminance diffusion plastic and potassium (Steve Bogash, Penn State), thenms.com/users/Haygro9a8/newsletters/13-05-2010_08:24:26.html

Louisiana State University, *Performance of Hot-set Tomato Varieties*, lsuagcenter.com/en/crops_livestock/crops/vegetables/Performance+of+Hot+Set+Tomato+Varieties+in+Louisiana+Summer+Fall+2006.htm

Alabama Cooperative Extension Service, *Blossom Drop in Tomatoes*, aces.edu/department/com_veg/blossom_drop.pdf

Chapter 50: Peppers

Uncle Steve's Hot Stuff, usHOTstuff.com. Includes a chart of over fifty hot peppers and their heat ratings.

Chile Pepper Institute, New Mexico State University chilepepperinstitute.org

Robert Farr, The Chile Man in Virginia, thechileman.com

Saving Our Seeds, *Organic Pepper Seed Production Manual*, savingourseeds.org/publications.html

Chapter 51: Eggplant

Alabama A&M and Auburn Universities, *Guide to Commercial Eggplant Production*, aces.edu/pubs/docs/A/ANR-1098/ANR-1098.pdf (currently under review)

UC IPM Online, *Weed Management for Organic Eggplant Production*, ipm.ucdavis.edu/PMG/r211700511.html

ATTRA, *Flea Beetle: Organic Control Options*, attra.ncat.org/attra-pub/summaries/summary.php?pub=135; *Colorado Potato Beetle: Organic Control Options*, attra.ncat.org/attra-pub/summaries/summary.php?pub=132

Chapter 52: Potatoes

ATTRA, *Potatoes: Organic Production and Marketing*, attra.ncat.org/attra-pub/summaries/summary.php?pub=96

Oregon State Potato Information Exchange (PIE), oregonstate.edu/potatoes Many links.

OSPUD Participatory Organic Potato Project, ospud.org

Oregon State University, *An Online Guide to Plant Disease Control*, plant-disease.ippc.orst.edu

Virginia Cooperative Extension Service, *Potato Seed Selection and Management*, pubs.ext.vt.edu/2906/2906-1391/2906-1391.html

University of Idaho Extension (many publications), extension.uidaho.edu/resources2.asp?title=CROP%20PRODUCTION&category1=Crops&category2=Potatoes&color=91A967&font=4B5F27

Organic and Alternative Methods for Potato Sprout Control in Storage, Mary Jo Frazier, Nora Olsen and Gale Kleinkopf, info.ag.uidaho.edu/pdf/CIS/CIS1120.pdf

Manitoba Agriculture, Food and Rural Initiatives, *Commercial Potato Production: Botany of the Potato Plant, Growth Stages*, gov.mb.ca/agriculture/crops/potatoes/bda04s02.html#Growth_Stages_

University of Maine, *Potato Facts: Selecting, Cutting and Handling Potato Seed*, umext.maine.edu/onlinepubs/pdfpubs/2412.pdf

Organic Seed Potatoes

Moose Tubers, Fedco Seeds, ME, fedcoseeds.com/moose.htm

Potato Garden, CO (formerly Ronniger Potato Farm and Milk Ranch Specialty Potatoes), potatogarden.com

Wood Prairie, ME, woodprairie.com/potato

Maine Potato Lady, mainepotatolady.com

Southern Exposure Seed Exchange, VA, southernexposure.com

Grand Teton Organics, ID, grandtetonorganics.com

Chapter 53: Okra

Kerr Center, *Growing Heirloom Okra*, kerrcenter.com/publications/heirloom-okra-report.pdf

Alabama Cooperative Extension Service, *Rejuvenating Okra: Producing a Spring Crop and a Bigger Fall Crop from the Same Planting* (ratooning), aces.edu/pubs/docs/A/ANR-1112/ANR-1112.pdf

Chapter 54: Sweet Corn

ATTRA *Sweet Corn: Organic Production*, attra.ncat.org/attra-pub/summaries/summary.php?pub=31

North Carolina State University, *Organic Sweet Corn Production*, ces.ncsu.edu/depts/hort/hil/hil-50.html

Cornell University, *Resource Guide for Organic Pest and Disease Management*. Search under Crop Management Practices for Sweet Corn, nysaes.cornell.edu/pp/resourceguide/cmp/corn.php

SARE, Ruth Hazzard and Pam Westgate, *Organic Insect Management in Sweet Corn*, mysare.sare.org/publications/factsheet/pdf/01AGI2005.pdf

Video, *Farmers and their Ecological Sweet Corn Production Practices*, produced by Ruth Hazzard and Vern Grubinger, uvm.edu/vtvegandberry/Videos/cornvideo.html

Crookham Company, *Sweet Corn Genotypes*, crookham.com/sweet-corn-genotypes

Clemson Cooperative Extension *Corn Production Guide, Growth Stages*, clemson.edu/extension/rowcrops/corn/guide/growth_stages.html

Chapter 55: Sweet Potatoes

ATTRA, *Sweet Potato: Organic Production*, attra.ncat.org/attra-pub/summaries/summary.php?pub=32

North Carolina Sweet Potato Commission, ncsweetpotatoes.com The commercial growing page has lots of information and photos, ncsweetpotatoes.com/sweet-potato-industry/growing-sweet-potatoes-in-north-carolina

North Carolina State University Commercial Horticulture Publication, *Presprouting Sweetpotatoes*, ces.ncsu.edu/depts/hort/hil/pdf/hil-23.pdf

Guidelines for Sweetpotato Seed Stock and Transplant Production, ces.ncsu.edu/depts/hort/hil/pdf/hil-23-c.pdf

Sweet Potato IPM (not organic, but includes good info on insect pests), *Crop Profile for Sweetpotatoes in Virginia*, ipmcenters.org/cropprofiles/docs/VAsweetpotato.html

Suppliers of slips

Sand Hill Preservation, IA, 136 varieties of USDA Organic heirlooms, available in small quantities, sandhillpreservation.com/pages/sweetpotato_catalog.html

Steele Plant Co, TN, eight varieties in any quantity, good prices, great service, not organic, sweetpotatoplant.com

Southern Exposure Seed Exchange, VA, nine organic varieties, southernexposure.com

Chapter 56: Seed Growing

Seed to Seed, Suzanne Ashworth, 2nd edition 2002, Sees Savers Exchange. Good for small-scale seed growers, information on each vegetable family for seven bioregions and information on what crosses with what.

Breed Your Own Seed Varieties, Carol Deppe, 2000, Chelsea Green. Technical details for those breeding new varieties.

Seed Savers Exchange, a nonprofit membership organization that saves and shares heirloom seeds: seedsavers.org

Organic Seed Alliance, Organic Seed Conference, Growers' Trade Association, news, research, advocacy, and more: seedalliance.org

Seedsaving and Seedsavers' Resources, an international resource for seed savers: homepage.tinet.ie/~merlyn/seedsaving.html or homepage.eircom.net/~merlyn/seedsaving.html

Saving Our Seeds, information on isolation distances, seed processing techniques, manuals on growing specific seeds and links to more information: savingourseeds.org

Vegetable Seed Saving Handbook, howtosaveseeds.com/basic.php

Fedco Seeds, CR Lawn, *Why Save Seed?* fedcoseeds.com/seeds/why_save_seeds.htm

Seed Saving for Beginners, isolation distance, self- or cross-pollination, seed longevity, fedcoseeds.com/seeds/seed_saving.htm

Eli Rogosa Kaufman, *From Generation to Generation: Seed Saving*. A school curriculum about seed growing, fedcoseeds.com/forms/seedschool.pdf None of us are too old for this!

Local Harvest, localharvest.org

Seed Saving Supplies and Equipment, Southern Exposure Seed Exchange (silica gel, seed vials), southernexposure.com

Abundant Life Seed Foundation (seed cleaning screens), abundantlifeseeds.com

Plans for making your own seed winnower and other seed cleaners, saveseeds.org/tools/tool_winnower_hand.htmlsustainableseedsystems.wsu.edu/nicheMarket/smallScaleThreshing.html

Chapter 57: The Business of Seed Crops

Small-scale Organic Seed Production, Patrick Steiner, FarmFolk/CityFolk. For farmers in British Columbia and beyond. Why and how to incorporate vegetable seed production into existing farming systems. Interviews several small-scale seed-growers from Canada and the US to get a glimpse of their experience over the years, bcseeds.org/seed-manual.php

The Wisdom of Plant Heritage, Organic Seed Production and Saving, Bryan Connolly, revised edition 2011, Chelsea Green. "A how-to for the small producer, the best technique for growing the seed that contains the knowledge of how to thrive, propagate and please the taste buds under local conditions," nofany.org/catalog/publications/wisdom-plant-heritage-organic-seed-production-and-saving

The Organic Seed Alliance promotes the value of seed and seed-*saving* skills, seedalliance.org

The Siskiyou Sustainable Cooperative, formed to strengthen the quality and consistency of organic products and improve grower training and support in the Siskiyou region of Oregon, siskiyoucoop.com

Index

alfalfa, 122, 192

allelopathy: brassicas, 121, 129, 145–6; rye, 118–9, 145, 147, 177, 346–7; sorghum-sudangrass, 130, 147; other crops, 125, 131, 132

Alternaria fungal diseases, 75–6, 140, 261, 266, 309, 330, 351 (*See also* early blight)

alyssum, sweet, 98, 151, 213, 214, 251, 354, 388

amaranth, 244–5

angular leaf spot, cucurbit (*Pseudomonas syringae*), 134–5, 140, 294, 301

annual ryegrass, 126, 127, 130

Anthracnose fungal diseases, 135, 177, 252, 285, 294, 301, 303, 309, 353, 360

aphids: botanical/biological controls, 154, 251; as disease vectors, 136, 285, 332, 353, 354, 366; insecticidal soap, 97, 154, 215, 230, 251, 278, 309, 344, 354, 366; physical controls, 189, 201, 214, 230, 251, 366; physical deterrents, 150, 285; predators for control of, 97, 151, 154, 213–4, 251–2, 309, 366; removing from harvested crop, 215, 253; susceptible varieties, 94, 209

armyworms, 318, 354–5, 360, 366, 388–9

Arnold, Paul and Sandy, 234

arugula, 5, 11, 45, 85, 86, 87, 92, 96, 98, 159, 228

Ashworth, Suzanne, 405

Asian greens, 45, 225–32; fall/winter production, 85, 86, 87, 92–3, 95, 98, 104, 231–2; harvesting, 159, 226, 231; pests, 97, 230–1; storage, 161, 166; transplanting/sowing, 36, 40, 229–30; varieties, 92–3, 220, 226–8 (*See also* Chinese (Napa) cabbage; pak choy; tatsoi)

asparagus, 11, 12, 45, 109

asparagus beans, 11, 13, 32, 182, 184–5

assassin bugs, 127, 284, 300

Austrian winter peas, 117, 118, 119, 122, 126, 305, 346–7

Bacillus thuringiensis (Bt), 153, 215, 230, 278, 284–5, 290, 355, 366, 388

bacterial blight of beans (various spp.), 177, 181, 204

bacterial speck (*Pseudomonas syringae*), 353, 360

Bahret, Melissa, 102

baking soda for disease management, 139–40, 201, 285, 301, 313, 351

basil, 29, 42, 87, 160

beans, 11, 13, 45, 175–81; frost sensitivity, 81, 87; harvesting, 155, 160, 181; hoop-house production, 101, 102, 103, 404; pests/diseases, 74, 138, 177, 179–81; seed saving, 399; sowing, 49, 62, 66, 176; succession planting/rotations, 5, 25, 70, 71, 74, 85, 177–9; varieties, 32, 175

Beauvaria bassiana (beneficial fungus), 230, 251, 252, 374

beets, 13, 45, 260; fall/winter production, 85, 87, 88, 92, 94, 98, 250; for greens, 11, 92, 98, 243–4; harvesting, 267; pests/diseases, 266–7; sowing/transplanting, 53, 62, 66, 68, 243, 262, 263–4; spacing, 48–9, 264, 266; storage, 6, 162, 164, 166, 268; varieties, 87, 261

beneficial insects. *See* farmscaping; parasitic wasps; *specific insects*

Bergmark, Christine, 85

Bezilla, Ken, 87

bird pests, 154, 388

black rot/gummy stem blight (cucurbits, *Didymella bryoniae*), 285, 291, 301, 309

black scurf (potatoes, *Rhizoctonia solani*), 369, 375

black-eyed peas. *See* Southern peas

blackleg (brassicas, *Leptosphaeria biglobosa*), 138, 278

Bleasdale, J. K. A., 47, 50, 69, 265, 370

blister beetles, 242, 355, 375, 381

blossom end rot, 343, 354, 360, 377

Bogash, Steve, 356

boron, 208–9, 304, 343, 360, 372; deficiency, 114, 262, 272, 277, 354, 384–5, 396

Botrytis grey mold, 136, 140, 181, 252, 318

Boutard, Anthony and Carol, 184

Brady, Nyle, 108

brassicas, 208–16; as cover crops, 106, 116, 121, 129, 145–6, 369, 387; crop requirements, 66, 109, 208–9; disease management, 134, 138, 231, 276, 278; farmscap-ing for, 213–4, 217, 218, 278; pests/pest management, 150, 152, 215–6, 217, 230–1, 278; planning for, 12, 218–9; in rotations, 211–2, 217–8, 223–4; for salad mix, 96–7, 98; seed saving, 232, 279; sowing/trans-planting, 36, 53, 55, 56, 57, 58, 80, 212–3, 217, 218, 219–23, 229–30; varieties, 209–11, 220; weed management, 142 (*See also* harlequin bugs; *specific crops*)

broccoli: cover crops before/after, 116, 117; fall/winter production, 87, 88, 94; ger-mination requirements, 66; growth/care of, 214–5; harvesting, 155, 158, 159, 216; pests, 150; spacing, 49, 51, 211, 213, 222; suggested planting quantities, 45, 211; transplanting, 213, 222; varieties, 209, 220 (*See also* brassicas)

broccoli raab, 228

brown marmorated stink bug, 150, 151, 284, 299–300, 355

brussels sprouts, 4, 87, 219, 223

Bubel, Mike, 161, 168, 171–2

Bubel, Nancy, 61, 65, 161, 168, 171–2, 271, 403, 407

buckwheat: as cover crop, 103, 116, 117, 119, 121, 122–3, 133, 229; for farmscaping, 98, 213, 284, 354, 374; as trap crop, 300

Burkholder, Marlin, 160

Byczynski, Lynn, vii–viii, 105

cabbage: cover crops before/after, 116, 117; fall/winter production, 85, 86, 87, 88; germination requirements, 66; growth/care of, 214–5; harvesting, 156, 158, 159, 162, 216, 224; pests, 150, 211, 215, 217, 278; spacing, 49, 211, 213, 222; spring cabbage greens, 96; storage, 161, 162, 164, 166, 167, 170; suggested planting quantities, 45, 211; transplanting, 36, 213, 222; varieties, 85, 86, 87, 88, 209, 220 (*See also* brassicas)

cabbage worms, 211, 215, 217, 278

calcium, 108, 114, 138, 208–9; adding, 113, 140, 203, 360; and buckwheat, 133; defi-ciency, 203, 252, 272, 343, 354, 360, 377

calendula, 213, 404, 405

Canadian field peas, 127

canker, parsnip (various species), 138, 261, 267

canola, 121, 129

cantaloupes. *See* muskmelons

carrot rust fly, 150, 153, 265, 266, 267, 272

carrot weevil, 153, 266, 272

carrots, 45; fall/winter production, 86, 87, 88, 94; germination requirements, 66, 67; harvesting, 119, 159, 267; pests/diseases, 75–6, 150, 153, 261, 266, 332; sowing, 40, 61, 63, 262, 263; spacing, 48, 49, 263, 265; storage, 6, 164, 166, 167, 268; succession planting, 70, 71, 75–6; varieties, 87, 260–1; weed management, 66, 142, 146, 263, 264, 265

cauliflower, 47, 66, 81, 87, 218, 219, 222, 223, 224

celeriac, 45; fall production, 87, 88; harvesting, 273; pests/diseases, 272; sowing/transplanting, 56, 270–2; storage, 6, 163, 166, 273; varieties, 87, 270

celery, 45; fall/winter production, 81, 87, 88, 96, 274; germination requirements, 66, 68; harvesting, 159; pests/diseases, 272; sowing/transplanting, 55, 56, 57, 61, 270–2; spacing, 49, 271–2; storage, 164, 166, 170, 273; varieties, 87, 269–70

Cercospora leaf spot (fungal), 140, 242, 261, 266, 267, 272, 309

chard, 45, 240–3; fall/winter production outside, 85, 86, 87, 88; harvesting, 89, 159, 242; sowing/transplanting, 55, 67, 241–2; varieties, 93, 240–1; winter hoophouse production, 92, 93, 95, 98

chervil, 85, 87

chicories, 85, 86, 87, 88, 93

Chinese (Napa) cabbage, 45, 226; fall/winter production outside, 81, 85, 87, 88; harvesting, 159; pests, 97; sowing/transplanting, 230; storage, 166; varieties, 87, 93; winter hoophouse production, 92, 93, 98

chives, 86, 88

cilantro, 86, 87, 213, 284

claytonia, 88, 94

cleome, 151, 213, 214

clovers, 127–9, 305; as spring cover crops, 123; as undersown cover crops, 5, 116, 117, 120, 121, 218, 387; as winter cover crops, 117–8, 119

clubroot (brassica, *Plasmodiophora brassicae*), 134, 138, 231, 276, 278

coldframes, 84, 88, 259

Coleman, Eliot, 21, 51, 59, 63, 88, 91, 99–100, 112, 172

collards: harvesting, 89, 159, 216, 224; suggested planting quantities, 45, 211; transplanting, 213, 223; varieties, 88, 210, 220; winter production, 4–5, 85, 86, 87, 88, 94, 219, 224, 250

Colorado potato beetles, 149, 150, 355, 366, 369, 374

compost teas, 60, 109, 112–3, 136–7, 137–8, 375

Connolly, Bryan, 413

copper, deficiency in corn, 384

copper fungicides, 140, 351, 353

corn (sweet), 45, 87, 383–90; cover crops/rotations, 116, 117, 118, 121, 387, 388; growth requirements, 384–5; harvesting/storage, 155, 156, 160, 389–90; pests/diseases, 388–9; sowing/transplanting, 49, 53, 55, 57, 61, 62, 385–6; succession planting, 5, 70, 71, 75, 387–8; varieties, 383–4, 387, 388, 389; weed management, 142, 143

corn earworms (tomato fruitworms), 152, 354, 360, 381, 388–9

cover cropping, 115–23, 124–33; all-year green fallow, 116, 120, 223–4; for disease management, 117, 252, 350, 369; in hoophouses, 103, 106; for summer, 119–20, 122–3; undersowing, 5, 86, 116, 117, 120, 121, 200, 218, 290; for weed management, 115, 117, 121, 144, 211, 281, 289–90; for winter, 115–6, 116–9 (*See also* allelopathy; no-till methods; *specific crops*)

cowpeas. *See* Southern peas

Cranshaw, Whitney, 152, 266

cresses, 85, 87–8, 96, 228

crop rotations, 4, 18–26, 365; and brassicas, 211–2, 217–8, 223–4; and disease management, 104, 134, 135, 138, 181, 268, 282, 350, 369; in hoophouses, 97, 103–4; and legumes, 177, 201; and pest management, 149, 177, 298–9, 375, 388, 389; and succession planting, 24–6; ten-year plan for major crops, 20–4 (*See also* cover cropping; *specific crops*)

cucumber beetles, 101, 282, 309, 381; as disease vectors, 135, 193, 285, 291, 299; physical controls, 150, 231, 284, 289, 298–9; resistant vegetables, 288, 289, 291, 295; sprays, 154, 299; sticky traps/pheromone lures, 152, 284, 291, 299; trap crops, 281, 291, 299

cucumbers, 45, 87; harvesting, 156–7, 160, 295, 301; pests/diseases, 135, 294, 299–301; season extension, 74, 81, 302; sowing/transplanting, 54, 57, 297–8;

spacing, 49, 296, 297; storage, 162, 302; succession planting, 5, 25, 70, 71, 73–4, 85, 299; varieties, 74, 294–5, 296

cutworms, 151, 252, 272, 318, 354, 360, 375, 389

daikon radishes, 66, 87, 88, 121, 129, 166

damping off (various fungi), 138, 140, 252, 266–7, 331, 351, 366

Davis, Jeanine, 387

Day, Eric, 266

deer/rabbit deterrents, 150, 154, 180, 193, 252

Denckla, Tanya, 270

Deppe, Carol, 107, 183, 189, 368, 372

Dermer, Lisa, 107

dibbles, 59, 62, 64

dill, 87, 213, 251, 284, 404, 405

disease management techniques, 134–40; avoiding soil splashback, 346, 350, 360, 381; biorational controls, 137, 139–40, 163, 201, 285, 291, 301, 313, 351; composting/foliar feeding, 112, 251, 252, 350, 375; cover cropping, 117, 252, 350, 369; crop rotation, 104, 134, 135, 138, 181, 268, 282, 350, 369; disease vectors, 134–5, 290, 291, 353, 360; field sanitation, 98, 318, 350, 375; harvest timing, 157, 177, 185, 362; integrated pest management, 135–7; microbial controls, 136–7, 140, 351, 352, 353, 375; resistant varieties, 294, 350, 353; seed saving, 351, 400; seed treatments, 138–9, 313–4, 351; soil solarization, 98, 106, 139, 151, 252; succession planting, 285, 350–1; transplant selection, 56, 57; and weed management, 212, 350 (*See also under specific crops; specific diseases*)

Docter, Michael, 155

downy mildew: appearance, 285, 291, 330; controls, 140, 252, 278, 291, 318, 330; forecasts, 136; resistant varieties, 234, 294, 341; weather conditions, 285, 301, 330

drip irrigation: advantages, 230, 298, 343; disease prevention, 135, 285, 304, 308, 325, 343, 350, 360, 381; and plastic mulch, 214, 298; and sowing, 68, 282; and transplanting, 222, 249, 395; in tunnels/hoophouses, 94, 259; weed management, 144

Drowns, Glenn, 303

early blight (*Alternaria solani*), 135, 140, 351, 352, 367, 375

Earthway seeders, 50, 59, 62, 63–4, 119
edamame, 13, 45, 191–5; hoophouse production, 101, 102, 103, 104, 195, 403; seed saving, 102, 105, 403–4; soil requirements, 97, 192; succession planting, 5, 25; varieties, 102, 191–2, 404
eggplant, 45, 87, 363–7; and cover crops, 121, 127; harvesting, 81, 156, 158, 367; pests/diseases, 19, 35, 101–2, 150, 242, 352, 365–7; seed viability/saving, 29, 367; sowing/transplanting, 51, 53, 55, 56, 57, 364; storage, 162, 367
endive, 85, 87–8, 93
Engeland, Ron, 311
erosion, 110, 115, 141
ethylene, 163–4, 167, 168, 268, 292, 333, 356, 370
European corn borers, 180, 183, 360, 388–9

farmscaping, ix, 120, 135, 145, 151, 213–4, 284, 354, 374, 388 (*See also specific beneficials; specific flowers; specific pests*)
fava beans, 13, 186–90; as cover crop, 119, 126, 187–8; harvesting, 11, 190; sowing, 10, 40, 188–9; spacing, 49, 188; varieties, 87, 187, 188; winter production, 87, 96, 190
fennel, 86, 87–8
fertility. *See* soil fertility
fish emulsion, 60, 138
flame weeding, 13, 144, 145, 147, 308; carrots, 66, 146, 263, 264, 265; other crops, 204, 318, 329, 374, 387
flaming for pest control, 149, 215, 300, 374
flea beetles: brassicas, 150, 152, 217, 278; nightshades, 19, 101–2, 150, 354, 360, 364, 365–6; other crops, 242, 245, 381
flea beetles, control of: biological/microbial, 230, 354; physical, 101–2, 150, 154, 218, 242, 277, 364, 365–6; trap crops, 215
Florida stake-and-weave system, 102, 347–50
fluid sowing, 69, 263
foliar feeding, 102, 281; to boost yield, 270, 314–5, 358, 360; for frost protection, 81, 84; for pest/disease management, 251, 252, 350; for seedlings, 344 (*See also* compost teas)
forage radish, 121, 129
forsythia, 193, 198
Frazier, Mary Jo, 377
frost seeding, 124
Fusarium fungal diseases, 314; infection route, 309; microbial controls, 140, 351;

prevention, 318, 330, 331, 381; resistant varieties, 234, 294, 303; soil and seed survival, 134; and soil fertility, 138, 314; symptoms, 201, 242, 269, 272, 318, 330, 331, 352, 396

galinsoga, 142, 143, 145, 146–7
garlic, 12, 45, 311–23; as antifungal agent, 137; curing/storing, 6, 161, 162, 166, 321–3; growth, 312, 319–20; harvesting, 156, 214, 320–1; pests/diseases, 313–4, 318–9; planting, 312, 313–4, 315–6; seed storage, 312–3, 323; soil requirements, 314–5; types, 311–2; weed management, 317–8; winter hardiness, 87, 88
garlic scallions, 12, 86, 313, 316, 317
genetically modified organisms, 4, 120, 125, 131, 281, 374, 383, 384, 409
germination testing, 29, 407, 409
ginger, 101, 102–3, 166
Goss, Jeffrey, 88
grasshoppers, 76, 154, 231, 242, 252, 277, 278, 381
green beans. *See* beans
Grieshop, Matt, 300
Grohsgal, Brett, 85, 88–9
groundhogs, 355
growing degree days, 152, 203
Grubinger, Vern, 64, 389

hairy vetch, 117, 119, 121, 123, 126, 305, 346–7
Hall, Bart, 119
harlequin bugs, 217, 223; and brassica cover crops, 106, 121, 183; physical controls, 60, 150, 211, 215, 230; trap crops, 151, 214
harvesting techniques, 89, 155–60, 162, 177, 185, 362 (*See also under specific crops*)
Hazzard, Ruth, 389
Hochmuth, George J., 47
Hon Tsai Tai (Asian green), 94, 227–8
honeybees, 135, 153, 289, 308
Hong Vit (leaf radish), 228
hoophouses and hoophouse growing: benefits, 5, 79; cover crops/rotations, 97, 103–4, 106; pests/diseases, 97–8, 101–2, 106; photos/descriptions, 6, 79, 90–1, 100–1; scheduling, 88, 103, 104; seed saving, 105–6, 403–4; summer hoophouse production, 74, 99–107, 207, 358, 360; winter/early spring hoophouse production, 84, 90–8, 259 (*See also under specific crops*)

hornworms, 151, 354
horseradish, 86, 88, 164, 166
horticultural oils, 140, 154, 251
hydrogen peroxide, 140, 351, 353

Ingham, Elaine, 109, 137
interplanting. *See* relay planting
iron, 114, 209, 343, 372

Jagger, Chris, 63
Japanese beetles, 105, 152, 193, 381
Jeavons, John, 109–10
Jerusalem artichokes, 86, 88, 164, 166
jicama, 101

kale, 45, 208, 211; harvesting, 89, 156, 159, 216, 224; overwintering, 4–5, 81, 84, 85, 86, 87, 88, 219, 224, 250; sowing/transplanting, 92, 212–3, 222–3; undersowing, 121; varieties, 88, 93, 94, 209–10, 223; winter hoophouse production, 92, 93, 98, 209–10
kaolin clay, 154, 300
Knudson, William A., 194
kohlrabi, 45; fall/winter production, 85–6, 87, 88, 219, 220, 223; harvesting, 158, 216, 224; pests/diseases, 210–1; storage, 6, 164, 166, 224; varieties, 223
Komatsuna, 85, 86, 87–8, 94, 227

lablab bean as no-till cover crop, 315–6
lacewings, 214, 251, 252, 309, 329, 374
ladybugs, 374; aphid control, 97, 154, 251, 309, 366; Colorado potato beetle control, 149; flea beetle control, 278; harlequin bug control, 215; thrips control, 252, 319, 329
lambsquarters, 143, 145
Lampkin, Nic, 114
late blight (*Phytophthora infestans*), 134, 252, 352, 369, 375
leaf beet, 404
leaf miners, 237, 242, 267
leaf radish, 85, 86
leaf-footed bugs, 284
leeks, 4–5, 45, 338–41; harvesting, 159, 341; sowing/transplanting, 49, 57, 339–40; storage, 161, 164, 166, 170, 341; varieties, 87, 88, 338; winter production, 86, 87–8
Leskey, Tracy, 300
lettuce, 45, 246–59; bolting/bitterness, 248, 253, 258; germination requirements, 66–7, 68; harvesting, 156, 159, 252–3; pests/diseases, 98, 140, 251–2,

332; relay planting, 204–5, 242, 250–1; season extension, 80, 254, 255–9; sowing/transplanting, 36, 49, 50, 53, 54, 55, 57, 61, 67, 69, 92, 247, 248–9, 255, 257; storage, 161; succession planting, 5, 8, 71, 74–5, 253–5, 256–7; types/varieties, 87, 88, 93, 246–7, 250; winter hoophouse production, 92, 93, 95, 98, 250, 259; winter production outside, 81, 85, 86, 87, 88
lima beans, 11, 13, 81, 87, 182, 183, 185
Lorenz, Oscar, 50
low tunnels, 79, 259
lupins, 127

mache (corn salad), 5, 85, 86, 88, 94
magnesium, 108, 113, 384
manganese, 209, 372
Maruba Santoh, 85, 87, 92–3, 98, 159, 226, 230
Maynard, Donald N., 47, 50
McCormack, Jeff, 295–6
McDonald, Richard, 215
Mebrahtu, Tadesse, 191
Mei Qing Choi (Asian green), 228
Mexican bean beetles, 74, 149, 152, 153, 177, 179–80, 183, 193
Miami peas, 117, 119, 131
millets, 106, 117, 119–20, 121, 122–3, 130–1, 300, 369
mites, 313–4, 318
Mizell, Russ, 300
mizspoona, 94, 228
mizuna, 45, 226–7; fall production, 85, 87, 88; harvesting, 159, 231; spring production, 44, 86; winter production, 92, 93, 95, 96, 98, 104, 195
Mohler, Charles (Chuck), 141–2, 143
molybdenum, 209, 343
Montgomery, David, 110
mosaic viruses. See under viral diseases
Moyer, Richard, 187
mulching: biodegradable plastic mulch, 5, 118, 177, 304, 402; black plastic for soil warming, 214, 304, 364, 395; landscape fabric, 347; to moderate soil temperature/moisture, 198, 214, 258, 270, 316, 343–4, 373; pest management, 150, 153–4, 298, 366, 373, 374; to reduce soil-borne diseases, 135, 346; and row spacing, 51, 359–60; and soil fertility, 110, 113; weed management, 144, 145, 146–7, 317, 329, 343, 346, 364, 374; winter cold protection, 84, 315, 316 (See also no-till methods)
muskmelons, 45, 294; frost sensitivity, 81,

87; harvesting, 156–7, 160, 301–2; pests/diseases, 101; sowing/transplanting, 54, 55, 56, 57, 297–8; storage, 163–4; succession planting, 25, 71; varieties, 295–6, 302
mustards, 45; as cover crops, 121, 122, 129, 139, 252; harvesting, 159; as trap crops, 215; varieties, 139, 228–9; winter production, 86, 87, 93, 96, 250

Natural Resources Conservation Service, 148
Navazio, John, 234, 412
neck rot (onions, Botrytis alii), 330, 332, 341
neem oil, 137, 140, 154, 230, 284, 299
nematodes (beneficial), 113, 230, 262, 285, 299, 366
nematodes (pathogenic), 139, 313, 318–9; and cover crops, 125, 126, 127, 129, 130, 132, 133, 369, 375; root knot nematodes, 125, 130, 242, 309, 354, 369, 379, 381; root lesion nematodes, 133, 180
NeSmith, Scott, 50, 307
Nielson, Anne, 300
nitrate accumulation in leafy greens, 98, 157, 250
nitrogen, 114, 115, 123, 142, 176, 328, 384; from leguminous cover crops, 117, 118, 120
nitrogen-fixing bacteria, 176, 192, 194, 199, 204
no-till methods, 117, 204, 218, 315–6, 387; benefits, 121–2; and weed management, 122, 144, 147, 346–7
nutrition, 167, 194, 202, 225–6, 244, 246–7, 276, 303, 362, 368, 391
nutsedges, 126, 132, 145, 147

oats, 125; as spring cover crop, 116, 119, 123; as undersown cover crop, 5, 117, 121, 200, 238, 387; as winter cover crop, 117–8, 119, 316
okra, 45, 378–82, 404; frost sensitivity, 81, 87; harvesting, 156, 158, 381–2; pests/diseases, 300, 355, 381; relay planting, 52, 380–1; seed viability, 29; transplanting, 38, 51, 53, 55, 379–80; varieties, 378–9
onion maggots, 314, 318, 319, 329
onions (bulbing), 45, 324–33; bolting, 326, 329; curing/storage, 6, 161, 162, 166, 332–3, 336–7; germination requirements, 66, 67; growth of, 324–7; harvesting, 156, 160, 329, 332, 336; overwintering, 86–7, 88, 92, 95, 96, 327; pests/diseases,

329, 330–2; record-keeping, 12; sowing/transplanting, 36, 40, 51, 53, 55, 56, 57, 327, 328–9; spacing, 49, 52, 328, 329; varieties, 88, 324, 326, 327–8, 330; weed management, 142, 329 (See also potato onions; scallions)

pak choy, 45; fall production, 81, 85, 87; harvesting, 159, 226, 231; spring production, 44, 86; varieties, 226, 228; winter production, 92–3, 98, 104
Paquin, Scott, 411
parasitic wasps: for aphids, 251, 366; Braconid wasps, 215, 230, 319, 354; for corn earworm, 388; for Japanese beetles, 193; for leaf hoppers, 374; for Mexican bean beetles, 74, 149, 153, 177, 179–80, 193
parsley, 66, 67, 85, 86, 88
parsley worms, 266, 272
parsnips, 45, 260; harvesting, 267–8; overwintering, 4–5, 86, 88; pests/diseases, 138, 261, 267; sowing, 61, 62, 66, 262, 264; spacing, 49, 264, 266; storage, 164, 166, 268; varieties, 261–2
parthenocarpy, 295, 300
peanuts, 6, 87, 102, 162–3, 203–7
peas (shelling), 49, 196, 198
peas (snap and snow), 45, 196–202; cover crops/rotations, 116, 121; for farmscaping, 374; harvesting, 156, 158, 160, 201; hoophouse production, 96, 103; pea shoots, 197, 202; pests/diseases, 138, 201; relay planting, 5, 197, 199–200, 235–6; sowing, 66, 67, 88, 198–9, 202; varieties, 196–8
Peet, Mary, 299, 374
peppers, 45, 357–62; cover crops/rotations, 117, 118, 121, 358; frost sensitivity, 81, 87, 361; harvesting/storage, 155, 156, 160, 162, 170, 361–2; hot peppers, 101, 102; pests/diseases, 353, 360, 362; season extension, 80, 360–1; sowing/transplanting, 53, 56, 57, 61, 358–60; varieties, 357–8
perpetual spinach/spinach beet, 243
pest management techniques, 148–54, 251–2; crop rotation, 149, 177, 298–9, 375, 388, 389; cultural controls, 56, 57, 149, 266, 388; flaming, 149, 215, 300, 374; insecticidal soap, 139, 154, 300, 309, 344, 354, 355, 366, 374; integrated pest management, 135–7, 148, 388; monitoring, 148–9, 151–2; mulching, 150, 153–4, 298, 366, 373, 374; pheromone lures, 152, 284, 290, 291, 299, 300; physical/

mechanical means, 148, 149–51, 201, 230, 242, 251, 284–5, 354, 355; pre-planting treatments, 313–4, 318–9; rowcovers, 215, 230, 231, 278, 300, 364, 365–6; soil solarization, 106, 151; sprays, 152–4, 215, 230, 278, 290, 354–5, 374, 388–9; succession planting, 74, 149; trap crops, 151, 214, 215–6, 281, 291, 299, 300, 355; for vertebrate pests, 150, 154, 164, 172, 180, 193, 252, 355, 388 (*See also* farmscaping; nematodes (beneficial); parasitic wasps; *specific crops*; *specific pests*)

phenology, 152, 198, 271, 283, 359, 369, 385

Phomopsis blight, 140, 366

phosphorus, 113, 114, 343, 360, 372, 384

Phytophthora diseases, 140, 267, 301, 351, 353, 366 (*See also* late blight)

pigweed (red-root), 142–3, 145, 245, 281

pirate bugs, 252, 319

Planet Junior seed drill, 62–3

planning, 3–4; crop quantities, 9, 12–3, 29–30, 41–2, 43–4, 45–6; crop review, 6, 7, 14–7, 31; for crop rotation, 20–6; and efficient harvesting, 155–6; field manual, 8, 12–3; and first frost date, 70–1, 85–6, 88, 119, 125–33, 398; outdoor planting schedule, 40; seed inventory/ordering, 29–32; sequence, 7–8; starting transplants, 35–6, 39–40; styles/options, 9–10; for succession planting, 8, 11–2, 70–3, 75, 178–9; summer hoophouse crops, 103, 104

Poisson, Leandre and Gretchen, 259

potassium, 108, 113, 114, 138, 208, 262, 328, 343, 372, 396

potato onions, 5, 40, 86–7, 88, 334–7

potatoes, 45, 87, 368–77; in crop rotation, 116, 117, 118, 218, 369; curing, 162, 169, 377; growth, 368, 372; harvesting, 156, 168, 376–7; hoophouse production, 94; pests/diseases, 153–4, 369, 374–6; physiological age, 169, 370, 377; planting, 369–72, 373–4; pre-sprouting, 370–2; spacing, 49, 52; storage, 6, 161, 164, 167, 168, 169, 170, 171, 377; varieties, 368–9

powdery mildew: cucurbits, 138, 285, 291, 294, 301; peas, 183, 197, 201; other crops, 252, 278, 318, 351

powdery mildew controls: baking soda, 139–40, 201, 285, 351; biological controls, 140, 351; milk, 136, 201, 291, 351; prevention, 138, 140, 252, 285; resistant varieties, 197, 294

praying mantids, 252, 300

pre-sprouting seeds, 68–9, 85

pumpkins. *See* winter squash and pumpkins

Pythium fungus, 138, 140, 266, 267, 331, 351, 396

radicchio, 86, 87, 88

radishes, 46, 122–3, 160, 285; fall/winter production outside, 85, 87, 88; winter hoophouse production, 92, 93, 95, 98, 250 (*See also* winter radishes)

ragweed, 125, 142

record-keeping, viii, 5, 14–5, 70–3, 142, 152

relay planting (interplanting), 5, 52, 144, 151, 197, 199–200, 204–5, 235–6, 242, 250–1, 380–1, 388

Reynolds, Grace, 315

Rhizoctonia rots, 140, 267, 272, 278, 331, 351, 369, 375

Roberts, Warren, 50

rodents, 164, 172

Rogers, Ted, 300

rolling dibbles, 59, 62

root cellars, 163, 165, 168–72

rotations. *See* crop rotations

rowcover temperature gain, 258, 274

rowcover types, 78–9, 150, 237–8

rust (allium, *Puccinia porri*), 318, 341

rutabagas, 6, 46, 87–8, 164, 167, 275–9

rye cover crops, 86, 117–9, 121, 125, 305; allelopathy, 118–9, 145, 147, 177, 346–7

sabadilla, 290

salsify, 86, 88, 164, 167

Salter, P. J., 47, 69, 265, 370

Samulis, Ray, 50, 386

scab (cucumber, *Cladosporium cucumerinum*), 294, 295, 301

scab (potato, *Spongospora scabies*), 138, 267, 278, 369

scallions, 12, 46, 86, 87, 88, 92, 93, 95, 98, 159–60

Schonbeck, Mark, 108, 119

Sclerotinia diseases, 98, 140, 181, 204, 252, 353, 375

scurf, sweet potato (*Monilochaetes infuscans*), 138, 396

season extension, 3–6, 77–82, 259, 308 (*See also* hoophouses and hoophouse growing; *specific crops*)

seed saving, 399–407; biennials, 243, 268, 274, 404, 405–6; brassicas, 232, 279; business aspects, 399–400, 408–13; corn, 390; cucurbits, 106, 286, 293, 302, 303, 310, 399, 400, 402–3, 404, 405, 412;

garlic, 312–3, 323, 400; legumes, 102, 105, 185, 190, 194, 202, 206–7, 399, 401, 403–4, 412, 413; lettuce, 106, 405, 413; nightshades, 351, 356, 358, 362, 367, 399, 400, 401–2, 404, 412, 413; okra, 382, 404, 412; spinach, 239

seed storage, 27–8, 247, 406–7

seed viability, 28–9, 66, 176, 188, 194, 234, 239, 262, 281, 327, 358

Senposai, 12, 85, 87, 92, 93, 98, 159, 227

Septoria leaf spot, 117, 272, 343, 352, 360, 400, 401

sesbania, 132

shadecloth, 60, 80–1, 100–1, 107, 160, 239, 257

Sisti, Rich, 316

slugs, 230–1, 242, 252

soil blocks, 37–8, 54, 305

soil fertility, 4, 108–14, 115; balancing soil chemistry, 108–9, 113–4; biointensive mini-farming, 109–10; composts/fertilizers, 97, 109, 110–3, 114, 137–8, 262; and disease prevention, 135, 137–8, 314; and pest management, 149; soil biology, 109–10, 112–3, 343; and weed management, 145, 147 (*See also* cover cropping; *specific minerals/nutrients*)

soil solarization, 98, 106, 136, 139, 145, 151, 252

soil temperature, 58, 61, 65, 66–7

sorghum, 300, 412

sorghum-sudangrass, 106, 117, 119, 130, 145, 147, 315–6

sorrel, 85, 87, 88

Southern blight (*Sclerotium rolfsii*), 140, 367

Southern peas (cowpeas), 45, 81, 87, 97, 101–3, 182–4, 403; as cover crops, 117, 119, 121, 122–3, 132

sowing (direct), 53, 61–4, 65–9, 88–9, 144, 282

soybean cover crops, 5, 103, 116, 117, 119–20, 121, 122–3, 131, 387

spider mites, 154, 272, 294, 309, 366

spinach, 46, 116, 233–9; harvesting, 89, 98, 156, 158, 159, 237; hot-weather substitutes, 244–5; overwintering, 4–5, 85, 86, 87–8, 98, 235, 237–8, 250; pests/diseases, 237, 332; relay planting, 5, 197, 199–200, 235–6; shadecloth use, 80, 239; sowing, 36, 55, 65, 66, 67, 69, 234–5, 238–9; storage, 161; transplanting, 40, 58, 92, 234–5, 236–7; varieties, 88, 93, 233–4; winter hoophouse production, 92, 93, 95, 98, 105, 236–7

Spinosad, 97, 153–4, 215, 230, 284, 299, 355, 366, 374, 388
squash bugs, 280, 283–4, 290
squash vine borer, 284–5, 290
stale seedbed technique, 131, 144, 145, 147
Stansbury, Patricia, 192, 194
Steiner, Patrick, 413
Stern, David, 316
stink bugs, 154, 183, 355, 381 (*See also* brown marmorated stink bug)
storage of crops, 6, 161–7, 168–72 (*See also under specific crops*)
strawberries, 25, 79, 94, 136, 158
string-weaving. *See* Florida stake-and-weave system
succession planting, 5, 70–6, 85; and crop rotations, 24–6; and disease management, 285, 350–1; for pest management, 74, 149; planning for, 8, 11–2, 70–3, 75, 178–9 (*See also under specific crops*)
sulfur, 113, 140, 154
summer squash and zucchini, 46, 280–6; crop rotations, 281–2; frost sensitivity, 81, 87; harvesting, 156–7, 158, 160, 285–6; hoophouse production, 74, 281, 283, 284; pests/diseases, 283–5; sowing/transplanting, 54, 57, 62, 282; succession planting, 5, 25, 70, 71, 72–4, 85, 283; as trap crops, 151; varieties, 74, 280–1, 283, 286; weed management, 142, 282–3
sunflowers, 106, 122–3, 213, 300, 354, 374, 404
sunn hemp, 122, 123, 132, 315–6
sunscald, 332, 354, 360, 397
sweet potatoes, 46, 87, 391–8; allelopathy, 147; cover crops/rotations, 118, 121, 391; curing/storage, 6, 107, 161, 167, 398; diseases/afflictions, 138, 395–6; harvesting, 156, 162, 396–8; planting, 58, 392, 394–5; selecting seed, 397; starting slips, 391–4

tarnished plant bugs, 180, 272
tatsoi, 46, 85, 86, 87–8, 92, 93, 95, 96, 98, 159, 227, 231
Tendergreen (Asian green), 87
Tenderleaf (Asian green), 228
thrips, 183, 318, 341; as disease vectors, 353–4, 360; microbial controls, 153, 252; mineral controls, 154; physical controls, 150, 252, 329; predators, 252, 319, 329
tip burn, 98, 252
Tipping, Don, 411
Tokyo Bekana, 85, 87, 92–3, 98, 226, 230
tomatoes, 80, 81, 87, 118, 342–56; harvest-

ing, 155, 158, 160, 355–6; no-till methods, 117, 122, 346–7; pests/diseases, 117, 135, 332, 343, 344, 346, 350–3, 354–5, 400; physiological disorders, 114, 343, 354; sowing/transplanting, 36, 46, 53, 54, 55, 56, 57, 58, 344–6, 351; spacing, 49, 51; storage, 163–4, 165, 167; succession planting, 342–3, 345, 350; trellising/supporting, 347–50; varieties, 342–3, 344, 356; weed management, 142, 346–7
Tozer, Frank, 186
transplanting and starting transplants, 33–40, 53–60; aftercare, 60, 306, 340, 364; age/size of starts, 56–8, 328–9, 339, 344, 359, 364; bare-root, 38–9, 53–4, 57, 58, 212–3, 218, 229, 249; benefits, 5, 53, 67, 77, 144, 150, 344; equipment/facilities needed, 33–5, 37–8; flats/cells, 36–7, 54–5, 57–8, 212, 282; hardening off, 55–6, 150, 212, 344–5, 359, 364; planting techniques/patterns, 50–1, 58–9, 298, 305–6, 339–40, 345–6; scheduling/planning, 35–6, 39–40, 298, 358; soil blocks, 37–8, 54, 282, 297–8, 305; transplant shock, 57, 60, 150 (*See also under specific crops*)
trap crops, 151, 214, 215–6, 281, 291, 299, 300, 355
Trichoderma species (beneficial fungi), 137, 140, 351, 352, 375
turnips, 46, 275; as cover crops, 122–3, 129; harvesting, 159, 160, 278–9; hoophouse production, 12, 92, 93, 95, 98, 250, 279; pests/diseases, 97, 278; sowing/transplanting, 66, 277–8; storage, 6, 164, 167, 279; varieties, 87, 93, 276; winter production outside, 85, 87, 88
Twin Oaks Community, ix, x–xii, 26
Tyfon Holland Greens (Asian greens), 87, 94, 228

varieties, 4 (*See also under specific crops*)
vegetable weevil larvae, 97, 154, 231
velvetbean, 132
velvetleaf, 142, 145
Verticillium wilt (*Verticillium dahliae*), 134, 138, 351, 352, 360, 366, 369, 381
viral diseases, 135, 183, 201, 272, 285, 294, 299, 301, 332, 350, 353–4, 360, 396
Volk, Gayle, 311–2

Wallace, Ira, 407, 408–13
wasps, parasitic. *See* parasitic wasps

watering/irrigation, 58, 60, 61, 68, 79–80, 94–5, 107 (*See also* drip irrigation)
watermelons, 46, 303–10; cover crops/rotations, 116, 117, 118, 305; crop requirements and yield, 304, 307; harvesting/storage, 155, 160, 309–10; pests/diseases, 135, 309, 400; sowing/transplanting, 55, 56, 57, 305–6; spacing, 49–50, 51, 304, 306–8; varieties, 303–4, 306; weed management, 308
Weaver, William Woys, 189, 304
weed management techniques, 141–7; cover cropping, 115, 117, 121, 144, 211, 281, 289–90; crop spacing, 47–8, 50; cultivation, 177, 249–50, 282–3, 289–90, 317, 329, 374; hand pulling, 157–8, 272, 283, 317; no-till planting, 122, 144, 147, 346–7; preventing germination, 143–4, 145; reduce weed seeding, 143, 144–5, 308; rowcover, 78–9; vinegar spray, 145, 317–8; weed identification, 142–3 (*See also* flame weeding; mulching; *specific crops*)
Wehner, Todd, 295
Weil, Ray, 108, 116
Westgate, Pam, 389
white rot (alliums, *Sclerotium cepivorum*), 318, 330
whiteflies, 154, 272
Wiediger, Alison and Paul, 58, 91, 284
wilt (*Erwinia tracheiphila*), 135, 285, 291, 294, 299, 301, 309
winter barley, 119, 125
winter purslane, 85
winter radishes, 66, 87–8, 121, 129, 164, 166, 167 (*See also* radishes)
winter squash and pumpkins, 46, 81, 87, 287–93; cover crops/rotations, 116, 118, 121, 290; curing/storage, 6, 107, 161, 162, 165, 167, 292–3; harvesting, 155, 156, 158, 291–2; pests/diseases, 290–1; sowing/transplanting, 288, 289; varieties, 287–8, 291, 292, 293; weed management, 142, 289–90
winter wheat (as cover crop), 119, 125, 177, 183, 398
wireworms, 266, 375

yarrow, 213, 215, 251
Yukina Savoy, 85, 92, 93, 95, 98, 159, 227, 231

zinc, 343, 384

About the Author

PAM DAWLING was born in London, England, and while at Cambridge University joined a group forming back-to-the-land communes. She saw organic farming as a healthy way to live and an alternative to industrialized agriculture. After gardening for about seventeen years in the UK, she moved to the USA.

Since 1991 she has been living in central Virginia, practicing farming with awareness of ecology, finite resources and the future of the planet. To make best use of every space in the garden, and every window in the calendar, Pam started to keep crop records and research better methods, varieties and timing.

She wrote up what she learned as monthly articles for *Growing for Market* magazine. After six years of articles she wrote *Sustainable Market Farming* for small-scale sustainable vegetable growers. Many other farmers have contributed their time and expertise to the information here, and the book stands in appreciation to them and as a way to help more growers.

Pam lives at Twin Oaks Community, a secular, income-sharing, work-sharing ecovillage established in 1967. There she helps grow food for around 100 people on three and a half acres and provides training in sustainable vegetable production for community members.

sustainablemarketfarming.com
facebook.com/SustainableMarketFarming

If you have enjoyed *Sustainable Market Farming*, you might also enjoy other

BOOKS TO BUILD A NEW SOCIETY

Our books provide positive solutions for people who
want to make a difference. We specialize in:

Sustainable Living ◆ Green Building ◆ Peak Oil
Renewable Energy ◆ Environment & Economy
Natural Building & Appropriate Technology
Progressive Leadership ◆ Resistance and Community
Educational & Parenting Resources

For a full list of NSP's titles, please call 1-800-567-6772 or check out our web site at:

www.newsociety.com

new society
PUBLISHERS